The Age of the Earth

Geological Society Special Publications

Society Book Editors

A. J. FLEET (CHIEF EDITOR)

P. DOYLE

F. J. GREGORY

J. S. GRIFFITHS

A. J. HARTLEY

R. E. HOLDSWORTH

A. C. MORTON

N. S. ROBINS

M. S. STOKER

J. P. TURNER

Special Publication reviewing procedures

The Society makes every effort to ensure that the scientific and production quality of its books matches that of its journals. Since 1997, all book proposals have been refereed by specialist reviewers as well as by the Society's Books Editorial Committee. If the referees identify weaknesses in the proposal, these must be addressed before the proposal is accepted.

Once the book is accepted, the Society has a team of Book Editors (listed above) who ensure that the volume editors follow strict guidelines on refereeing and quality control. We insist that individual papers can only be accepted after satisfactory review by two independent referees. The questions on the review forms are similar to those for *Journal of the Geological Society*. The referees' forms and comments must be available to the Society's Book Editors on request.

Although many of the books result from meetings, the editors are expected to commission papers that were not presented at the meeting to ensure that the book provides a balanced coverage of the subject. Being accepted for presentation at the meeting does not guarantee inclusion in the book.

Geological Society Special Publications are included in the ISI Science Citation Index, but they do not have an impact factor, the latter being applicable only to journals.

More information about submitting a proposal and producing a Special Publication can be found on the Society's web site: www.geolsoc.org.uk.

GEOLOGICAL SOCIETY SPECIAL PUBLICATION NO. 190

The Age of the Earth: from 4004 BC to AD 2002

EDITED BY

C. L. E. LEWIS
History of Geology Group, Macclesfield, UK

&

S. J. KNELL
Department of Museum Studies, University of Leicester, UK

2001

Published by

The Geological Society

London

THE GEOLOGICAL SOCIETY

The Geological Society of London (GSL) was founded in 1807. It is the oldest national geological society in the world and the largest in Europe. It was incorporated under Royal Charter in 1825 and is Registered Charity 210161.

The Society is the UK national learned and professional society for geology with a worldwide Fellowship (FGS) of 9000. The Society has the power to confer Chartered status on suitably qualified Fellows, and about 2000 of the Fellowship carry the title (CGeol). Chartered Geologists may also obtain the equivalent European title, European Geologist (EurGeol). One fifth of the Society's fellowship resides outside the UK. To find out more about the Society, log on to www.geolsoc.org.uk.

The Geological Society Publishing House (Bath, UK) produces the Society's international journals and books, and acts as European distributor for selected publications of the American Association of Petroleum Geologists (AAPG), the American Geological Institute (AGI), the Indonesian Petroleum Association (IPA), the Geological Society of America (GSA), the Society for Sedimentary Geology (SEPM) and the Geologists' Association (GA). Joint marketing agreements ensure that GSL Fellows may purchase these societies' publications at a discount. The Society's online bookshop (accessible from www.geolsoc.org.uk) offers secure book purchasing with your credit or debit card.

To find out about joining the Society and benefiting from substantial discounts on publications of GSL and other societies world-wide, consult www.geolsoc.org.uk, or contact the Fellowship Department at: The Geological Society, Burlington House, Piccadilly, London W1J 0BG: Tel. +44 (0)20 7434 9944; Fax +44 (0)20 7439 8975; Email: enquiries@geolsoc.org.uk.

For information about the Society's meetings, consult *Events* on www.geolsoc.org.uk. To find out more about the Society's Corporate Affiliates Scheme, write to enquiries@geolsoc.org.uk.

Published by The Geological Society from:
The Geological Society Publishing House
Unit 7, Brassmill Enterprise Centre
Brassmill Lane
Bath BA1 3JN, UK

(*Orders*: Tel. +44 (0)1225 445046
Fax +44 (0)1225 442836)
Online bookshop: *http://bookshop.geolsoc.org.uk*

The publishers make no representation, express or implied, with regard to the accuracy of the information contained in this book and cannot accept any legal responsibility for any errors or omissions that may be made.

© The Geological Society of London 2001. All rights reserved. No reproduction, copy or transmission of this publication may be made without written permission. No paragraph of this publication may be reproduced, copied or transmitted save with the provisions of the Copyright Licensing Agency, 90 Tottenham Court Road, London W1P 9HE. Users registered with the Copyright Clearance Center, 27 Congress Street, Salem, MA 01970, USA: the item-fee code for this publication is 0305-8719/01/$15.00.

British Library Cataloguing in Publication Data
A catalogue record for this book is available from the British Library.

ISBN 1-86239-093-2
ISSN 0305-8719

Typeset by Aarontype Ltd, Bristol, UK

Printed by Cambrian Press, Aberystwyth, UK

Distributors

USA
AAPG Bookstore
PO Box 979
Tulsa
OK 74101-0979
USA
Orders: Tel. +1 918 584-2555
Fax +1 918 560-2652
E-mail: *bookstore@aapg.org*

Australia
Australian Mineral Foundation Bookshop
63 Conyngham Street
Glenside
South Australia 5065
Australia
Orders: Tel. +61 88 379-0444
Fax +61 88 379-4634
E-mail: *bookshop@amf.com.au*

India
Affiliated East-West Press PVT Ltd
G-1/16 Ansari Road, Daryaganj,
New Delhi 110 002
India
Orders: Tel. +91 11 327-9113
Fax +91 11 326-0538
E-mail: *affiliat@nda.vsnl.net.in*

Japan
Kanda Book Trading Co.
Cityhouse Tama 204
Tsurumaki 1-3-10
Tama-shi
Tokyo 206-0034
Japan
Orders: Tel. +81 (0)423 57-7650
Fax +81 (0)423 57-7651

Contents

Preface	vi
KNELL, S. J. & LEWIS, C. L. E. Celebrating the age of the Earth	1
FULLER, J. G. C. M. Before the hills in order stood: the beginning of the geology of time in England	15
VACCARI, E. European views on terrestrial chronology from Descartes to the mid-eighteenth century	25
TAYLOR, K. L. Buffon, Desmarest and the ordering of geological events in *époques*	39
RUDWICK, M. J. S. Jean-André de Luc and nature's chronology	51
TORRENS, H. S. Timeless order: William Smith (1769–1839) and the search for raw materials 1800–1820	61
MORRELL, J. Genesis and geochronology: the case of John Phillips (1800–1874)	85
SHIPLEY, B. C. 'Had Lord Kelvin a right?': John Perry, natural selection and the age of the Earth, 1894–1895	91
WYSE JACKSON, P. N. John Joly (1857–1933) and his determinations of the age of the Earth	107
LEWIS, C. L. E. Arthur Holmes' vision of a geological timescale	121
YOCHELSON, E. L. & LEWIS, C. L. E. The age of the Earth in the United States (1891–1931): from the geological viewpoint	139
BRUSH, S. G. Is the Earth too old? The impact of geochronology on cosmology, 1929–1952	157
KAMBER, B. S., MOORBATH, S. & WHITEHOUSE, M. J. The oldest rocks on Earth: time constraints and geological controversies	177
DALRYMPLE, G. B. The age of the Earth in the twentieth century: a problem (mostly) solved	205
HOFMANN, A. W. Lead isotopes and the age of the Earth – a geochemical accident	223
CALLOMON, J. H. Fossils as geological clocks	237
MANNING, A. Time, life and the Earth	253
STRINGER, C. B. Dating the origin of modern humans	265
REES, M. J. Understanding the beginning and the end	275
Index	285

Preface

CHERRY L. E. LEWIS & SIMON J. KNELL

This book is dedicated to the memory of John Christopher Thackray (1948–1999), archivist at the Natural History Museum (NHM) and Geological Society. It evolved from *Celebrating the Age of the Earth*, the Geological Society's William Smith Millennium Meeting, convened in June 2000 by Cherry Lewis on behalf of the History of Geology Group (HOGG), of which Simon Knell was then a committee member. John had been chairman of HOGG since its foundation in 1992 and without him this project would never have happened.

The idea began with a meeting between John and Cherry, as she sought material for her forthcoming book on Arthur Holmes.

Late in the sunny afternoon of 18th June 1998, John Thackray and I were walking from the Natural History Museum towards South Kensington tube, discussing possible topics for future HOGG meetings. At that time John was Chairman of HOGG and I was a new member, of but a few months standing. It was only the second time I had met him, but already he had been immensely helpful in my search for material relating to Holmes.

John had found a large number of Holmes' letters in the NHM archives and I had spent the day reading them, so I was full of all the exciting things I had uncovered. 'What about a meeting on Holmes and his contemporaries?' I suggested, as we hovered on the pavement waiting to cross the road. John looked doubtful, 'Would that be enough to fill a whole day?' he asked, as we dodged the traffic, 'But we could have one on the age of the Earth' he continued, while we refuged on the island. Instantly, standing in the middle of Thurloe Street, we looked at each other and both knew that this was 'it'. With the Millennium approaching *The Age of the Earth* seemed to have a particular resonance. We threw a few ideas around until we reached the station, but I heard nothing more until I received this email about two weeks later:

Dear Cherry,
we have just had a HOGG committee meeting, and you will be alarmed to hear that your name was mentioned. We were talking about what meeting to have in the autumn of 1999 – the end of the millennium ... [It was agreed] that we should hold a one-day meeting on the history of geochronology, age of the earth, geological timescale, all that sort of thing ...

Then we had to think of a convenor.

All the committee coughed and looked at their feet, until someone remembered that you are working on AH and the age of the earth, and might be just the perfect person. ... it is not a huge amount of work.

Perhaps we could talk about it. Or perhaps you wish you had never met us!
BWs,
John

How could Cherry refuse? Over the next few months they exchanged frequent emails as it became the main topic of much HOGG discussion. The idea grew: a one-day conference, then two; a provisional line-up of British speakers soon became international; the end of 1999 became mid-2000, to give us more time to organize it.

While Cherry organized speakers, John talked to people behind the scenes: the Royal Society liked the idea so much they suggested we apply for sponsorship, while the Geological Society elected it for the Millennium's William Smith meeting. With funds pledged we could now invite speakers from abroad, and to our delight the Geological Society of America also promised to sponsor two airfares. By the end of that year it was all looking rather good, but it was then the terrible shock came. John was seriously ill with cancer. Incredibly, he lived only a few months longer. He died on 6 May 1999 at the age of fifty-one.

In the mid-1980s John had worked in the Geological Museum where Simon also had an office. John had been one of Simon's PhD examiners and it was John who encouraged him to join the HOGG committee. John really was the lifeblood of that group but in remembering him here we celebrate more than his breadth of knowledge, professionalism and scholarship. Unassuming, ever helpful, and with a wonderfully dry wit, John's most important quality was the way he always put things into perspective. With great affection we dedicate this book to his memory.

We had a clear vision of what we wanted the meeting to achieve, inspired by William Sollas who in 1900, as President of Section C – Geology, had opened his address to the British Association for the Advancement of Science (BAAS) in the following terms: 'The close of one century, the dawn of another, may naturally suggest some brief retrospective glance over the path along which our science has advanced ... from which we may gather hope of its future progress.'

Sollas' theme was the age of the Earth, then perhaps *the* topic of interest to scientists from many disciplines. And this was the point – scientists from *many* disciplines. Not only did we want to take a retrospective look at our science over the last hundred years or so, but here was a rare opportunity to involve people from a range of scientific disciplines interested in the age of the Earth as a topic, and geochronology as a tool – geologists, biologists, archaeologists, astronomers and, of course, historians. In the early part of the twentieth century it had been common practice for scientists from the different BAAS Sections to get together to share their knowledge on subjects of interest to them all. Today it rarely happens – we are all so 'specialized' – but uniquely we had this chance.

Over two days (28 and 29 June 2000) the audience was taken from Elizabethan 'mind control' to the very cutting edge of work now being done to date the Earth and its rocks. The first day was largely given over to historians who revealed the many ways in which the Earth's age was first ignored and then grew to become one of the most important scientific questions of all time. The second day emerged from these historical perspectives to that of scientists reflecting upon a golden age of radiochemistry, as geochronological techniques were developed with which to establish a geological time scale, and from which all those used today have descended. The meeting also included the William Smith lecture given by Hugh Torrens, as well as contributions from other distinguished individuals from a range of scientific disciplines, on how a knowledge of geology

Speakers and contributors to *Celebrating the Age of the Earth*, held at the Geological Society of London, Burlington House, London, on 28 and 29 June 2000. Starting from the back row (row 4), left to right:

Row 4: John Calloman, Patrick Wyse Jackson, Ken Taylor, Martin Rudwick.
Row 3: Martin Rees, Stephen Moorbath, John Fuller, Ezio Vaccari.
Row 2: Aubrey Manning, Gerald Friedman, Chris Stringer, Joe Burchfield.
Row 1: Richard Wilding, Stephen Brush, G. J. Wasserburg, Al Hofmann, Cherry Lewis.

and geochronology contributes to, and overlaps with, their subjects.

But this volume is not just a record of the conference. There are a number of contributions here that were not presented at the meeting and, with time to reflect since the event, all papers have been revised. As editors we certainly made our authors work and in this regard we must thank all the referees who contributed so much to helping us. We would also like to thank the publishing staff for their work in producing such a splendid book; HOGG for sponsoring some of the photos in the volume, as well as the Geological Society of London, the Royal Society and the Geological Society of America for enabling us to *Celebrate the Age of Earth* in the first place. John Thackray would have been proud of us all.

It is recommended that reference to all or part of this book should be made in one of the following ways:

LEWIS, C. L. E. & KNELL, S. J. (eds) 2001. *The Age of the Earth: from 4004 BC to AD 2002*. Geological Society, London, Special Publications, **190**.

MANNING, A. 2001. Time, life and the Earth. *In*: LEWIS, C. L. E. & KNELL, S. J. (eds) *The Age of the Earth: from 4004 BC to AD 2002*. Geological Society, London, Special Publications, **190**, 253–264.

Celebrating the age of the Earth

SIMON J. KNELL[1] & CHERRY L. E. LEWIS[2]

[1] *Department of Museum Studies, 105 Princess Road East, Leicester LE1 7LG, UK*
 (*email*: sjk8@leicester.ac.uk)
[2] *History of Geology Group, 21 Fowler Street, Macclesfield, Cheshire SK10 2AN, UK*
 (*email*: clelewis@aol.com)

Abstract: The age of the Earth has been a subject of intellectual interest for many centuries, even millennia. Of the early estimates, Archbishop Ussher's famous calculation of 4004 BC for the date of Creation represents one of the shortest time periods ever assigned to the Earth's age, but by the seventeenth century many naturalists were sceptical of such chronologies. In the eighteenth century it was Nature that provided the record for Hutton and others.

But not all observers of geology enquired about time. Many, like William Smith, simply earned a living from their practical knowledge of it, although his nephew, John Phillips, was one of the first geologists to attempt a numerical age for the Earth from the depositional rates of sediments. For more than fifty years variations of that method prevailed as geology's main tool for dating the Earth, while the physicists constrained requirements for a long timescale with ever more rigorous, and declining, estimates of a cooling Sun and Earth.

In 1896 the advent of radioactivity provided the means by which the Earth's age would at last be accurately documented, although it took another sixty years. Since that time ever more sophisticated chronological techniques have contributed to a search for the oldest rocks, the start of life, and human evolution. In the attempt to identify those landmarks, and others, we have greatly progressed our understanding about the processes that shape our planet and the Universe, although in doing so we discover that the now-accepted age of the Earth is but a 'geochemical accident' which remains a contentious issue.

Establishing the Earth's age and chronology has been something of an intellectual baton passed from one academic creed to another as each seemed to offer new hope of solution or understanding. In the seventeenth century the Earth's age was subject to the strictures of religious and social orthodoxy. The product of Creation, theological study offered one logical path to its understanding, but also to its containment. This pattern would repeat itself, as forces for an expanded timescale repeatedly did battle with those who thought the Earth relatively young. This was never simply a conflict between 'progressive' science and a resistant society, but rather a reflection of the path travelled by an idea through a complexity of beliefs, dogmas, theories, measurements and methodologies.

In the eighteenth century, Theories of the Earth marshalled existing evidence, scientific practice and analogy, to construct models for the origin and development of the planet. At the beginning of the nineteenth century, modern geology emerged as a rigorous and empirical field of study then centred on stratigraphy with its implied notions of relative age. This, when combined with the wide adoption of uniformitarianism, gave the saicence new and seemingly rational ways to think about time. But the new geologists were often reticent about discussing time in terms of years. By the 1860s, however, the subject added the great evolutionary debate to its intellectual baggage as T. H. Huxley defended Darwin's billion-year estimate for natural selection. Now both biologists and geologists joined forces to demand a longer timescale which permitted the playing out of those evolutionary and uniformitarian processes necessary to explain the condition of the Earth.

However, the age of the Earth had acquired a new interest group. Applying theoretical ideas developed in the arly years of the century, a number of respected physicists began to offer hope of at last calculating the Earth's age in years. But if their calculations were correct, then those great underpinning theories of the natural sciences could not be. With the discovery of radioactivity in 1896, the future of the Earth's age was firmly in the court of that emerging interdisciplinary subgroup, the geophysicists. In the twentieth century these extended the timescale

From: LEWIS, C. L. E. & KNELL, S. J. (eds). *The Age of the Earth: from 4004 BC to AD 2002*. Geological Society, London, Special Publications, **190**, 1–14. 0305-8719/01/$15.00 © The Geological Society of London 2001.

remarkably. With Einstein's general relativity field equations, the topic was also returning to cosmology, to explanations of the Earth's age in the context of the age of the universe, to the realm of astronomers.

What unifies these different approaches and their practitioners is their extrapolation of time from the study of process, whether that process was observable or modelled by theory. Early chronologers rigorously applied contemporary historiographic methods to chart the 'progress' of history. Later geologists would extrapolate the time necessary for shaping the surface of the Earth or for the deposition or erosion of observable strata. Biologists too envisaged a time-dependent process of evolution. Physicists made predictions and measurements from the physics of heat and radioactive decay, and astronomers deduced process from planetary construction and arrangement. In the relative calm of the early twenty-first century, in a field which has seen centuries of deep controversy, we should perhaps reflect upon the confidence our predecessors had in *their* age for the Earth. They too thought they had 'got it right'.

Chronologies and theories

In one of his popular articles, Stephen Jay Gould (1993, pp. 181–193) responded to James Barr's (1985) definitive study of a great historical chronologer with a diatribe against previous 'Whiggish' interpretations of this early form of scholarship. The historiography of science had long rejected the notion of 'progress', of the past having been a succession of errors, but Gould was here talking to scientists who necessarily remained wedded to, and driven by, this concept of progress towards 'truth'. Though not the first to do so, Barr saw chronologers not as religious dogmatists but as rigorous scholars working within the social and intellectual constraints of their time. They were participating in the earliest of historiographic practices – the construction of a linear timeline of events – which, in reconceived form, would later underpin the great stratigraphic enterprise of the nineteenth century.

Rees's early-nineteenth-century *Cyclopaedia* separated 'sacred chronology' from other historical chronologies. Here chronology was 'one of the eyes of history' and not simply the study of ancient theological texts. Its potential sources were many:

> As its use is extensive, the difficulty of acquiring it is not inconsiderable. It derives necessary assistance from astronomy and geography, also from arithmetic, geometry, and trigonometry, both plain and spherical; and likewise from a studious and laboured application of various sources of information, supplied by the observation of eclipses, by the testimonies of credible authors, and by ancient medals, coins, monuments and inscriptions (Anon. 1819).

Chronologies were constructed as an analytical tool so as to distinguish 'cause' and realize a 'chain of events' (Anon. 1837, p. 131). To achieve this, their content was selective and thematic.

To discern the age of the Earth, the only 'credible authors' were those relating creation stories. The Bible, or rather the texts from which it was written, then became the primary historical texts, key sources for the interpretation of time. Historians generally point to Theophilus of Antioch, of the second century AD, as perhaps the first to use the Biblical record as a source for the construction of a chronology. The art of chronology, however, extended back long before the Christian era, often deriving 'fabulous ages' of hundreds of thousands of years for particular civilizations (Anon. 1819, 1833, 1849). In the sixth century BC Zoroaster, a religious teacher who lived in Persia (now Iran), believed that the world had been in existence for over 12 000 years; the Roman writer Cicero relates that the venerable priesthood of Chaldea in ancient Babylonia held the belief that the Earth emerged from chaos two million years ago, while the old Brahmins of India regarded Time and the Earth as eternal (Holmes 1913).

By the sixteenth century the Bible formed a core resource for the writing of histories in Europe. Of particular significance was the interpretation of the Second Letter of Simon Peter, which states 'One day is with the Lord as a thousand years, and a thousand years as one day' (2 Peter 3:8) (see also Dean 1981*a*; Oldroyd 1996*a*). It draws on Psalm 90 of the Old Testament, 'The human condition': 'Lord ... To you, a thousand years are a single day'. Chronologers read this literally, and made each of the six days of creation equivalent to 1000 years. The Biblical motivation for discussing the Earth's limited age was to encourage the living of 'saintly lives'. For here was not simply the birth but also the death of the Earth, the Judgement Day, which will 'come like a thief, and then with a roar the sky will vanish, the elements catch fire and fall apart, and the Earth and all it contains will be burnt up' (2 Peter 3:10). If this were not enough to encourage piety, Psalm 90 also pondered human mortality, 'over in a trice, and then we are gone.'

Fig. 1. Oil Painting of Archbishop James Ussher in 1658. Reproduced by kind permission of the governors and guardians of Armagh Public Library.

Scholars and particularly the clergy – long an intellectual elite – used these 6000 years as a framework. The relative merits of the Hebrew Massoretic text, the Samaritan Pentateuch and Greek Septuagint were much debated, while later studies of other civilizations revealed even longer chronologies and other creation stories. Given the range of sources and scope for interpretation, it is no wonder that no definitive chronology was produced.

Undoubtedly the most famous chronology known to us today is that published in the seventeenth century by James Ussher (Fig. 1) but, as Fuller (2001) shows, he was not alone nor was he the most revered in his time (see also Anon. 1819, 1833). A brilliant scholar, Ussher had completed a draft whilst at Trinity College, Dublin, in 1597. He was then only 16 years old. Barr (1985) disentangled the complexity of Ussher's work, of which his Creation date of 4004 BC is best known. This curiously high-resolution date is derived from traditional interpretations he had inherited, which placed 4000 years between the Creation and the coming of Christ, and an adjustment to take account of evidence that Herod (a contemporary of Christ) died in 4 BC. However, only one-sixth of Ussher's chronology was concerned with the Bible, the rest was a history of classical times. That we know of Ussher, however, rather than one of the hundreds of other chronologers, is simply an accident of history (for which, see Fuller 2001).

For one group of scholars it seemed the answer might be found by undertaking comparative studies of published chronologies in order to locate sources of error and correction. William Hales' three-volume *New Analysis of Chronology* (1809–1812) is an example which became very familiar to early-nineteenth-century British audiences, but he had predecessors and contemporaries across Europe (Anon. 1837, p. 133; Davies 1969, p. 13; Fuller 2001). From his own studies Hales preferred a creation date of 5411 BC. But by the time he published, chronologies, however useful they were for the construction of civil histories, were no longer restricting views of the age of the Earth. Indeed scholars had, for almost a century, taken the debate beyond the religious confines of orthodoxy. Davies (1969, p. 16), for example, suggests that scepticism of established chronologies existed amongst seventeenth-century naturalists. Though such views were rarely committed to paper, he felt sure that the London coffee houses must have been alive with such talk. The devout John Ray (1627–1705), for instance, reflected upon the oganic origin of fossils and considered the implications for time of the depths of sediments and rates of denudation – a recurrent theme in the centuries to come (Davies 1969, p. 59; Dean 1981a, p. 444).

Despite the enforcement of religious orthodoxy by the state (Davies 1969, p. 11), freedom of expression about the origins and history of the Earth developed rapidly in the eighteenth century. In 1694, for example, Edmond Halley considered the implications of a comet impact on the Earth. He imagined an Earth repeatedly reborn following chaotic destruction. It seemed to support a view of an eternal Earth undergoing periods of dissolution and recreation – a notion then considered 'one of the most dangerous of heresies' (Kubrin 1990, p. 65; Davies 1969, p. 12). It was also a challenge to the British establishment, much like that presented by theories of biological transmutation a century later (Desmond 1989). Such ideas did not simply undermine religious teaching but proclaimed as a falsehood the natural social order into which the established church and wider society were divided. To counter such opinions of himself, which were likely to affect his career prospects, Halley then spent a good deal of his time pursuing evidence for a finite age of the Earth. For him, there were good socio-political reasons for wanting to discover a particular answer which reinforced accepted views (Kubrin 1990, p. 65). Gould pondered one such of Halley's observations (Gould 1993, pp. 168–180). Understanding that the salts in the world's oceans were delivered to the sea via rivers, Halley suggested that if the rate of contribution could be calculated then here would be a measure for the age of the Earth. It was an idea that was to resurface in the late nineteenth century in the hands of John Joly (Wyse Jackson 2001). What surprised Gould was that Halley was considering a maximum age for the Earth, an argument against eternity, rather than a minimum. Thirty years later, following considerable social change, Halley was at last able to publish his paper discussing an Earth reborn without fear of social or religious retribution.

Halley's contemporary, Sir Isaac Newton, also considered the Earth's age and chronology, though without disrupting the traditional views of the Church. In his posthumous *Chronology of Antient Kingdoms Amended* he used astronomical calculations to correct those Greek and other ancient chronologies which were frequently used as historical sources (Newton 1728).

Rationalist Deism was an inevitable reaction to the strictures and assumptions of religious orthodoxy in seventeenth and eighteenth century Britain. It proved highly influential in intellectual circles across Europe and North America. For James Hutton, its most famous geological disciple, Nature provided a true record: the Earth's timescale was indefinite. Hutton was born only a few years after it had been finally possible for Halley to publish his paper inferring a longer timescale, perhaps an eternity. Hutton, however, said nothing of eternity. On this, his great admirer, Archibald Geike, was emphatic (Geike 1899, p. 719; also Burchfield 1975; Dean 1981*a*, p. 454). Notable amongst Hutton's critics was Jean André de Luc who, from the same record of Nature, found only concordance with Biblical chronology. For de Luc (discussed by Rudwick 2001) geomorphological processes created features in the landscape that were indisputably the product of time. If process rates were known then a mechanism for measuring time existed. It was this same kind of thinking which Georg Louis Leclerc, the Comte de Buffon, famously used to extrapolate a longer timescale of 75 000 years (discussed by Taylor 2001), a timescale which he still felt too short. Buffon applied measurement and reason, and promoted a theory based on contemporary empiricism (Buffon 1807).

Jean Louis Giraud, Abbé Soulavie (1752–1813), like Nicolas Desmarest (also Taylor 2001), interpreted temporal sequence from a study of volcanoes. In his book of 1791, he compared volcanic deposits to the Control of Acts, a recent piece of legislation which ensured that the dates of contracts would be registered such that originals could be verified. He also predicted the future role of fossils in determining the Earth's chronology:

All these volcanoes and their products may be placed today in a certain order by reason of these curious 'registries of control' of nature. The study of these monuments is, in the mineral kingdom, the veritable art of verifying the dates of nature, as the study of plant or shell-bearing rocks is the art of verifying the dates and eras of the ancient history of organized beings in the kingdom of the living world (Soulavie 1791).

He used superposition of volcanic deposits, their relative state of preservation and comparative elevation above the sea as indicators of succession, but he was not suggesting the Smithian principle of characteristic fossils.

Throughout the eighteenth century, across the whole of Europe, Theories of the Earth mixed conjecture, religious orthodoxy and observation (Vaccari (2001) gives a wide overview). That the Earth had undergone a succession of 'revolutions' was a popular concept. It was critical to George Cuvier's (1817) *Essay on the Theory of the Earth* which interpreted the succession of extinct animals preserved in the fossil record and the wild contortions of rocks which, following the arguments of Steno, were believed to have originally been deposited horizontally. But by the time Cuvier's *Theory of the Earth* was published, the genre was attracting much criticism from a new breed of empiricist geologists.

Relative age, uniformitarianism and time

As geology emerged as a rigorous and self-contained science in the early nineteenth century, there remained a belief that a unifying theory might permit explanation of the geological world. Time, as a useful geological concept, however, did not emerge from theory but from observation. The stratigraphic work of William Smith (Fig. 2), Cuvier and Alexandre Brongniart, Thomas Webster and John Farey in the first two decades of the nineteenth century proved fossils as useful time markers. It was an idea that would develop rapidly in terms of its understanding and application throughout the first half of that century. Unlike Buffon and Cuvier (and indeed Webster), Smith's goal was not intellectual or social (Knell 2000) but entirely practical (as Torrens (2001) demonstrates). The knowledge he developed resulted from attempts to solve the practical problems of coal prospecting. As Torrens shows, this was one of the major economic challenges of the period. What Smith actually thought of the role of time in causing the arrangement of fossils and strata is a matter for debate (Knell 2000; Torrens 2001; Fuller 2001). For the most part he was talking about the 'natural order' of strata, though we should have no doubt that his views changed. Certainly by 1824 Smith's work on the Yorkshire coast, was giving 'strong confirmation to the geological axiom that "Deposits of equal antiquity enclose analogous fossils" ', as his nephew and co-worker, John Phillips (Fig. 2), reported to the Yorkshire Philosophical Society (Knell 2000, p. 148). That Smith should have co-authored such a statement is something most historians of Smith have previously denied. But as is apparent from Torrens' paper, Smith's ideas were never fully formed. When he and Phillips began their detailed study of the Yorkshire coast in the 1820s they were testing their knowledge of sequence and indicators, and the reliability of the Smithian method. What Torrens most clearly demonstrates is that in an era of frequent coal prospecting scandals it was easy to understand why Smith was revered by some and yet simply not believed by others. The risks were great and the Smithian technique was neither infallible nor at the height of its sophistication. In 1813, Ernst Friedrich, Baron von Schlotheim, like others on the frontline of the new science, still felt fossil-based chronology a possibility rather than a reality (Schlotheim 1813). George Greenough (Fig. 2) and a few others doubted it had any particular value at all. Schlotheim, like John Farey and later Von Buch, called for higher resolution studies of organic remains, as he felt these might reveal the epochs of revolutions in terms of spans of years. By the 1830s, chaotic fossil nomenclature was, more than anything, leading to stratigraphic imprecision. The British Association for the Advancement of Science charged John Phillips with the task of sorting this out (Fig. 2).

Phillips epitomized the empiricism which, it was claimed, had turned geology into a rigorous science. He made an entire career out of time, the kind of relative time which Callomon (2001)

Fig. 2. The British Association for the Advancement of Science at Newcastle in 1838. Oil painting on canvas by Thomas Henry Gregg. The British Association for the Advancement of Science first met in York in 1831. In 1838 the Duke of Northumberland was the President of the Meeting.
Key: 1 Sir Roderick Impey Murchison; 2 Richard Owen; 3 Henry Thomas De la Beche; 4 ?; 5 John Phillips; 6 Adam Sedgwick; 7 George Bellas Greenough; 8 Charles Lyell; 9 William Buckland; 10 ?; 11 William Smith.

now pursues. In the middle years of the nineteenth century, Phillips was producing endless tables of facts, believing, initially at least, that errors arose from omission and that comprehensive data would yield more accurate results. He was later to understand the error of this reasoning as the wealth of data he had generated was sometimes beyond human powers to process (Knell 2000). Phillips was so swept up in the mid-century fashion for statistics that one may feel somewhat surprised when Morrell (2001) tells us of Phillips' reticence to commit himself to a number for the age of the Earth.

Though reticent, Phillips understood the human desire to create a scale for the chronology given by relative time. He knew all attempts were bound to fail due to countless factors that would corrupt and distort the figures. One could only expect 'plausible inferences concerning the time elapsed in the production of stratified rocks', nothing more (Phillips 1849, p. 795). However, ever willing to apply a little mathematics to philosophical puzzles, he made an attempt 'in a strictly philosophical spirit' in the *Encyclopaedia Metropolitana*. He considered a number of ways in which the problem might be resolved, including using upright fossil plant remains as evidence of rapid depositional events (such as the locally famous fossil tree trunks in the coal measures of Altofts, near Wakefield, which Phillips knew well). However, the most logical line of enquiry remained that which had attracted the attention of Desmarest, Soulavie, Buffon and de Luc, and which was now characterized by the uniformitarian principles of Charles Lyell (Fig. 2), John Playfair and Hutton: that is, the deduction of time from the products of geomorphological processes. Here he undertook a calculation of the time taken to deposit Coal Measure flagstones within which layers of rolled grains one-twentieth of an inch thick could be detected. A 40 ft thickness of rock implied 9600 layers. If each layer represented a tide, then about 700 layers would result from annual deposition. Thus 40 ft of rock could be deposited in approximately 13.5 years. Half of the Yorkshire Coal Measures consisted of sandstones, some 1500 ft of strata in all. By this reasoning, deposition of the whole of the Coal Measures may have lasted 1000 years. Such figures, combined with Phillips' belief that all strata ultimately originated as molten rock, made him think the Earth did have a finite age, a view he felt disagreed with Lyell (Phillips 1849, p. 798). However, by this time Phillips had read Joseph Fourier's (1822) *Théorie Analytique de la Chaleur* (*The Analytical Theory of Heat*) and was, along with other 'theorists'

(as Morrell (2001) reveals), foreseeing the conflict between the infinite timescale of uniformitarianism and the restricted needs of those theorizing on the physics of the Earth (as discussed by Brush 2001; Dalrymple 2001; Lewis 2001; Shipley 2001; Yochelson & Lewis 2001; Wyse Jackson 2001).

Some have suggested that, as the century wore on, geology continued to be threatened by conservative and speculative philosophy. Though Fuller (2001) sees in Leonard Horner's 1861 address to the Geological Society of London remnants of religious orthodoxy snapping at the heals of science, Davies (1969, p. 210) was more dismissive. Was Horner 'tilting at windmills' or detecting a real threat in orthodox religion? More than a century later, geologists continue the same campaign while observers ponder the same question. By the middle of the nineteenth century, Victorian society was undergoing a crisis of faith, to which geology with its notions of endless time had contributed (see Dean 1981*b*). However, attempts were made to understand the Bible in ways acceptable to contemporary interpretations of the geological evidence. One came in the 1850s when Hugh Miller, the Scottish geologist and writer, attempted to reconcile the Genesis and the fossil record in *The Testimony of the Rocks* (1857) (Oldroyd 1996*b*). He explained Genesis Ch. 1 as a literally visual revelation, each 'day' corresponding to a geological period. Although popular in its day, it was ultimately insufficiently naturalistic for the scientists and too liberal an interpretation for the theologians.

However, by this time it was not religion that was restraining the age of the Earth, but the physics of heat and its implications for a cooling Earth. Now Lyell's most fundamental principle – that of uniformitarianism – and all that rested upon it, seemed under threat. With Archibald Geikie in the ascendancy as Britain's pre-eminent geologist, an unbridled supporter of Lyell and Hutton, the geologists were not going to take any restrictions on the age of the Earth lying down. Amongst his many arguments Geikie used denudation rates, rather than rates of sedimentation, to discuss the scale of geological time. Based on empirical data for the mean height of the continents, and fluvial sediment and solution loads, he predicted the complete wasting of Europe in just 4 million years, North America in 4.5 million and South America in 7 million. It demonstrated that rates of denudation were not as slow as usually stated and that vast quantities of time were not required to produce considerable change to the surface of the Earth (Geikie 1868). It was not that Geikie wanted a young Earth but that a young Earth did not undermine uniformitarianism. Evolutionary biologists were talking in terms of Darwin's estimate of 1000 million years for the record of natural selection, time the geologists were willing to give. As Huxley remarked in defence of geology, 'Biology takes her time from geology' (Huxley quoted by Desmond 1994, p. 370). The physicists, led by William Thomson – later Lord Kelvin – and his collaborator Peter Guthrie Tait, were not followers of Lyell or Darwin.

The storehouse of creation

The closing decade of the nineteenth century saw the start of a scientific revolution that was to impact on geology in a way that would transform it forever and prepare it for the definitive 'revolution' that occurred in the 1960s. By 1893 debate between geologists and physicists about the age of the Earth was at its peak. The ages determined by geologists, using empirical methods and observing Earth processes, ranged from 3 million years 'for the whole incrusted age of the world' (Winchell 1883, p. 378), to Wilber John McGee's 1892 value of 15 *billion* years for the whole age of the Earth, of which 7 billion had elapsed since the end of the Palaeozoic (Yochelson & Lewis 2001). These extreme values were largely unacceptable to both geologists and physicists, and did nothing to bring the two sides into closer agreement, although the urgent need for a reliable timescale was becoming more and more evident: 'How immeasurable would be the advance of our science could we but bring the chief events which it records into some relation with a standard of time!' (Sollas 1900, p. 717).

Lord Kelvin's interest in the age of the Earth had been prompted by his work on the age of the Sun's heat. He considered that the Sun, 'assumed to be an incandescent liquid', must obey the laws of thermodynamics, therefore it could not continue to radiate sunlight for ever unless it was being provided with energy. Since there was no evidence for this energy, there was no alternative – it *must* be cooling down – and the inevitable result of cooling was that the Sun had a finite life. In this conclusion Kelvin was supported by astronomers.

Given these apparently irrefutable circumstances, it is not surprising that when Charles Darwin, in the first edition of *On the Origin of Species*, published his estimate of 300 million years for erosion of the Weald, a section relatively high up in the geological sequence, that Kelvin's response was dismissive: 'What then are we to think of such geological estimates as 300 000 000 years for the "denudation of the Weald?"' (Thomson 1862, p. 391; also Morrell 2001). Since the age of the Earth was constrained by the age of the cooling Sun, it was only to be expected that its inhabitants '... cannot continue to enjoy the light and heat essential to their life, for many million years longer, unless sources now unknown to us are prepared in the great storehouse of creation' (Thomson 1862, p. 393). Within a month of questioning Darwin's estimate in public, Kelvin delivered his now famous attack on geologists and their methods for determining an age for the Earth that ignored the laws of thermodynamics.

At that time Kelvin proposed the Earth's age to be between 20 and 400 million years (Thomson 1864), but he found the problem so interesting that over the following three decades he continued to pursue it, confessing that he 'would rather know the date of the *Consistentior Status* than of the Norman Conquest' (Thomson 1895a, p. 227). The *Consistentior Status* was considered to be the time when the molten Earth had first acquired a solid crust. As more data became available Kelvin reduced his estimate first to 100 million years (Thomson 1871), and then, in 1893, to only 24 million (Thomson 1895b) when new results on the melting temperature of rocks became available (King 1893). This new age concurred well with the age of the Sun, reckoned to be 20 million years. To the physicists, the inevitability of the Sun's decline was a compelling reason for a young Earth, and the fact that the two ages, Earth and Sun, now gave consistent results was conclusive proof that both were right. There was little for the geologist to argue against, except to suggest to the physicist that 'there may be an error somewhere in his data or the method of his treatment' (Walcott 1893, p. 639).

The effort to quantify geological time affected both the theory and practice of geology and was largely responsible for it maturing into a professional scientific discipline. Up to that point, geologists had worked out their estimates for the age of the Earth from their empirical observations of sedimentary processes, based on the uniformitarian premise that the rates of these processes did not vary over time. It was largely the work of Kelvin that made clear to geologists the need for *quantification* and it was he who set them on a path which observed the laws of physics, from which they have subsequently deviated little. A rigid adherence to Lyell's uniformitarian interpretation of geology was no longer possible, and although not all physicists were united in their interpretation of Kelvin's data, as is evident from the views expressed by his former student John Perry (Shipley 2001), they were, on the whole, agreed on the need for geologists to abandon their dogmatic attitudes. It was Perry, in fact, who complimented Kelvin in *Nature* for his exposé of uniformitarianism: 'Lord Kelvin completely destroyed the uniformitarian geologists, and not one now exists. It was an excellent thing to do. They are as extinct as the dodo or the great auk' (Perry 1895, p. 227).

That same year, 1895, William Röntgen discovered X-rays and the following year Henri Becquerel detected similar rays being emitted from uranium. The decade of discovery that ensued was a time unparalleled in the history of science (Lewis 2001). The speed at which new discoveries were made about the atom unravelled the whole fabric of physics, and a completely new design was needed to build it back up again. This new form of energy at last explained the discrepancy between the long timescales observed by geologists, and the short one predicted by Kelvin. For even if the Earth was still cooling from a time when it had been a molten globe, the heat generated within the Earth by radioactive decay had counteracted that cooling for hundreds of millions, even billions, of years.

Based on the accumulation of helium, a by-product of the decay process, in 1904 Ernest Rutherford (Fig. 3) determined the very first radiometric date: a fergusonite mineral gave an age of 500 Ma. More than forty years after Kelvin's pronouncement that the Sun would not keep shining unless new sources of energy 'now unknown to us are prepared in the great storehouse of creation', Rutherford was to opportunistically turn this phrase to his advantage in order to placate Kelvin, whose age for the Earth was now demonstrably in error: 'That prophetic utterance refers to what we are now considering tonight, radium!' (Eve 1939, p. 107). But Kelvin died three years later, having never fully accepted that a new source of heat had indeed lain in the great storehouse of creation – creation of the Solar System. As Martin Rees (2001) explains, most elements, particularly the heavy ones like uranium, form during a supernova explosion, the gaseous debris from which ultimately resulted in formation of our Solar System. Thus stellar nucleosynthesis could be viewed as Kelvin's 'great storehouse of creation' in which all elements on Earth are prepared.

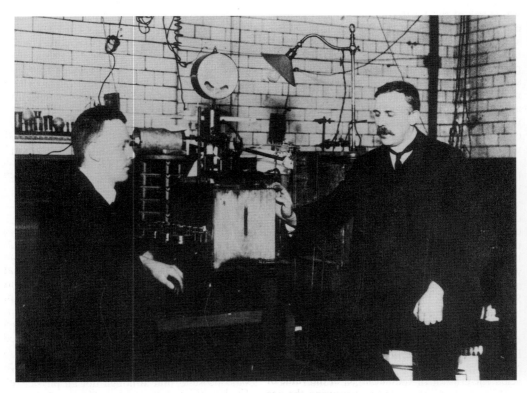

Fig. 3. Ernest Rutherford (1871–1937) (right), who determined the first radiometric date, with Hans Geiger, in their laboratory at Manchester University, around 1912. Credit: Science Museum/Science and Society Picture Library.

Throughout history, there have always been those slow to accept changes to the *status quo*, and who fiercely resist innovation, particularly when, like radioactivity, it is rapidly introduced and means the discarding of many strongly held beliefs. Having desired, in Lyell's day, an Earth of almost infinite duration, geologists fought Kelvin for fifty years to be allowed an Earth that was even a hundred million years old. Within a few years of Rutherford's first radiometric date, it became clear that, after all, geologists could again have an Earth of almost limitless extent, but now the demand was for *less* time. They were being asked to accept a ten-fold increase – an Earth that was more than a billion years old: just where were all the sediments needed to fill such enormous amounts of time?

During this period of transition, two men were particularly prominent in developing the science of geochronology. Today John Joly is famous for his attempts to date the age of the Earth from the sodium content of rivers, but he was also a man with an extraordinary range of interests and abilities (Wyse Jackson 2001). Joly was quick to grasp the principles of radioactivity and to apply them to geological problems, but his inability to accept the long timescales being indicated by the method was typical of many of his generation. It required someone younger, with an open mind unfettered by the need to maintain political prestige, to solve the new problems. Arthur Holmes' additional advantage over many of his geological contemporaries, was that he had a background in physics, but being familiar with all the old methods used to estimate the age of the Earth, he also understood the difficulties being encountered by geologists:

> The surprises which radioactivity had in store for us have not always been received as hospitably as they deserved. With the advent of radium geologists were put under a great obligation, for the old controversy [the need for a long time scale] was settled overwhelmingly in their favour. But the pendulum has swung too far, and many geologists feel it impossible to accept what they consider the excessive periods of time which seem inferred (Holmes 1913, p. 167).

For more than four decades Holmes was in the vanguard of geochronology, and during that time he contributed more to understanding that science than any other single individual. He pursued an almost evangelical crusade to persuade geologists and the world at large, of the value of radioactivity as a means for measuring the age of the Earth, and an Earth of great antiquity (Lewis 2001).

In the United States, rejection or acceptance of ideas about radioactivity and its application to geological problems progressed at much the same rate as it did in England, although in the first quarter of the twentieth century America lagged behind by a year or so. Since most of the work on radioactivity was then being done in Europe, time was needed for ideas to travel by boat across the Atlantic and for the more important papers to be reproduced in American journals (Yochelson & Lewis 2001). But by the early 1930s, most scientists in both countries essentially accepted the geological timescale provided by Holmes, and little reference was made to any timescale provided by sedimentation or denudation rates. Now only the astronomers seemed to question the validity of radiometric dates (Brush 2001).

During the Second World War, development of the atom bomb meant that almost all available physicists in America were co-opted to work on the Manhattan Project. The outcome for geochronology was a greatly improved understanding of uranium, refined mass spectrometers, and a number of people who had become highly skilled in the *art* of mass spectrometry (Lewis 2000). The superior dates now available not only pushed back the age of the Earth as older minerals were found, but they also led three people in Europe to recognize that the new methods for dating the minerals also held the key to dating the Earth. Argument as to the timing of events and the merits of each model led to it being known for many years as the 'Holmes–Houtermans' model for dating the Earth, as both Arthur Holmes and Fritz Houtermans published their very similar versions in 1946, within months of each other. But when Gerling's 1942 paper was translated from Russian, it seemed that he had had the idea first, so the name became even more cumbersome the 'Gerling–Holmes–Houtermans model' (Dalrymple 2001).

In fact, it has now come to light that the principles for the model originated with Holmes back in 1932 when he first proposed the concept of 'initial ratios' (Lewis 2001), but the argument as to who was 'first' is not important here. What is interesting is the apparently simultaneous development of the same idea by a number of people, not obviously in contact with each other, although all working in the same domain. It is a common theme in science. Robert Strutt claimed that he had thought of how to date minerals before Rutherford did, or certainly before Rutherford published (Egerton 1948), and although Frederick Soddy is credited with the concept of isotopy because he coined the term 'isotope' in 1913, the idea of several species of the same element had been around for a decade previously (Kauffman 1982). The process of 'discovery' is in fact rarely a 'Eureka' moment, being almost always one of a gradual evolution of ideas – the replication of memes (Dawkins 1989) that occasionally mutate. Scientists, however, do like to attribute discoveries to an individual who was *first*. After all, one of the basic priorities of research, particularly these days, is to be able to claim credit in the intellectual rat race. The Theory of Plate Tectonics should rank alongside Darwin's Theory of Evolution and Einstein's Theory of Relativity, but perhaps because it cannot be attributed to any one individual, it is somehow assigned to a lower rank.

Regardless of who developed the model for dating the age of the Earth, once it was in place all that remained was for techniques to progress to a point where accurate data could be obtained and plugged into the model to give the correct result. Claire Patterson (Fig. 4), so named on his birth certificate but who frequently called himself Clair in an attempt to avoid the gender confusion (L. Patterson, pers. comm. 1999), was another whose mass spectrometry skills had been honed on the Manhattan Project. In 1956, after a decade spent developing techniques to measure minute amounts of lead, this skill enabled him to finally date the 'time since the earth attained its present mass' (Patterson 1956, p. 236). He resolved it to be 4550 ± 70 Ma. After centuries of controversy, it may appear that the final event went almost unnoticed, but it is only with hindsight that we now know Patterson was 'right'. At the time, it might only have been one more of the many recent attempts to get it right.

Ages of the 'oldest'

There is something in the human psyche that drives us to look for extremes – the first and last, the highest and lowest, largest and smallest, youngest and oldest. With the search for the age of the Earth ticked off that list, we have looked for other extremes. Biologists search for the first signs of life and anthropologists for the first humans; in astronomy it is the furthest stars; in geology the oldest rocks.

Since development of the Theory of Plate Tectonics, we have progressed to a remarkable understanding about the way the Earth has evolved. Despite this, we still have a rather hazy view of those first 500 million years,

Fig. 4. Claire Patterson (1922–1995) who dated the age of the Earth (courtesy of the Archives, California Institute of Technology).

although the hope is that the oldest rocks of the planet will carry a 'memory' of the processes that occurred hundreds of millions of years before their formation. Evidence from zircons that give an age of 4.4 Ga, suggests that crust was forming on the Earth's surface soon after accretion (Kamber et al. 2001), which is considered to have lasted some tens of millions of years (Hofmann 2001). But 'accretion' seems to have persisted in the form of 'bombardment' for at least another 500 Ma, as massive impacts, known to have bombarded the Moon until 3.9 Ga, presumably also hit the Earth. No evidence is found on Earth for this bombardment, but the apparent lack of survival of any in situ crust older than 3.8 Ga (Kamber et al. 2001) strongly supports the theory. The race to find older and older crust appears to be limited by this bolide barrier. However, history reminds us that for thirty years (1911–1941) the age of the Earth remained static at 1.6 Ga because, during all that time, no minerals were found that indicated it might be any older. But in 1941 discovery of the Manitoba pegmatite, that gave an age of more than 2.2 Ga (Nier et al. 1941), made it apparent that the Earth was actually much older than anyone had hitherto realized. Thus the search for older and older crust should not be discouraged!

The biologist's search for the start of life on Earth, asks not only the question 'when?', but also 'how?'. Was it in the murky primordial soup, or on the back of one of those massive bolide impacts? Evidence for life is now found in a remarkable variety of environments (Manning 2001), and it therefore does not seem unreasonable to anticipate that life survived even the meteorite bombardment. In West Greenland ^{13}C-depleted graphite microparticles, claimed to be biogenic in origin, have been found in the oldest known sedimentary rocks that are 3.7–3.8 Ga (Kamber et al. 2001). As flexible and durable as life seems to be, a continued search may well push that boundary back even further.

Searching for our ancient hominid ancestors is a topic of particular interest to us all, and perhaps it is a mark of the success of archaeology to capture the public's imagination, that when asked the question 'What technique is used to date age of the Earth?' the vast majority will answer 'carbon dating'. In the last fifteen years, refined archaeological dating techniques have immeasurably improved our ability to understand the time-scale of human evolution, particularly for the last 200 000 years (Stringer 2001). The new dating techniques that facilitate this understanding are the direct descendents of those established by the early pioneers in the last century who worked so hard to launch geochronology as a science. Now it is impossible to imagine how we would manage science without them.

Chronology is a crucial tool for the understanding of many disciplines. Being able to date events tells us much about the *processes* that enabled them. For example, the key to unlocking geology's unifying theory was development of a *timescale* of magnetic reversals seen on the floor of the oceans. Immediately it became clear that the youngest rocks were nearest to the central ridge, while the oldest were furthest away and adjacent to the continents – so the oceans really were opening, and the continents really were drifting apart! Sea-floor spreading, hitherto just a theory, became a fact.

Similarly, as Brush (2001) discusses, when the value for Hubble time, which provides an age for the Universe, was seriously underestimated in the 1930s and '40s, it resulted in a Universe that was younger than the Earth and stars it contained. Astronomers then sought a range of models, or processes, in an attempt to explain this discrepancy. For example, the fact that the 'timescale problem' did not exist in the Steady-State Theory was one of its advantages. But because the timescale was wrong, so too were the processes used to explain formation of the Universe. Until the 1950s, Holmes and his colleagues, with their persistent and consistent dating of minerals from Earth, continued to be a thorn in the side of astronomers, in much the same way that geologists had been a thorn in the side of physicists during the previous century. It was not until the late 1950s that the Hubble constant was revised and provided a more realistic age for the Universe, which then facilitated development of current models for its evolution. Today, some astronomers claim that the Big Bang Theory explains evolution of the Universe to such an extent that they can 'place 99 percent confidence in an extrapolation back to the stage when the Universe was one second old' (Rees 2001). In the light of history, and all the bold statements discussed in this volume that have subsequently proved incorrect, this is a brave assertion.

Conclusion

The value that Patterson determined for the age of the Earth has not changed in nearly fifty years, beyond that allowed for by his error range. Compatible with ages of meteorites and the Moon, it now seems a consistent and stable number, but is it? Like an explosive volcano, the age of the Earth debate did not die, it just lay dormant for fifty years until science had progressed sufficiently to question the underlying assumptions on which it was based. Our improved understanding of the extreme chemical differentiation that occurred during separation of the crust from the mantle reveals that although the mechanisms for enriching uranium and lead in the crust are quite different, the result is that, on average, the overall enrichment is much the same for both elements (Hofmann 2001).

As Hofmann explains, although this 'geochemical accident' served Patterson well and apparently provided him with the 'right' answer, terrestrial lead isotopes, such as those that contributed to the ocean sediments used by Patterson to represent 'average Earth lead', can only provide an approximate age. The ages of other events, such as formation of meteorites now considered to be 4.56 Ga, the timing of core differentiation around 4.40 Ga and discovery of 4.40 Ga zircons, place the 'age of the Earth' into a wide band of 160 Ma. But even so, only the lower limit of this 'wide band' falls outside Patterson's error range of ±70 Ma, on his original estimate of 4.55 Ga. According to Hofmann, however, it may be possible to reduce the error to perhaps as little as 10 Ma.

The extinct decay products of short-lived radioactive nuclides are now being used to interpret the chronology of the earliest history of the solar system, so perhaps they will also provide further constraints on the formation history of the Earth. Until then, the age of the Earth remains controversial and the search for a refined value continues. It could make for interesting discussion at the turn of the next century.

We would like to thank all the authors who have contributed to making this volume the rich and varied book it has become, and on whom we have extensively drawn for this paper. As editors, we would like to thank not only the referees of this paper, M. Taylor and M. Stoker, but also all those referees who have willingly given of their time and expertise to comment on the wide range of papers put before them. Thank you all. It made our task a lot easier!

References

ANON. 1819. Chronology. *In*: REES, A. (ed.) *Cyclopaedia*, Vol. 7 [unpaginated].
ANON. 1833. Chronology. *The London Encyclopaedia*, Vol. 5, 663–685.
ANON. 1837. Chronology. *Penny Cyclopaedia*, Vol. 7. Knight, London, 130–134.
ANON. 1849. Chronology. *In*: SMEDLEY, E., ROSE, H. J. & ROSE, H. J. (eds) *Encyclopaedia Metropolitana*, Vol. 16. Griffin, London, 639–655.
BARR, J. 1985. Why the world was created in 4004 BC: Archbishop Ussher and Biblical chronology. *Bulletin of John Rylands University Library Manchester*, **67**, 575–608.
BRUSH, C. 2001. Is the Earth too old? The impact of geochronology on cosmology, 1929–1952. *In*: LEWIS, C. L. E. & KNELL, S. J. (eds) *The Age of the Earth: from 4004 BC to AD 2002*. Geological Society, London, Special Publications, **190**, 157–176.
BUFFON, LECLERC, G. L. COMTE DE. 1807. *Époques de la Nature*, Paris. [Translated extract in MATHER, K. F. & MASON, S. L. (eds) 1939. *A Source Book in Geology*. McGraw-Hill, New York, 65–73].
BURCHFIELD, J. D. 1975. *Lord Kelvin and the Age of the Earth*. Macmillan, London.
CALLOMON, J. H. 2001. Fossils as geological clocks. *In*: LEWIS, C. L. E. & KNELL, S. J. (eds) *The Age of the Earth: from 4004 BC to AD 2002*. Geological Society, London, Special Publications, **190**, 237–252.
CUVIER, G. 1817. *Essay on the Theory of the Earth*, [translated by Robert Jameson], Edinburgh.
DALRYMPLE, G. B. 2001. The age of the Earth in the twentieth century: a problem (mostly) solved. *In*: LEWIS, C. L. E. & KNELL, S. J. (eds) *The Age of the Earth: from 4004 BC to AD 2002*. Geological Society, London, Special Publications, **190**, 205–221.
DAVIES, G. L. [HERRIES] 1969. *The Earth in Decay: A History of British Geomorphology 1578–1878*. Macdonald, London.
DAWKINS, R. 1989. *The Selfish Gene* (second edition). Oxford University, Oxford.
DEAN, D. R. 1981a. The age of the earth controversy: beginnings to Hutton. *Annals of Science*, **38**, 345–456.
DEAN, D. R. 1981b. 'Through science to despair': geology and the Victorians. *In*: PARADIS, J. & POSTLEWAIT, T. (eds) *Victorian Science and Victorian Values. Annals of the New York Academy of Sciences*, **364**, 111–116.
DESMOND, A. 1989. *The Politics of Evolution*. University of Chicago, Chicago.
DESMOND, A. 1994. *Huxley: The Devil's Disciple*. Michael Joseph, London.
EGERTON, A. C. 1948. Lord Rayleigh 1875–1947. *Obituary Notices of Fellows of the Royal Society for 1947*, 503–538.
EVE, A. S. 1939. *Rutherford*. Cambridge University, Cambridge.
FOURIER, J. 1822. *Thèorie Analytique de la Chaleur*. Paris.
FULLER, J. G. C. M. 2001. Before the hills in order stood: the beginning of the geology of time in England. *In*: LEWIS, C. L. E. & KNELL, S. J. (eds) *The Age of the Earth: from 4004 BC to AD 2002*. Geological Society, London, Special Publications, **190**, 15–23.
GEIKIE, A. 1868. On denudation now in progress. *Geological Magazine*, **5**, 249–254.
GEIKIE, A. 1899. Presidential address to Section C. *Report of the British Association for the Advancement of Science*, 719–730.
GOULD, S. J. 1993. *Eight Little Piggies*. Penguin, London.

HALES, W. 1809–1812. *A New Analysis of Chronology* (three vols). London.

HOFMANN, A. W. 2001. Lead isotopes and the age of the Earth – a geochemical accident. *In*: LEWIS, C. L. E. & KNELL, S. J. (eds) *The Age of the Earth: from 4004 BC to AD 2002.* Geological Society, London, Special Publications, **190**, 223–236.

HOLMES, A. 1913. *The Age of the Earth.* Harper, London and New York.

KAMBER, B. S., MOORBATH, S. & WHITEHOUSE, M. J. 2001. The oldest rocks on Earth: time constraints and geological controversies. *In*: LEWIS, C. L. E. & KNELL, S. J. (eds) *The Age of the Earth: from 4004 BC to AD 2002.* Geological Society, London, Special Publications, **190**, 177–204.

KAUFFMAN, G. B. 1982. The atomic weight of lead of radioactive origin: A confirmation of the concept of isotopy and the group displacement laws. *Journal of Chemical Education,* **59**, 3–8, 119–123.

KING, C. 1893. The Age of the Earth. *American Journal of Science,* **45**, 1–20.

KNELL, S. J. 2000. *The Culture of English Geology, 1815–1851: A Science Revealed Through Its Collecting.* Ashgate, Aldershot.

KUBRIN, D. 1990. 'Such an impertinently litigious lady': Hooke's 'Great Pretending' vs. Newton's Principia and Newton's and Halley's Theory of Comets. *In*: THROWER, N. J. (ed.) *Standing on the Shoulders of Giants.* University of California, Berkeley, 55–90.

LEWIS, C. L. E. 2000. *The Dating Game.* Cambridge University, Cambridge.

LEWIS, C. L. E. 2001. Arthur Holmes' vision of a geological timescale. *In*: LEWIS, C. L. E. & KNELL, S. J. (eds) *The Age of the Earth: from 4004 BC to AD 2002.* Geological Society, London, Special Publications, **190**, 124–138.

MANNING, A. 2001. Time, life and the Earth. *In*: LEWIS, C. L. E. & KNELL, S. J. (eds) *The Age of the Earth: from 4004 BC to AD 2002.* Geological Society, London, Special Publications, **190**, 253–264.

MCGEE, W. J 1892. Comparative chronology. *American Anthropologist,* **5**, 327–344.

MORRELL, J. 2001. Genesis and geochronology: the case of John Phillips (1800–1874). *In*: LEWIS, C. L. E. & KNELL, S. J. (eds) *The Age of the Earth: from 4004 BC to AD 2002.* Geological Society, London, Special Publications, **190**, 85–90.

NEWTON, I. 1728. *The Chronology of Antient Kingdoms Amended.* London.

NIER, A. O., THOMPSON, R. W. & MURPHY, B. F. 1941. The isotopic constitution of lead and the measurement of geological time. III. *Physical Review,* **60**, 112–116.

OLDROYD, D. R. 1996a. *Thinking about the Earth: A History of Ideas in Geology.* Athlone, London.

OLDROYD, D. R. 1996b. The geologist from Cromarty. *In*: SHORTLAND, M. (ed.) *Hugh Miller and the Controversies of Victorian Science.* Clarendon, Oxford, 76–121.

PATTERSON, C. C. 1956. Age of meteorites and the Earth. *Geochimica et Cosmochimica Acta,* **10**, 230–237.

PERRY, J. 1895. On the Age of the Earth. *Nature,* **51**, 224–227.

PHILLIPS, J. 1849. Geology. *Encyclopaedia Metropolitana,* Vol. 6.

REES, M. J. 2001. Understanding the beginning and the end. *In*: LEWIS, C. L. E. & KNELL, S. J. (eds) *The Age of the Earth: from 4004 BC to AD 2002.* Geological Society, London, Special Publications, **190**, 275–283.

RUDWICK, M. 2001. Jean-André de Luc and nature's chronology. *In*: LEWIS, C. L. E. & KNELL, S. J. (eds) *The Age of the Earth: from 4004 BC to AD 2002.* Geological Society, London, Special Publications, **190**, 51–60.

SCHLOTHEIM, E. F., BARON VON. 1813. *Taschenbuch für die Gesammte Mineralogie*, vol. VII, 2–134, [Translated extract in MATHER, K. F. & MASON, S. L. (eds) 1939. *A Source Book in Geology.* McGraw-Hill, New York, 174–175].

SHIPLEY, B. 2001. 'Had Lord Kelvin a Right?': John Perry, natural selection and the age of the Earth, 1894–1895. *In*: LEWIS, C. L. E. & KNELL, S. J. (eds) *The Age of the Earth: from 4004 BC to AD 2002.* Geological Society, London, Special Publications, **190**, 91–105.

SOLLAS, W. J. 1900. Presidential address, Section C – Geology: Evolutional Geology. *Report of the British Association for the Advancement of Science, Bradford 1900,* 711–730.

SOULAVIE, J. L. G., ABBÉ. 1781. *La Chronologie Physique des Volcans,* Paris. [Translated extract in MATHER, K. F. & MASON, S. L. (eds) 1939. *A Source Book in Geology.* McGraw-Hill, New York, 155–157].

STRINGER, C. B. 2001. Dating the origin of modern humans. *In*: LEWIS, C. L. E. & KNELL, S. J. (eds) *The Age of the Earth: from 4004 BC to AD 2002.* Geological Society, London, Special Publications, **190**, 265–274.

TAYLOR, K. 2001. Buffon, Desmarest, and the ordering of geological events in *époques*. *In*: LEWIS, C. L. E. & KNELL, S. J. (eds) *The Age of the Earth: from 4004 BC to AD 2002.* Geological Society, London, Special Publications, **190**, 39–49.

THOMSON, W. (LORD KELVIN). 1862. On the Age of the Sun's Heat. *Macmillan's Magazine,* **5**, 288–393.

THOMSON, W. (LORD KELVIN). 1864. On the Secular Cooling of the Earth. *Transactions of the Royal Society of Edinburgh,* **23**, 157–169.

THOMSON, W. (LORD KELVIN). 1871. On Geological Time. *Transactions of the Geological Society of Glasgow,* **3**, 321–329.

THOMSON, W. (LORD KELVIN). 1895a. Copy of a letter from Lord Kelvin, *Nature,* **51**, 227.

THOMSON, W. (LORD KELVIN). 1895b. The Age of the Earth. Nature, **51**, 438–440.

TORRENS, H. S. 2001. Timeless order: William Smith (1769–1839) and the searching for raw materials 1800–1820. *In*: LEWIS, C. L. E. & KNELL, S. J. (eds) *The Age of the Earth: from 4004 BC to AD 2002.* Geological Society, London, Special Publications, **190**, 61–83.

VACCARI, E. 2001. European views on terrestrial chronology from Descartes to the mid-eighteenth century. *In*: LEWIS, C. L. E. & KNELL, S. J. (eds) *The Age of the Earth: from 4004 BC to AD 2002.*

Geological Society, London, Special Publications, **190**, 25–37.

WALCOTT, C. D. 1893. Geological time as indicated by the sedimentary rocks of North America. *Journal of Geology*, **1**, 639–676.

WINCHELL, A. 1883. *World Life, or Comparative Geology*. Chicago.

WYSE JACKSON, P. N. 2001. John Joly (1857–1933) and his determinations of the age of the Earth. *In*: LEWIS, C. L. E. & KNELL, S. J. (eds) *The Age of the Earth: from 4004 BC to AD 2002*. Geological Society, London, Special Publications, **190**, 107–119.

YOCHELSON, C. & LEWIS, C. L. E. 2001. The age of the Earth in the United States (1892–1931): from the geological viewpoint. *In*: LEWIS, C. L. E. & KNELL, S. J. (eds) *The Age of the Earth: from 4004 BC to AD 2002*. Geological Society, London, Special Publications, **190**, 139–155.

Before the hills in order stood: the beginning of the geology of time in England

JOHN G. C. M. FULLER

2 Oak Tree Close, Rodmell Road, Tunbridge Wells, Kent TN2 5SS, UK

Abstract: That order should govern the nature of the world is an idea not confined to England, though the history of science in this country demonstrates again and again that a conception of Divine Order lay at its heart. To people of earlier days, want of order implied confusion, displacement, derangement, time out of joint, even the presence of malevolent power – 'when the planets in evil mixture to disorder wander'. The divine scheme revealed by scripture was a frame and support for Earth science. It told an indisputable story of an ordered beginning, a diluvial reordering, and a future end in dissolution. It was a story backed by secular law, and no thinking person could have been unaware of it. Yet 'when' and 'how' were legitimate questions, answered in detail by hexaemeron writers. It is an educational curiosity in England that a particular Biblical chronology drawn up in 1650 accompanied scriptures printed for use in schools until 1885 – a matter of consequence to the history of all geological thought in this country.

This contribution to the subject of time in geology offers some background to Elizabethan and early-seventeenth-century conceptions of the Earth's history, and explains how central control by Government enabled one particular version of chronology to pervade all common perceptions of English and American Earth science to the end of the nineteenth century. Freethinkers and other hapless writers on the history of science commonly introduce Archbishop Ussher's date for the beginning of the world, 4004 BC, as if it were a joke, yet fail to explain it, and never say why they find it amusing. Remember, as Elias Ashmole said, 'Posterity will pay us in our own coin, should we deride the behaviour and dress of our ancestors' (Ashmole 1652, Preface).

Also examined are Leonard Horner's problems with the Oxford Bibles in 1861. Then President of the Geological Society, he sought to discover 'by what specific authority' was Archbishop Ussher's supposed date for the Creation placed in English Bibles, where schoolchildren might take it to be part of Holy Writ (Horner 1861). Horner's question has lain unattended ever since.

Order, the Great Chime and Symphony of Nature

> Whatsoever things are of Natures tempering and dighting, either in the earth his closet or entrayles, or within the water, may well be devided and sorted into these foure kinds: Earthes, Liquors or Juices, Stones, Mettales.

Thus begins the Preface to John Maplet's book of 1567, *A Greene Forest*. Maplet was one of several leading naturalists then at Cambridge, and was a fellow of Caius. He described an array of stones and earths according to their usefulness and rarity, first sorting them into a three-fold hierarchy. Of lowest rank were those deemed common and base; above them came stones less common though still base; and occupying higher places were those judged rare and precious. Like the rest of Creation, everything belonging to the Earth had its proper place in the natural order, and Maplet's mineralogy illustrated a characteristic mode of thinking among Elizabethan writers (Tillyard 1943). All things had been ordered by Providence:

> And it may not be called ordre, excepte it do contayne in it degrees, high and base, accordynge to the merite or estimation of the thyng that is ordered. Where there is any lacke of ordre nedes must be perpetuall conflicte. More over take away ordre from all thyngs what shulde then remayne? Certes nothynge finally (Sir Thomas Elyot 1531, p. 3).

Want of order meant derangement of nature, a threat of malevolence, and opposition to the beneficent providence of God. Within the Church too, one could have heard the same message promulgated by formal exhortation and homily, delivered with Government authority and read aloud to the people:

> Almighty God hath created and appointed all things in heaven, earth, and waters, in a most excellent and perfect order. In earth he hath assigned and appointed Kings, Princes, with other Governors under them, in all good and necessary order (Thomas Cranmer 1547, 'An exhortation concerning good order').

Twelve such homilies or sermons were issued in 1547, with royal injunction to read them at services. Their purpose was in part to augment the meagre preaching skills available, but more importantly they were to strengthen central control of doctrine, for the local church in both village and town was for everyone the principal agency of social regulation. Queen Elizabeth reissued the first (1547) *Book of Homilies* in 1559, revised 'for the better understanding of the simple people', and issued a second one in 1563. She regarded these books as a powerful means of control: 'I wish such men as may be found not worthy to preach, be compelled to read homilies' (quoted by Neale 1953–1957, vol. 2, p. 70).

Elizabethan thought control

But order and hierarchy by themselves could say little about past events and the passage of time. The only history of the Earth on general offer spoke of an initial Chaos regulated and transformed into a finished world within seven days. According to this Biblical account, the inhabited world suffered catastrophic inundation by water; and further narratives foretold that it was destined for annihilation by fire, possibly soon. Passages in the Bible describing these three events – Creation, catastrophic Flood, and final Conflagration – were formally read aloud at regular intervals in churches throughout England. No thinking person capable of hearing and understanding could have been unaware of them. In England they formed the common ground and shared language for any debate on Earth's history.

At her accession in 1558, Queen Elizabeth inherited a nation afflicted by disorder, bankruptcy and ecclesiastical reversion. To restore the Royal Supremacy, uniformity of doctrine, and the authority of central government in all matters concerning attendance of the people at Church, Elizabeth in 1559 signed into law her first Act of Uniformity, and issued a newly revised *Book of Common Prayer* for public use in all churches. Both actions had deep and lasting influence on the future course of English Earth science.

By powers in the 1559 Act of Uniformity, attendance at Church on Sundays and Holy Days was enforced on pain of a fine for absence, amounting for a workman to a full day's wage. In later years the fines were hugely increased. At services, important readings tied to particular calendar days by the lectionaries accompanying previous books of Common Prayer were in Elizabeth's revision of 1559 to be read additionally on selected Festival Sundays. For example, chapter 1 of Genesis, containing the Creation narrative, was entered in the old lectionaries to be read on 2 January, which would fall on a Sunday only fourteen times in a hundred years. Elizabeth's new lectionary accompanying the 1559 *Book of Common Prayer* called additionally for Genesis 1 to be read on Septuagesima Sunday every spring. Similarly, the Flood narratives in chapters six, seven and eight of Genesis, listed for 4 and 5 January, were read additionally at Sexagesima. The Conflagration prophecy in the Second Letter of Peter, being part of the New Testament, was appointed for reading three times a year on fixed dates in April, August and December.

The Queen's insistence on uniformity of doctrine, the newly augmented lectionary, and her coercive pressure on everyone to attend the established Church, shaped and controlled English thinking about the nature of the Earth for the next two hundred years. By injunction and monopoly she took control of printing in London, and allowed only two presses elsewhere, one at Oxford and another at Cambridge. It is hardly an exaggeration to say that the history of modern geology in England and North America cannot be readily explained without acknowledging the role of central government during the reign of Elizabeth I.

The General Epistle of Barnabas

Propagation of a doctrine that the world was destined to a fiery end, the time *when* of its ending being withheld, aroused prophetic divinations of all kinds. The core speculation was this: if the world has a predetermined lifetime, and the amount of that lifetime now past could be discovered, then the time still remaining could be forecast. Two references in canonical Scripture were thought to be relevant: Psalm 89 ('Vulgate', v. 4): 'For a thousand years in thy sight are as yesterday, which is past'; and (2) the Second Letter of Peter (3.8): 'But of this one thing be not ignorant, my beloved, that one day with the Lord is as a thousand years, and a thousand years as one day.' In Elizabethan times, as now, these thousand-year days were coupled to the six Creation days, though considered either separately or together they did not offer any measure of the world's expected lifetime. On the other hand, the General Epistle of Barnabas, using the same data, was very specific on this (12: 3b-5):

> And God made in six days the works of his hands; and he finished them on the seventh day, and he rested the seventh day, and sanctified it. Consider, my children, what that signifies, he finished them in six days. The meaning of it is this; that in six thousand years the Lord God will bring all things to an end. For with him one day is a thousand years; as he himself testifieth, saying, Behold this day shall be as a thousand years. Therefore, children, in six days, that is, in six thousand years, shall all things be accomplished.

This Epistle dates to the fourth century *Codex Siniaticus*, and its forecast of six thousand years for the ultimate duration of time was familiar to Elizabethans, as Shakespeare noted in *As You Like It* (Act 4, Scene 1) in 1599: 'The poor world is almost six thousand years old.' He would have known that he was living in the sixth millennium of the world, for even the folklore of the countryside said that the birth of Christ had freed old Adam from four thousand years in captivity:

> Adam lay ybounden, bounden in a bond:
> Four thousand winter thought he not too long ...
> Ne had the apple taken been,
> Ne had never our Lady abeen heavene queen.[1]

For Shakespeare, that event lay nearly sixteen hundred years in the past. Doomsday forecasting was commonplace. Whether English, Latin or Greek, the pool of data varied little. Prognosis by numbers entertained and appalled those who listened to its secrets. Take, for example, this receipt for the burning of the world, *mundi conflagratio*. Add together the Roman numeral letters hidden in those two words, and from them read the year MDCLVII, 1657. The year of that number *Anno Christi* will be fatal, for *Anno Mundi* 1657 was also fatal. In that year of the world, some said, Noah's Flood had come.

> We at the present see Time's changing state,
> And Nature's fearful alterations,
> As if Time now did preach the Heavens' debate,
> And Stars to band in dismal factions.
> (John Norden 1600).

Speculum Mundi, 1635

All things on Earth and in the seas, or belonging to the celestial spheres above, came into being on one of the six days of Creation. So reasonably it followed that each day's work by the Creator, as recorded in the opening verses of the book of Genesis, might be made the subject of an individual study, a herbal or a bestiary perhaps, or a discourse on stones and metals such as the first part of John Maplet's *A Greene Forest*. In that way, natural history came to be incorporated within an ancient literary form called a *hexaemeron* or *hexameron*, meaning a collected treatise on the works done during the six days of Creation. Such books were in effect encyclopaedias of natural history, arranged in a chronological sequence.

In 1635, John Swan, Rector of Sawston near Cambridge and formerly a scholar at Trinity College, published a lengthy hexameron entitled *Speculum Mundi, or a Glasse Representing the Face of the World*. This work attracts attention by combining, possibly for the first time in England, descriptions of things in Nature with an estimation of time elapsed since Creation; and further, that the passage of time manifestly had brought decay. Swan's book thus set itself apart from the works of Bible chronologers, for it treated time as a matter of physical consequence within a world of natural objects and phenomena – celestial bodies, air, sea, dry lands, minerals, plants and animals. Swan's hexameron went far beyond a treatment of time solely as a problem of scriptural interpretation. For example he wrote:

> All things which are to us conspicuous, consisting of matter and form, are of themselves frail and fading, having such a nature, that they either are or may be subject to corruption; but such is the world: and therefore as in respect of its essence it is finite; so likewise in respect of time it cannot be infinite ... so that if the parts of the world be subject to corruption, then must likewise the whole world also (Swan 1635, p. 3).

On the finite character of the world, Swan rejected the idea of a history in six evenly-timed eras or stages of one thousand years. 'Yet nevertheless', he wrote, 'I will not deny but that the world may stand six ages before it endeth.' By virtue of their source, the six ages whose durations he chose to calculate were exactly the same as those named and described in Robert Grosseteste's *Hexaemeron* of 1225 (Table 1).

---
[1] English partsong carol, 15th century. Sloane MS No. 2593.

Table 1. *The six ages of the world according to the hexaemera of Robert Grosseteste and John Swan*

As there were Six days of Creation So there are Six Ages of the World	
Robert Grosseteste *Hexaemeron*, c.1225	**John Swan** *Speculum Mundi*, 1635
1. Adam to Noah	1656
2. Noah to Abraham	422
3. Abraham to David	866
4. David to Captivity	448
5. Captivity to Christ	605 ±
	3997
6. The Christian Era	1635

Swan's elapsed time before Christ totalled 3997 years. He counted the years of the fifth age as 605 'or thereabouts', thus placing the overall sum of the first five ages potentially at an exact figure of 4000 years. For Swan, the years of the sixth age would have been 1635, though still unfinished: 'These I grant to be the six ages of the world: but who is so mad as to say or think that there were just [equal] thousands of yeares betwixt each or any of them?' (Swan 1635, p. 16).

The reason for Swan's uncertainty over the duration of the fifth age was that there were no canonical scriptures between Malachi and the New Testament, and time had to be allowed for the supposed Chaldean, Persian, Grecian and Roman Monarchies. Swan's book went through three further editions – 1643, 1665 and 1670 – and it appeared in William London's 1657 *Catalogue of the Most Vendible Books in England*. Besides the main catalogue, *Speculum Mundi* was also listed in a select group of seven natural histories, deemed by London to be 'properly usefull for schooles.' In 1653, Swan brought out another work, *Calamus Mensurans: or the Standard of Time*, containing further computations of scriptural chronology.

Scriptural chronology

Given that the present world could be expected to exist for six thousand years, more or less, according to popular belief based on Biblical and Apocryphal tradition, calculations might determine how much of that time was past, and how much lay ahead. There were three main scriptural sources.

(1) The Septuagint, or Greek Old Testament, adopted by Jewish people of the Diaspora at Alexandria. This dated back at least to the second century before the Christian era, and remains in Greek use.

(2) The Vulgate or Latin Bible, the Old Testament of which Jerome translated from ancient Hebrew texts toward the end of the fourth century of the Christian era. This was the original text for English Bibles current in Queen Elizabeth's time.

(3) The Massoretic text agreed by a group of Hebrew scholars in the second century of the Christian era, which is preserved in copies dating from the tenth century. These were the chief Old Testament sources for the Authorized or King James' translation of 1611.

As well as internal difficulties in the setting of dates for a literature not intended as a timetable of events, the Greek Septuagint and Hebrew Massoretic Scriptures differed somewhat from one another in the elapsed time recorded between important early events, particularly in the years from Creation to Abraham's entry into

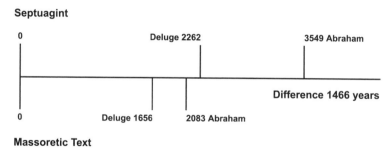

Fig. 1. Differences in the elapsed times between early events in the Septuagint and Massoretic texts, from Creation to the arrival of Abraham in Canaan.

Canaan. This is best illustrated by a diagram showing the years of the world, counting from Creation (Fig. 1).

There are two further causes of major difficulty. Firstly, there was a lengthy time-gap unfilled by canonical writings between the end of the Hebrew Scriptures and the opening of the New Testament. Secondly, historically verifiable dates, rather than commonly accepted Biblical ones, grew more troublesome to accommodate in the later parts of the Old Testament. Everyone reading the Massoretic text seemed to agree that the Great Deluge began in the year of the world 1656, for this was not a date that could be safely challenged in England after publication of the King James' Bible in 1611. Nevertheless, dates of events in the Old Testament later than the Deluge were increasingly liable to be interpreted and controlled by other historical records.

Government control over publishing in England broke down in the early 1640s, and a Civil War ravaged the country between 1642 and 1649. The King was defeated by Parliamentary forces, and bloodily murdered. For eleven years the country was in the hands of a Puritan dictatorship. It was a period marked by huge proliferation of Bible-oriented literature, including masterworks of scriptural chronology by John Lightfoot (1602–1675) and James Ussher (1581–1656).

By Government appointment and preferment, Lightfoot became Vice-Chancellor of Cambridge University in 1654. He is now remembered chiefly as a chronicler of the Scriptures, and his dates in the Old Testament accord exactly to the Massoretic text, which was the basis of the 1611 King James' Bible. He quoted the year of the world 3519 (about 480 years before the Christian era) as the point where 'the Chronicle of the Old Testament ends' (Lightfoot 1647).

Unlike Lightfoot, Ussher was a professional Irish cleric, though resident in England. He is chiefly remembered for his *Annales Veteris Testamenti* ('Annals of the Old Testament') in which he propounded once more the chronology of the Massoretic text (Ussher 1650–1654). Ussher's significance in the history of science, and of geology in particular, derives entirely from an association between his name and the dates later printed in thousands of English Bibles. By itself, his chronology added nothing to the Massoretic dataset employed by scores of other chronologists.

Lastly, on Biblical chronology, there was disagreement about placing a switch-point where the Old Testament years ended and the Christian era began. The need to find some credibly fixed point in time arose because the Christian era was not invented until the sixth century, and did not pass into general use in Europe until the eleventh century. A date for the Nativity using the old Roman calendar put it within a ten-year spread, interpretable as lying somewhere between years that are now called 7 BC and AD 3.

After a period of general acceptance, the Nativity year chosen was found to be about four years later than it should have been, the error being due to an original misunderstanding of the Roman calendar. This four-year error is the source of the oddly incremental 4 in the 4004 BC date of Creation introduced into English Bibles, where one can read against the verse 'When Jesus was born in Bethlehem' (Matthew 2:1) these words: 'The Fourth Year before the Common Account called Anno Domini'. That is to say, the years of the world at the Nativity were 4000, as everyone had said, but the Christian calendar had been started four years late (as it still is).

Had our ancestors dealt less circumspectly with Holy Writ they might have realized that their dataset was incomplete. Who in England before the Civil Wars would have dared say that the Scriptures were insufficient to reveal the Divine will, and were unable to reveal the time *when* of the world's Creation?

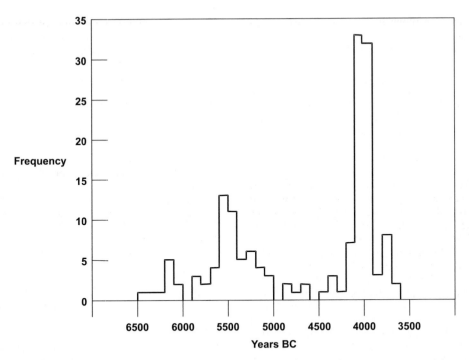

Fig. 2. Frequency histogram of 156 estimated ages of the Earth, drawn from data published by William Hales (1809, vol. 1 pp. 211–214). These various estimates were based largely, though not exclusively, on Biblical sources, and ranged in date of origin from AD 94 to AD 1800. Histogram peaks at 4000 BC and 5500 BC reflect clusters of concentration on the Massoretic and Septuagint texts. Archbishop Ussher's date of Creation hides indifferently among sixty-five others, including John Swan's, and the countryside wisdom of medieval folklore, all of them in the peak between 3900 and 4100 BC.

Small wonder that Samuel Pepys in 1661, after discoursing with Jonas Moore on the possibility that England and France were formerly joined, wrote in his *Diary* on 23 May that Moore: 'Spoke very many things, not so much to prove the Scripture false, as that the time therein is not well computed nor understood'. Of course, Pepys was writing after the Civil Wars, when constitutional government had been restored, and natural sciences were flourishing. Not least among them were geological enquiries into the fabric of the Earth, and various semi-theological proposals seeking to fit Scripture and natural philosophy into coherent 'Theories'. So fierce were partisan political controversies clustering around the words of Holy Writ that the recently formed Royal Society ruled to abstain from any discussion of them, yet when Burnet's *Telluris Theoria Sacra* claimed to solve all phenomena of Noah's Flood consonant with the Scriptures, the writings of the ancients, and the Cartesian philosophy, a meeting of the Society in 1681 desired him to bring in an account of it (Birch 1756–1757, vol. 4, p. 69n).

Chronologies continued to multiply through the next century, and by 1809 William Hales was able to enumerate 156 separate attempts to fix a date of Creation (Hales 1809). This was proof, if any were needed, that James Ussher's estimate was not of itself specially remarkable (Fig. 2).

Leonard Horner and the annotated Bible

To latter-day ears, a proposition that the Earth sciences in England came into being and were successfully developed without reference to geological time might seem unreasonable, yet is demonstrably true. No appeal to the passage of time was needed to arrange Elizabethan stones and minerals into orders and hierarchies; physical congruities among fossil things found in the Earth had no Caroline calendars; even whole strata could be measured and described according to their distinctive contents, whatever their age. For example, it is a fact that during the early years of the nineteenth century, William Smith mapped the strata of England without reference to geological time, largely distinguishing each stratum by means of its contained variety of organic remains. The strata, as Smith delineated them, were parts of a rotating system set in motion at the beginning of the world – 'the motion of the earth, which probably commenced while these strata were in a soft state' (Smith 1801 in Cox 1942, p. 85). The stratigraphic relationships among them were all of a piece, and owed little to the subsequent passage of time, other than superficial damage caused by the Deluge.

Why was stratigraphic geology at this time being conducted by its chief practitioner without mention of any relative antiquity among the strata? Two reasons emerge: firstly, that strata could be separately identified by their constitution and physical properties alone; and secondly, that the strata had come into existence together, as a whole, by Divine fiat on the third Creation day. That Biblical fact had been taught in schools since the days of Queen Elizabeth, and was still being taught when William Smith was at school. The central tenet for Smith or any other schoolboy was that the Earth had been set to turn on its axis *after* the strata had formed, not before.

Leonard Horner was in his second presidency of the Geological Society when, in his Anniversary Address of 1861, he felt compelled to ask whether it was wise to teach children that mankind appeared less than six thousand years ago, when that was manifestly at odds with 'modern discoveries in ethnology and philology' (Horner 1861):

> It will be useful to look into the history of ... that marginal note in our Bibles over against the first verse of the first chapter of the Book of Genesis, that "In the beginning God created the heaven and the earth" [four thousand and four years *before* the birth of Christ].

This date Horner rightly ascribed to a chronology derived from the works of Archbishop Ussher, from which he quoted two passages of Ussher's original Latin. The first was from *Annales Veteris Testamenti*, and the second from *Chronologia Sacra*. Horner did not have translations printed in the published version of his Address, leaving some geologists of later generations ignorant of what Ussher actually propounded in the quoted passages, so they are here translated:

> Annals of the Old Testament calculated from the first origin of the world. In the beginning God created the heaven and the earth, and this beginning of time, following our chronology, occurred at the beginning of that night which preceded the 23rd of October in the year 710 of the Julian period. So on the first day of the age, October 23rd, the first weekday, God created the angels together with the highest heaven: then, having completed the highest roof of the work, the wondrous artist moved on to the lowest foundations of this world's creation, and established this lowest globe forged from the earth and the abyss [primal Chaos].

> Sacred Chronology. So from the evening opening the first day of the world, up to the midnight proving the start of the day (in fact 25th December) on which we suppose Christ was born, we conclude that 3999 Julian years, 2 months of 30 days, 4 days and 6 hours had passed. In fact we conclude that at the 1st January of the Julian period year 4714 (from when we calculate the beginning of the commonly-accepted Christian era) 4003 years, 2 months, 11 days and 6 hours had passed.

Horner's Address continued:

> I have endeavoured, by enquiries at Oxford, Cambridge, Edinburgh, and at the Queen's printers in London, to ascertain by what specific authority, Royal or ecclesiastical, the date of 4004 was added to the first verse in Genesis in the authorized version [of the Bible], and I have not been able to discover that any record exists of such an authority.

His chief concern, aside from the apparent absence of authority for placing dates in Bibles, was the effect that such instruction had on the education of children. He noted for instance Old Testament histories, as used in schools, stating: 'First Lesson ... the Creation, B.C 4004 ... drawn from the Sacred Volume'.

Behind these concerns, and driving the careful language of the Society's *Proceedings*, was a realization that education in England, and geological education in particular, was being constrained and distorted by an ancient numerical device, older even that medieval folklore, and unaltered since the time that 'Adam lay ybounden four thousand winter long'.

Bishop Lloyd and Dr Fell

Whether such a specific authorization as Leonard Horner sought did actually exist – one that gave Royal or ecclesiastical warrant for placing Ussher's dates in Bibles – remains elusive. But evidence can now be found to show how chronological annotations first came to be added to printed Bibles, and that the practice went on for many years after Horner had questioned it.

The main story began in 1672 at Oxford, when Dr John Fell proposed an annotated Oxford Bible. Fell's influence upon the early history of Oxford printing was immense. He was Dean of Christ Church from 1660, Bishop of Oxford from 1675, and earlier had been treasurer of accounts for building the Sheldonian Theatre. It was his idea that printing should have a place at the Sheldonian, and there the Press was first sited.

Here is the opening of Fell's *Proposals for an Annotated Oxford Bible*, 1672:

> It is designed to print an Edition of the H. Scriptures in the University of Oxford, with all possible care & accuratenesse in reference both to the correctnesse of the Text, & beauty of the Character & all other extrinsick ornaments: with Annotations also plainly & practically rendring the mind of the Text, so as to be understood by the unlearned reader: to which will be added Arguments to the several books, Chronological observations, Geographical tables &c as seem necessary to the whole (Madan 1925b, p. 143).

So in 1672 at Oxford, the inclusion of 'chronological observations' in a Bible was proposed, and in 1679 Fell's sublessees at the Press introduced chronological dates in the margins of their edition. These dates were calculated from zero at Creation, and measured as years of the world (not the later BC–AD system); and were in accordance with Archbishop Ussher's chronology. They were reproduced later in other Oxford Bibles (Carter 1975). The 1679 Oxford edition:

> Is perhaps the first English Bible to print marginal dates as well as references and notes: they are *Anni Mundi*, not *Anni Christi*. The first is at Gen. c.5, v.3, 130; Abraham's birth is 2008; Job is dated 2400; the Battle of Megiddo is 2394; the Birth of Christ is 4000; the Crucifixion 4036 (Madan 1925a, p. 377).

Leonard Horner's Anniversary Address made specific reference to a Bible published in 1701, to which William Lloyd, Bishop of Worcester, had added chronological dates:

> It is commonly admitted that William Lloyd (1627–1717), created Bishop of Worcester in 1700, undertook, apparently on the motion of Archbishop Tenison and at the request of Convocation in 1699, to prepare an improved edition of King James' version; and that his Bible appeared in 1701 (Darlow & Moule 1968, p. 233).

Known as the *London Folio of 1701*, this edition by the King's Printer exhibited a marked difference in its chronological notation from the Oxford Bibles. Bishop Lloyd worked out a new method that first appeared in this edition, not as previously counting forward from Creation, *Anni Mundi*, but counting backwards and forwards from the Birth of Christ, BC and AD. Thus Ussher's chronological dates, which had been added to the Oxford Bible of 1679, reappeared in 1701, modified by Bishop Lloyd, and in the new system of notation.

From that time forward, the chronology of Archbishop Ussher as presented by William Lloyd enjoyed almost universal distribution: it gained near-sacred authority, and a record of citation infinitely beyond challenge. If the number of editions of a book is taken as some measure of its influence, nearly a thousand separate editions of the English Bible had been published even before the nineteenth century began. A revised version of the Old Testament appeared in 1885, without marginal chronology, but the authorized or King James' version continued to be printed. Even present-day 'Reference Editions' from Oxford have marginal notes 'based largely on Ussher's chronology', beginning at chapter 11 of Genesis with '21st c. B.C.'.

Concluding note

Endless repetition, association with Holy Writ, and worldwide distribution in English were the reasons for popular acceptance of Ussher's fabled date of 4004 BC for the Creation of mankind and the world. The figure enjoyed an unchallengeable citation frequency and a guarantee of immaculate quality, yet it revealed no practical difference from hundreds of other calculations made from the same body of data. In fact, the date 4004 BC was not really Ussher's at all: William Lloyd was the one who introduced to English Bibles the BC–AD system of dating with a fixed-point Nativity. Previously, dates had been presented as years of the world, Creation being time-zero.

Geological time, in the sense of secular duration and measured ages of rocks, did not become a matter of popular account until the middle part of the nineteenth century. Minerals and fossils needed no time of duration on Earth to be arranged and classified in cabinets. Stratigraphic mapping of matchless originality and accomplishment had no need of age, relative or otherwise, until the 1820s (Fuller 1995).

Yet nearly two centuries previously, in 1616, an entry in Samuel Pepys' *Diary* revealed a seventeenth-century intellectual interest in computing more accurately the age of the world. Of course, it was scriptural interest as much as geological; and that was to be expected, for Queen Elizabeth had earlier introduced a system of central thought-control and printing-management by Government and the Church, rendering it certain that in England there would be Biblical models and a common language of biblical understanding for building future theories of the Earth. These features, consequent on sixteenth-century legislation, persisted in English Earth science, and bore heavily on geological thought well into the second half of the nineteenth century. For example, at the 1865 meeting in Birmingham of the British Association for the Advancement of Science, a public declaration or manifesto was put up in defence of Biblical geology, seeking approval among the delegates. The preamble included these words: 'It is impossible for the Word of God as written in the book of Nature, and God's Word written in Holy Scripture, to contradict one another' (Kinns 1882, p. 5). Of the 716 signatories, 111 were Fellows of the Geological Society, and 72 were Fellows of the Royal Society. Leonard Horner had good reason for asking questions about English education and Oxford Bibles.

I would like to thank particularly R. Gosling for translating passages of J. Ussher's Latin, Canon P. A. Welsby for help with enquiries at Lambeth Palace Library, and A. Cushing for producing an expertly finished text.

References

ASHMOLE, E. 1652. *Theatrum Chemicum Britannicum*. Grismond, London.
BIRCH, T. 1756–1757. *The History of the Royal Society of London* [reprinted 1968, 4 vols, Johnson, USA].
CARTER, H. 1975. *A History of the Oxford University Press: 1 To the Year 1780*. Clarendon, Oxford.
CRANMER, T. 1547. *Sermons or Homilies, Appointed to be Read in Churches*. London.
DARLOW, T. H. & MOULE, H. F. 1968. *Historical Catalogue of Printed Editions of the English Bible*. British and Foreign Bible Society, London.
ELYOT, T. 1531. *The Governour*. London [reprinted 1907, J. M. Dent].
FULLER, J. G. C. M. 1995. *'Strata Smith' and his Stratigraphic Cross-Sections, 1819*. AAPG, Tulsa, and the Geological Society, London.
GROSSETESTE, R. c.1225. *Hexaemeron* [translated by MARTIN, C. F. J. 1996. *On the Six Days of Creation*, Oxford].
HALES, W. 1809. *A New Analysis of Chronology and Geography*, 3 vols. London.
HORNER, L. 1861. Anniversary address of the president. *Quarterly Journal of the Geological Society*, **17**, xxvii–lxxii.
KINNS, S. 1882. *Moses and Geology; or, the Harmony of the Bible with Science*. Cassell, London.
LIGHTFOOT[E], J. 1647. *The Harmony, Chronicle and Order of the Old Testament: the Years Observed and Laid Down Chronically*. London.
LONDON, W. 1657. *Catalogue of the Most Vendible Books in England*. London.
MADAN, F. 1925a. *Oxford Books: A Bibliography*, Vol. 3. Clarendon, Oxford.
MADAN, F. 1925b. The Oxford press, 1650–75. *The Library* (4th series), **6**, 113–147.
MAPLET, J. 1567. *A Greene Forest, or a Naturall Historie*. Denham, London.
NEALE, J. E. 1953–57. *Elizabeth I and Her Parliaments, 1584–1601*, 2 Vols. Cape, London.
NORDEN, J. 1600. The approaching end. *Vicissitudo Rerum*.
SMITH, W. 1801. Uncompleted preface to *Natural Order of Strata*. Reproduced in COX, L. R. 1942. New light on William Smith and his work. *Proceedings of the Yorkshire Geological Society*, **25**, 81–90.
SWAN, J. 1635. *Speculum Mundi ... Whereunto is Joined an Hexameron*. Cambridge.
TILLYARD, E. M. W. 1943. *The Elizabethan World Picture*. Chatto & Windus, London.
USSHER, J. 1650–1654. *Annales Veteris et Novi Testamenti* (Part 1 1650; Part 2 1654). London.

European views on terrestrial chronology from Descartes to the mid-eighteenth century

EZIO VACCARI

Centro di studio sulla Storia della Tecnica – CNR, c/o Università di Genova, via Balbi 6, 16126 Genova, Italy (email: ezio.vaccari@lettere.unige.it)

Abstract: The Theories of the Earth formulated by the English scholars Thomas Burnet, William Whiston and John Woodward at the end of the seventeenth century circulated widely within the continent of Europe during the first decades of the eighteenth century. These theories established a sequence of physical conditions of the Earth according to the chronology outlined in the Book of Genesis, emphasizing two main stages: the Creation and the Deluge. Although the authority of the Biblical account of the age and early history of the Earth was normally accepted at the beginning of the eighteenth century, the continental reception of English Theories of the Earth varied. This was due to the complexity of the European context which since the 1660s had produced the theories of René Descartes, Gottfried Wilhelm Leibniz and Athanasius Kircher, as well as Nicolaus Steno's dynamic view on the development of the Earth's surface. Steno emphasized the importance of the interpretation of rock strata in the field for reconstruction of the Earth's history. He also carefully avoided contradicting the Biblical account and associated the Deluge with one of the geological stages identified in his history. Nevertheless, the Stenonian heritage stimulated some Italian scientists – such as Antonio Vallisneri, Luigi Ferdinando Marsili, and later Giovanni Targioni Tozzetti and Giovanni Arduino – to presuppose, within the results of their researches, an indefinitely great antiquity of the Earth. Theoretical models linked to Biblical chronology included those of Emanuel Swedenborg in Sweden and Johann Jakob Scheuchzer in Switzerland, while in France, Benoît De Maillet proposed a Theory of the Earth which was censured by the Church because of its possible implications regarding the eternity of matter. Among European scholars of the first decades of the eighteenth century, the Stenonian heritage (notably the necessity of fieldwork in a regional context) and the global Theories of the Earth were equally influential.

Over the last half century there have been several studies of early speculation on time and the age of the Earth. These include works by Haber (1959), Rossi (1979), Albritton (1980, 1984), Rudwick (1976, 1986, 1999), Dean (1981), Gould (1987), Guntau (1989), Dalrymple (1991), Rupke (1998), Rowland (1998) and Richet (1999). In particular, Rappaport's (1997) well-documented and stimulating analysis of the development of geological ideas between 1665 and 1750 has now become a key starting point for research into this period. Studies by Toulmin & Goodfield (1965), North (1977), Whitrow (1977, 1988), Wilcox (1987) and Holland (1999) further investigate the idea of time and its measurement. This paper offers a synthesis of ideas on the age of the Earth between the late seventeenth and the early eighteenth centuries within continental Europe. In particular it examines the reception given to the Theories of the Earth proposed by the English scholars Thomas Burnet (1635–1715), William Whiston (1667–1752) and John Woodward (1665–1728), whilst also remarking on the Italian–Swiss geological network of the early eighteenth century.

The Book of Genesis and the ancient chronologies

When several Theories of the Earth appeared in England at the end of the seventeenth century (Burnet 1681; Warren 1690; Ray 1692; Woodward 1695; Whiston 1696), the majority of European scholars saw the history of the Earth as being framed by the account given in the Book of Genesis. That history was conceived as a process of several phases, beginning with God's Creation of the world and passing through the universal catastrophe of the Deluge (Rudwick 1976, pp. 68–72; Rappaport 1997, pp. 189–199). However, a small number of scholars rejected Biblical time, adopting instead ancient philosophies of eternalism, often adapting the Aristotelian concept of endless cyclical time. This position was

From: LEWIS, C. L. E. & KNELL, S. J. (eds). *The Age of the Earth: from 4004 BC to AD 2002.* Geological Society, London, Special Publications, **190**, 25–37. 0305-8719/01/$15.00 © The Geological Society of London 2001.

supported, at the beginning of the eighteenth century, by freethinkers who, based on chronological evidence of the history of Egyptian and Chinese peoples, believed in the very remote antiquity of the world. This evidence seemed to prove an age of the Earth up to one hundred times greater than the Biblical chronology and reduced the Deluge to a merely local event exclusively linked to the history of the Hebrew people.

From the early sixteenth century, European commentaries on Genesis normally stated that time began with God's Creation of the world (Dean 1981, p. 441). Although the Biblical chronologists, who were theologians, often did not agree amongst themselves about the results of their calculations, the Earth was generally considered to be from 5000 to 6000 years old, while the Deluge was judged to have occurred 1600–2300 years after the Creation. Nevertheless, around the middle of the seventeenth century, particularly with the appearance of some treatises accused of impiety – such as Isaac de la Peyrère's (1594–1676) *Praeadamitae* (1655) – the context of chronological speculation on the age of the Earth drastically changed and became something of a 'minefield' (Rossi 1969, pp. 137–151).

An Earth without time: the Cartesian and Kircherian models

The Biblical chronologists and freethinkers were indulging in theoretical speculations or genealogical computations which contained no reference to a possible 'geological' interpretation of the Earth. However, around this time, French philosopher René Descartes (1596–1650) proposed a theoretical model for the formation of the Earth's crust in his treatise *Principia Philosophiae* (1644) (Fig. 1). Descartes theorized that the cooling Earth – an extinguished star – settled out its differentiated matter into layers. Later, vapours escaping from these lower layers passed through the external crust to leave various cavities beneath. The resulting crustal collapse created mountains and basins which became the oceans.

According to François Ellenberger (1988, pp. 216–224), Descartes did not display a particular interest or competence in geological matters. Unlike other contemporary scholars, such as Kircher and Steno, he did not consider fossils and strata or devote a complete work to the features of the terraqueous globe. Nevertheless, historians generally agree that Descartes' model, although confined within a few pages, had significant influence throughout Europe (Roger 1982). Descartes did not specify a precise timescale for his theory

and carefully avoided any possible chronological reference to the Scriptures. It has been suggested that 'Descartes supposed that major geological changes were induced over a period of (seemingly) just a few years' (Oldroyd 1996, p. 47), although he probably left open the possibilities of a short or long timescale, compatible with the Biblical chronology. On this topic Roger (1982, p. 105) remarks that 'the depth of time through which the succession of events developed itself in Descartes's cosmogony was a theoretical one, able to be actually reduced to the very instant of the creation'. But Descartes was not interested in the age of the Earth, only in its formation. His model – like that of Gottfried Wilhelm Leibniz (1646–1716) at the end of the century (Leibniz 1693) – concerns the mechanical process of the Earth's formation, quite removed from the speculative concerns of the chronologists.

The German Jesuit Atahanasius Kircher (1602–1680) held a different position. His ponderous treatise *Mundus Subterraneus* (1664–1665), aimed not to describe the formation of the Earth's surface but to illustrate its permanently functioning processes, conceived as an unchangeable 'Geocosm, the wonderful handiwork of God' (translated by Mather & Mason 1939, p. 18). Consequently, his visual representations of the terraqueous globe – the first of this kind in the early scientific literature (Fig. 2) – show the rigid Kircherian model, based on a system of internal circulation with contacts between water and fire, the 'pyrophylacia' and the 'hydrophylacia'. Kircher made no references to the age of the Earth, as his theory was in perfect harmony with the Book of Genesis and Biblical chronology. In a later work, the *Arca Noë* (Kircher 1675), the elucidation of the Biblical story of the Flood is supported by several illustrations and by a chronological table. A crucial episode of Genesis, the Deluge became central to debates on the age of the Earth between the late seventeenth and mid-eighteenth centuries (Rappaport 1978; Dean 1985; Cohn 1996).

It is still not easy to evaluate Kircher's influence in Europe. Some studies, especially of his correspondence and on the circulation of his books, have provided new light on this topic (Fletcher 1968, 1988; Gorman 1997), but it seems that the size and the production costs of *Mundus Subterraneus* – a work well known to Martin Lister, Robert Moray, John Locke, Nicolaus Steno, Baruch Spinoza, Henry Oldenburg, Christian Huygens and others – prevented its translation and consequently wider readership. The exception was a partial edition in English, of about 70 pages, limited to the treatment of volcanoes (Kircher 1669).

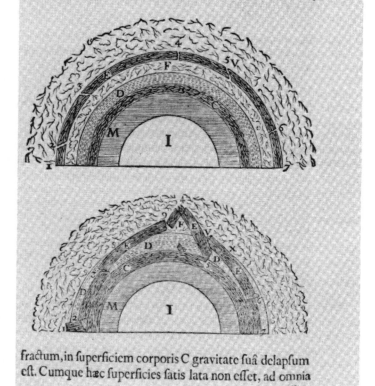

Fig. 1. The formation of the Earth's surface according to René Descartes (1644).

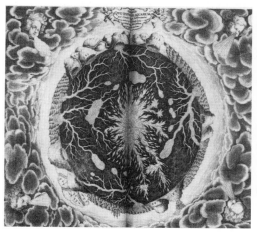
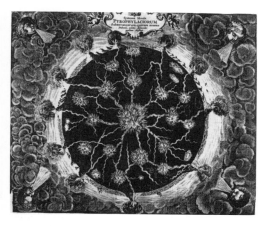

Fig. 2. The 'Geocosmus' of Athanasius Kircher (1664–1665), with the 'hydrophylacia' (left) and the 'phyrophylacia' (right).

Kircher's representation of the Earth, linked to the Mosaic tradition of the Bible, probably inspired German alchemist and mining expert Johann Joachim Becher's (1635–1682) *Physica Subterranea* (1669). This explained the Earth's internal operations with chemical reactions and productive stages caused by the combination of elements (Oldroyd 1996, pp. 38–40). In an account of the formation of the face of the Earth published in 1702 and 1706, Swedish chemist Urban Hiärne (1641–1724) also proposed the structural model illustrated by Kircher (Frängsmyr 1969, pp. 17–53, 346) (Fig. 3). However, Hiärne believed in the perpetual change of the face of the Earth and did not accept the Deluge as the exclusive geological cause of the widespread distribution of fossils. By this time fossils were generally considered to be of organic origin. The influence of Kircher was also particularly strong in Spain up to the middle of the eighteenth century (Glick 1971; Pelayo 1996, pp. 42–54). Here the power of the Jesuits encouraged the persistence of a very orthodox Catholic vision of the history of the Earth based on the Biblical chronology and the fundamental role of the Deluge, as outlined particularly in the writings of José Antonio González de Salas (1588–1651).

Steno's dynamic view of the development of the Earth's surface

In contrast, Danish anatomist Nicolaus Steno (1638–1686) was more influenced by the Cartesian model, although he placed it within the reconstruction of a local geological history substantially based on the effects of alternate forces of fire and water within the traditional Biblical timescale. François Ellenberger (1988, p. 272) has written that Steno actually saw in the field what Descartes had simply imagined and theorized in his *Principia Philosophiae*.

Between 1667 and 1669, after several explorations and much research on the geomorphology

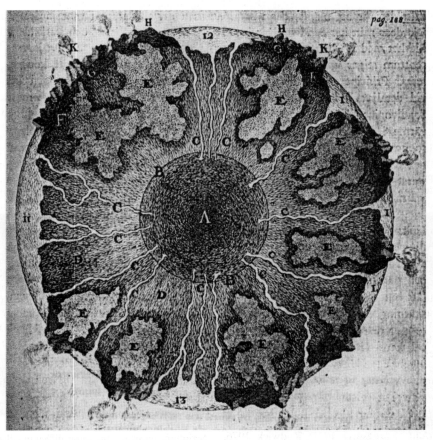

Fig. 3. The terraqueous globe according to Urban Hiärne (from *Actorum chymicorum Holmiensium parasceve, Acta et tentamina chymica*, Holmiae, 1712).

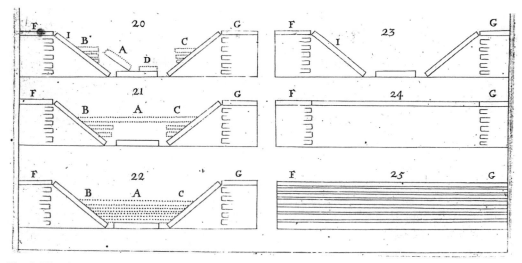

Fig. 4. The six stages of the geological history of Tuscany (in Steno 1669).

and the lithology of Tuscany and other parts of northern Italy, Steno emphasized the importance of the accurate interpretation of rock strata in the field as a sequence of records for the reconstruction of the history of the Earth (Herries Davies 1995). In the treatise *De solido intra solidum dissertatio Prodromus* (1669), which was intended to be the introduction of a greater geological and palaeontological study of Tuscany, Steno published a very significant plate representing the historical sequence of geological events. The six figures of the plate included in the *Prodromus* (Fig. 4), which may be read in chronological order from 25 to 20, introduced a new way of interpreting and illustrating the rock strata of the Earth's surface. Figure 25, which illustrates the oldest stage, shows the result of sedimentary deposition of strata within the primaeval ocean, while figure 24 shows the effects of subsequent underground erosion of the same strata supposedly due to the activity of volcanic fires and earthquakes. In the following stage, the collapse of the upper strata and the formation of valleys and mountains are described. Then, as illustrated in figure 22, new depositions of strata due to a big flood (identified with the Universal Deluge) determined the new morphology of the valley, later again modified by further episodes of underground erosion (Steno's figure 21) and finally by other collapses accompanied by the formation of a new type of relief (figure 20).

As a scholar and very religious man, Steno considered it absolutely natural and necessary to frame this geological history within the biblical account (Frängsmyr 1971; Rudwick 1976, p. 68). However, the most important stage of the Biblical timescale after the Creation, the Deluge, seems not to be the central geological event of his dynamic theory. According to Steno, in the past at least two floods covered Tuscany with water. The older strata were deposited by the primeval ocean after the second day of Creation, while the younger strata were deposited by the Deluge, which however appears to be like a normal flood, certainly not unique in its effects and instead part of a complex sequence of geological events which may be repeated in the future. Reading the *Prodromus*, one sees that in the last part of his treatise Steno (1669, pp. 69–76) introduces a series of confrontations between 'Natura' and 'Scriptura', with references to the Deluge ('universale diluvium') and to the Biblical chronology. His aim seemed to be to show the general harmony between Nature and the Bible as well as to 'correct' a geological system which appeared, rather dangerously, capable of functioning outside the chronological framework of the Book of Genesis. In fact, he states: 'ne vero a novitate periculum quisquam metuat, Naturae cum Scriptura consensus paucis exponam, recensendo praecipuas difficultates, quae circa singulas terrae facies moveri poterunt'[1] (Steno 1669, p. 69).

As with the theoretical models of Descartes and Kircher, the European influence of the 'Stenonian heritage' also deserves more detailed investigation. The *Prodromus*, which was well-known among French, German, Italian and

---

[1] 'but lest anyone be afraid of the danger of novelty, I set down briefly the agreement between Nature and Scripture, reviewing the main difficulties that can be raised about individual aspects of the Earth.'

Swiss scholars involved in geological studies since the late seventeenth century (Ellenberger 1988, pp. 246–248; 1994, pp. 130–131), also had a significant influence in England (Eyles 1958) where it was translated by one of the secretaries of the Royal Society of London, Henry Oldenburg (Steno 1671), and probably stimulated John Woodward's interest in the Earth's strata.

The Theories of the Earth and their influence

The abbreviated outline of the hypothesis on the origins of the Earth proposed by Leibniz, inspired by the Cartesian and Stenonian models, probably also reached a wide scholarly audience at the beginning of the eighteenth century, being published in three pages of the famed German periodical *Acta Eruditorum* in January 1693 (Oldroyd & Howes 1978). Later, a much fuller version was published in the posthumous *Protogaea* (1749), and this contributed decisively to the diffusion of Leibniz's geological theory (Roger 1968; Ellenberger 1994, pp. 137–148). Leibniz, like Descartes, made no direct references to the age of the Earth, although an initial mention of the Mosaic narration is balanced by the statement that 'autor arbitratur, globum terrae multo majores passum mutationes quam quisquam facile suspicetur'[2] (Leibniz 1693, p. 40). In fact, the *Protogaea* presented a sequence of geological events, from the original molten Earth to the formation of a cooled glass-like crust, which was then covered by the sea and finally collapsed shaping the mountains, and depositing different kinds of rock and fossiliferous strata.

The Biblical framework and Mosaic timescale were widely supported by the Theories of the Earth of Thomas Burnet (1681), John Woodward (1695), William Whiston (1696) and other contemporary British scholars during the 1680s and 1690s (Herries Davies 1969, pp. 63–94; Roger 1973; Porter 1979; Oldroyd 1996, pp. 48–58). Notwithstanding their basic similarities, their positions did often diverge, as in the case of the controversy between Burnet and the Newtonian theorists (Force 1983). However, these physico-theological authors generally did not follow the Biblical text literally and were not particularly concerned with calculating the Earth's exact age. On the contrary, they aimed to reconstruct a sequence of physical states of the Earth within the general framework of Genesis, not least to reinforce the authority of the Biblical timescale against the impiety of eternalism and the possible risks of an anti-Christian mechanism. Significantly, in 1690–1691, Thomas Burnet criticized the chronologists stating that 'we do not know what the true Age of the World is', because 'there are three Bibles, if I may so say, or three Pentateuchs, the Hebrew, Samaritan, and Greek: which do all differ very considerably in their Accounts, concerning the Age of the World: And the most learned men are not yet able to determine with certainty, which of the three Accounts is most authentick' (quoted by Dean 1981, pp. 443–444).

During the first decades of the eighteenth century the Theories of the Earth formulated in particular by Burnet, Whiston and Woodward circulated widely throughout Europe. Until the 1740s they received a mixed reception, but they did greatly stimulate debates on the role of biblical chronology and especially of the Deluge within the history of the Earth. The Latin editions of Burnet's *Sacred Theory of the Earth* (*Telluris Theoria Sacra*: first edition 1681, second edition 1689) were widely read throughout Europe, and a third edition was printed in Holland (Burnet 1699). Interest in the Protestant German-speaking countries in Theories of the Earth based on Biblical chronology, led to this book's translation into German (Burnet 1698), together with Whiston's (1713) *New Theory of the Earth*, which found wide acceptance in Germany by the middle of the eighteenth century (Force 1985, p. 28), and Woodward's (1744) *Essay Toward a Natural History of the Earth*. In Spain, on the other hand, the persistence of a strong Kircherian influence significantly contributed to the rejection of Burnet's theory, which was regarded as not sufficiently orthodox (Piquer 1745), and also stimulated negative comments on the writings of Woodward and Whiston (Capel 1985, pp. 120–123).

In Italy, Woodward received a more positive reception than Burnet or Whiston (Morello 1989, pp. 611–615). The cause may be found in Burnet's diluvialism, which was considered too radical by several eighteenth-century Italian scholars, and in the excessively astronomical speculations of Whiston. Woodward was instead a great field investigator and collector, who paid constant attention to strata and fossils. For this reason his *Natural History of the Earth* was of particular interest to scholars who continued the 'Stenonian heritage'.

Woodward and the Italian–Swiss geological network

John Woodward's *Essay* (1695) was first translated into Latin in Switzerland (1704) and later

---

[2] 'the author believes that the terrestrial globe has undergone much greater changes than might readily be supposed.'

into French (1735), Italian (1739) and German (1744). Consequently, the diffusion of this book into Europe is particularly important (Jahn 1972). The Latin edition was produced by Johann Jakob Scheuchzer (1672–1733), a great Swiss naturalist and Alpine traveller, who in 1701 had started a correspondence with Woodward which continued for 25 years. Scheuchzer, rightly characterized by Melvin Jahn (1974, p. 19) as 'Woodward's principal European mentor', published various treatises to demonstrate that fossil fishes and fossil plants were all organic remains of the Deluge (Scheuchzer 1708, 1709). And in the four volumes of *Physica Sacra* (1731–1735), Scheuchzer reinforced the link with Biblical chronology and illustrated the main episodes of Genesis (Fig. 5), within a general theory of the formation of the

Fig. 5. Noah's Ark before the Deluge (in Scheuchzer 1731–1735).

Earth's surface which adopted the Woodwardian idea of the universal dissolution of the original Earth's crust (with its primitive mountains) by the waters of the Deluge. According to Woodward this catastrophic event had deposited regular strata, which later cracked to form new relief and mountain chains, and after that all the water had flowed back into the central abyss of the Earth.

In Italy, the Latin translation of Woodward's *Essay* was widely discussed during the first two decades of the eighteenth century in the Academy of Inquieti (later the Academy of Sciences) of Bologna. Johann Jakob Scheuchzer and his brother Johann (1684–1738) were both members of the Inquieti (Cavazza 1990, pp. 65, 74–75). Between 1704 and 1708 they travelled from Zürich to Bologna several times to present lectures on geological subjects and to promote Woodward's writings at the Academy's meetings (Tega 1986, pp. 62, 67–70, 85–87). Consequently, an active group of Bolognese naturalists started to explore the Apennines, looking for fossil evidence of the Deluge as proof of the Woodwardian Theory of the Earth and its short timescale (Minuz 1987, pp. 46–49; Sarti 1992). A distinguished figure among them was Giuseppe Monti (1682–1760), author of *De Monumento Diluviano* (1719), in which he supported Woodward's idea of the universal dissolution of the Earth's crust, and interpreted some fossil bones found in the Bolognese hills as remains of a diluvial walrus.

Within this Swiss–Italian geological network there was a third supporter of the Woodwardian theory (Ellenberger 1994, pp. 124–134). This is Louis Bourguet (1678–1742), a French Huguenot scholar active in Neuchâtel, who published a *Mémoire sur la Théorie de la Terre* in 1729 (Bork 1974). Another Bolognese scholar, Luigi Ferdinando Marsili (1658–1730), lived in Switzerland and travelled there from early 1704 to the end of 1705 with Johann Scheuchzer, whose geological work was influenced by Woodward as well as by Steno's *Prodromus* (Ellenberger 1995). Marsili studied the morphology of the strata of the mountains around the alpine lakes, partially explaining their formation through the action of the Deluge and referring to 'many phenomena' inside and outside the terraqueous globe, as being due to the action of the waters (Marsili 1725, pp. 38–39; Gortani 1930). Marsili was more interested in the different structures of the Earth's surface than in their explanations within a Biblical timescale. Although he had initially followed the Woodwardian theory, he became substantially critical of the supposed singularity of the Deluge's geological role and eventually supported Vallisneri's theory of a succession of several floods in the Earth's history (Marsili 1728; Vaccari in press).

Investigating the possible origins of springs, using some data provided by Johann Scheuchzer and by Marsili himself, the naturalist Antonio Vallisneri (1661–1730) stated that the strata of the mountains had been due to the action of several floods: 'Appariscono fondati i monti, cioè di strati, o di tavolati, ma sollevati di sopra il piano della terra, come una crosta sovra un'altra, ognuna delle quali sia stata lasciata in forma di posatura da varie inondazioni, in tempi a noi ignoti seguite, eccettuata quella dell'universale Diluvio'[3] (Vallisneri 1715, p. 25). Vallisneri developed this position in his book *De' Corpi Marini* (1721) where, besides the floods, he considered in a Stenonian way the effects of earthquakes and subterranean fires ('fuochi sotterranei') in shaping of the Earth's surface. Here he attacked the rigidity of traditional Earth histories based on the Biblical chronology by reducing the role of the Deluge. Using evidence collected in the field, Vallisneri distinguished a series of events of various intensities (local and general floods) which possibly operated over a very long, although not yet of quantifiable, timespan.

This notion of the Earth's great, indefinite (but not boundless), antiquity is also found in the writings of Giovanni Targioni Tozzetti (1712–1783) and Giovanni Arduino (1714–1795), two major figures of Italian geology in the second half of the eighteenth century (Vaccari 1991, 1993, 1996). Vallisneri paved the way for the future establishment of a relative chronology for the formation of the mountains (Greene 1996, pp. 67–70), no longer strictly linked to the two Biblical stages of Creation and Deluge. It is perhaps not surprising then that Woodward sent Scheuchzer a very negative judgment of Vallisneri's *De' Corpi marini*: 'I have within these few days, received Dr Vallisneris book: and find him little or not at all Vers'd in Minerals, and the State of the Earth, the subject he writes upon. All here, and particularly the Italians, agree that he merits no Notice. There is nothing new in what he writes' (letter written between 1722 and 1724 and quoted by Jahn 1974, p. 25).

The great variety of different concentrations of fossils observed in the field also convinced Marsili, like Vallisneri (Fig. 6), of the succession

---

[3] 'The mountains appear formed of strata, or layers, but which are raised above the level of the Earth, like superimposed crusts, each of them left behind in the form of sediments by various floods, which occurred in unknown times, apart from that of the Deluge.'

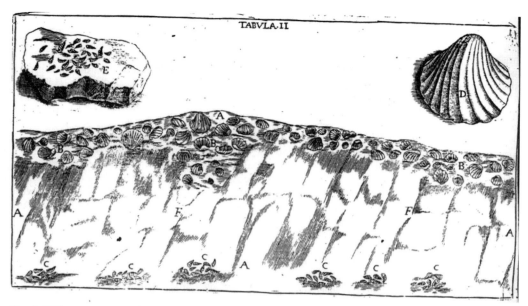

Fig. 6. Different concentrations of fossils near the coast of Tuscany (in Vallisneri 1721).

of several sea-floods of different intensity, which had shaped the Earth's surface over a long timespan. Again, the same problem of the variety and complexity of fossils found on the mountains at different levels prompted Anton Lazzaro Moro (1687–1764) to reject the diluvial theory. He instead proposed an 'ultra-plutonist' Theory of the Earth entirely based on the action of fire and heat (the mountains are all volcanoes), within a chronological framework far longer than that allowed by Genesis (Moro 1740). Reactions against the writings of Moro and Vallisneri, from Italian defenders of the role of the Deluge and Biblical chronology (for example, Costantini 1747), clearly show the tension and energy of the Italian geological debates in the middle of the eighteenth century.

Biblical chronology and the perception of longer time

In other European countries the context of geological studies during the early eighteenth century showed similar development towards the hypothesis of extended timescales for the age of the Earth. In Sweden, for example, the scientific authority of Emanuel Swedenborg (1688–1772) gave strong support to Biblical chronology and to the decisive geological role of the Deluge (Swedenborg 1721), here situated within a history of an Earth in gradual decay and marked by the constant diminution of the seas (Frängsmyr 1969, pp. 131–179). However, acceptance of this latter phenomenon allowed Carl von Linné (1707–1778), in his *Oratio de Telluris Habitabili Incremento* (1744), to dismiss the Deluge as the central geological cause because of its short duration. Later he suggested the possibility that the age of the Earth might be much greater than that sanctioned by conventionally accepted Biblical chronology, as all the geological changes which had occurred to the Earth's surface needed an immense period of time (Frängsmyr 1983, pp. 145–155).

During the first half of the eighteenth century, 'stressing the inadequacy of Biblical time, French writers increasingly recognized that geological processes had been of immense duration' (Dean 1981, p. 447). Benoît de Maillet (1656–1738) wrote the *Telliamed* (1748) between 1692 and 1718, in which he proposed a Theory of the Earth which would later be censured by the Church because of its possible implications regarding the eternity of matter. According to Maillet the Earth was once entirely covered by waters which had caused all the geological changes in the surface, including the deposition of strata and the formation of mountains (Carozzi 1969). Calculating the present rate of the retreat of the sea he concluded that such diminution has been going on for more than two billion years. Significantly, while Maillet was promoting the clandestine circulation of *Telliamed*'s manuscript in the 1720s (Cohen 1993), the philosopher Montesquieu (1689–1755) wondered, in his *Lettres Persanes* (1721), how it could be possible for those who understand nature and have a reasonable idea of God to believe that matter and created things are only 6000 years old.

The extent to which the normally accepted Biblical timescale for the age of the Earth constrained some scholars and scientists between the late seventeenth and the early eighteenth century is a crucial question. Was it really possible to fit the estimated duration of all geological processes and phenomena, which were increasingly being observed and analysed in the field, into such a rigid timeframe? If the various interpretations of the Biblical accounts – with their search for harmony between Scriptures, Nature and scientific ideas – determined a gradual shortening of the timescale during the seventeenth century, did some scholars find themselves caged into the 6000 years of Genesis despite their perceptions of a longer timescale? Or was the Biblical timescale perceived as sufficient and reasonable?

Rhoda Rappaport (1997, p. 175) suggests that the short geological timescale was also reinforced by observations and experiments. The most readily observable geological processes (earthquakes, erosion, floods, volcanic eruptions), which were also documented in such works as Strabo's *Geographia*, Mercator's *Atlas* and Varenius's *Geographia Generalis*, were rapid events. The nature of these geological events was fully compatible with the Biblical timescale and with the notion of the Deluge, which was widely thought to be 40 days long.

This was certainly part of the general context. But some scholars also perceived of the possibility that the Earth may be much older. British scholars, such as John Ray (1627–1705), Edward Lhwyd (1660–1709) and Robert Hooke (1635–1703), who carefully observed sediments and geomorphological features in the field, tended to consider the timescale of Genesis too short. Although when they 'thought of pushing back the date of Creation beyond Ussher's 4004 BC, they were certainly not proposing the addition of millions of years to the Scriptural age of the Earth' (Herries Davies 1969, p. 17). Indeed, Hooke's position on this matter has been the subject of various historiographical interpretations (cf. Drake 1996, pp. 100–103; Rappaport 1997, pp. 192–193).

In Italy, Marsili and Vallisneri followed Steno's footsteps and prepared the ground for the great lithostratigraphical fieldwork undertaken in the 1750s by Giovanni Arduino (1760, 1774). Arduino adopted a concept of relative chronology for his theory of the formation of the Earth's crust, which was subdivided into four rock units ('ordini'); he completely ignored the Biblical timescale. However, other mid-eighteenth-century scholars, such as Johann Gottlob Lehmann (1719–1767) in Germany (1756) or Torbern Olof Bergman (1735–1784) in Sweden (1766), had greater difficulty in emancipating themselves from the account of Genesis, in spite of their accurate lithostratigraphical analyses.

The study of strata and rock formations, particularly in mountainous regions, provided the main keys for the establishment of a greater antiquity of the Earth. The records of a sequence of geological events of long duration, as for example in the building of a mountain, were no longer the subsidiary data of a unique supernatural event, such as the Deluge. The problem was not to reject completely Genesis or even Creation itself, but to reduce the role of the Deluge. Modification and extension of the Earth's age and chronology were underway and followed a pathway which required reappraisal of the Deluge, and the possibility that its significance as a geological event would need to be dismantled.

I would like to thank S. Knell (University of Leicester), M. Rudwick (University of Cambridge) and K. Taylor (University of Oklahoma) whose comments and suggestions have greatly improved this paper.

References

ALBRITTON, C. C. JR. 1980. *The Abyss of Time. Changing Conceptions of the Earth's Antiquity after the Sixteenth Century*. Freeman Cooper, San Francisco.

ALBRITTON, C. C. JR. 1984. Geologic Time. *Journal of Geological Education*, 32, 29–37.

ARDUINO, G. 1760. Due lettere, [...] sopra varie sue osservazioni naturali. Al Chiaris. Sig. Cavalier Antonio Vallisnieri. *Nuova Raccolta di Opuscoli Scientifici e Filologici, Venezia*, 6, xcix–clxxx.

ARDUINO, G. 1774. Saggio Fisico-Mineralogico di Lythogonia e Orognosia. *Atti dell'Accademia delle Scienze di Siena detta de' Fisiocritici*, 5, 228–300.

BECHER, J. J. 1669. [...] *Physicae Subterraneae Libri Duo*. J. D. Zunneri, Frankfurt.

BERGMAN, T. O. 1766. *Physisk beskrifning öfver jordklotet, pa Cosmographiska Sällkapets vägnar författad*. Werlds-Beskrifning, Uppsala.

BORK, K. B. 1974. The geological insights of Louis Bourguet (1678–1742). *Journal of the Scientific Laboratories, Denison University*, 55, 49–77.

BOURGUET, L. 1729. *Lettres Philosophiques sur la formation des sels et des cristaux, [...] avec un Mémoire sur la Théorie de la Terre*. Honoré, Amsterdam.

BURNET, T. 1681. *Telluris Theoria Sacra: orbis nostri originem et mutationes generales, quas iam subiit aut olim subiturus est, complectens. Libri duo priores de Diluvio & Paradiso.* Kettilby, London.

BURNET, T. 1698. *Theoria Sacra Telluris: d. i. Heiliger Entwurff oder biblische Betrachtung des Erdreichs.* Liebernickel, Hamburg.

BURNET, T. 1699. *Telluris Theoria Sacra: originem et mutationes generales orbis nostri* [...]. Wolters, Amsterdam.

CAPEL, H. 1985. *La física sagrada. Creencias religiosas y teorías científicas en los orígenes de la geomorfología española.* Serbal, Barcelona.

CAROZZI, A. V. 1969. De Maillet's Telliamed (1748). An Ultra-Neptunian Theory of the Earth. *In:* SCHNEER, C. J. (ed.) *Toward a History of Geology.* MIT, Cambridge (Mass.) and London, 80–99.

CAVAZZA, M. 1990. *Settecento inquieto. Alle origini dell'Istituto delle Scienze di Bologna.* Il Mulino, Bologna.

COHEN, C. 1993. La communication manuscrite et la genèse de 'Telliamed'. *In:* MOUREAU, F. (ed.) *De bonne main. La communication manuscrite au XVIIIe siècle.* Universitas, Paris, 59–69.

COHN, N. 1996. *Noah's Flood. The Genesis Story in Western Thought.* Yale University, New Haven and London.

COSTANTINI, G. A. 1747. *La verità del Diluvio Universale vindicata da' dubbi e dimostrata nelle sue testimonianze.* Bassaglia, Venice.

DALRYMPLE, G. B. 1991. *The Age of the Earth.* Stanford University Press, Stanford.

DEAN, D. R. 1981. The Age of the Earth Controversy: Beginnings to Hutton. *Annals of Science,* **38**, 435–456.

DEAN, D. R. 1985. The Rise and Fall of the Deluge. *Journal of Geological Education,* **33**, 84–93.

DESCARTES, R. 1644. *Principia Philosophiae.* Elzevirium, Amsterdam.

DRAKE, E. T. 1996. *Restless Genius. Robert Hooke and his Earthly Thoughts.* Oxford University, Oxford and New York.

ELLENBERGER, F. 1988. *Histoire de la Géologie.* Tome 1. *Des Anciens à la première moitié du XVIIe siècle.* Lavoisier, Paris.

ELLENBERGER, F. 1994. *Histoire de la Géologie.* Tome 2. *La grande éclosion et ses prémices 1660–1810.* Lavoisier, Paris.

ELLENBERGER, F. 1995. Johann Scheuchzer, pionnier de la tectonique alpine. *Mémoires de la Societé Géologique de France,* nouvelle série, **168**, 39–53.

EYLES, V. A. 1958. The influence of Nicolaus Steno on the development of geological science in Britain. *In:* SCHERZ, G. (ed.) *Nicolaus Steno and his Indice.* Acta historica scientiarum naturalium et medicinalium – vol. 15, Copenhagen, 167–188.

FLETCHER, J. 1968. Athanasius Kircher and the distribution of his books. *Library,* **23**, 108–117.

FLETCHER, J. (ed.) 1988. *Athanasius Kircher und seine Beziehungen zum gelehrten Europa seiner Zeit.* Harrassowitz, Wiesbaden.

FORCE, J. E. 1983. Some Eminent Newtonians and Providential Geophysics at the turn of the Seventeenth Century. *Earth Science History,* **2**, 4–10.

FORCE, J. E. 1985. *William Whiston: Honest Newtonian.* Cambridge University, Cambridge.

FRÄNGSMYR, T. 1969. *Geologi och skapelsetro. Föreställningar om jordens historia från Hiärne till Bergman.* Almqvist & Wiskell, Stockholm.

FRÄNGSMYR, T. 1971. Steno and geological time. *In:* SCHERZ, G. (ed.) *Dissertations on Steno as Geologist.* Odense University, Odense, 204–212.

FRÄNGSMYR, T. 1983. Linnaeus as a geologist *In:* FRÄNGSMYR, T. (ed.) *Linnaeus: the Man and his Work.* University of California, Berkeley, 110–155.

GLICK, T. F. 1971. On the influence of Kircher in Spain. *Isis,* **62**, 379–381.

GORMAN, M. J. 1997. The correspondence of Athanasius Kircher: The world of a seventeenth century Jesuit. An international research project. *Nuncius,* **12**, 651–658.

GORTANI, M. 1930. Idee precorritrici di Luigi Ferdinando Marsili su la struttura dei monti. *In:* Comitato Marsiliano (ed.) *Memorie intorno a Luigi Ferdinando Marsili.* Zanichelli, Bologna, 257–275.

GOULD, S. J. 1987. *Time's Arrow, Time's Cycle. Myth and Metaphor in the Discovery of Geological Time.* Harvard University, Cambridge (Mass.) and London.

GREENE, J. C. 1996. *The Death of Adam. Evolution and Its Impact on Western Thought* [revised printing of the first edition 1959]. Iowa University, Ames.

GUNTAU, M. 1989. Concepts of Natural Law and Time in the History of Geology. *Earth Sciences History,* **8**, 106–110.

HABER, F. C. 1959. *The Age of the World. Moses to Darwin.* The Johns Hopkins Press, Baltimore.

[HERRIES] DAVIES, G. L. 1969. *The Earth in Decay: a History of British Geomorphology 1578–1878.* Macdonald, London and Elsevier, New York.

[HERRIES] DAVIES, G. L. 1995. The Stenonian Revolution. *In:* GIGLIA, G., MACCAGNI, C. & MORELLO, N. (eds) *Rocks, Fossils and History.* Festina Lente, Florence, 45–49.

HOLLAND, C. H. 1999. *The Idea of Time.* Wiley, Chichester and New York.

JAHN, M. E. 1972. A Bibliographical History of John Woodward's 'An Essay Towards a Natural History of the Earth'. *Journal of the Society for the Bibliography of Natural History,* **6**, 181–213.

JAHN, M. E. 1974. John Woodward, Hans Sloane, and Johann Gaspar Scheuchzer: a re-examination. *Journal of the Society for the Bibliography of Natural History,* **7**, 19–27.

KIRCHER, A. 1664–1665. *Mundus subterraneus, in XII Libros digestus.* J. Janssonium & E. Weyerstraten, Amsterdam, 2 vols.

KIRCHER, A. 1669. *The Vulcano's: or, Burning and Fire-vomiting Mountains, Famous in the World: with their Remarkables. Collected from the most part out of Kircher's Subterraneous World.* Darby for Allen, London.

KIRCHER, A. 1675. *Arca Noë, in tres libros digesta quorum I. De rebus quae ante Diluvium II. De rebus, quae ipso Diluvium ejusque duratione III. De iis, quae post Diluvium à Noëmo gesta sunt.* Janssonium, Amsterdam.

LEHMANN, J. G. 1756. *Versuch einer Geschichte von Flötz-Gebürgen, betreffend deren Entstehung, Lage darinne besindliche Metallen, Mineralien und Fossilien.* Lange, Berlin.

LEIBNIZ, G. W. 1693. Protogaea. AUTORE, G. G. L. *Acta Eruditorum [...] Mensis Januarii Anno MDCXCIII, Lipsiae,* 40–42.

LEIBNIZ, G. W. 1749. *Protogaea sive de prima facie telluris et antiquissimae Historiae vestigiis in ipsis naturae monumentis dissertatio ex schedis manuscriptis viri illustris in lucem edita a Christiano Ludovico Scheidio.* Schmidii, Gottingen.

LINNÉ, C. VON 1744. *Oratio de Telluris habitabilis incremento.* C. Haak, Leiden.

MAILLET, B. de 1748. *Telliamed, ou Entretiens d'un philosophe indien avec un missionaire françois sur la diminution de la mer, la formation de la terre, l'origine de l'homme, etc.* Guer, Amsterdam, 2 vols.

MARSILI, L. F. 1725. Osservazioni fisiche intorno al Lago di Garda, detto anticamente Benaco. *In*: MARSILI, L. F. 1930. *Scritti inediti.* Zanichelli, Bologna, 1–126.

MARSILI, L. F. 1728. Lettera al nostro Autore [...]. *In*: VALLISNERI, A. 1728: *De' Corpi Marini che su' Monti si trovano* [...] (second edition). Lovisa, Venice, 141–150.

MATHER, K. F. & MASON, S. L. 1939. *A Source Book in Geology 1400–1900.* McGraw Hill, New York.

MINUZ, F. 1987. 'Ad Naturae Historiam Spectantia'. *In*: TEGA, W. (ed.) *Anatomie Accademiche II. L'Enciclopedia scientifica dell'Accademia delle Scienze di Bologna.* Il Mulino, Bologna, 43–58.

MONTI, G. 1719. *De Monumento Diluviano nuper in Agro Bononiensi detecto.* Rossi, Bologna.

MONTESQUIEU, C. DE SECONDAT, BARON DE 1721. *Lettres Persanes.* Marteau, Cologne [i.e. J. Desbordes, Amsterdam].

MORELLO, N. 1989. La geologia in Italia dal Cinquecento al Novecento. *In*: MACCAGNI, C. & FREGUGLIA, P. (eds) *Storia sociale e culturale d'Italia,* vol. 5/II. *La storia delle scienze.* Bramante, Busto Arsizio, 587–632.

MORO, A. L. 1740. *De' crostacei e degli altri marini corpi che si truovano su' monti.* Monti, Venice.

NORTH, J. D. 1977. Chronology and the Age of the World. *In*: YOURGRAU, W. & BRECK, A. D. (eds) *Cosmology, History, and Theology.* Plenum, New York and London, 307–333.

OLDROYD, D. R. 1996. *Thinking about the Earth: A History of Ideas in Geology.* Athlone, London.

OLDROYD, D. R. & HOWES, J. B. 1978. The first published version of Leibniz's 'Protogaea'. *Journal of the Society for the Bibliography of Natural History,* **9**, 56–60.

PELAYO, F. 1996. *Del Diluvio al Megaterio. Los orígenes de la Paleontología en España.* Consejo Superior de Investigaciones Científicas, Madrid.

PEYRÈRE, I. DE LA. 1655. *Praeadamitae.* Elsevier, Amsterdam.

PIQUER, A. 1745. *Fisica Moderna Racional, y Experimental.* Tomo Primero. Garcia, Valencia.

PORTER, R. 1979. Creation and Credence: The Career of Theories of the Earth in Britain, 1660–1820. *In*: BARNES, B. & SHAPIN, S. (eds) *Natural Order. Historical Studies of Scientific Culture.* Sage, Beverly Hills and London, 97–123.

RAPPAPORT, R. 1978. Geology and orthodoxy: the case of Noah's Flood in eighteenth-century thought. *The British Journal for the History of Science,* **11**, 1–18.

RAPPAPORT, R. 1997. *When Geologists were Historians, 1665–1750.* Cornell University, Ithaca and London.

RAY, J. 1692. *Miscellaneous Discourses concerning the Dissolution and Changes of the World. Wherein the primitive chaos and Creation, the general Deluge, fountains, formed stones, sea-shells found in the earth, subterraneous trees, mountains, earthquakes, vulcanoes, the universal conflagration and future state, are largely discussed and examined.* Smith, London [second edition corrected and enlarged: *Three physico-theological Discourses.* Smith, London, 1693].

RICHET, P. 1999. *L'âge du monde. À la découverte de l'immensité du temps.* Seuil, Paris.

ROGER, J. 1968. Leibniz et la théorie de la terre. *In*: Centre International de Synthèse, *Leibniz 1646–1716: Aspects de l'homme et de l'oeuvre.* Aubier-Montaigne, Paris, 137–144.

ROGER, J. 1973. La Théorie de la Terre au XVIIe siècle. *Revue d'histoire des sciences,* **26**, 23–48.

ROGER, J. 1982. The cartesian model and its role in the eighteenth-century 'Theory of the Earth'. *In*: LENNON, T. M., NICHOLAS, J. M. & DAVIS, J. W. (eds) *Problems of Cartesianism.* McGill-Queen's University Press, Kingston and Montreal, 91–125.

ROSSI, P. 1969. *Le sterminate antichità. Studi vichiani.* Nistri-Lischi, Pisa.

ROSSI, P. 1979. *I segni del tempo. Storia della terra e storia delle nazioni da Hooke a Vico.* Feltrinelli, Milan [English edition: *The Dark Abyss of Time.* Chicago University, Chicago and London 1984].

ROWLAND, S. M. 1998. Age of the Earth, before 1800. *In*: GOOD, G. A. (ed.), *Sciences of the Earth: An Encyclopedia of Events, People and Phenomena.* Garland, New York and London, 7–13.

RUDWICK, M. J. S. 1976. *The Meaning of Fossils. Episodes in the History of Palaeontology* (second edition). Science History, New York.

RUDWICK, M. J. S. 1986. The Shape and Meaning of Earth History. *In*: LINDBERG, D. C. & NUMBERS, R. L. (eds) *God and Nature: Historical Essays on the Encounter between Christianity and Science.* University of California, Berkeley, 296–321.

RUDWICK, M. J. S. 1999. Geologists' Time. A Brief History. *In*: LIPPINCOTT, K. (ed.) *The Story of Time.* Merrell Holberton, London, 250–253.

RUPKE, N. 1998. 'The end of history' in the early picturing of geological time. *History of Science,* **36**, 62–90.

SARTI, C. 1992. Giuseppe Monti and palaeontology in the eighteenth century Bologna. *Nuncius,* **8**, 443–455.

SCHEUCHZER, J. J. 1708. *Piscium Querelae et Vindiciae.* Typis Gessnerianis, Zurich.

SCHEUCHZER, J. J. 1709. *Herbarium Diluvianum collectum.* Gesneri, Zurich.

SCHEUCHZER, J. J. 1731–1735. *Physica Sacra*. J. A. Pfeffel, Augsburg and Ulm, 4 vols.

STENO, N. 1669. *De solido intra solidum naturaliter contento dissertationis prodromus*. Ex Typographia sub signo Stellae, Florence.

STENO, N. 1671. *The Prodromus to a Dissertation concerning Solids Naturally Contained within Solids*. Winter, London.

SWEDENBORG, E. 1721. [Indications of the Deluge. Letter to Jacobus A Melle]. *Acta Literaria Sveciae* 192–196.

TEGA, W. (ed.) 1986. *Anatomie Accademiche I. I Commentari dell'Accademia delle Scienze di Bologna*. Il Mulino, Bologna.

TOULMIN, S. & GOODFIELD, J. 1965. *The Discovery of Time*. Harper & Row, New York.

VACCARI, E. 1991. Storia della Terra e tempi geologici in uno scritto inedito di Giovanni Arduino: la 'Risposta Allegorico-Romanzesca' a Ferber. *Nuncius*, **6**, 171–211.

VACCARI, E. 1993. *Giovanni Arduino (1714–1795). Il contributo di uno scienziato veneto al dibattito settecentesco sulle scienze della Terra*. Olschki, Florence.

VACCARI, E. 1996. Cultura scientifico-naturalistica ed esplorazione del territorio: Giovanni Arduino e Giovanni Targioni Tozzetti. *In*: BARSANTI, G., BECAGLI, V. & PASTA, R. (eds) *La politica della scienza. Toscana e Stati italiani nel tardo Settecento*. Olschki, Firenze, 243–263.

VACCARI, E. (in press). The study of folds in the early eighteenth century: Luigi Ferdinando Marsili and Antonio Vallisneri. *Eclogae Geologicae Helvetiae*.

VALLISNERI, A. 1715. *Lezione Accademica intorno all'origine delle Fontane*. Ertz, Venice.

VALLISNERI, A. 1721. *De' corpi marini che su' monti si trovano, della loro origine, e dello stato del mondo davanti il Diluvio, nel Diluvio e dopo il Diluvio. Lettere critiche*. Lovisa, Venice.

WARREN, E. 1690. *Geologia: or, A Discourse concerning the Earth before the Deluge, wherein the form and properties ascribed to it, in a book intituled the Theory of the Earth, are expected against; and it is made appear, that the dissolution of the Earth was not the cause of the Universal Flood*. Chiswell, London.

WHISTON, W. 1696. *A New Theory of the Earth, from its Original to the Consummation of All Things, wherein the Creation of the World in six days, the universal deluge, and the general conflagration, as laid down in the Holy Scriptures, are shewn to be perfectly agreeable to reason and philosophy*. Roberts, London.

WHISTON, W. 1713. *Nova Telluris Theoria, das ist, Neue Betrachtung der Erde*. Ludwigen, Frankfurt.

WHITROW, G. J. 1977. The Role of Time in Cosmology. *In*: YOURGRAU, W. & BRECK, A. D. (eds) *Cosmology, History, and Theology*. Plenum, New York and London, 159–177.

WHITROW, G. J. 1988. *Time in History. The Evolution of our General Awareness of Time and Temporal Perspective*. Oxford University Press, Oxford and New York.

WILCOX, D. J. 1987. *The Measure of Times Past. Pre-Newtonian Chronologies and the Rethoric of Relative Time*. University of Chicago, Chicago and London.

WOODWARD, J. 1695. *An Essay toward a natural history of the Earth and terrestrial bodies especially minerals as also the seas, rivers and springs. With an account of the universal Deluge: and of the effects it had upon the Earth*. Wilkin, London.

WOODWARD, J. 1704. *Specimen Geographiae Physicae quo agitur de Terra, et corporibus terrestris speciatim mineralibus: nec non mari, fluminibus, et fontibus. Accedit Diluvii universalis effectuumque eius in Terra descriptio*. Gessneri, Zurich.

WOODWARD, J. 1735. *Geographie Physique ou essay sur l'histoire naturelle de le terre, trad. de l'anglais par M. Noguez ... avec la réponse aux observations de M. le docteur Camerarius, plusieurs lettres écrites sur la même matière; & la distribution methodique des fossiles*. Briasson, Paris.

WOODWARD, J. 1739. *Geografia Fisica, ovvero Saggio intorno alla storia naturale della terra, con la giunta dell'Apologia del Saggio contro le Osservazioni del Dottor Camerario e d'un Trattato de' fossili d'ogni specie*. Pasquali, Venezia.

WOODWARD, J. 1744. *Physicalische Erd-Beschreibung, oder Versuch einer natürlichen historie des Erdbdens*. Weber, Erfurt.

Buffon, Desmarest and the ordering of geological events in *époques*

KENNETH L. TAYLOR

Department of the History of Science, University of Oklahoma, 601 Elm Street, Room 622, Norman, OK 73019, USA (email: ktaylor@ou.edu)

Abstract: During the eighteenth century many naturalists and philosophers became persuaded of the great antiquity of the Earth, and of the promise that knowledge of the Earth's past and development could be built up through investigations of natural terrestrial features. In common with most geological issues of the time, these opinions rose to prominence to a considerable degree in connection with the so-called Theories of the Earth. This paper discusses some interconnections between Theories of the Earth and the emerging enterprise of geological field investigation, as they related to efforts toward establishing relative ages of geological phenomena. It considers in particular the two rather different Theories of the Earth offered by Buffon in 1749 and 1778, respectively. While the earlier one (*Théorie de la Terre*) emphasized principles for extracting physical knowledge of the Earth's configuration through empirical investigation, the latter theory (*Époques de la Nature*) drew attention to the project of organizing knowledge about the Earth around a directional sequence of periods. The central impulses of Buffon's two conceptions of the Earth were combined in actualistic field investigations by geologists of the late eighteenth century, Nicolas Desmarest in particular, which contributed significantly to the establishment of methods for determining distinct stages or sequences of the Earth's past.

Between 300 and 200 years ago, questions about the extent of time through which the Earth's natural operations have functioned were often on the minds of articulate naturalists. However, many of these thinkers seem *not* to have set explicit formulations about such questions at the centre of their work. So, to oversimplify the research problem a little, one faces a choice between two possible approaches to historical study of scientific ideas about terrestrial time. One way is to search through what authors said directly about the matter. The other is to focus on what the authors appear to have thought was most important in their efforts to comprehend the Earth (or perhaps what they believed they could speak sensibly about as responsible savants), and then see what inferences might be drawn out regarding how they dealt with time.

Both approaches are important. It is not surprising, if only because of the need to keep the problem within manageable proportions, that many of the most prominent historical studies concerned with the Earth's age are organized around the first of these conceptions (Haber 1959; Albritton 1980; Dean 1981; Hallam 1989, ch. 5; Dalrymple 1991). This paper, however, takes the second, less direct path. At least two related reasons justify this approach. First, we know that there were cultural and institutional constraints on what many reputable naturalists of the eighteenth century felt they could say about the extent of time. This is one consideration, although not the only one, in understanding why some of the scientific writers of the period who disavowed a religiously orthodox timescale of several thousands of years were cautious in expressing their views, often resorting to the use of ambiguity. (Quite effective for this purpose, in the French language, was the double meaning of the word *siècle*, which in addition to 'century' might alternatively signify 'age' or 'period'.) So time is one of those topics where we have reason to suppose certain eighteenth-century writers did not say all that was on their mind. Second, the constraints on candid expression about time need not always have been external; they might result from personal intellectual integrity and self-discipline. Scientific thinkers who did not want to give utterance to what they saw as mere speculation might have good reason to be circumspect on an issue where they were keenly aware it was so difficult to find solid footing. Time conceptions, or explorations of new possibilities for moving beyond conjecture with regard to the Earth's duration, often might be only implicit or latent in the ways astute writers addressed what they thought to be the central questions, or issues where there seemed reasonable prospect of making headway.

An empirical spirit was highly valued, and was still treated as something rather new, in the natural sciences of the eighteenth century. This was an age when great enthusiasm was expressed about the revitalization of philosophy owing to revived consultation of the book of nature, in place of reading texts inherited from traditional authorities. It was a time when Francis Bacon's reputation was at its peak, as an originator and publicist of inductive and experimental methods, on the Continent as well as in England. In part as a result of applying this Baconian spirit to features observable over the Earth's surface or enclosed within its accessible parts, many eighteenth-century naturalists had begun to conclude that the Earth in its present condition must be far more greatly changed from its earlier state than had been generally thought by their predecessors. If the Earth had truly undergone vast changes, the time required for their accomplishment might very well be much more than was allowed by conventional chronologies. Leading natural philosophers, guided by precepts emphasizing experience as the source of knowledge, were aware that serious difficulties lay in the way of efforts to know the past, particularly a remote past which no human being could have witnessed. It was apparent to those committed scrupulously to empirical methods that the best chances for acquiring secure knowledge of nature's past lay in cultivating the means of interpreting natural 'monuments', natural relics of times and events inaccessible to direct observation. Among those best informed about such methods were antiquarian and historical scholars, many of whom were just as keen about pursuing historical understanding of 'natural antiquities' as they were about human history (Rappaport 1982, 1997). By the close of the eighteenth century it was widely felt that a far stronger grip had been gained on the means to extend knowledge of the Earth's past, if not actually to determine its age, than had been the case at the century's beginning. I will try here to trace a part of the developments within geological science that lent justification to that opinion.

Buffon, Desmarest and natural *époques*

During the last third of the eighteenth century several scientific thinkers and investigators advanced ideas for interpretation of terrestrial features in terms of distinct stages or sequences of the Earth's past. A concept that came to play a very significant role in efforts to construct periods in the Earth's past was expressed through the French word *époque*, or epoch. It is a story

Fig. 1. Portrait of Buffon, 1761, by Drouais, engraved by Baron. Frontispiece, *Histoire Naturelle*, vol. 1 (1749), from the set held in the History of Science Collections, University Libraries, University of Oklahoma. (Evidently this frontispiece, identified as a 1761 Drouais portrait, was added before binding of the 1749 volume.)

that can be told largely through the work of two notable French naturalists, Georges-Louis Leclerc, the Comte de Buffon (1707–1788), and Nicolas Desmarest (1725–1815) (Figs 1 and 2). The first linked his interest in *époques* directly with notions about the Earth's age, while the second evidently preferred to avoid making pronouncements on a problem he saw as unavoidably conjectural. While their approaches to establishing a vocabulary of *époques* reflect great differences in their scientific sensibilities, I suggest here that behind such differences there were important intellectual affinities that can be understood through their mutual attachment to ideas found within the prevailing geological idiom, the Theories of the Earth. I will further suggest that Desmarest's important contribution toward the establishment of scientific procedures for placement of geological phenomena in relative order is best understood as arising out of guiding principles found within the tradition of Theories of the Earth, rather than as opposed to them.

Fig. 2. Portrait of Desmarest, from an engraving by Tardieu. This presumably shows Desmarest at an age of at least 70 years, since he is shown wearing the apparel of the Institute (founded 1795) designed by David. From a photograph given to the author in 1966 by Desmarest's descendant, Marthe Chartrin.

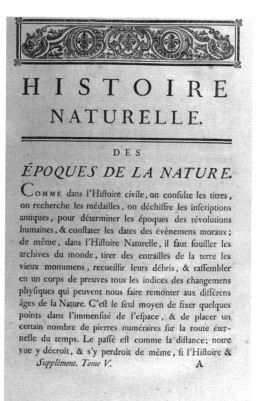

Fig. 3. Title and opening text of Buffon's *Époques de la Nature*, from *Histoire Naturelle, Supplément* vol. 5 (1778), p. 1.

Buffon and Desmarest each put the term *époque* forward publicly during the 1770s. For both, the term had reference to successive periods and geological events in the Earth's history. Similarly, each of them attached to the notion of *époques* an expansive conception of the time through which the Earth's processes have operated. But an initial review of their respective uses of an *époques* terminology reveals a number of striking differences that seem to outweigh the similarities between them.

The Burgundian Buffon, by far the more famous of the two, centred his 1778 masterpiece *Époques de la Nature* around a series of stages in the Earth's development, from its supposed generation around 75 000 years ago (Buffon 1778, p. 67) (Fig. 3). Desmarest, considerably less renowned than Buffon both in his own day and later (but certainly not an obscure figure to those familiar with geology's history), utilized a conception of distinct geological epochs in an analysis of volcanic features he had been studying since 1763 in the region of Auvergne, in south-central France. Although he had made brief indication of such an analysis into epochs in his earlier work on the Auvergne basalts (Desmarest 1774, 1777),[1] it was in a 1775 memoir, read for a special meeting of the Paris Academy of Sciences, that Desmarest formally presented his views on the epochs of nature in these volcanic remnants. A much less grandiose exposition than that of Buffon, Desmarest's paper was only published in truncated form a few years later, in 1779, the same year that Buffon's great treatise became available for purchase (Desmarest 1779).

---

[1] Desmarest's initial perceptions about the volcanic terrain in Auvergne – specifically his determination that columnar basalt is a volcanic product – were presented briefly in 1765. His subsequent and more thorough investigations of the region, published in two parts during the 1770s, were presented in 1771.

A longer version was presented anew for the French Institute a quarter of a century later, in 1804, and published in 1806 (Desmarest 1806).

The close coincidence in timing between the theory of Buffon and Desmarest's summary has fostered suspicions among some commentators (e.g. Roger 1962, pp. xl–xli; Ellenberger 1994, p. 240) of contention between Buffon and Desmarest over rights to the term *époque* in scientific treatment of the Earth. If there was in fact a contest over rightful possession of the word, it was a competition between two men of quite unequal status in the French scientific world. Buffon, nearly two decades older than Desmarest, was no less greatly his social and cultural senior, as a patrician who had long been the superintendant of the King's garden and keeper of the royal natural history collections (Roger 1989). Desmarest, by contrast, was of humble origins, and in the 1770s was a newcomer to the Paris Academy, to which he had been elected in 1771 with patronage assistance from reform-minded ministers and officials who appreciated his work in support of rational improvements in crafts and industry (Taylor 1969, 1971).

Jacques Roger, the foremost interpreter of Buffon's life and work, found that the term *époque* was used with reference to the Earth's history by other naturalists besides either Buffon or Desmarest at least as early as 1750 (Roger 1962, pp. xl–xli). Yet at the time that Buffon and Desmarest began to cultivate their respective conceptions of *époques*, such occasional antecedent uses of the term had not been incorporated within any systematic schema for analysing the Earth. Buffon himself first used the word in print, as a means to designate a distinct period of the Earth's history, in a 1766 passage of his monumental *Histoire Naturelle* where he envisioned setting the problem of animal life's history within the broad framework of the terrestrial globe's creation and development (Buffon 1766, p. 374; Roger 1962, p. xxvi). Buffon's first presentation of his *Époques de la Nature* came in August 1773, when he read a chapter of the text at a meeting of the Academy of Dijon (Roger 1962, p. xxxi). While the *Supplément* volume of *Histoire Naturelle* containing the *Époques* has an imprint of 1778, it was not placed on sale until April 1779 (Roger 1962, p. xxxvii). If Desmarest became annoyed, as may well have been the case, at Buffon's apparent appropriation of the terminology of *époques* as the learned world was abuzz about the royal naturalist's splendid new treatise, his pique may be understandable in that he had been using that term in his private notes and correspondence since the early 1760s.[2]

Regardless of their respective entitlements to priority regarding use of the term *époque*, in the final two decades of the eighteenth century both Buffon and Desmarest were acknowledged as significant exponents of an intensified sense of sequential developments in the Earth, and of the operations of natural processes over periods greatly surpassing the time of human record and memory. Both friends and enemies of Buffon agreed that his *Époques de la Nature* did much to strengthen conviction in a natural timescale dwarfing the extent of human history, and to recast scientific thinking in terms of development or a periodized series of phases of change (Gohau 1987, ch. 7; Ellenberger 1994, pp. 215–217; Oldroyd 1996, pp. 89–92). Although with a much more restricted and less varied audience, Desmarest too gained respect, within a comparatively exclusive community of scientific scholars, as a pioneering advocate for interpreting geological phenomena in a way that allowed determination of successive operations and effects.[3]

Attentive observers saw differences as well as similarities in the deployment of an 'epochs' concept, on the part of Buffon and Desmarest. For Buffon, epochs represented vast stages in an unfolding story boldly narrated, situating a panorama of natural events and effects in a temporal order, the overall plausibility of which was not meant to depend critically on any narrow body of data. Indeed, Buffon's story rationalized the facts, rather than being constructed out of them (Taylor 1992*a*). For Desmarest, however, the account of the Auvergne volcanic epochs involved no dramatically comprehensive narrative. In fact, Desmarest's regionally restricted story was expressly incomplete. It made no

---

[2] Desmarest wrote about viewing 'les traces de chacune des époques des trois révolutions principales que le globe a essuyées' – evidence of three main epochs of global revolution – in his notes from observations in Périgord in 1761: 'Remarques de Mr. Desmarest (de l'Académie des Sciences) sur la géographie physique, les productions & les manufactures de la généralité de Bordeaux, lors de ses tournées dupuis 1761 jusqu'en 1764' [the period concerned actually ended in 1762], Archives Départementales de la Dordogne, Périgueux, MS 26, part I, p. 27. On Desmarest's further private use of the term *époque* in the later 1760s see Taylor 1992*a*, p. 374, n. 10.

[3] Ezio Vaccari has kindly pointed out to me that during the same period under discussion here, several Italian naturalists were also adopting a terminology of *epoca* to define periods of global alteration. In particular, Dr Vaccari informs me that a form of geological periodization or change was discussed in terms of *epoca* by Giambattista Passeri in 1753, by Alberto Fortis in 1766, and by Giovanni Arduino in the early 1770s.

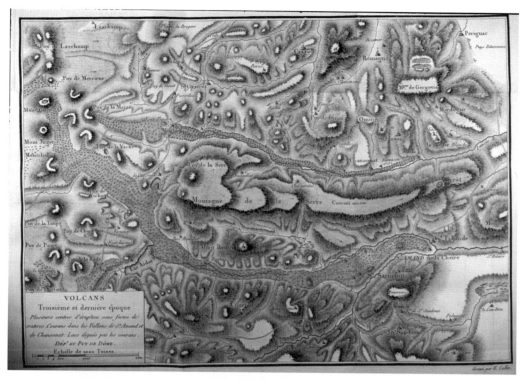

Fig. 4. Desmarest's map (plate VII) showing volcanic remnants from distinct epochs, from his 1806 publication, 'Mémoire sur la détermination de trois époques de la nature par les produits des volcans'. The Auvergne locality centres on the valleys of Chanonat (N) and Saint-Amand (S), separated by the long, sloping plateau of the Montagne de la Serre. This map features the third and last epoch, the products of which are the volcanic cones on the left and the stippled lava flows in the two valleys. The Montagne de la Serre was identified as the remnant of a lava flow from the much earlier second epoch, left high and dry by subsequent erosion.

claims to give an account of changes on a global scale, and it was confined to discussion of effects deriving from discrete periods, separated from one another by time gaps about which he indicated nothing could be said, at least for the moment (Fig. 4).

Each of these Enlightenment naturalists plainly required a natural timescale that disregarded traditional chronologies. But they differed markedly in how overtly this departure from convention was expressed. Buffon's chronological framework was explicitly stated in terms of a developmental sequence where the present moment stands some 75 000 years from the Earth's origins as a piece of incandescent matter torn from the sun by a passing comet (Fig. 5). This computation was ostensibly informed by Buffon's experiments, already reported in earlier volumes of the *Histoire Naturelle*, on the cooling of molten iron in the forges at his Burgundy estate (Buffon 1774, 1775). Buffon's knowledge of how radically variable the results of such an extrapolation might be is suggested in the fact that his private calculations of the Earth's duration were measured in millions of years instead of a few scores of thousands as in his published account (Roger 1962, pp. lx–lxvii; 1989, pp. 537–543).

From a later perspective it is easy to underestimate the novelty of a deliberate narration of the Earth's history through a period of less than a hundred thousand years. But we should consider that a ten-fold enlargement of orthodox figures for the Earth's age could be a genuine challenge to the imagination. Even if many astute readers discounted the pretended precision of Buffon's chronology, they saw the unalloyed naturalism in his disregard for orthodox time-reckoning. There is reason to believe that a number of Buffon's progressive contemporaries, even among naturalists who disdained Buffon as an armchair naturalist and an undisciplined system-builder, registered grudging respect because this prominent office-holder had pushed the envelope confining open expression of what serious thinkers might say about the

Fig. 5. Plate from *Histoire Naturelle*, vol. 1 (*Théorie de la Terre*), 1749 (opposite p. 127, at the start of 'Preuves de la Théorie de la Terre'). By N. Blakey, engraved by St. Fessard. An allegorical representation of the origins of the Earth from solar stuff, a result of a collision of a comet with the Sun. While this image accompanied Buffon's *Théorie de la Terre*, it was in the later *Époques de la Nature* (1778) that Buffon elaborated more extensively on the idea of the planets' generation from solar matter.

natural time frame. They of course knew that Buffon was read – not only because he held a scientific position of unexcelled prestige and authority, but above all because he was readable.

Desmarest was more discreet than Buffon, less rashly explicit about time. In his account of the Auvergne epochs, and for that matter in all his writings public or private, to my knowledge Desmarest never ventured to attach numbers to the periods of time his geological interpretations required. It seems probable that he did not believe such reckoning to be possible. Nonetheless there was no escaping the fact that Desmarest's distinctions among the volcanic epochs of Auvergne depended on recognition of the progressive destruction of the volcanic productions by erosive processes working at more or less constant rates. For Desmarest, the positions of volcanic remnants from the earliest epoch, capping crests overlooking sizeable valleys, spoke of the operation of geomorphic processes – excavation by moving water – in slow and regular fashion. So while Desmarest in his public declarations avoided sensationally direct or precise assertions about the duration of geologial processes, neither did he shrink from explaining his views on how natural operations produced the observed effects in ways clearly necessitating enormously long periods of time.

To summarize, then, Buffon popularized an 'epochs' concept in a form something like that of a likely story about the Earth's formation and progressive alteration. And Desmarest helped to introduce the possibility – or at least strengthened the possibility – of working by empirical analysis of geological phenomena to a genetic or developmental account of particular geological operations and effects.

Most of what I have said so far about Buffon and Desmarest and their place in late-eighteenth-century developments regarding geology and time is fairly well agreed upon – an abbreviated form of a standard narrative, one might say. Notice that while this story entails the approximately simultaneous presentation by Buffon and Desmarest of accounts of sequential geological *époques* conceived as periods of substantial duration – and thus a form of geological explanation organized around placement of events in a chronological series – it also emphasizes differences in the approaches taken by the two naturalists in establishing the events they respectively addressed, and in the contrasting degrees of continuity and completeness of the resulting sequences. It also makes no reference to possible interdependence between Buffon and Desmarest as regards adoption and use of a vocabulary based on *époques*.

Contemporary theories of the Earth, and an undifferentiated past

I propose now to enlarge on this picture, to sketch out a relevant part of the culture of geology in the period when Buffon and Desmarest were preparing themselves to make public presentations about Epochs of Nature. The result should not require any major retractions from the story that has just been rehearsed, as far as it goes. It will, however, draw attention to some common ground held by Buffon and Desmarest, as one looks more closely at their apparently contrasting applications of the *époques* concept. I will maintain that Desmarest's mode of thinking about geological problems drew significantly from some Buffonian doctrines set forth long before either Buffon or Desmarest talked in public about natural epochs. And I will suggest also that this means we must be cautious about drawing any stark contrast between Buffon as exclusively a system-building theorist and Desmarest as an altogether field-based empiricist thoroughly disengaged from Theories of the Earth.

Buffon's *Époques de la Nature* is viewed as one of the culminating entries in the genre of scientific writing called Theories of the Earth. Even as it was being discussed and debated, in the years around 1780, there were voices saying that this style of treatment was of limited use, that it was in the process of being replaced by less comprehensive, more empirically based studies of restricted areas without pretensions of global application (Taylor 1992*a*). But for us to heed *only* those voices risks allowing the term 'Theory of the Earth' to be hijacked as a pejorative by a few late Enlightenment advocates of scientific reform. Through most, if not the whole, of the eighteenth century, the expression 'Theory of the Earth' was generally used rather flexibly, and by and large *not* pejoratively, to refer to scientific understanding of the Earth – including ongoing but so far largely inconclusive efforts at such understanding (Roger 1973; Taylor 1992*b*; Magruder 2000).

This usage is exemplified in a standard reference work of the third quarter of the century: the *Rational Universal Dictionary of Natural History* produced by the French mineralogical naturalist Jacques-Christophe Valmont de Bomare (1731–1807). Valmont de Bomare, from Rouen, first put out this multi-volume compendium in the 1760s, following years of travelling at Royal expense throughout Europe – including Iceland – as an official *naturaliste-voyageur du gouvernement*. The *Dictionnaire* went through five successively enlarged editions by the end of the century. As a teacher of public courses, as well as through his encyclopedic exposition, Valmont de Bomare was a successful digester and popularizer of scientific knowledge (Burke 1976).

In a moderately lengthy article in the *Dictionary* entitled 'Theory of the Earth', Valmont de Bomare addressed geological knowledge broadly (Valmont de Bomare 1768–1769, vol. 11, pp. 222–245). The article exemplifies a formula valuable for historians of geology proposed several years ago by our late colleague François Ellenberger, to the effect that we can profitably distinguish *specific* or *individual* Theories of the Earth from the *ideal* or *generic* Theory of the Earth enterprise (Ellenberger 1994, pp. 12–16). Specific Theories, according to this distinction, were systems of explanation offered with more or less assurance as an author's True Answer to problems relating to the Earth's configuration, origins, development, purposes or future prospects. Generic theories by contrast represented the large and frequently critical and sceptical enterprise in which attainment of a true Theory, or at least elements of it, was viewed as highly desirable but a work still in progress. In other words, a particular use of the phrase 'Theory of the Earth' might refer to aspirations for a complete system, or alternatively it might refer to something close to what later came to be thought of as geological investigation. One needs to look at each case to know where it fits along the spectrum between these two points.

Valmont de Bomare's account of the Theory of the Earth belongs in this latter, generic category. It exhibits doubts about any possibility for serious resolution of questions about the Earth's beginnings, declaring that rather than try to determine origins what we should do instead is to study the Earth in its present state and arrangement. It pursues such themes as: the contrast between sudden, violent causes, and slow and general ones; the inadequacy of any single cause to account fully for the evident transformations the Earth has undergone; and disparity between superficially apparent confusion and disorder in the Earth, as distinct from order and regularity discernible on close inspection.

Quite significantly, relative to the concerns of the present paper, Valmont de Bomare's article on the theory of the Earth treats as more or less conventional a distinction between the *present* or *new* state of the world and an *old* or *former* condition from which the present is demarcated by one or multiple *revolutions*. Valmont is somewhat unclear about the accessibility of satisfactory knowledge regarding the revolution or revolutions separating present from past, as he is also about the extent of time that may be at stake. For our purposes, however, the notable feature of Valmont's treatment is the largely opaque divide between the old and new worlds, as regards any prospect for articulation of the past into discrete parts. Past conditions in the Earth are tantalizingly suggested by various sorts of terrestrial evidence – and indeed the Theory of the Earth enterprise here has, as perhaps the major part of its agenda, to explore that evidence and illuminate that past. But Valmont de Bomare puts forward no protocol for setting past conditions in any kind of temporal order or sequence, let alone for developing a chronology. Valmont was apparently a scrupulous empiricist, uninterested in merely speculative discussion of time and the past. As far as he could see, there existed no reliable means to establish chronological sequences in the Earth's past, which is to say for doing history. Thus what might be called the 'chronological duality' in Valmont's demarcation, distinguishing simply between the *present* world and a *former* one, is perhaps mostly the result of a lack of perceived options. The former world, from this perspective, is in a sense monolithic, undifferentiated, impervious to scientific analysis into a refined set of parts placed in relative sequence. This is a view of nature's remote past that would persist in nineteenth-century popular conceptions, as Martin Rudwick has shown, long after geological specialists had done a great deal to establish a highly differentiated series of periods in the Earth's history (Rudwick 1992).

Desmarest *vis-à-vis* Buffon's two Theories of the Earth

A decade after publication of this article in Valmont's *Dictionary*, anyone giving the summary of Desmarest's 1775 memoir close attention, or anyone reading Buffon's *Époques de la Nature*, would find in one or the other of these pieces denials of an opaque divide between present and past. Buffon's scheme of continuous epochs offered an appealingly complete account, but little in the way of tools for confident determination that the story was told correctly. Desmarest, on the other hand, laid his emphasis on precisely those analytical tools for access on a limited basis to an ordered past. An observer at the end of the 1770s whose attitudes were formed by – or conformable to – a dependable reference like Valmont de Bomare's *Dictionary* would be justified in regarding both Desmarest and Buffon as breaking through a barrier between present and past, although in rather different ways. And that observer would probably consider each of their two achievements as part of the Theory of the Earth endeavour.

In general, modern historians of science have not wanted to categorize Buffon's and Desmarest's 'epochs-undertakings' alike, in contrast to a contemporary outlook for which I have just made a claim. Such reluctance is readily understandable, indeed in some measure I share it. Unlike Buffon, Desmarest in the 1770s outlined interpretive procedures for information generated by investigation that showed a way to translate observations of a landscape into a historical narrative. It represented a fulfilment, or rather a promising beginning of fulfilment, of an aspiration expressed by forebears such as Leibniz and Fontenelle, to establish a reliable history of nature derived out of its analogues to archival documents in civil history – natural monuments and inscriptions (Roger 1968; Rossi 1984; Rappaport 1991, 1997; Leibniz 1993; Cohen 1998).

All the same, without retreating on maintaining the difference between Buffon's narrative *époques* and Desmarest's analytical ones, I support a respectful observance of their contemporaries' habit of linking both with Theories of the Earth. Beyond the fact that, as I believe, users of Valmont de Bomare's *Dictionary* would speak that way, an additional consideration is that important elements in the path that brought Desmarest to his historic formulation of *époques* had roots in Theories of the Earth, in fact especially in the original *Théorie de la Terre* with which Buffon opened his monumental *Histoire Naturelle* in 1749.

That Buffon produced not just one but two theories of the Earth – the first so titled and the second the *Époques de la Nature*, three decades later – has elicited much comment. The two theories had in common mainly the same hypothetical origin of planet Earth as generated from a chunk of solar matter drawn out by a passing comet. Perhaps because of Buffon's reputation as a broad-canvas naturalist and literary synthesizer, Buffon's Earth-theorizing in both instances has sometimes been treated as spun out of the imagination, and thus as an encouragement to a conjectural kind of thinking. (If many contemporaries had reservations about Buffon's standing as an active scientific observer, few of his readers are likely to have been misled on this point. They know he did not personally perform all the things he said should be done in the name of good science.) But in fact the two theories had quite different characters. Of the two theories, the *Époques de la Nature* is a good deal better known and understood, certainly in part because it has long been regarded as a masterpiece of French literature, which the *Théorie de la Terre* has not. Jacques Roger gave us a finely commentated edition of *Époques*, and nothing of the sort has ever happened with Buffon's 1749 *Théorie*.

Buffon's 1749 *Théorie de la Terre* – from which by the way it is apparent Valmont de Bomare drew for his article of the same name – was in considerable measure an appeal for a program of judicious observational study as the correct path toward understanding the Earth (Fig. 6). The greater part of the 1749 *Théorie* reflects a spirit of 'empiricism in principle', for which Buffon was a master spokesman. It deserves notice that the *Théorie* was identified as the 'second discourse' of the *Histoire Naturelle* project (Buffon 1749*b*), immediately after the initial discourse entitled 'On the manner of studying and expounding natural history' (Buffon 1749*a*). A consistency in methodology, as well as didactic tone, is sustained through both discourses. Buffon presented himself to be sure as an interpreter of the Earth, but no less as a teacher of how to interpret the Earth. The central interpretive principles are empirical in character and tend to emphasize assimilation of observations to laws, not to history (Gohau 1990, 1992).

In the *Théorie* Buffon advocated careful and patient observation especially of a wide range of configurational regularities in terrestrial features at various levels, ranging from continental in scale to a matter of inches. The underlying idea,

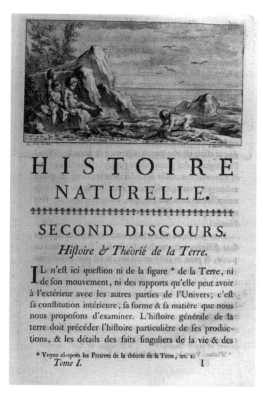

Fig. 6. Opening page (65) of Buffon's *Théorie de la Terre*, in the first volume (1749) of *Histoire Naturelle*. Vignette by DeSeve, engraving by St. Fessard.

of which Buffon was not so much originator as eloquent expositor, was that generalizations arising out of combinations and comparisons of these dispositional regularities are the most promising avenues to as-yet unknown pieces of a proper Theory of the Earth (Taylor 1988). There is a kind of epistemological modesty in this Buffonian programme that may surprise those familiar only with Buffon's later theory, the *Époques de la Nature*. Although certainly not without its own evidences of confidence – indeed overconfidence – in the extent of human knowledge about certain terrestrial features and their meanings, Buffon's outlook in the *Théorie* was strongly investigatory. The *Théorie de la Terre* was not a result in hand, it was an empiricist's adventure in progress. Perhaps not incidentally, in his *Théorie* Buffon evaded direct discussion of the question of how long terrestrial operations have been working.

Buffon's emphasis on examination of dispositional regularities was part of a style much respected among francophone naturalists throughout the eighteenth century – and I believe others outside the French-speaking world as well. The promise it seemed to hold out for its adherents was in discovery of useful laws, generalizations in the spirit of natural philosophy, rather than in the form of historical formulations. In retrospective view, we may be tempted to think that investigators of the late eighteenth century would have been well advised to seek historical formulations, in preference to philosophical generalizations; we know that during the early nineteenth century geological science was to achieve an unprecedented coherence in part because of a focus on stratigraphy, oriented precisely around a concern for sequence rather than law. But we must remember that it was not until the nineteenth century that scientists began to be comfortable with the idea that placement of events in historical sequence constituted a wholly legitimate form of scientific explanation. For our eighteenth-century protagonists, it came naturally to expect that a proper resolution of the puzzles of Earth science – attainment of a satisfactory Theory of the Earth – must take the shape of laws or fixed generalizations.

It is perhaps ironic that the principal hero of this discussion, Nicolas Desmarest, whose achievement it was to articulate a method for translating observations of apparently static structures of a volcanized landscape into empirically determined chronological phases or epochs, was brought up geologically on a diet of timeless configurational generalizations. For so he was formed as a beginning investigator, as one finds on close examination of his early manuscripts and publications (Taylor 1997). In both indirect and direct ways, Desmarest acknowledged that his principles of geological investigation were largely the same as those found in Buffon's 1749 *Théorie de la Terre*.

But maybe this is not quite so ironic after all. For much more was involved in Desmarest's Auvergne research than the insight that the volcanic structures he studied there could be distinguished in accord with discrete epochs. Desmarest had first to gauge the configurational limits of these structures, by fieldwork. In what is no doubt an oversimplified formula, Desmarest's geochronological achievement with his *époques* required that he use two distinguishable sorts of cognitive faculties in an integrated manner: one of them spatial, and the other temporal or historical. For the first of these, the school to which Desmarest belonged, the approach he adopted in seeking to chart out dispositional regularities, was – as he acknowledged – nowhere better expressed than in Buffon's 1749 *Théorie de la Terre*.

In preparing this paper I have benefited from the kindness of M. J. S. Rudwick in allowing me to read in draft parts of his forthcoming book, *Bursting the Limits of Time*. I am grateful to K. V. Magruder for his invaluable assistance with the illustrations. I thank S. Knell, M. Rudwick and E. Vaccari for a number of constructive suggestions for revision of the manuscript. Figures 1, 3, 4, 5 and 6 courtesy of the History of Science Collections, University of Oklahoma Libraries.

References

ALBRITTON, C. R. JR. 1980. *The Abyss of Time: Changing Conceptions of the Earth's Antiquity after the Sixteenth Century*. Freeman Cooper, San Francisco.

BUFFON, G. L. LECLERC, COMTE DE. 1749a. Premier Discours. De la manière d'étudier & de traiter l'histoire naturelle. *In: Histoire Naturelle, générale et particulière*. Vol. 1. De l'Imprimerie Royale, Paris, 3–62.

BUFFON, G. L. LECLERC, COMTE DE. 1749b. Second Discours. Histoire & Théorie de la Terre. *In: Histoire Naturelle, générale et particulière*. Vol. 1. De l'Imprimerie Royale, Paris, 65–124 [Preuves de la théorie de la terre: 127–612].

BUFFON, G. L. LECLERC, COMTE DE. 1766. De la dégénération des animaux. *In: Histoire Naturelle, générale et particulière*. Vol. 14. De l'Imprimerie Royale, Paris, 311–374.

BUFFON, G. L. LECLERC, COMTE DE. 1774. Expériences sur le progrès de la chaleur. *In: Histoire Naturelle, générale et particulière*. Supplément, Vol. 1. De l'Imprimerie Royale, Paris, 145–300.

BUFFON, G. L. LECLERC, COMTE DE. 1775. Recherches sur le refroidissement de la Terre & des planètes. *In: Histoire Naturelle, générale et particulière*. Supplément, Vol. 2. De l'Imprimerie Royale, Paris, 361–515.

BUFFON, G. L. LECLERC, COMTE DE. 1778. Des Époques de la Nature. *In: Histoire Naturelle, générale et particulière*. Supplément, Vol. 5. De l'Imprimerie Royale, Paris, 1–254 [Additions et corrections aux articles qui contiennent les preuves de la Théorie de la Terre, 255–494; Notes justificatives des faits rapportés dans les Époques de la Nature, 495–599].

BURKE, J. G. 1976. Jacques-Christophe Valmont de Bomare. *In*: GILLISPIE, C. C. (ed.) *Dictionary of Scientific Biography*, Vol. 13. Charles Scribner's Sons, New York, 565–566.

COHEN, C. 1998. Un manuscrit inédit de Leibniz (1646–1716) sur la nature des 'objets fossiles'. *Bulletin de la Société Géologique de France*, **169**, 137–142.

DALRYMPLE, G. B. 1991. *The Age of the Earth*. Stanford University, Stanford.

DEAN, D. R. 1981. The Age of the Earth Controversy: Beginnings to Hutton. *Annals of Science*, **38**, 435–456.

DESMAREST, N. 1774. Mémoire sur l'origine & la nature du basalte à grandes colonnes poygones, déterminées par l'histoire naturelle de cette pierre, observée en Auvergne. *Mémoires de l'Académie*

Royale des Sciences, Paris, Année 1771, 705–775 [presented to the Academy 3 July 1765 (in part) and 11 May 1771].

DESMAREST, N. 1777. Mémoire sur le basalte. Troisième partie, où l'on traite du basalte des anciens; & où l'on expose l'histoire naturelle des différentes espèces de pierres auxquelles on a donné, en différens tems, le nom de basalte. *Mémoires de l'Académie Royale des Sciences*, Paris, Année 1773, 599–670 [presented to the Academy 11 May 1771].

DESMAREST, N. 1779. Extrait d'un mémoire sur la détermination de quelques époques de la nature par les produits des volcans, & sur l'usage de ces époques dans l'étude des volcans. *Observations sur la physique, sur l'histoire naturelle et sur les arts*, **13**, 115–126 [summary of paper presented to the Academy at the *Séance publique* at its official reconvening on St. Martin's day (11 November) 1775].

DESMAREST, N. 1806. Mémoire sur la détermination de trois époques de la nature par les produits des volcans, et sur l'usage qu'on peut faire de ces époques dans l'étude des volcans. *Mémoires de l'Institut des Sciences, Lettres et Arts. Sciences mathématiques et physiques*, **6**, 219–289 [presented to the Institut *1 Prairial an XII* (21 May 1804)].

ELLENBERGER, F. 1994. *Histoire de la géologie. Tome 2: La grande éclosion et ses prémices, 1660–1810.* Technique et Documentation (Lavoisier), Paris, London, New York [1999 English translation by M. Carozzi, *History of Geology: The Great Awakening and its First Fruits, 1660–1810*, Balkema, Rotterdam].

GOHAU, G. 1987. *Histoire de la géologie.* Editions La Découverte, Paris [1990 English translation and revision by A. V. Carozzi, & M. Carozzi, *A History of Geology*. Rutgers University, New Brunswick and London].

GOHAU, G. 1990. *Les sciences de la terre aux XVIIe et XVIIIe siècles: Naissance de la géologie.* Albin Michel, Paris.

GOHAU, G. 1992. La 'Théorie de la Terre', de 1749. *In*: GAYON, J. (ed.) *Buffon 88: Actes du Colloque international pour le bicentenaire de la mort de Buffon.* Vrin, Paris, 343–352.

HABER, F. C. 1959. *The Age of the World: Moses to Darwin.* The Johns Hopkins, Baltimore.

HALLAM, A. 1989. *Great Geological Controversies.* (second edition). Oxford University, Oxford and New York.

LEIBNIZ, G. W. VON. 1993. *Protogaea: De l'aspect primitif de la Terre et des traces d'une histoire très ancienne que renferment les monuments mêmes de la nature.* Translated by B. de Saint-Germain, edited with introduction and notes by J.-M. Barrande, Presses Universitaires du Mirail, Toulouse.

MAGRUDER, K. V. 2000. *Theories of the Earth from Descartes to Cuvier: Natural Order and Historical Contingency in a Contested Textual Tradition.* PhD dissertation, University of Oklahoma.

OLDROYD, D. R. 1996. *Thinking About the Earth: A History of Ideas in Geology.* Athlone, London.

RAPPAPORT, R. 1982. Borrowed Words: Problems of Vocabulary in Eighteenth-Century Geology. *British Journal for the History of Science*, **15**, 27–44.

RAPPAPORT, R. 1991. Fontenelle Interprets the Earth's History. *Revue d'histoire des sciences*, **44**, 281–300.

RAPPAPORT, R. 1997. *When Geologists Were Historians, 1665–1750.* Cornell University, Ithaca and London.

ROGER, J. (ed.) 1962. *Buffon, Les Époques de la Nature, Édition critique.* Mémoires du Muséum National d'Histoire Naturelle, Série C, Sciences de la Terre, Tome 10. Éditions Muséum, Paris.

ROGER, J. 1968. Leibniz et la théorie de la Terre. *In: Leibniz 1646–1716: Aspects de l'homme et de l'oeuvre.* Aubier-Montaigne, Paris, 137–144.

ROGER, J. 1973. La théorie de la Terre au XVIIe siècle. *Revue d'histoire des sciences*, **26**, 23–48.

ROGER, J. 1989. *Buffon: Un philosophe au Jardin du Roi.* Fayard, Paris [1997 English translation by S. L. Bonnefoi, edited by L. P. Williams, *Buffon, A Life in Natural History*, Cornell University, Ithaca and London].

ROSSI, P. 1984. *The Dark Abyss of Time: The History of the Earth & and the History of Nations from Hooke to Vico.* Translation from the Italian by Cochrane, L. G. University of Chicago, Chicago and London.

RUDWICK, M. J. S. 1992. *Scenes from Deep Time: Early Pictorial Representations of the Prehistoric World.* University of Chicago, Chicago and London.

TAYLOR, K. L. 1969. Nicolas Desmarest and Geology in the Eighteenth Century. *In*: SCHNEER, C. J. (ed.) *Toward a History of Geology.* MIT Press, Cambridge (Mass.) and London, 339–356.

TAYLOR, K. L. 1971. Nicolas Desmarest. *In*: GILLISPIE, C. C. (ed.) *Dictionary of Scientific Biography*, Vol. 4. Charles Scribner's Sons, New York, 70–73.

TAYLOR, K. L. 1988. Les lois naturelles dans la géologie du XVIIIème siècle: Recherches préliminaires. *Travaux du Comité Français d'Histoire de la Géologie*, 3ème série, **2**, 1–28.

TAYLOR, K. L. 1992a. The *Époques de la Nature* and Geology during Buffon's Later Years. *In*: GAYON, J. (ed.) *Buffon 88: Actes du Colloque international pour le bicentenaire de la mort de Buffon.* Vrin, Paris, 371–385.

TAYLOR, K. L. 1992b. The Historical Rehabilitation of Theories of the Earth. *The Compass: The Earth-Science Journal of Sigma Gamma Epsilon*, **69**, 334–345.

TAYLOR, K. L. 1997. La genèse d'un naturaliste: Desmarest, la lecture et la nature. *In*: GOHAU, G. (ed.) *De la géologie à son histoire: ouvrage édité en hommage à François Ellenberger.* Comité des Travaux Historiques et Scientifiques, Paris, 61–74.

VALMONT DE BOMARE, J. C. 1768–1769. *Dictionnaire raisonné universel d'histoire naturelle. Edition augmentée par l'auteur* (second edition), 12 vols, Yverdon.

Jean-André de Luc and nature's chronology

MARTIN J. S. RUDWICK

Department of History and Philosophy of Science, Free School Lane, Cambridge CB2 3RH, UK

Abstract: Jean-André de Luc (or Deluc) (1727–1817), who first proposed the term 'geology' almost in its modern sense, was one of the most prominent geologists of his time. His 'theory of the Earth', published in several versions between 1778 and 1809, divided geohistory in binary manner into two distinct phases. The fossiliferous strata had been formed during a prehuman 'ancient history' of immense but unquantifiable duration. Then the present continents had emerged above sea-level in a sudden physical 'revolution', at the start of the Earth's 'modern history' of human occupation. De Luc argued that the rates of 'actual causes' or observable processes (erosion, deposition, volcanic activity, etc.) provided 'natural chronometers' that proved that the 'modern' world was only a few millennia in age; and he identified the natural revolution at its start as none other than the Flood recorded in Genesis. So 'nature's chronology' could be constructed from natural evidence, to match the well-established historical science of chronology based on textual evidence from ancient cultures. De Luc's natural chronology was restricted to the recent past, but it provided a template for later geologists to develop a geochronology extending into the depths of geohistory. The historical importance of de Luc's work has only been obscured by the myth of intrinsic conflict between science and religion.

The theme of this volume is the 'age of the Earth'. The familiar phrase often implies a single numerical figure, which has certainly been of great and longstanding interest to cosmologists. But to Earth scientists what matters much more is to know what happened when, within that literally global timespan: in other words, to have a geochronology. However, early geologists got a very long way without quantifying deep time at all: they could and did work with an unquantified or so-called relative timescale, reconstructing events at least in the right temporal sequence. They clearly regarded it as desirable to put numbers on such a sequence, in order to give temporal precision to causal processes and to Earth history. But they were often reluctant to do so, for one very good reason: they knew they had little evidence for any such quantification, and they did not want to discredit their infant science by indulging in mere speculation.

Traditionally, however, this reluctance to quantify geohistory has been given a quite different explanation. It has been seen as a manifestation of a perennial conflict between science and religion, focused in this case on a conflict between geology and Genesis. Early geologists, it was said, had avoided specifying the Earth's timescale for fear of the Church, and those who had dared to do so had been persecuted for their temerity. But this is a historical myth, in the true sense of that word: not so much a false story, though it is that too, but a story that persists because it carries heavy ideological loading and is useful to those with various secularist agendas. The myth of intrinsic conflict between science and religion must be demythologized, as it has been by modern historians of the sciences (for example, Brooke 1991; Brooke & Cantor 1998). It was largely constructed in late-nineteenth-century Europe (including Britain) and North America. It was a weapon in the hands of those who, often in the cause of professionalizing the sciences, were trying to wrest cultural and political power away from older social and intellectual elites, among which ecclesiastical elites had been prominent (Gieryn 1999, pp. 37–64; Moore 1979, pp. 19–122; Turner 1978). The myth survives here and there in Europe (including Britain) even today, especially in the rhetoric of some atheistic popular science writers. It was also revived in the United States during the twentieth century, owing to specific cultural conditions that gave as much political power to religious fundamentalists as to their secularist opponents (Numbers 1985, 1992).

Once the historically contingent character of the myth of intrinsic conflict is recognized, the way is open to be liberated from it. Only then can we get a clear view of what was going on in the past history of the sciences, including the Earth sciences. The traditional brief timescale for the Earth, epitomized by James Ussher's famous date of 4004 BC for the Creation, has generally been regarded as utterly opposed to the immensely long timescale that Earth scientists now take for granted (Albritton 1980). In this paper I argue, on the contrary, that the short timescale was the direct progenitor of the long one, and facilitated its adoption. Historians no longer play the futile game of assigning praise or blame to people of the past for helping or hindering 'the progress of science'. But as a matter of history, the attempt to date human events even back to the Creation was one of the conceptual resources that were used in constructing an analogous scale for deeply prehuman time and geohistory. It was not the only such resource, but it was certainly an important one, and its role deserves to be recognized.

The science of chronology

A first clue to the affinity between the two timescales is given by the modern word 'geochronology' itself. It applies to the Earth the older concept of chronology, just as the word 'geohistory' applies the concept of history. Chronology was, and still is, a science within the practice of human historiography (the word 'science' is used here in the original pluralistic sense that only the anglophone world has abandoned, to denote *any* body of disciplined knowledge, about either nature or humanity). Chronologers analysed historical records, both texts and material artefacts (inscriptions, coins, etc.), in order to reach accurate and reliable dates for historical events: first within some specific culture, but then correlating those dates on to a standard cross-cultural timeline. Chronology has ancient roots, but it first developed rigorous internal standards in the seventeenth century (Grafton 1975, 1991; North 1977); Ussher was just one of many scholarly chronologers, and not even the most distinguished. The goal of most early chronologers was indeed religious: it was to locate the great events of divine action, from Creation to Incarnation, in their context of world history. But in the course of doing so chronology itself was gradually secularized, as the histories of other cultures came to be treated in the same way as that of the ancient Jews and the Christian church. Chronology was eventually eclipsed in prestige by other kinds of historical work, but it never went extinct: its modern results are visible whenever, for example, dates BC are assigned to ancient Egyptian or Chinese events and artefacts.

Chronologers strove above all for precision based on good evidence critically evaluated (this admirable goal could readily be transposed into the later science of geochronology). However, working with fragmentary and often problematic texts, they found in practice plenty of scope for argument and controversy. For as they probed the records back in time from the familiar Romans and ancient Greeks, their sources became ever more sparse and enigmatic. Those chronologers, such as Ussher, who claimed to extend the timeline even back to a primal Creation recognized that they were pushing the science to its limits. It was here that their findings were most uncertain and controversial. Although they were all using much the same sources, there was no consensus, and Ussher's 4004 BC was just one of dozens of rival figures for the date of Creation. Given the central place of the Bible in the culture of Christendom, it is not surprising that the records extracted from Genesis were often given a privileged status, over and above what little was known about the early history of other cultures. But to justify that preference chronologers were obliged to compare and evaluate all the ancient texts, and hence to develop techniques of textual criticism that were later applied to the Biblical documents themselves.

However, most work on chronology was focused not on the date of Creation but on later history, where there was much more documentary evidence for constructing a reliable timescale. The great bulk of Ussher's work, like that of any other chronologer, dealt with the ordinary history of the last few centuries BC, with that of Greeks and Romans as well as Jews, not with the origin of the cosmos or even that of the human race (Ussher 1650–1654).

This remained true of chronology a century and a half after Ussher's death. Chronology continued to flourish in the time of, say, Hutton, but as a scholarly discipline based on an array of texts that by then included those of ancient India and China. It was a science that tried to correlate the records of *all* human cultures and to condense them onto a single timeline of universal history. By around 1800, however, most chronologers had abandoned the attempt to extend the timeline back to the origin of the cosmos, restricting their science instead to the few millennia of recorded *human* history. For the growth of Biblical criticism, particularly in the German universities, had led them to recognize that the Creation story in Genesis was the product of a culture profoundly different from their own, and that to treat it as a scientific text was misleading and inappropriate. So when naturalists became convinced on quite different grounds that there must have been a vast *prehuman* geohistory, there was no intrinsic conflict between them and the chronologers, or between geology and Genesis. Occasionally there was indeed forceful and even vehement argument; but only at specific times, in specific places, and above all in specific social settings. There was, for example, more argument in Britain than elsewhere in Europe, and more between savants and popularizers than among the scholars and naturalists themselves. Sweeping generalizations about perennial conflict, or alleging endemic ecclesiastical opposition, are not supported by the historical record.

Theories of the Earth

By around 1800, most naturalists with first-hand field experience of the relevant natural features shared a general sense of the vast though unquantifiable magnitude of deep time. They had no inhibitions about expressing that sense as concretely as their evidence allowed. They alluded almost casually to a million years or to thousands of '*siècles*' (the key French word could denote either centuries or indefinitely longer 'ages'). Such phrases expressed the time that seemed to be needed to account for the deposition of thick limestones and other stratified sediments, the erosion of deep valleys in mountain regions, the accumulation of piles of lava flows on the flanks of volcanoes, and so on (Ellenberger 1994, pp. 35–39; Taylor 2001). These 'guesstimates' usually remained vague and barely quantified, for the good and sufficient reason that naturalists knew they had no reliable evidence on which to base any firmer figures.

Significantly, the most striking exception proved the point by being highly controversial. Buffon's famously precise figures for the dating of his successive '*époques de la nature*' depended wholly on his hypothesis that the Earth had originated as an incandescent globe in space, and that it had cooled thereafter in the manner indicated by his experiments with small model balls (Buffon 1778; Roger 1962, 1989). In effect, the geophysics and the timescale came as a package: if Buffon's theory of global cooling was rejected, as it usually was on other grounds, his timescale necessarily collapsed with it. In fact his work often received the ultimate scientific dismissal of the time: it was a mere '*roman*', a novel, a piece of fanciful science fiction. But in any case Buffon's timescales, published and private, fell within the same range (from tens of thousands to a few millions of years) as the estimates of other naturalists. Even at the lower end, such figures went far beyond the traditional timescale of Ussher's century; Buffon kept the higher end to himself not out of fear of ecclesiastical authorities but because he knew he could not justify it scientifically with any concrete evidence.

Buffon's model of global cooling was rightly treated as just one example of what was usually called 'Theory of the Earth' (Taylor 2001). The phrase referred not to any specific theory about the Earth, but to a scientific *genre*, just as the novel, landscape and opera were artistic genres. 'Geotheory', as I suggest it should be called, was a genre that offered a Theory Of Everything about the Earth, a terrestrial TOE; geotheorists proposed 'systems' or models of how the Earth works, which aimed at accounting for all its major features and all its causal processes. Geotheory always had a temporal dimension, of course; but that did not make it intrinsically *historical*, any more than, say, the theory of gravitation gave a historical explanation of planetary orbits. In the genre of geotheory, some specific 'systems' postulated a directional sequence, yet they were still scarcely geohistorical: in the case of Buffon's *Époques*, for example, the development was in effect determined or programmed from the start by the physics of global cooling. Alternatively, some models proposed a cyclic or steady-state 'system', as in the case of Buffon's (1749) earlier one, or more famously Hutton's. These were even less geohistorical, because the whole point of such models was to show that under unchanging laws of nature no period in past, present or future could be unique or even distinctive in character. Hutton made this explicit in his famous parallel between the endless changes or 'revolutions' of the Earth's surface and deep interior, and the equally endless revolutions of the planets around the Sun (Hutton 1788, p. 304; Gould 1987, pp. 60–97). In fact, on the magnitude of time, Hutton was far from being an innovator; he was a typical 'natural philosopher' of the Enlightenment, who like Buffon took it for granted that in nature time is available without limit for explanatory purposes.

De Luc as a European intellectual

It has been necessary to recall savants such as Buffon and Hutton, and far earlier ones such as Ussher, at some length, in order to provide an adequate context for understanding the origin of geochronology in the decades around 1800. For that origin is not to be found in Hutton's grand verbal gestures about infinite time and a 'system' without perceptible beginning or end. Nor is it to be found in the sense, which was quite widespread among Hutton's contemporaries, of an unquantifiably vast but *finite* geohistory. Instead, the future science of quantitative geochronology had its roots just where the word itself should lead us to expect: in transposing the pre-existing science of chronology from human history onto the Earth (Rappaport 1982, 1997). In initiating that process, one naturalist stands out: without lapsing into the discredited heroic style of history, Jean-André de Luc (or Deluc) (1727–1817) can be seen to have played a pivotal role. (The French naturalist Jean-Louis Giraud-Soulavie (1752–1813) also used the idea of 'nature's chronology' extensively in his natural history of southern France (Soulavie 1780–1784), which was well-known throughout Europe, but on this point his work had less impact than de Luc's.)

De Luc was well qualified to become Hutton's most formidable and persistent critic (Fig. 1). They were of the same generation, indeed they were born less than a year apart. De Luc's interests, like Hutton's, ranged far beyond what was later to become the science of geology; they were both Enlightenment 'philosophers', and to call either of them a scientist is deeply misleading and anachronistic. De Luc was a native and citizen of the Protestant city-state of Geneva (not yet incorporated into Switzerland). He gained a fine scientific reputation throughout Europe, particularly for his meteorology and not least for devising an accurate portable barometer that was widely used in plotting physical geography (De Luc 1772; Feldman 1990). In 1773, in middle age, he migrated to England, where he was elected to the Royal Society and appointed 'Reader' or intellectual mentor to the German-born Queen Charlotte, the wife of King George III (Tunbridge 1971). Like the royal family he lived at Windsor, just across the Thames from another immigrant, royal protégé and Fellow of the Royal Society, the great astronomer William Herschel. Throughout his long life de Luc wrote in his native French, which was then also the international language of the sciences, and indeed of the arts and diplomacy, just as much as English is today.

De Luc was certainly not a marginal figure in intellectual life, either in Europe as a whole or even just in England. But if he was so important at the time, it is legitimate to ask why he is not better known among historians and modern geologists. One reason is that his published work is voluminous, verbose and repetitious, and takes much effort to read and digest; but the same is often said of Hutton, and with as much truth. A more powerful reason is that de Luc was quite explicit about the religious motivation that underlay his scientific work; but this too is also true of Hutton (a fact that often dismays geologists who take the trouble to read the work of their father-figure, if they do so with secularist presuppositions). However, the crucial difference between them is that Hutton's theology was deistic, whereas de Luc's was quite orthodoxly Christian. Queen Charlotte called him approvingly a 'proper philosopher' (*philosophe comme il faut*), because unlike many others he was not a sceptic in religious matters; he called himself unambiguously a 'Christian philosopher' (*philosophe Chrétien*) (Tunbridge 1971). And so, under the baneful influence of the myth of intrinsic conflict, de Luc has often been dismissed, even by historians who ought to have known better, as if he was in effect a modern American fundamentalist. He was not, and he deserves to be treated as seriously by modern geologists and historians as he was by his contemporaries, even if they do not share his religious beliefs.

Fig. 1. Jean-André de Luc: a portrait drawn around 1798 by Wilhelmine de Stetten and engraved by Friedrich Schröder (courtesy of the Bibliothèque Publique et Universitaire, Geneva).

De Luc's geotheoretical model

Like Buffon, Hutton and many others, de Luc devised his own Theory of the Earth or geotheoretical model. It first began to appear in 1778, when he published some letters that he had sent to his patron while he was travelling in the Alps. In a famous footnote, he tentatively proposed the term '*géologie*', although not in its modern sense, for it was to denote the genre of geotheory as the terrestrial analogue of cosmology (De Luc 1778, pp. vii–viii). He then published no fewer than 150 more of his letters to the Queen, in six volumes, supporting his model with a mass of evidence based on his extensive fieldwork in central Europe and the Low Countries (De Luc 1779). All this work was published in Holland, which was a major centre of the international scientific book trade, and of course in French.

Over a decade later, as the political Revolution began to erupt in France, de Luc revised and amplified his earlier work. He sent a series of 31 papers to Paris, almost one a month, for publication in *Observations sur la Physique* (soon afterwards renamed *Journal de Physique*), which was one of the leading periodicals in Europe for the natural sciences (De Luc 1790–1793). At the same time a set of four long letters to Hutton, criticizing the Scotsman's recent geotheoretical 'system', appeared in the English intellectual *Monthly Review*, making de Luc's own model well-known in his adopted country (De Luc 1790–1791). Then, after the leading German naturalist Johann Friedrich Blumenbach of Göttingen had visited him in Windsor, de Luc sent him a set of seven letters, in which he promised to expound his 'system' more succinctly (in the event he was almost as prolix as before). Blumenbach had these essays translated for an important periodical edited in Gotha, the *Magazin für das Neueste aus der Physik* (De Luc 1793–1796; Dougherty 1986). All but one were then retranslated, from German into English, for the new *British Critic* (De Luc 1793–1795), before the whole set finally appeared in Paris, despite the state of war between France and Britain, in book form and in their original French (De Luc 1798). Much later, but still in wartime, his *Elementary Treatise on Geology* (1809) was published in Paris and London, in French and in English translation, making his ideas available at last in a final and somewhat briefer form.

This lengthy publication history underlines de Luc's prominence in the European scientific elite of his time, which remained highly international in outlook in spite of the wars sparked by the Revolution in France. His geotheoretical model changed and developed over the years, in part as a result of his continuing fieldwork, but for the purpose of this paper some enduring themes are of greatest importance. Unlike Buffon, Hutton and most other geotheorists, de Luc showed relatively little interest in offering causal explanations for events in the deep past. This was not because he doubted that they had had natural causes of some kind, but because he was much more concerned to establish the character of the Earth's *history*. He wanted to integrate the natural world, as it was being explored by naturalists such as himself, into the strongly historical perspective of his Christian beliefs, making the recent 'physical history' of the Earth the natural backdrop to the great events of human history.

Specifically, de Luc claimed that the Earth had undergone a radical 'Revolution' in the quite recent past, during which the previous sea floors had become dry land, and the previous land areas had sunk beneath the waves: in effect, an almost total interchange of continents and oceans. He thought this had probably been caused by sudden crustal collapse, rather than by the kind of deep-seated upheaval proposed by Hutton: the present continents would have emerged as a result of a global (eustatic) fall in sea level, after the deep collapse of the former continents. But exactly how it had happened was much less important to de Luc than the fact that it had happened. It was, he believed, a matter of contingent geohistory, an event for which there was compelling field evidence of many kinds. If such an event failed to fit the grand causal theories put forward by others such as Hutton, so much the worse for those theories. He criticized Hutton for spending too much time indoors and not enough in the field (De Luc 1790–1791, **2**, p. 601). Certainly his own field experience was far wider than Hutton's; the latter had, for example, never seen the Alps or any other high mountains, or travelled elsewhere on the Continent (except to get his medical degree at Leiden).

De Luc's geotheoretical model can, I suggest, be usefully termed *binary* in character. His radical Revolution divided geohistory into two sharply contrasted periods, the Earth's '*histoire moderne*' and its '*histoire ancienne*', the familiar world of the present and an ancient or former world (De Luc 1779, **5**, pp. 489, 505–506). Furthermore, since the earlier continents had collapsed out of sight, any evidence for whatever human life they might have sustained had become inaccessible. So in effect de Luc's binary model also divided geohistory into a prehuman and a human period. However, his two periods of geohistory were contrasted not only in relation to human life, but also in their timescales.

Timescales of modern and ancient geohistory

De Luc argued that the 'ancient' or 'former' world (the key French word *ancien* bears both meanings) was of immense but unquantifiable antiquity. The literalism of the earlier tradition in chronology, in its application to the Creation story, was repudiated unambiguously. De Luc insisted on 'the enormous antiquity of the earth', adding that 'naturalists who have thought otherwise were not attentive observers' (De Luc 1790–1793, **39**, p. 334). He reminded his readers how he had long rejected any attempt to compress the whole early history of the Earth into six literal days, as being 'in effect as much contrary to natural history as [it is] to the text that is to be explained'. Like other naturalists of his generation, de Luc had a good sense of how rock formations bore witness to successive periods of deposition over vast spans of time. He seems to have appreciated this increasingly over the years, probably as a result of seeing for himself the huge piles of fossiliferous strata exposed in various parts of Europe. About those along the south coast of England, for example, he exclaimed, 'What time must there have been for the formation of this pile of beds!' The timescale was 'indeterminate' or unquantifiable, but the successive periods of the former world could at least be put confidently into the correct order; the dating of this deep geohistory could be relative, but not absolute (De Luc 1790–1793, **40**, pp. 282, 455).

Like other naturalists across Europe, de Luc also had a good sense of how the different formations were characterized by distinctive sets of fossils, suggesting a complex history of life. The limestones around Paris, he noted, 'contain shells that often vary from bed to bed ... Their [the beds'] identity is recognized as much by these [fossil] bodies as by the nature of the rock' (De Luc 1790–1793, **40**, pp. 281–282). On a larger scale he described how, on the south coast of England, shales were overlain by limestones, which were capped in turn by the Chalk; he noted that ammonites were confined to the lower and therefore older formations; and he interpreted that fact in terms of their extinction in the course of geohistory (De Luc 1790–1793, **39**, pp. 458–460; Ellenberger and Gohau 1981). It was all rather crude by comparison with William Smith's more detailed stratigraphical work a decade or two later, and of course even more so in modern terms. Nonetheless it shows that de Luc had a clear conception of how an 'ancient history' of the world might be constructed from field evidence. Anyway, his strong Christian convictions certainly did not inhibit him from thinking in terms of humanly unimaginable spans of time. He did reconstruct geohistory as a sequence of seven vast periods which, like Buffon's 'epochs', corresponded explicitly to the seven 'days' of the Creation story in Genesis. He certainly ascribed important truth value to that story, but not in a literal sense, because he claimed that its primary purpose was religious and its significance theological. His seven periods of geohistory did not mark literal days, any more than Buffon's seven 'epochs'.

Compared to these vast tracts of deep time, de Luc's present world was quite young, and spanned only a brief period (it is worth noting that his final metaphorical 'day', like Buffon's, was not a divine sabbath rest as in the Genesis story, but the period of *human* dominance on earth). The major event that had brought the present world into being was only a few millennia in the past. De Luc identified it as none other than the event recorded in Genesis as Noah's Flood. That equation has been enough to provoke a knee-jerk reaction from some historians, who have promptly ridiculed or dismissed him (for an early and influential example see Gillispie 1951, pp. 56–66). But in fact his interpretation of the Flood story (De Luc 1798, pp. 287–337) was quite subtle, and far from the crass kind of literalism favoured by modern American fundamentalists. What mattered to him most was just one point, its date. He was concerned above all to show that the chronologers' date for the Flood, based on ancient textual records, was not contradicted by evidence from the world of nature. Only if the Biblical documents were reliable *as human history*, he argued, could their authors be treated as trustworthy guides in moral, social and religious matters (De Luc 1779, **5**, pp. 630–646). Like many other intellectuals throughout Europe, de Luc believed that it was the repudiation of traditional moral norms, by fashionably deistic or even atheistic philosophers, that had led to the social catastrophe of the Revolution in France, which was lurching into its most radical, violent and regicidal phase even as he wrote his letters to Paris.

For de Luc's purposes it was not necessary for naturalists to achieve the kind of precision to which chronologers aspired, dating past events to a precise year. It was sufficient to show that there was natural evidence for a major physical change at around the right time, no more than a few millennia in the past. To establish this he made a close study of what he called *'causes actuelles'*, physical causes that were 'actual' in the sense that is now obsolete in English but still current in French: they were *present* processes of change, processes that could be observed at work in the present world, such as erosion, deposition, volcanic activity and so on (Ellenberger 1987). On the basis of a mass of fieldwork he argued that many such actual causes cannot have been operating for an indefinitely long time, but must have started at a finite and relatively recent time in the past. He identified this as the time of the revolution that was marked by the emergence of the present continents *as land areas*, for only at that point could the varied terrestrial processes have begun to act on them.

One of de Luc's favourite examples was the delta of the Rhine, which he studied repeatedly during his travels in the Netherlands. The delta was manifestly growing in size by new deposition, at a rate that could be judged roughly from historical records of its former limits in Roman and medieval times. But its size was finite, so its historical rate of growth could be extrapolated back to an origin at some finite time in the past. Other examples were provided by lakes such as Lac Léman (the Lake of Geneva), where, at the far end from his native city, the Rhône was clearly extending its delta year by year; yet if it had been doing so indefinitely it would long since have converted the whole lake into alluvial pastures. The screes below the nearby Alpine peaks were also growing yearly, by rock falls from above, but they too were of finite size. On the heathlands of the north German plain (which included both the homeland of his patron Charlotte and the domain of her Hanoverian husband George III), the thin cover of peat overlying sediments with marine fossils suggested likewise that the accumulation of peat from plant decay could not have been going on, and the region could not have been above sea level, for an indefinitely long time. (It will be clear from a modern perspective that many of de Luc's processes, in the regions he knew at first hand, had indeed started, or restarted, at a geologically recent time, namely after the last Pleistocene glaciation; but neither he nor any of his contemporaries had any reason to consider glacial ice as a possible causal agent outside the high mountain regions.)

Nature's chronometers

Features such as these seemed to de Luc to provide several independent lines of evidence that 'the present world', the physical regime now observable on the present continents, dated from a relatively recent time in the past. They provided him with the possibility of constructing a 'physical chronology' based on *natural* evidence, to parallel and supplement the chronologers' narrative based on human textual evidence. More than that, they suggested how nature's chronology might even be quantified into nature's *chronometry*, at least for the period since the decisive revolution in which the present continents had emerged, if not for the far longer periods of the Earth's 'ancient history' before that event.

In his earlier writings, de Luc invoked simple and even homely analogies to express this quite novel concept of geochronometry. For example, he argued that the heathlands of Hannover preserved the nearest there was to the pristine state of the continents:

> There are still many uncultivated lands there, which, like the teeth of a young horse, can give us some idea of the age of the world; I mean, of the date when the present surface took the form in which we know it today (De Luc 1779, **3**, 11–12).

The analogy, more familiar in de Luc's day than in ours, was apt and illuminating. A horse's teeth, steadily worn down by its grassy diet, were an infallible and even quantitative guide to its age, making it proverbially inappropriate to 'look a gift horse in the mouth' to assess its age and its value before accepting it. So likewise the thickness of the peat might be used to estimate the date of the continent's emergence as a land mass; that the former was slight implied that the latter was recent.

The analogy with timekeeping became more explicit in a later letter to the Queen, in which de Luc referred to the growth of a delta such as that of the Rhine:

> This is the true *clepsydra* of the centuries, for dating the Revolution: time's zero is fixed by the unchanging sea-level and its degrees are marked by the accumulation of the deposits of the rivers, just as they are by the piling up of sand in our ancient instruments of chronometry (De Luc 1779, **5**, p. 497).

Here was another apt analogy, although de Luc confused the dripping of a primitive water-clock or clepsydra with the trickling sand in an hourglass; the latter was as familiar in his day as horses' teeth, being used on board ship and in the parson's pulpit as well as in the kitchen (where in the form of the humble egg-timer it survived well into the twentieth century).

In de Luc's view the slowly accumulating deposits in a delta marked the passage of time since the continent emerged, just as reliably as the amount of sand in the lower half of an hourglass marked the time that had elapsed since it was last up-ended.

De Luc later enhanced his argument by calling his physical features *'nature's chronometers'* (De Luc 1790–1791, **2**, p. 580). Here the analogy was with the celebrated clockwork 'chronometers' (for which the word had been coined) that had recently enabled John Harrison to solve the outstandingly important practical problem of determining longitude at sea. De Luc conceded that by comparison with this supreme high-tech achievement of the eighteenth century his dating of natural features was far from precise. Nonetheless, the vivid metaphor made his goal of temporal accuracy unmistakeable. As he himself put it, referring again to the growth of the Rhine delta:

> There then is a true chronometer: one finds there the total operation [of the process of deposition] since the birth of our continents; one can see its causes and their progress, and one can distinguish the parts of all the products of known times. Doubtless there are too many causes of irregularity in this process for one to be able to count the centuries; but it is evident that their number cannot be considerable (De Luc 1790–93, **41**, p. 344).

So nature's own chronometers could provide at least the rudiments of a quantified chronology for the Earth itself, based on the observable rates of actual causes or ordinary natural processes.

De Luc himself applied his 'chronology of nature' only to the one problem that was his greatest concern: the dating of the boundary event that separated the present world, the world of human history, from the incalculably more lengthy former world of 'ancient history' represented by the formations of fossiliferous strata. Nonetheless it was a decisive move, for it showed by example how nature itself could provide the basis, however roughly, for its own quantitative chronometry, which in turn could give precision to nature's chronology, that is, for geochronology. It provided a template and a precedent for those who would later try to extend nature's chronology back in time beyond human history into prehuman geohistory.

De Luc's legacy to geochronology

De Luc's binary model for geotheory was immensely influential in the early nineteenth century, above all because it was adopted by the great French naturalist Georges Cuvier (Ellenberger & Gohau 1981). Cuvier used his skills as a comparative anatomist to show that most fossil bones belonged to terrestrial animals distinct from living species. He interpreted the contrast as the result of a mass extinction or 'catastrophe' in the quite recent past, a 'revolution' that corresponded closely to de Luc's (Rudwick 1997). Like de Luc, Cuvier remained uncertain about the cause of this event, being far more concerned to establish its sheer historicity. He agreed with de Luc that it had happened only a few millennia ago, but he widened the evidence for such a date by drawing on a vast multicultural range of human records. With a cultural relativism typical of the Enlightenment, he reduced the Biblical narrative of the Flood to just one story among many, all of them more or less garbled but still perhaps with a kernel of historical truth (Cuvier 1812, pp. 94–106; Rudwick 1997, pp. 239–246). By treating them all as potentially valid evidence, Cuvier consolidated de Luc's tying of geohistory into human history, geochronology into chronology.

Ironically, however, de Luc's notion of nature's own chronometers, or of nature's chronology, was developed most effectively by a geologist who used it to refute the idea that there had been any catastrophic event whatever in the geologically recent past. A decade after de Luc's death (at the age of ninety), and more than two decades after Cuvier's innovative research on fossil bones, the London geologist Charles Lyell reacted against the Delucian ideas of his former mentor William Buckland at Oxford, and used de Luc's 'actual causes' to argue that no catastrophic event had disturbed the steady pace of physical change on Earth, as far back in deep time as the record of the rocks could go. He could hardly adopt de Luc's phrase as it stood, without drawing unwelcome and embarrassing attention to his close affinity with the earlier naturalist, at least on this point. But under the guise of 'modern causes', as Lyell called them, de Luc's method for analysing and calibrating geohistory got a second wind, and became the basis for Lyell's own geotheoretical model, later dubbed uniformitarianism.

Lyell argued that a statistically uniform rate of organic change at the level of species could be used to construct a new natural chronometer. As in the case of 'actual causes', Lyell avoided adopting de Luc's term, but his idea was the same. His chronometer was to be based on the steadily increasing percentage of extant species, and the decreasing proportion of extinct ones, in the fossil molluscan assemblages of the successive Tertiary formations, which on this basis he termed Eocene, Miocene and Pliocene (Lyell 1830–1833, **3**, pp. 45–61; Rudwick 1978). But his grand theory of constant faunal change soon foundered in the face of awkward empirical detail, and with it his bid to find a natural chronometer that could be applied to even earlier formations.

Conclusion

The aspiration to construct a quantitative geochronology remained, however, at least among a few geologists; and it can be traced through the rest of the nineteenth century, as in the work of John Phillips, and into the radiometric dating of the twentieth (Lewis 2001; Morrell 2001; Wyse Jackson 2001).

In taking that longer view, however, we should not overlook the decisive role of Jean-André de Luc, despised and rejected though he often has been, in transposing the ideas of chronology and chronometry from the human realm into the world of nature, and in working out how to interpret physical features as indices of geochronology. Furthermore, we should not ignore the historical evidence that the intellectual roots of this crucial transposition from culture into nature lay in de Luc's commitment to a religious tradition that had historicity at its heart. It was the science of chronology, powered originally by the desire to locate revelatory divine action within universal human history, that provided the model for de Luc's parallel 'chronology of nature'; and it was this in turn that others could later extend, as geochronology, into the depths of geohistory. Thus in the long run the traditional brief timescale exemplified by Ussher's 4004 BC was the historical template for the vastly longer timescale of modern geology. The perverse revival of a short timescale (or 'young Earth') by religious fundamentalists, in the radically changed epistemic circumstances of the modern world, should not blind us to its decisive importance in the earlier development of the earth sciences.

This paper is based on a talk given at the Geological Society in London in June 2000, during the millennial William Smith meeting, 'Celebrating the Age of the Earth'. Some of the research was done under grant no. SBR-9319955 from the (US) National Science Foundation.

References

ALBRITTON, C. C. 1980. *The Abyss of Time: Changing Conceptions of the Earth's Antiquity after the Sixteenth Century*. Freeman & Cooper, San Francisco.

BROOKE, J. H. 1991. *Science and Religion: Some Historical Perspectives*. Cambridge University, Cambridge.

BROOKE, J. H. & CANTOR, G. N. 1998. *Reconstructing Nature: The Engagement of Science and Religion*. Clark, Edinburgh.

BUFFON, G. LECLERC, COMTE DE. 1749. Histoire et théorie de la terre. *In*: Buffon, *Histoire Naturelle, Générale et Particulière, avec la Description du Cabinet du Roi*, Vol. 1, 63–203.

BUFFON, G. LECLERC, COMTE DE. 1778. Des Époques de la nature. *In*: Buffon, *Histoire Naturelle*, supplément, Vol. 5, 1–254.

CUVIER, G. 1812. Discours préliminaire. *In*: CUVIER, G. *Recherches sur les Ossemens Fossiles de Quadrupèdes, où l'on Rétablit les Caractères de plusieurs Espèces d'Animaux que les Révolutions du Globe paroissent avoir Détruites*, 4 vols. Déterville, Paris.

DE LUC, J.-A. 1772. *Recherches sur les Modifications de l'Atmosphère*, 2 vols. Geneva.

DE LUC, J.-A. 1778. *Lettres Physiques et Morales sur les Montagnes et sur l'Histoire de la Terre et de l'Homme: Addressées à la Reine de la Grande-Bretagne*. De Tune, The Hague.

DE LUC, J.-A. 1779. *Lettres Physiques et Morales sur l'Histoire de la Terre et de l'Homme, Addressées à la Reine de la Grande-Bretagne*, 5 vols in 6. De Tune, The Hague, and Duchesne, Paris.

DE LUC, J.-A. 1790–1791. Letters to Dr James Hutton, F.R.S. Edinburgh, on his Theory of the Earth. *Monthly Review or Literary Journal Enlarged*, **2**, 206–227, 582–601; **3**, 573–586; **5**, 564–585.

DE LUC, J.-A. 1790–1793. Lettres à M. de La Métherie. *Observations sur la Physique, sur la Chimie, sur l'Histoire Naturelle et sur les Arts*, **36**, 144–154, 193–207, 276–290, 450–469; **37**, 54–71, 120–138, 202–219, 290–308, 332–351, 441–459; **38**, 90–109, 174–191, 271–288, 378–394; **39**, 215–230, 332–348, 453–464; **40**, 101–116, 180–197, 275–292, 352–369, 450–467; **41**, 32–50, 123–140, 221–239, 328–345, 414–431; **42**, 88–103, 218–237; **43**, 20–38.

DE LUC, J.-A. 1793–1795. Geological letters, addressed to Professor Blumenbach. *British Critic: A New Review*, **2**, 231–238, 351–358; **3**, 110–118, 226–237, 467–478, 589–598; **4**, 212–218, 328–336, 447–457, 569–578; **5** 197–207, 316–326.

DE LUC, J.-A. 1793–1796. Geologische Briefe an Hrn. Prof. Blumenbach. *Magazin für den Neueste aus der Physik und der Naturgeschichte*, **8**(4), 1–41; **9**(1), 1–123; (4), 1–49; **10**(3), 1–20; (4), 1–104; **11**(1), 1–71.

DE LUC, J.-A. 1798. *Lettres sur l'Histoire Physique de la Terre, addressées à M. le Professeur Blumenbach, Renfermant de Nouvelles Preuves Géologiques et Historiques de la Mission Divine de Moyse*. Nyon, Paris.

DE LUC, J.-A. 1809. *Traité Élémentaire de Géologie*. Courcier, Paris.

DOUGHERTY, F. W. P. 1986. Der Begriff de Naturgeschichte nach J. F. Blumenbach anhand seiner Korrspondenz mit Jean-André DeLuc. Ein Beitrag zur Wissenschaftsgeschichte bei der Entdeckung der Geschichtlichkeit ihres Gegenstandes. *Berichte zur Wissenschaftsgeschichte*, **9**, 95–107.

ELLENBERGER, F. 1987. Les causes actuelles en géologie: origine de cette expression, la légende et la réalité. *Bulletin de la Société Géologique de France*, **3**(8), 199–206.

ELLENBERGER, F. 1994. *Histoire de la Géologie*, Vol. 2: *La Grande Éclosion et ses Prémices, 1660–1810*. Technique et Documentation, Paris.

ELLENBERGER, F. & GOHAU, G. 1981. A l'aurore de la stratigraphie paléontologique: Jean-André de Luc, son influence sur Cuvier. *Revue d'Histoire des Sciences*, **34**, 217–257.

FELDMAN, T. S. 1990. Late Enlightenment meteorology. *In*: FRÄNGSMYR, T., HEILBRON, J. L. & RIDER, R. E. (eds) *The Quantifying Spirit in the Eighteenth Century*. University of California, Berkeley and Los Angeles, 143–177.

GIERYN, T. F. 1999. *Cultural Boundaries of Science: Credibility on the Line*. University of Chicago, Chicago.

GILLISPIE, C. C. 1951. *Genesis and Geology: a Study in the Relations of Scientific Thought, Natural Theology and Social Opinion in Great Britain, 1790–1850*. Harvard University, Cambridge (Mass.).

GOULD, S. J. 1987. *Time's Arrow, Time's Cycle: Myth and Metaphor in the Discovery of Geological Time*. Harvard University, Cambridge (Mass.).

GRAFTON, A. T. 1975. Joseph Scaliger and historical chronology: the rise and fall of a discipline. *History and Theory*, **14**, 156–185.

GRAFTON, A. T. 1991. *Defenders of the Text: The Traditions of Scholarship in an Age of Science, 1450–1800*. Harvard University, Cambridge (Mass.).

HUTTON, J. 1788. Theory of the Earth; or an investigation of the laws observable in the composition, dissolution and restoration of the land upon the globe. *Transactions of the Royal Society of Edinburgh*, **1**, 209–304.

LEWIS, C. L. E. 2001. Arthur Holmes' vision of a geological timescale. *In*: LEWIS, C. L. E. & KNELL, S. J. (eds) *The Age of the Earth: from 4004 BC to AD 2002*. Geological Society, London, Special Publications, **190**, 121–138.

LYELL, C. 1830–1833. *Principles of Geology: Being an Attempt to Explain the Former Changes of the Earth's Surface by Reference to Causes now in Operation*, 3 vols. Murray, London.

MOORE, J. R. 1979. *The Post-Darwinian Controversies: A Study of the Protestant Struggle to Come to Terms with Darwin in Great Britain and America, 1870–1900*. Cambridge University, Cambridge.

MORRELL, J. 2001. Genesis and geochronology: the case of John Phillips (1800–1874). *In*: LEWIS, C. L. E. & KNELL, S. J. (eds) *The Age of the Earth: from 4004 BC to AD 2002*. Geological Society, London, Special Publications, **190**, 85–50.

NORTH, J. D. 1977. Chronology and the Age of the World. *In*: YOURGRAU, W. & BRECK, A. D. (eds) *Cosmology, History and Theology*. Plenum, New York, 307–333.

NUMBERS, R. L. 1985. Science and religion [in American history]. *Osiris*, (2) **1**, 59–80.

NUMBERS, R. L. 1992. *The Creationists*. University of California Press, Berkeley and Los Angeles.

RAPPAPORT, R. 1982. Borrowed words: problems of vocabulary in eighteenth-century geology. *British Journal of the History of Science*, **15**, 27–44.

RAPPAPORT, R. 1997. *When Geologists were Historians, 1665–1750*. Cornell University, Ithaca.

ROGER, J. 1962. Buffon, Les Époques de la Nature: édition critique. *Mémoires du Muséum National d'Histoire Naturelle* sér. C, **10**.

ROGER, J. 1989. *Buffon: Un Philosophe au Jardin du Roi*. Fayard, Paris.

RUDWICK, M. J. S. 1978. Charles Lyell's dream of a statistical palaeontology. *Palaeontology*, **21**, 225–244.

RUDWICK, M. J. S. 1997. *Georges Cuvier, Fossil Bones and Geological Catastrophes*. University of Chicago, Chicago.

SOULAVIE, J.-L. GIRAUD-. 1780–1784. *Histoire Naturelle de la France Méridionale ou Recherches sur la Minéralogie du Vivarais [etc.]*, 7 vols. Belle, Nîmes, and Quillau, Paris.

TAYLOR, K. L. 2001. Buffon, Desmarest and the ordering of geological events in *époques*. *In*: LEWIS, C. L. E. & KNELL, S. J. (eds) *The Age of the Earth: from 4004 BC to AD 2002*. Geological Society, London, Special Publications, **190**, 39–49.

TUNBRIDGE, P. A. 1971. Jean André de Luc, F.R.S. (1727–1817). *Notes and Records of the Royal Society of London*, **26**, 15–33.

TURNER, F. M. 1978. The Victorian conflict between science and religion: a professional dimension. *Isis*, **69**, 356–376.

USSHER, J. 1650–1654. *Annales Veteris Testamenti, a Prima Mundi Origine Deducti [etc.]*, 2 vols. Flesher, London.

WYSE JACKSON, P. N. 2001. John Joly (1857–1933) and his determination of the age of the Earth. *In*: LEWIS, C. L. E. & KNELL, S. J. (eds) *The Age of the Earth: from 4004 BC to AD 2002*. Geological Society, London, Special Publications **190**, 107–119.

Timeless order: William Smith (1769–1839) and the search for raw materials 1800–1820

HUGH S. TORRENS

Lower Mill Cottage, Furnace Lane, Madeley, Crewe CW3 9EU, UK
(*email*: gga10@keele.ac.uk)

Abstract: Smith first described himself as 'land surveyor and drainer' in his 1801 *Prospectus* but then as 'engineer and mineralogist' in his first book of 1806. His several careers are discussed with an attempt to shed new light on his pioneering career as 'mineral surveyor' (a term invented by his pupil, John Farey, in 1808). The trials for coal with which he was involved can be divided into two: those in which he used his new stratigraphic knowledge in positive searches for new coal deposits; and those where his stratigraphic science could often negatively demonstrate that many such searches were doomed to failure. These latter attempts were being made in, and misled by, repetitious clay lithologies, which resembled, but were not, Coal Measures. Smith was the first to show how unfortunate it was for such coal hunters that the British stratigraphic column abounded in repetitious clay lithologies. It was also unfortunate for Smith that many of the founding fathers of the Geological Society were unconvinced of the reality or the utility of Smith's discoveries. Its leaders at first did not believe he had uncovered anything of significance and then simply stole much of it. The development of Smith's stratigraphic science in the world of practical geology remains poorly understood, but the legacy of his method for unravelling relative geological time and space was one of the most significant of the nineteenth century.

The history of geology cannot merely concern its past academic study. It must include past practical aspects, like finding the materials on which mankind depends. If only for this reason, a fresh look at William Smith (Fig. 1) is needed. We should first think of him as a man who *did* things (in today's terms, a technologist) rather than one who *knew* things (today, a scientist). He was successively land and canal surveyor and engineer, land drainer and irrigator, sea erosion and harbour engineer, and, finally, mineral surveyor (or prospector) during the period while he was creating his geological reputation. This was in large part before there were journals for, or societies or chairs of, English geology.

Smith's geological knowledge had by 1815 grown from the small region round Bath and the Somerset Coalfield until it encompassed nearly the whole of England and much of Wales. His stratigraphic grasp was even more complete, from the London Clay down to the, then vital, Coal Measures and beyond, leaving us a legacy reflected in many stratigraphic names in universal use today, like Callovian. His geological achievements range from his early ordering, and identification, of Bath strata between 1794 and 1799, to his great and single-handed (and single-minded) *Geological Map* of much of Britain of 1815. He duly passed these skills on to his several pupils. If the claim is made that no one could have achieved so much in a country with so slight a previous tradition in geology, one might (even if this claim were true) remind people of Elgar's similar achievement in music a hundred years later. One might also reflect how badly the practice of geology, in which Smith was involved, gets recorded as history, particularly in the country which first underwent an Industrial Revolution.

Smith should be seen as the man who first brought science to the clandestine world of mineral prospecting. He seems to have first used this skill, in 1801, to try to find new coal supplies near Bath on properly informed stratigraphic grounds, although his success then is uncertain. He soon used his new scientific knowledge to urge in 1805 that an expensive trial for coal elsewhere in Somerset would inevitably prove abortive, and should be abandoned. But his advice was ignored. Smith was, above all, a pioneer in the practice of geology.

William Smith in the history of geology

William Smith was the first recipient of what later became the premier award of the Geological

From: LEWIS, C. L. E. & KNELL, S. J. (eds). *The Age of the Earth: from 4004 BC to AD 2002*. Geological Society, London, Special Publications, **190**, 61–83. 0305-8719/01/$15.00 © The Geological Society of London 2001.

Fig. 1. William Smith (aged 68) as painted on one day in the early summer of 1837 and given to him by his fellow lodger in London, the French artist Hughes Forau (1803–1873). This portrait was presented to the Geological Society of London in 1881 by Smith's younger brother's son, William Smith (1814–1882) of Cheltenham.

Society of London, founded in November 1807. This is the Wollaston Medal awarded to him in 1832 (Sedgwick 1832). The second award of the Wollaston Fund went in support of one who followed in Smith's footsteps, William Lonsdale (1794–1871) (Murchison 1832). This has made Smith a revered figure among anglophone geologists who now regard his position as established: 'only rarely nowadays can it be said that a new concept originates with one man alone, but the science of biostratigraphy was founded by William Smith, and he owed nothing to earlier writers' (Hancock 1977, p. 3).

Among historians he is instead problematic. Some question his role in the advancement of geology. Rupke saw his award as merely one given to a man who was at 'the centre of a cult ... which professed that he was "the father of English Geology"'. In his view George Greenough, the first President of the Geological Society, or James Parkinson, an early palaeontologist and then stratigrapher inspired by Smith, 'would be most deserving of the title' (Rupke 1983, pp. 191–193). Knell has recently, and rightly, preferred to see the cult to which Smith was then central as one of jealousy among Geological Society members (Knell 2000, p. 30). Laudan had gone further than Rupke, noting that 'Smith can be pitied for his isolation and admired for his determination in the face of it, but it is hard to imagine that the development of geology would have been much different if he had never published' (Laudan 1987, p. 168). Knell again countered this view by exposing the extent to which Smith's researches revolutionized the culture of English geology after 1815 (Knell 2000).

Some problems are exposed here. First, what is the historian's main duty? I agree with the French historian who wrote 'le devoir de l'historien est de comprendre et de justifier avant d'accabler' (Ellenberger 1988, p. 92). Too many recent statements from historians have crushed Smith rather than understood him. Second, historians must regard information contemporary with the events they try to reveal as of greater value than later, secondary, information. This paper tries to make use of such primary material.

But problems remain in trying to understand Smith and his discoveries, made, as they were, during the Napoleonic Wars. We lack all diaries from his formative years before 1802, except that for 1789, and there are few other papers. Second, there is no adequate biography. His nephew, John Phillips (1800–1874), who later became Professor of Geology at Oxford University (Knell 2000, p. 33), attempted one in 1844 (Phillips 1844). But this, published soon after Smith's death, is understandably nepotic, if only because of its manifest sins of omission, ignoring Smith's wife and his fraught finances and eventual imprisonment for debt. Four subsequent attempts at an alternative view have all been abandoned before publication. These were by William Stephen Mitchell (1840–1892) in the 1870's, A. G. Davis (1892–1957) from 1943 on, L. R. Cox (1897–1965) from 1938 and Joan Eyles (1907–1986) from 1962.

This lack of basic data about Smith has resulted in some extraordinary claims being made of him. These range from the wording on his monument at his birthplace put up in 1891, which still reads 'In memory of "The Father of British Geology"' to Steve Jones' claim that Smith's 'first geological map of Britain, made in 1815, managed to trace only one thin seam of limestone called the Cornbrash across southern England. Now after years of exploration, the logic of the whole world has been laid bare' (Jones 1999, p. 210).

Smith's early careers around Bath 1791–1803

Smith was born at Churchill, Oxfordshire, son of the village blacksmith who died when he was seven. From 1787 to 1791, Smith became pupil and assistant to land surveyor Edward Webb (1751–1828) at Stow-in-the-Wold, Gloucestershire. Here Webb taught Smith how to observe the land, a skill vital to his future geological studies. In 1790, Smith was struck by the soils he saw during an attempt to find coal in a field opposite the Shoe Inn, Plaitford, New Forest, Hampshire (Phillips 1844, p. 5).[1] In 1791 Smith was sent by Webb to survey an estate in the North Somerset coalfield. This was the start of Smith's first career, as *land surveyor*. He lodged at Rugbourne Farm, High Littleton, and became involved in many underground surveys for Lady Elizabeth Jones (d. 1800), widow of Sir William Jones (d. 1791) of Ramsbury, Wiltshire. Here much of the same data that John Strachey (1671–1743) had earlier used and published, based on the local knowledge of the stratification in the area as uncovered by colliers, was used by Smith (Fuller 1992).

Smith's breakthrough came in March 1793. He had sufficiently impressed local landowners with his surveying that he was asked to make preliminary surveys for a Somerset coal canal. This was to help bring to market coal which was otherwise land-locked. It was to involve deep surgery of the Earth's strata. This was Smith's entrance to his second career, as *canal surveyor*. Excavation of this canal, which started in 1794, allowed Smith to extend the geological observations he had made within the coalfield to areas outside it. He soon realized that the regularity of stratification inside the coalfield was maintained outside it, contrary to what colliers claimed (Phillips 1844, p. 6). Smith now started to document this. Between August and October 1794, the Somerset Coal Canal management, very progressive as far as new technology was concerned (Doughty 1978), encouraged Smith to go on a canal fact-finding tour through England with two local coal owners, Sambourne Palmer (1758–1814) and Richard Perkins senior (1753–1821) (Torrens 1997a, p. 242). Their tour took an easterly route up to Yorkshire and returned more westerly through Shropshire – where the Industrial Revolution had falteringly started nearly a century before. Here, in October, they saw the scaled-down Caisson canal lift in operation, which had been recently completed at Oakengates (*Shrewsbury Chronicle*, 3 October 1794).

In April 1795, William Benett (fl. 1782–1825) was appointed resident engineer to the Somerset Coal Canal (but based in distant Beckington, near Frome, Somerset) with Smith as on-the-spot surveyor, or sometimes sub-engineer (Mitchell 1872). Uniquely, much of the canal was cut along two nearly parallel valleys. This allowed strata in one valley, as they were progressively exposed from west to east, to be compared with those exposed in the other a few miles away. This was a crucial feature in helping Smith's growing understanding of local stratigraphy. Contracts to dig the first, western, canal cuts were settled by June 1795. The canal reached fossiliferous, Lias rocks [Jurassic] above unfossiliferous Red Earth [Trias] during the second stage of excavations, from September 1795.

Smith moved to a new residence, more befitting an aspiring canal surveyor/engineer, from November 1795. He lived in the centre of Cottage Crescent (today Bloomfield Crescent) way above, and south of, the city of Bath. Here he later recorded how:

> from this point [my] eye roved anxiously over the interesting expanse which extended before me to the Sugar-loaf mountain in Monmouthshire, and embraced all in the vicinities of Bath and Bristol; then did a thousand thoughts occur to me respecting the geology of that and adjacent districts continually under my eye, which have never been reduced to writing (Phillips 1844, p. 15).

Smith's canal superintendence was already making it hard to find time for other, to him less pressing, concerns. But enough written evidence survives from this period to assess how Smith's knowledge of local strata grew. On 5 January 1796 he noted at the Swan Inn, Dunkerton, no doubt after a hard day's work on the nearby Somerset Coal Canal:

> fossils have been long studdied as great Curiosities collected with great pains treasured up with great Care at a great Expense and shown and admired ... because it is pretty. And this has been done by Thousands who have never paid the least regard to that wonderful order and regularity with which Nature has disposed of these singular productions and assigned to each class its peculiar Stratum (Cox 1942, p. 12).

In December 1796, Smith was elected a member of the Bath Agricultural Society (*List of Members* 1797, p. 75), notably, among so many Esquires, as 'Mr William Smith, near Bath'. He took his first apprentice/assistant in 1795, Thomas Bartley (1780–1819), son of the then

---
[1] Grid Reference SU 2719.

secretary of that society, whom he trained both as land and mineral surveyor and drainer (*Cambridge Chronicle*, 13 July 1810). In August 1797 Smith wrote down his earliest ordering of Bath strata to survive (Douglas & Cox 1949). This starts with No. 1 'Chalk Strata' and descends to No. 28, 'Limestone' (Carboniferous Limestone) below the Coal Measures. By 1799, Smith was less certain of the position of this last limestone, which suggests he had now inherited the earlier, mistaken tradition of John Strachey, who had confused Mendip Limestone with Lias limestone (Fuller 1992, p. 77). Cox could only show that Smith was aware of Strachey's work by the end of 1796 (Cox 1942, p. 10).

Torrens (1997a) discussed Smith's geological predecessors in the Bath area. He confirmed the inordinate number of enthusiasts for collecting fossils there, but noted that none of them yet ordered these stratigraphically. They thus pay mute tribute to Smith's originality. Smith's influence on the local literati can be further inferred from the printed *Proposals*, dated April 1797,[2] which Rev. Richard Warner (1763–1857) issued for his *The History of Bath* (1801). These include the note that one of Warner's intended chapters was to deal with 'the Fossilogy of Bath and its neighbourhood'. Smith later recorded that Warner's fossil collection had been the first that he arranged in stratigraphic order (Torrens 1979). Warner was one of those who employed Smith to stop landslips at his Widcombe cottage, near Bath.[3]

By March 1798, Smith felt sufficiently secure financially to purchase, mostly on mortgage, a beautiful, but fateful, 17-acre estate beside the Somerset Coal Canal at Tucking Mill (Eyles 1974). This, with his expensive attempts to work Bath Stone here, from 1808, from adjacent quarries via a newly laid railway which led down to a new saw-mill (Farey 1814, p. 30; Pollard 1983), proved a financial disaster for Smith and was one major reason for his imprisonment for debt in London in 1819 (Eyles 1969, p. 157).[4]

Smith's next elaboration of the Bath stratigraphic column, in 1799, went from No. 1 'The Chalk' down to only No. 23, 'The Coal'. It went no lower because there was less incentive for him to do so. Coal, *the* raw material of the industrial revolution (Wrigley 1962), was being increasingly sought as the country came to terms with the threat of Napoleonic invasion. Smith's new 'Order of Strata' listed at least four separate strata of blue clays. These are (from below):

(1) No. '14, Marl Blue' (today Lias Clay)
(2) Nos. '8–11, Blue and Yellow Clays with Fullers Earth' (Fullers Earth Clay)
(3) No. '5, Clay' (Forest Marble)
(4) No. '3, Clay' (Clunch or Oxford Clay).

These repetitious lithologies were in addition to the more famous additional example, which was the only double repetition that Smith recorded, and equally carefully separated, as

(a) No. '7, Upper Freestone' (Bath Freestone)
(b) No. '12 Lower Freestone or Bastard Freestone' (Inferior Oolite).

Smith's 1799 'Order' was reproduced by Smith (1815, table 1) and Fitton (1833a, pp. 38–39), among others.

Smith's note of January 1796, quoted above, shows that he was also well-informed of the distributions of fossils that were to be found in many, but *not* all, Bath strata and how he had realized their utility. These realizations enabled Smith to start separating repetitious lithologies against his standard stratigraphic order. This he seems to have been the first to achieve. He was also able to start colouring stratigraphic maps of the area, only some of which survive (Cox 1942, p. 25). In addition, Smith was being consulted as a water engineer, by late 1796, both on problems of water supply, as at Heytesbury, Wiltshire (Rogers 1976, p. 245) or as a land drainer. His first commission in this field was for Somerset Coal Canal chairman, James Stephens (1748–1816), on his Camerton estates, Somerset (Smith 1806, p. 54; Torrens 1997a, p. 241). These commissions mark the start of his third career, as *water drainer/irrigator*.

There is abundant evidence from Smith's own collection of fossils (which largely survives in the Natural History Museum, London) that he had been active as a collector of fossils while working as surveyor for the coal canal. Some of the best examples are the suites of fossils he collected during coal canal excavations, including those from the ill-fated Caisson lock between 1796 and 1799. These fossils were later vital in helping to locate its present site (Torrens 1975, pp. 7–8). Collecting fossils was certainly not something Smith only turned to 'apparently between 1801 and 1808 when he wrote to James Sowerby' (Fuller 1969, p. 2272).

---

[2] Gough Gen. Top. 365, Bodleian Library, Oxford.
[3] Oxford University Museum, Smith MSS (hereafter OUMSM), Box 40.
[4] In view of the central importance of this house to Smith's turbulent career, and to publications of the Society (see Duff & Smith 1992, frontispiece) one must hope the Geological Society will see that the plaque they helped re-erect on the wrong house in 1932 is moved to the correct site.

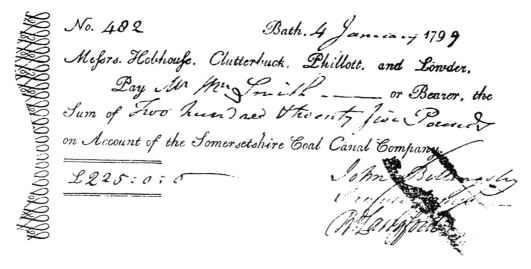

Fig. 2. Smith's final 1799 pay cheque from the Somerset Coal Canal.

The watershed of Smith's career proved to be 1799. On 5 April that year he ceased to be surveyor to the coal canal (Mitchell 1874, p. 7). This dismissal was confirmed by the General Meeting of the Somerset Coal Canal company on 5 June 1799 (Mitchell 1869). Smith, aged 30, who had been earning a regular income, now had to earn a new living by his own endeavours. His last pay cheque, for the six months January to June 1799, is reproduced here (Fig. 2).[5]

Smith's new, now geological, aspirations are revealed by the *timing* of his 'Dictation of the Order of the Strata in the Bath area', as named on the tablet later fixed to 29 Pulteney Street, Bath, one of the residences of his new supporter, Rev. Joseph Townsend (1739–1816). This was on 11 June 1799, days after his final dismissal, not in December as the tablet claims. This 'Dictation' detailed the 23 strata from Chalk to Coal. It could provide, once it had been disseminated, a vital advance in both English geology and in mineral prospecting by providing a standard against which rocks elsewhere in England and abroad could be compared down to the Coal. In the same year Smith produced his first, now lost, geological map of the country around Bath (Phillips 1844, p. 27; Cox 1942, p. 25). In 1802 he moved to a new office taking a new land-surveying partner, Jeremiah Cruse (1758–1819) (Torrens 1983). This was in the centre of Bath, one of polite English society's main arenas for socializing. It was ideally situated to help bring Smith's geological ideas to a wider public. This was aided by the unexpectedly wide distribution of his 1799 'Order of Strata' (Phillips 1844, p. 31). The possible role of the new (second) Bath Philosophical Society, also founded in 1799, in this diffusion must also be considered. Forty per cent of its known members (its minute book is lost) were associates of Smith (Torrens 1990, pp. 183–184).

Smith's career as itinerant land drainer and irrigator (and sea-breach engineer) 1799–1809

The year 1799 also proved a watershed for Smith in a more direct way, since the amount of water shed that year was phenomenal:

> From its commencement to its close [1799] was, perhaps, as ungenial to the productions of the earth and to the animal creation as any upon record, and the inclemency extended over a great part of Europe... On 18 March great land floods had overflowed the river at Bath and from July 8 an extremely wet summer [followed by] a very bad harvest (Baker 1911, pp. 234–235).

This had a critical effect on Smith's career, just after his dismissal, as it forced him to put his expertise in land drainage, rather than his geological knowledge, to use. The almost certain cause of his dismissal was the controversy over the future of their ill-fated Caisson (Torrens 1975). Its failure had been a direct result of this unexpectedly wet season, which damaged the critical geometry of its cistern beyond economic

---

[5] From the original found pasted into the copy of Woodward (1895) at Bath Reference Library.

repair. Smith soon reported on, and solved, other serious subsidence problems as at Combe Grove, Bath, in February 1800 (Phillips 1844, pp. 33–34). In May 1800, he was draining for Thomas Crook (1746–1823), 'one of the best farmers of the Bath district' (Young 1798; Phillips 1844, p. 34), at Tytherton Kellaways, Wiltshire. Excavations here revealed to him a new rock unit, the Kelloways Stone. This was the first of many he was able to add to his 1799 stratigraphic column.

Crook's record as an improving farmer drew many people to visit his farm as it lay close to Bath. In 1800, Thomas William Coke (1752–1842), Earl of Leicester, brought his sick wife to Bath, where she died on 2 June 1800 (Stirling 1908, vol. 1, p. 445). During his stay at Bath Coke visited Crook's agricultural improvements at Tytherton Kellaways and saw Smith's work there (Fitton 1833a, p. 32). Coke invited Smith to come to Holkham, in far-away Norfolk in October 1800 (Phillips 1844, p. 35) to bring his new drainage methods into use there. Coke was the major agricultural improver in that area and it was his influence which brought Smith's draining and irrigation methods all the way to Norfolk, rather than any supposed 'hype and salesmanship of men like "Mr Brooks" and William Smith' (Wade Martins & Williamson 1994, p. 37). Coke was also a major landowner in coastal areas here facing the brutal erosion of the North Sea, and from 1801 he employed Smith in yet another new direction, as seabreach engineer, which work kept Smith variously, but often busily, employed in Norfolk and Suffolk for many years (Cox 1942, p. 20).

So it was as a water/drainage engineer that Smith first made his living after losing his employment with the coal canal. In 1808, he drew up lists of those who had so far employed him. These were mostly as drainer or irrigator.[6] Among these he drained the Phelips family's estates at Montacute, Somerset, and the Earl of Ilchester's estates at Melbury, Dorset, over 1800–1801. He drained 100 acres at Longleat, Wiltshire, for the Marquis of Bath in the spring and summer of 1801 at 50 shillings an acre. At Woburn, Bedfordshire, in 1801, while draining for the Duke of Bedford, to whom he had been introduced by Coke, he met the polymath John Farey (1766–1826), to whom he also demonstrated his geological discoveries (Torrens 1993). Farey became an enthusiastic pupil and proselytiser for Smith's discoveries and originality, both in geology and draining, so far as to earn the later epithet 'Smith's Boswell' (Mitchell 1869). Farey immediately introduced Smith to Benjamin Bevan (1773–1833) of Leighton Buzzard (Fitton 1833a), who used this new geological knowledge in his own new career as canal engineer (Eyles 1985). Smith's geological results were now starting to be publicized in print, as by William Matthews in 1800 (Torrens 1997a, pp. 245–246) and by Richard Warner from 1801 (Cox 1942, pp. 31–34), but this publicity was certainly not mainly through the printed word.

Smith's geological results were equally disseminated at agricultural meetings, like the annual Woburn Sheep Shearings or Smithfield Cattle Shows, at which Smith and Farey became regular attenders. At the 1804 Woburn meeting, the president of the Royal Society, Sir Joseph Banks (1743–1820), opened a subscription to help Smith publish his geological results (Torrens 1993) but Smith was now so busy travelling, covering annual mileages of up to an extraordinary 10 000 miles a year (Phillips 1844, p. 50), that he found this more and more difficult.

In addition there is evidence, in an English 'laissez faire' society operating under war-time conditions, that Smith felt his financial rewards would come much more readily from water drainage and irrigation work than from his, to us only with hindsight, more important geological discoveries (Smith 1806). As he explained in 1806 to his new disciple William Cunnington (1754–1810), wool merchant, mercer and draper of Heytesbury, Wiltshire, and one of Smith's earliest geological converts outside Bath (Farey 1818; Cunnington 1975):

> the Subject of Strata ... now appears to be a most gigantic Undertaking, the accomplishing of which would almost alarm me to despair ... I had no Idea of the depth to which I am imersed. The time and attention necessary to the completion of such a small work as that on Irrigation has fully convinced me of the much greater difficulties attending the publication of a voluminous Work on the Strata.[7]

The difficulty of raising interest in such a new subject as geology had only been confirmed in Smith's mind by the double bankruptcy, in 1801 and 1804, of John Debrett (1752–1822) who was to have published Smith's first work on strata. The 1801 *Prospectus* for this described Smith as 'Land Surveyor and Drainer' (Sheppard 1920, p. 110). Only when his first book was published in 1806 did Smith make reference to his geological attainments, although to describe himself as a mineralogist only proves how little the word

---

[6] OUMSM, Box 40.

[7] OUMSM, letter from Smith to Cunnington, 14 December 1806.

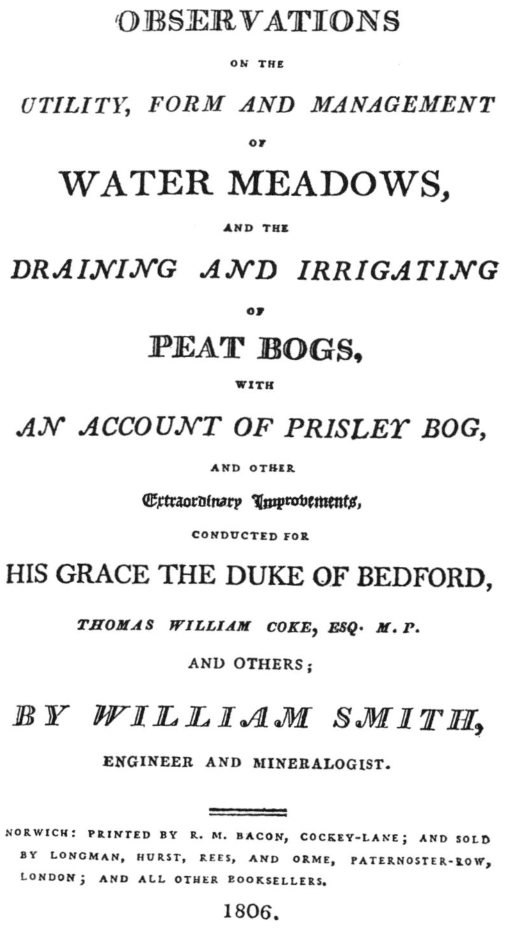

Fig. 3. Title page from Smith (1806) describing himself as engineer and mineralogist.

> OBSERVATIONS
> ON THE
> UTILITY, FORM AND MANAGEMENT
> OF
> WATER MEADOWS,
> AND THE
> DRAINING AND IRRIGATING
> OF
> PEAT BOGS,
> WITH
> AN ACCOUNT OF PRISLEY BOG,
> AND OTHER
> Extraordinary Improvements,
> CONDUCTED FOR
> HIS GRACE THE DUKE OF BEDFORD,
> THOMAS WILLIAM COKE, ESQ. M.P.
> AND OTHERS;
> BY WILLIAM SMITH,
> ENGINEER AND MINERALOGIST.
>
> NORWICH: PRINTED BY R. M. BACON, COCKEY-LANE; AND SOLD BY LONGMAN, HURST, REES, AND ORME, PATERNOSTER-ROW, LONDON; AND ALL OTHER BOOKSELLERS.
> 1806.

geology had yet penetrated the English language (Fig. 3).

Smith's fourth career as mineral prospector, 'by far the most valuable part of my work', from 1801

When Smith met Crook at Kellaways in 1800 he noted how Crook here:

> had conceived from the great resemblance to the slaty black marls which he had seen dug in Somersetshire that this stuff might be serviceable to his land. He accordingly set about an experiment which cost him a deal of money without doing the least good. 5 or 6 pits were sunk in different parts of his estate which uniformly produced the same stuff. Some of the pits were near 40 feet deep and the stuff came out in broad, flat scales or flags which fell to pieces by exposure. The miners could work in it with very little timber but the water was troublesome and smelled strong of sulphur. Mr C like many others, was nearly led into an experiment for coal by the flattering encouragement of the workmen employed upon the business, who frequently made fires of this stinking stuff and positively insisted upon it — that it would come to coal below.[8]

Such a costly experiment for coal in the Oxford Clay here had already been made (at an unrecorded date before 1800) by Sir Edward Baynton-Rolt (1710–1800) of Spy Park, Wiltshire, and the first Marquis of Lansdown, William Petty (1737–1805) of Bowood, Wiltshire, at the nearby village of Tytherton, Tetherton or Titherington, Wiltshire. Its only record is by Townsend, who noted that:

> the appearances of coal in this bed [Clunch or Oxford Clay] has given rise to numerous trials, encouraged by ignorance or fraud. Among these, I remember one at the expense of Sir Edward Baynton and the Marquis of Lansdown, to the S.E. of Tetherton ... Near Titherington, at a high level above the Kelloway rock, Sir Edward Baynton was prevailed upon to try for coal. Here he was amused and flattered with the hope of an extensive colliery, and from time to time the workmen showed him indubitable signs of coal, till the subscription funds and his patience were exhausted, and then they reluctantly departed ... [here] the depth must have been very considerably more than three hundred yards, with seven strata of water (Townsend 1813, pp. 127, 427–428).

As Townsend said, of this and the many other such trials made up till then, 'the knowledge of geology is of importance to coal adventurers, that they may neither deceive themselves in their expectations nor become a prey to designing men, who may endeavour to deceive them' (Townsend 1813, p. 427). Smith, and his pupils, including Townsend, now clearly understood the real value of knowing that strata were regularly ordered. This standard succession of strata could now be used to guide the previously haphazard hunt for raw materials in a rapidly industrializing Britain.

We should also note how difficult these times were politically. War was declared again after the

---
[8] OUMSM, Box 30, Folder 3.

abortive Peace of Amiens, in May 1803. Napoleon's grand army of 90 000 lay encamped at Boulogne waiting to embark on the invasion of England, as Martin Benson, rector of Merstham, Surrey, warned his congregation of the horror that faced them:

> We are fallen upon times of no ordinary complexion ... savage discord once more has opened upon the world. Buonaparte is invariable, and will be here, if opportunity favour him. The merciless tyrant of the unprotected, bloody exterminator of friend and foe, breathes his pestilential malice on our shores (Fuller 2000, p. 51).

Such attitudes seem to have inspired an enthusiasm for coal trials, some of which are outlined below.

Smith seems to have started to use his knowledge to prospect for minerals early in 1801, before he moved office to London in 1803. His 1808 list of those who had employed him to find 'Coal and other Subterranean Valuables'[9] only recorded one search for a material other than coal. This was 'limestone for Mr Holland, Oakhampton', Devon, probably the architect Henry Holland (1746–1806). The other ten trials he had supervised and listed here were all for coal, in Gloucestershire, Lancashire, Somerset, South and North Wales and Yorkshire. Many, as usual with such mining attempts, are badly recorded. Smith was later to emphasize that:

> the wealth of a country primarily consists in the industry of its inhabitants and in its vegetable and mineral productions ... whatever therefore tends to facilitate the discoveries of the one or the other, may with just propriety be considered a national concern ... the immense sums of money imprudently expended in searching for coal and other minerals, out of the regular course of the strata which constantly attend such productions..., prove the necessity of better general information on this extensive subject ... on my principles the most proper soil will be known for ... miners and colliers, in searching for metals and coal; builders for freestone, limestone, and brickearth; the inhabitants of dry countries, for water; the farmer, for fossil manures; will all be directed to proper situations, in search of the various articles they require; and will be prevented from expensive trials, where there can be no prospect of success (Smith 1815, pp. 2–5).

Positive coal trials: 'on my principles the most proper soil will be known for miners in searching for coal'

Three examples of scientifically based trials made by Smith away from then-known coalfields are noted here: those at Pucklechurch, Batheaston and Compton Dundon. All used Smith's observation of the unconformable relationship which brought Coal Measures to underlie the Red Earth, Ground or Marl (Triassic) in the Bath area (Farey 1813, p. 363). Smith's first underground survey at Mearns Colliery, High Littleton, Somerset, in 1791 was in just this situation (Phillips 1871, p. 499). Smith was again advantaged by a local circumstance, as here coal was commonly worked beneath the Red Ground. He wrote in 1815:

> The mass of strata usually called coal-measures, is known to be deprived of much of the superficial space which it would occupy by the overlapping of the red earth. When this unconformability of the red earth shall be more generally known, and its irregular thicknesses more correctly proved, it is highly probable that much more coal may be discovered (Smith 1815, p. 49).

Exactly when Smith realized the full significance of this unconformity is unclear. But he certainly recorded such unconformities, as that which affected Upper Jurassic strata near Bath, on his 1819 Bath to Southampton geological cross-section (Fuller 1995, section 2). This inconformity had meant that Smith's first 'Order of Strata' was incomplete in the vicinity of Bath – a problem he progressively rectified by adding those rock units which had been cut out. The evolution of Smith's standard here is well explained by Donovan (1966, pp. 12–16, figure 2). Smith showed another, Greensand, unconformity on his Taunton to Christchurch cross-section (Fuller 1995, section 5). In later notes Smith devoted a special section to 'Unconformity in the Stratification', dating it to 1807 or 1808.[10]

Pucklechurch, Gloucestershire, 1801–c. 1802

Smith's earliest attempt to use stratigraphic knowledge to find coal here, just outside a known coal field, seems to have been that beneath Lias and Red Ground cover on a property which Smith surveyed for George Whitmore (1775–1862) at Pucklechurch, Gloucestershire, in April

---

[9] OUMSM, Box 40.

[10] OUMSM, 23 June 1839 autobiographical notes.

1801.[11] At least two 'sections of strata sunk through for coal' which he then made survive at Oxford.[12] Fitton (1833a, p. 34) dated these as much too early, and confused even Phillips (1844, p. 8), as Cox (1942, p. 9) pointed out. Townsend (1813, pp. 130–131, 169) confirmed the stratigraphy uncovered here. Any lack of success in finding much coal can be explained by complicated structural repetitions of Coal Measures here. These are due to lateral faults which were not understood until Handel Cossham's (1824–1890) discoveries in 1884 (Vintner 1964, p. 41).

The Batheaston Mining Company, Somerset, 1804–1813

In March 1804 other trials for new collieries nearer Bath were being urged anonymously (*Bath Chronicle*, 29 March 1804). This was led by a woman who had come from the Shropshire Coalfields. On 16 April 1804, this new 'Batheaston Mining Company' was announced, to work land owned there, east of Bath, by entrepreneur Thomas Walters (1757–1847), who was to lease it to a Mrs Browne, to be worked by a company of shareholders on the model of a canal company (*Bath Journal*, 16 April 1804). Mrs Mary Lane Browne (1764–1838) had been born at Preston-on-the-Weald-Moors, two miles north of Ironbridge. She was the recent widow of Gilbert Browne (1741–1798), solicitor of Shifnal, Shropshire. They had moved to Bath in about 1792, where they must have met Walters.

Smith knew nothing of this Batheaston scheme until June 1804 when a friend (surely John Farey) showed him the *Monthly Magazine* for that month (Anon. 1804) in which his name was mentioned. Smith wrote to Cruttwell, the company solicitor, who expressed 'much surprize at your having been unacquainted with [our scheme] so long, having understood from Mrs Brown that she had written to you fully on the subject'.[13] Smith's official involvement was announced on 16 July after he had met the committee and reported on the feasibility of finding coals at Batheaston (*Bath Journal*, 16 July 1804). A site was soon chosen there, constrained both by having to be on Walters' property and by needing a supply of water for working machinery (*Bath Journal*, 30 July 1804).

In view of the sort of coal prospecting Mrs Browne was already concerned with at Haselor Farm, near Evesham (see below), Smith's involvement was of critical significance. This only started *after* Mary Browne had persuaded the landowner, Thomas Walters, of the supposedly great profits to be made by supplying war-torn England with coal, and *after* she had started her own misguided search near Evesham. As soon as Smith was consulted at Batheaston, this trial became one based on the scientific, if still rudimentary, stratigraphic information which Smith had uncovered. Batheaston is thus a significant site. It is one of the first major mineral prospects in England to have been scientifically grounded.

Smith could use his stratigraphic knowledge to prospect for stratified minerals (which he knew occupied determined positions) only after he had worked out the 'Natural Order' of rocks in the Bath area. This he elucidated between 1791 and 1799. Well aware of the strongly unconformable relationship existing below the Red Ground (Triassic) here, Smith's knowledge of this (sub-Triassic unconformity) allowed him also to predict that productive Coal Measures might be found here by sinking or boring into rocks *below* it.

At first two shafts were started at Batheaston,[14] but later borings in their floors hit major problems with influxes of water, which led to bigger problems, as the flow of the nearby Bath Hot Springs (on which the thriving Bath tourist industry depended) was suddenly affected by them. Attempts to overcome these problems with 30 horse-power and then 80 horse-power steam engines were made; and boring here continued with great difficulty down to a total depth of 671 feet before the attempt was finally abandoned in 1813, when the money ran out (Kellaways 1991).

The strata penetrated at such cost, below the Triassic unconformity here, proved unproductive, unluckily proving to lie immediately *below* productive Coal Measures. These seem certain to have been the Upper Cromhall Sandstones, in the uppermost part of the Carboniferous Limestone sequence immediately *below* any coal. If so, Smith's trials here – unlike Mrs Browne's attempt near Evesham – only just 'failed'. Because this Batheaston attempt was unsuccessful, it should not deceive us into thinking that it was in any way unscientific. To describe this attempt as 'incredible', as two historians have (Buchanan & Buchanan 1980, p. 31), is bad history, because in petroleum industry terms this was just a dry hole failure,

---

[11] Pucklechurch, Grid. Ref. ST 6976. Gloucestershire Record Office, Whitmore Estate papers, D45 P45.
[12] OUMSM, Box 26, folder 1 and out of sequence 1.
[13] OUMSM, draft letter, Smith to T.M. Cruttwell, 2 June 1804; letter, T.M. Cruttwell to Smith, 8 June 1804.

[14] Grid. Ref. ST 782677.

drilled for all the right reasons but without any commercial discovery being made.

Compton Dundon, Somerset, 1813–1815

Smith made later attempts to find coal below this same Triassic unconformity at Compton Dundon, near Glastonbury, Somerset, between 1813 and 1815. These were made for the Earl of Ilchester (Moore 1867, p. 457).[15] Borings here penetrated 609 feet of Triassic marls and it proved uneconomic to continue into rocks further beneath the unconformity. Smith knew that elsewhere in Somerset coal was often to be found just below such Red Ground cover but he had no way of knowing how extraordinarily thick such Triassic cover could prove to be.

Negative or abortive coal trials: 'Trials, where there can be no prospect of success'

The three trials described above involved Smith in using the stratigraphic knowledge he had uncovered, in attempts to predict where coal was likely, or might conceivably be found. But his new skills could equally be put to negative use, by predicting where it was impossible that coal could be found. Only the lack of attention so far given to the history of the practice of geology has allowed it to ignore such aspects of the history of geology and mineral prospecting. Buckland wrote in 1837:

> before we had acquired some extensive knowledge of the contents of each series of formations which the Geologist can readily identify, there was no a priori reason to expect the presence of coal in any one Series of strata rather than another. Indiscriminate experiments in search of coal, in strata of every formation, were therefore desirable and proper in an age when Geology was unknown (Buckland 1837, p. 524).

It was Smith who provided such 'extensive knowledge'.

---

[15] In 1977, I visited this locality and found a 90 year old informant who showed me where the main coal trial here had been made, despite this being over 70 years before his own birth. I made a mental note not to believe much of what he had told me. When I later discovered Smith's original map and boring log (Dorset Record Office D/F51) showing the site, it was exactly where he had said it was, in a field beside a helpfully named Coal Pit lane (Grid. Ref. ST 496325). I have since been more sympathetic to the possibilities of oral history.

Regardless of whether such attempts were desirable or proper, they were certainly common and it is only the terrible historical record of failure, here often inevitable, that denies us adequate records. Such trials are, all too often, impossibly hard to document. John Farey correctly claimed in 1806 that:

> almost every common, moor, heath, or piece of bad land in parts where coals are scarce, have at one time or other been reported by ignorant coal-finders to contain coal..., our inquiries, and those of Mr Smith, have brought to light hundreds of instances, where borings and sinkings for coals have been undertaken in such situations, and on such advice, in the southern and eastern parts of England, attended with heavy, and sometimes almost ruinous, expences to the parties, though a source of profit to the pretended coal finders, who, or some of their never-failing race of successors, equally sapient, have in many instances been able to return to the same spot or neighbourhood, and persuade a new proprietor to act again the same farce, and squander his money on an unattainable object (Farey 1806).

Haselor Farm, near Evesham, Worcestershire, 1804–1806

'Haselor Colliery' is a typical example of the sort of coal search then being undertaken, before people were aware of Smith's results. This attempt was made on Smith's Blue Marl (Lower Lias clays). Here, in a field beside the A44 road,[16] the same Mrs Mary Lane Browne, the initial stimulus for the Batheaston attempts, organized her own, misguided, attempt to open a colliery. This was funded by subscription, opened in February 1804, which attracted 214 shareholders initially paying £5 each. Mary Browne had lived here since the death of her husband in Bath in October 1798 (*Bath Journal*, 22 October 1798). Her attempt was by a sunk, walled shaft, now known to have reached the enormous depth of 840 feet at a minimum cost of £2300.[17]

The shaft first penetrated the Blue Lias surface clays, now known to be Jurassic. These had misled Mrs Browne by their lithological similarity to clays associated with true Coal Measures. But, however similar, these black/

---

[16] Grid. Ref. SP 011435.
[17] Sinkers' log 1805–6, Evesham Public Library, L338.272.

blue clays were stratigraphically way above the Coal Measures. Smith had documented the true stratigraphic position of 'The Coal' from 1797. So any attempt here, at such a high geological level, was doomed the moment the first sod was turned, if guided only by lithological similarity. It did not try to exploit the unconformable relationship that Smith had tried to use near Bath. It was reported of this trial:

> a similar [but only in also seeking coal!] experiment [to that intended at Batheaston] is already commenced in the vale of Evesham, and is persevering in by a powerful company in 200 shares under the direction of the land proprietor [Mrs Browne], who is an experienced mineralogist, and who has discovered to a gentleman at Batheaston that Little Salisbury [Solsbury] hill contains lead ore (*Bath Chronicle*, 29 March 1804).

Gilbert Browne, Mrs Browne's late husband, had been a successful coal hunter and entrepreneur within exposed Coal Measures, which subcropped at lower depths, on Cannock Chase, Staffordshire.[18] His wife must have taken on some of his mineralogical aspirations. But she was misled by the completely different stratigraphic situation she faced in Worcestershire. The Brownes had moved to Bath from Shifnal, Shropshire, in the midst of the 'birthplace' of the Industrial Revolution and clearly, in her case, of false economic expectations. Mary Browne might have learned to have more informed dealings with economic geology while she was based at Bath during 1792–1798, if she could have become more involved with Smith, based there until 1803. He was all too well aware of the problem of repetitious clay lithologies, listed in his 1799 'Natural Order of Strata', which were so often mistaken for Coal Measures.

As a testament to the costs of Mrs Browne's coal-hunting efforts near Evesham, on her death in 1838, she left an estate of less than £200.[19] She might have saved the considerable estates of her recently deceased Bath attorney husband, Gilbert Browne, who left her circa £5000 in his will and her equally recently dead father, Thomas Hampton (1724–1797), who left her at least another £1000 of his £5000 estate in his will.[20] But she, and her many subscribers, were to waste well over £2300 seeking non-existent coal at Haselor.

Her Haselor shaft, after penetrating Lias clays, passed down into very thick Red Marls (Triassic). The only published, very incomplete, log of this attempt comes from Hugh Edwin Strickland (1811–1853) thirty years later (Strickland 1835), who then lived at nearby Cracombe House, and was busy gathering geological data. He was already unable to gather a complete record of the rocks sunk through here. The only complete record comes from the sinkers' log, which shows that the sinking was abandoned in September 1806. This same source also records the bland and imprecise prospecting advice which had inspired the attempt:

> Sands, marles..., Freestone, Sandstone, Limestone, Slate, and various other species of stone, have arisen. These are placed over each other, in alternate and regular beds parallel to each other, and being commonly mixed with marine exuviae or other animal or vegetable remains ... In these, and in these only, (or in plains formed of the same materials) Coals is found and there are scarcely any of them that do not contain it. It is to hills of this sorts, therefore that we must direct our researches after Coal ... If alternate Strata of indurated Clay, Sand, Slate or Sandstone occur with Iron ore or mica we may be certain that Coal will be found at a greater Depth. The beds nearest to the surface are generally either earthy, slaty, or sulphureous and commonly thin and scanty, but under these, different beds of greater, thickness and of a better sort are found.[21]

This advice is of the standard, non-specific, sort then offered. It was all that was available to mineral prospectors before the stratigraphic advances made by Smith. Coals were then wrongly thought to be both common, and regularly distributed, throughout what we today call sediments. It was equally wrongly thought that the most common associate of coal was iron ore and that better quality coals always lay deeper, to falsely encourage people to persevere (Torrens 1997b). Smith's breakthrough was to have given new, reliable and specific advice. This would have showed that any search for coal here was stratigraphically impossible, as were so many of the attempts then made, because they were made at too high levels in his 'Natural Order of Strata'.

Some notes follow on five other, misguided, attempts in which Smith, or his pupils, were involved between 1805 and 1815.

---

[18] Brereton Colliery accounts, Staffordshire Record Office, D 240/E/C/1/20/1-36, 1791–1795.
[19] Will, Worcestershire Record Office.
[20] Prerogative Court of Canterbury.

[21] Sinkers' log, Evesham Public Library.

Shaft of the 'Intended Colliery' at South Brewham, Somerset. 1803–1810

The first significant occasion on which Smith is known to have given such negative advice, and at the same time enhanced his already large collection of fossils from the spoil thrown out of a deep shaft, occurred on 24 March 1805, at South Brewham village in Somerset, some miles south of Bath. Smith, on the basis of these fossils, then advised those seeking coal here that they were misled by a similar lithological repetition to that which was misleading Mrs Browne at Haselor. Here instead, it was the dark shales of Smith's Clunch (= Oxford) Clay which confused coal hunters, by their superficial resemblance to Coal Measures. The full story is told by Torrens (1998b).

This attempt started in 1803 with a similarly considerable initial capital to that needed at Haselor. It was equally oversubscribed. It was even claimed that the 'similarity' of soil, site and aspect here with known coal areas elsewhere in Somerset was 'scientific' justification for the attempt. In March 1805, Smith visited the site on his way back to Bath from the West Country. The shaft[22] had reached a depth of 120 feet and in the spoil Smith found specimens of the characteristic *Gryphaea*, which he knew identified the Kellaways Rock in Wiltshire.[23] These immediately proved to him that this search was being made too high in the stratigraphic column. But Smith's freely given advice was ignored. The sinking continued until 1810, by which time it had reached a depth of 652 feet, at a cost of thousands of pounds.

The Brewham trial proved significant for Smith because it seems to be the first occasion on which he identified his stratigraphic position, using knowledge of the stratigraphic distribution of fossils, from deep-mined material. He wrote on 1 March 1807 how he had:

> never [been] sure of fixing the Course of this rock [Kelloways Stone] to the westward until the Sinking of the Pit for Coal at Brewham when it was there discovered lying at a depth of 40 yards beneath the Shale, see *Journal to the West*.[24]

Smith's difficulty in 'fixing the Course' of this non-feature-forming rock should remind us of another problem which faced workers involved in the early establishment of the stratigraphy of South Britain. Smith was immediately able to confirm the presence of the correct stratum here, but he was less confident when these same fossils were found away to the north in Yorkshire in 1821 (Knell 2000, pp. 141–142). The greater the distance any correlations were made from the location of their original standard, here Wiltshire, the greater the doubt (because of uncertainties of the actual ranges of fossils involved) and the possibility of error. As another example involving the Coal Measures, we need only note John Farey's assumption, when faced with the far-away Jurassic Brora coals in Sutherland, Scotland, in 1813, that these too were Coal Measures (Waterston 1982). Smith left a revealing note, dated 10 November 1832, showing he was well aware of the problems here: 'Mr Farey who surveyed the Brora district ... used to puzzle me with his account of Belemnites there.'[25]

Shaft at Furzen Lease, Cirencester, Gloucestershire, 1806

Smith was involved with a number of other abortive trials for coal but, since they failed, details are scant. One was near Cirencester, Gloucestershire, in 1806. Here Smith recorded that when the Thames and Severn canal was dug in 1788 a lot of black shale (Clunch or Oxford Clay) thrown out of a lock pit at Latton had made people think would be coal.[26] At Furzen Lease,[27] south of Cirencester, the proximity of the same canal inspired a full trial, funded by the America merchant, and disowned Quaker, Daniel Roberts (1753–1811) but now on the blue clays of the Forest Marble.

Roberts wrote to Smith how 'I have proceeded with a shaft [here] and expect to get to a working seam at 80 yards depth'.[28] Smith immediately replied:

> I am confident the facts recorded therein [his planned but never published 1806 book: see Cox 1942, p. 17] would have deterred you from attempting to find coal at Furzen Lease. I had till yesterday no idea you was engaged in a trial upon such deceitful ground.[29]

Nothing more is heard of this attempt.

---

[22] Probable site Grid. Ref. ST 737367.
[23] The specimens survive in Smith's collection, NHM L 1529.
[24] OUMSM, Box 30, folder 4.
[25] OUMSM, Marlstone File.
[26] OUMSM, autobiographical note, 16 November 1838.
[27] Grid. Ref. SU 018986.
[28] OUMSM, letter, Roberts to Smith, 16 June 1806.
[29] OUMSM, draft letter, Smith to Roberts, 18 June 1806.

Shafts and borings by the Sussex Mining Company at Bexhill, Sussex, 1805–1811

Smith's pupils also soon took up his stratigraphically based methods in mineral prospecting. These most notably included John Farey who, on 12 September 1806, visited some of these Sussex sites at one of the most significant attempts to find new coal in British history, at Bexhill, a village facing France during 'the great terror' of feared-for French invasion, after the collapse of the Peace of Amiens. At four separate locations here,[30] coal-hunters were again misled by yet another lithological similarity: that between the lignite-yielding Wealden clays exposed along the coast, which they mistook for the similar clays associated with Coal Measures. William James (1771–1837) from the West Midlands was the leading proponent of coal searches here, inspired by some supposed coal found here when water wells were dug to supply troops temporarily stationed along the Sussex coast during the threatened invasion. The story of the several attempts of the Sussex Mining Company at Bexhill, between 1805 and 1811, has been told by Torrens (1998c). Their most remarkable aspect was the extraordinary expense in which James and his group became involved.

Farey knew immediately, from his work in Sussex, that, wherever exactly these beds were to be placed in the British stratigraphic series, they were much too high above any Coal Measures. Any uncertainty of exactly where below the Chalk these beds should be placed among Sussex strata was because of the rarity of fossils here and their unusual aspect, in non-marine rocks, which were soon to yield large (dinosaur) bones to Smith in 1809. Whatever their exact position, it was clear these beds lay not far below the Chalk. Farey put an advertisement in the local paper warning people that money would be wasted searching for coal at so high a stratigraphic level (Farey 1808, reproduced in Torrens 1998c, p. 185).

Attempts near Leominster, Herefordshire, 1809 to ?1814

At Leominster, Herefordshire, attempts were made in rocks which underlay Smith's Coal Measures. They were inspired by a local coal-hunting mania which broke out in May 1809 (*Hereford Journal*, 10 May 1809, p. 3 cols. 2 and 3) and continued all that year. In 1807, Lt-Col. John Joseph Atherton (1761–1809) had sold Walton Old Hall, near Liverpool, Lancashire, and bought the Street Court estate, Herefordshire.[31] Here he became an improving landowner and joined the Herefordshire Agricultural Society. Sadly he died in August 1809 (*Hereford Journal*, 16 August 1809), leaving his widow Marianne to run the estates. She was soon involved in coal hunting here. We first hear of this in a letter to Smith dated 4 January 1811 from Philip Meadows Taylor (1779–1868), who also came from Walton. He asked Smith's opinion: 'I write by the desire of Mrs Atherton ... to request you will excuse the liberty she has taken in having sent you by this Evening's Mail ... a paper parcel, containing specimens of different Strata which have been found in Boring for Coal here.'[32] These specimens must be those to which Phillips referred, when speaking of a period when 'such forbidden experiments for coal were common; on one occasion [Smith] paid twenty-four shillings for a parcel of micaceous sandstone, taken from the old red formations of Herefordshire, and put *in the mailbag*, as specimens of the matter sunk through in a trial for coal!' (Phillips 1844, p. 66). Taylor was a recent doubly bankrupted merchant who had been based in Liverpool (*Staffordshire Advertiser*, 15 September and 17 November 1810). He later moved to Dublin where he was again involved in ore mineral prospecting (Taylor 1838). His letter continued that he had heard from his two cousins, by letter from Samuel Taylor of Banham, farmer, and by visit from his brother Richard Cowling Taylor, surveyor, 'both of whom have assured us that you would favor us with your opinion as to the probable success of the present undertaking'. Richard Cowling Taylor was soon to become 'an admiring pupil of Smith, this extraordinary man and original genius in 1811' (Taylor 1848, p. 28).

P. M. Taylor's letter concludes on the sites of the coal trials here. They were near a 'low hill we consider secondary or tertiary and we have specimens of shells in the Stone in the Valley below. The soil a reddish clay. Mrs Atherton differs with me in opinion respecting the *originality* of this hill. She considers it *primitive*.' This letter demonstrates the total uncertainty facing those who tried to use more Wernerian stratigraphic terms so far from their Saxon

---

[30] Grid. Refs. TQ 754077 and 737083, and two other unknown locations.

[31] Prerogative Court of Canterbury Will, proved November 1809, PRO PROB 11/1505/797. Street Court estate, Grid. Ref. SO 424603.

[32] OUMSM, letter, P. M. Taylor to Smith, 4 January 1811.

roots. The stratigraphic position of these rocks was simply inexpressible in such terms.

How far Smith became involved here is unclear. That he replied is clear from his diary entry, 'wrote on the Street Court Trial for Coal',[33] but his reply has not survived. He was surely already clear that these trials were being made in the Old Red Sandstone, well below Coal Measures. This was a fact he clearly recorded in his 1817 *Geological Cross Section from London to Snowdon*, 'one of the most perfect sections of any portion of the globe so complex, which ever had been produced' (Fitton 1833a, p. 45, fig. 4).

The extent of coal borings here is clear from a memorandum left by the man who made them, a Mr Cowley. This, dated 23 July 1814, records that three separate boreholes had been made here, one of which was still in progress.[34] John Watson (d. 1832), a leading colliery viewer of Willington, Northumberland, was next called in. His letter dated 15 November 1814 reveals that the first two borings had reached 375 and 408 feet. Watson suggested, on merely lithological grounds, that there was still 'a great probability of Coal Seams being found in your estate'.[35]

The outcome of these coals trials, as so often with failure, is unclear. G. B. Greenough noted: 'Herefordshire – £17 000 subscribed expended in ?1812 in trials in this county',[36] although it is unclear if these were only those at Street Court. The extent of Marianne Atherton's financial plight is however revealed by the civil action for debt which Robert Lang brought against her in 1816. The Court of Exchequer at Westminster ordered, on 18 January 1817, that she repay her debt of £3000 to him, with damages.[37] It is at least clear how very expensive such abortive coal hunting was.

Shaft at Bagley Wood, near Oxford, 1812–c. 1815

One final attempt is worth discussing, if only because of its geographical situation. Phillips (1844, p. 66) has written how 'at Bagley Wood, near Oxford, within sight of the university halls, which then resounded with the fame of the attractive and useful lectures of [William] Buckland [1784–1856], an absurd experiment for coal was begun in the Kimmeridge Clay, and ended in a deplorable sacrifice of Fortune.' This attempt started in 1812 on the north edge of Bagley Wood, near Boar's Hill. Earlier attempts had clearly been made here, as so often, since *Jackson's Oxford Journal* for 1766 and 1783 refers to a Coalpit Bottom here (Townsend 1914, pp. 59, 96). The 1812 attempt was made in Smith's Oaktree Clay, his amalgam of Kimmeridge and Wealden Clays, yet other horizons which, due to their lithology, proved misleading to coal hunters (and to each other for Smith). This attempt was funded by the local Abingdon MP, Sir George Bowyer Bart (1783–1860), and inspired by an Oxford brewer, Sutton Thomas Wood (1748–1827). Some details are given by Cardwell (1987). Supposed 'coal', in fact isolated pieces of lignite, had been found here initially – again during the sinking of a well – at Goose Cottage, Sunningwell.[38]

This inspired Bowyer to plan, and stupidly in part to construct, a new canal to link up with the River Thames to take the yet-to-be-found coal away. By May 1815 his shaft had penetrated 70 feet of Kimmeridge Clay and 30 feet of Corallian Coral Rag below.[39] It was visited by the Oxford Geology Club and by Smith who collected a number of fossils, which again survive in London.[40] Bowyer was forced by this financial disaster to escape his many creditors, first to Italy. He died an 'outlawed exile in Dresden' 40 years later (Thorne 1986, pp. 235–236).

The aftermath of these trials in London

The gentlemanly Geological Society of London was formed in November 1807, while the Brewham and Sussex shafts were still being sunk. It is clear that many were unconvinced of the value of Smith's discoveries, as another letter written by Smith to William Cunnington demonstrates. Cunnington was duly elected an honorary member of the Geological Society in May 1808 (Woodward 1907, p. 271), just as Smith expected:

> Please to write down [the strata found at Brewham] in the order that were sunk through & the thickness of each with a reference to the organic remains imbedded and the parts of the shaft where the water came in and any other

---

[33] OUMSM, Diary, 11 January 1811.
[34] Northumbria Record Office, Watson collection, Wat/1/5/57.
[35] Northumbria Record Office, Watson collection, Wat/2/9, pp. 342–343.
[36] University College London (UCL), Greenough MSS, 167/29c.
[37] Herefordshire Record Office, BM 56/30, 1817.

[38] Grid. Ref. SP 498012.
[39] UCL, Greenough MSS, letter, W. D. Conybeare to Greenough [1815 post May].
[40] Oxford Club visit, UCL, Greenough MSS, letter, John Kidd to Greenough, 8 May 1815.

particulars you can obtain ... I hope you will excuse me for troubling you when you know that the new Geological Society purposely established to ascertain the actual stratification of the British Isles may be better satisfied with the truth of my System if the account of the above sinkings comes from an indirect person ... This [Brewham Trial] and the Road Common Experiment [yet another earlier abortive coal trial near Bath] have done much for the Science of Geology. By them, and also by the Batheaston Pit, the practicality of foretelling what may be found is clearly established.[41]

There were indeed many, as Smith claimed, who doubted the accuracy of his stratigraphic results, or that rocks could be so 'foretold'. James Brogden (1765–1842), MP and merchant whom Smith advised on setting up a working colliery at Trimsaran, Carmarthenshire, wrote to Smith in 1809:

> As I am anxious to promote your well founded claims to some national remuneration for the time and labor you have given to a great national object, I found an opportunity to see Mr [Humphry] Davy [1778–1829, an original member of the Geological Society 1807] (not however saying anything to him of your or my intentions in this respect) thinking he might promote them. He said he had never seen your Collection [of fossils] but that he had heard much about it. He confesses he is not a believer in a regular succession of strata; composed of the same materials and containing the same fossils through the whole of their course. He says if this discovery is made, you have certainly the merit of it ... I had almost forgotten to add that he said when you showed your map to the Board of Agriculture [1806] they thought that your System was not sufficiently proved.[42]

Similarly, when Lockhart Muirhead (1765–1829), first professor of natural philosophy at Glasgow University, reviewed John Farey's volume on the minerals of Derbyshire in 1811, he wrote:

> Sections IV and V which treat of the soil and strata, are prefaced by a sketch of the author's favourite notions of stratification, and particularly of the series of British strata, as indicated by the ingenious Mr. Smith of Mitford, near Bath ... That the members of this series have the continuous and extensive range assigned to them by Mr. Smith, and his pupil Mr Farey, we much doubt ([Muirhead] 1812, p. 153).

The view from Oxford University was similar. John Kidd (1775–1831), physician and professor of chemistry there, wrote:

> more accurate observation had also given reason to doubt some other positions respecting the mutual connection of the strata and the organic remains contained in them; that connection is at least not so exclusive as has been asserted: for the remains of animals concluded to be characteristic of the newest formations have been found in some of the earlier; and vice versa (Kidd 1815, p. 37).

Greenough, the Society's first President, was to reiterate this view as late as 1819:

> An opinion has for some time past been entertained in this country, that every rock has its own fossils ... That the fossils contained in secondary strata, are, of all empyrical and accidental characters, the most useful, in enabling us to follow the direction of these strata, no one can dispute: but their utility has been greatly over-rated (Greenough 1819, pp. 284–287).

These quotations provide clear evidence that there was considerable doubt, in learned circles at least, about the validity of Smith's results.

London reactions to the Sussex trials were more intriguing. Leonard Horner (1785–1864) wrote to the president, G. B. Greenough, on 4 April 1809, warning how his brother-in-law Dr Stephen Winthrop (1767–1819) had:

> at Warwick ... introduced me to a man [William James] who he thought might be a useful honorary member of the Geological Society – he having a great reputation among the gentry in the neighbourhood – he is a land-doctor, a solicitor and calls himself mineral surveyor. I found him to possess a great deal of conceit & very little knowledge – what he had was so confused that I should be afraid to place any reliance upon the truth of his observations. I understand he is the man who has set the good people of Sussex in search of Coal. I said nothing of the G.S. as I am satisfied he could be of no use.[43]

Despite this appraisal, James and another of the chief proponents of this financial disaster

---

[41] OUMSM, letter, Smith to Cunnington, March 1808.
[42] OUMSM, letter Brogden to Smith, 30 October 1809.
[43] UCL, Greenough MSS.

at Bexhill, which could have been avoided if notice had been taken of Smith's new science of stratigraphy, were soon elected Ordinary Members of the London Geological Society. William James, chief instigator of these trials, was elected on 3 January 1812, and John Forster, solicitor to his Sussex Mining Company, was elected on 4 December 1812 (Woodward 1907, p. 274).[44] These were privileges never accorded to either Smith or Farey, whose results were doubted. They were thought to be too practical, and ungentlemanly, for election (Woodward 1907, p. 53).

The first judgement that these Bexhill trials had been doomed to failure came from another of William Smith's pupils, Rev. Joseph Townsend (1739–1816). In his pioneering work on British stratigraphy, he noted: 'the much more rash adventure ... set at foot in Sussex in the same bed [which he wrongly thought was the Clunch Clay] at the expense of Thirty Thousand Pounds, without the possibility of finding any thing more than thin laminae of coal, from which no profit can arise' (Townsend 1813, p. 127). This, and Townsend's identical assessment of the Brewham trial, were soon quoted by Fitton, when commenting on the practical value of William Smith's discoveries ([Fitton] 1818, p. 334), in one of the first attempts made to assess Smith's achievement by a central member of the Geological Society of London.

When Smith was forced to sell his fine fossil collection to the British Museum from 1815, one of those who helped negotiate the sale was another of the Bexhill adventurers, who was by then Chancellor of the Exchequer; Nicholas Vansittart (1766–1851). In his case it is little wonder, after wasting so much of his own money, in ignorance of Smith and Farey's results, that Vansittart should have recommended the sale proceed. Smith made specific reference to the most recently abortive trial, that at Bagley Wood, when he wrote the memorandum which offered his collection to the British Museum in 1816:

> These specimens ... are of the utmost Geological importance as they completely prove the false principles on which a worthy Baronet was lately induced to sink for Coal in these Strata [at] the ruinous Expense of £20,000. The specimens which I can produce will for ever settle the false notions of Coal in the Strata [immediately] beneath the Chalk which have long prevailed and prevent such wanton expenditures of money in search of it as have been the ruin of many (Eyles 1967, p. 204).

---

[44] MSS Council Minutes, Geological Society, London.

Smith's fossil collection was by no means the only one then extant by which enquirers could see actual fossils arranged in stratigraphic order. The whole philosophy of James Sowerby's *Mineral Conchology* project which commenced publication in 1812 was, as the title page announced, to record and illustrate the stratigraphic distributions of those British 'shells which have been preserved at various times and depths in the Earth' (Sowerby 1812).

William Phillips (1775–1828), printer of the Geological Society's *Transactions*, gave an early assessment of the basic stratigraphic situation facing coal hunters, when he wrote of Smith's stratigraphic ordering:

> if we imagine all these strata to be compressed beneath the sand which covers the chalk into one twentieth part of what their outgoings occupy on the surface, we should even then be compelled to suppose that the strata of coal are more than two miles beneath the bottom of the London clay (Phillips 1816, p. 219).

This was one of the first assessments by an original member of the Geological Society that Smith's mineral prospecting results were reliable.

Smith, who had so single-handedly pioneered mineral prospecting in Britain, noted the Sussex trials in 1819, 'at Bexhill the extremity of the Forest Ridge against the Sea was the late very expensive and useless search for coal' (Sheppard 1920, p. 150). Conybeare and Phillips added their confirmation that, despite lithological similarities between Smith's Ironsand Formation and Coal Measures, such attempts were bound to prove abortive as soon as the relative stratigraphic relations of the two had been revealed. However much the former may 'much resemble, in some places ... the great coal formation, these circumstances have led to expensive but abortive attempts to procure [coal] from these beds near Bexhill' (Conybeare & Phillips 1822, p. 137). Thomas Webster confirmed the similarity in 1826, noting that:

> it was from the abundance of the iron, the beds of clay and shale with vegetable impressions and the fragments of charcoal in the sandstones, that the expectations had been formed of finding coal in this formation, before the difference between lignites and true coal was generally understood (Webster 1826, p. 34).

This assessment highlights Farey's contribution to the debate. Just as Farey had said in 1808, it was indeed the difference between isolated pieces of lignite and properly bedded true coal which was one of the points at issue. But the main point was both more general and more crucial: the very different stratigraphic positions of Smith's Ironsand Formation from the Coal Measures way below it.[45]

A final judgement can be that of John Herschel, writing in 1831, of the Bexhill trials. In his *The study of Natural Philosophy*, a book which influenced Charles Darwin 'to try and add my mite to the accumulated store of natural knowledge' (Burkhardt 1991, pp. 370–371), Herschel wrote:

> it is not many years since an attempt was made to establish a colliery at Bexhill, in Sussex. The appearance of thin seams and sheets of fossil-wood and wood-coal, with some other indications similar to what occur in the neighbourhood of the great coal-beds in the north of England, having led to the sinking of a shaft, and the erection of machinery on a scale of vast expense, not less than eighty thousand pounds are said to have been laid out on this project, which, it is almost needless to add, proved completely abortive, as every geologist would have at once declared it must ... (Herschel 1831, p. 145).

This influential judgement only reveals how soon the historical record of such practical, and papyrophobic, activities gets distorted. Herschel was writing within a generation of Farey's warning that large amounts of money were then being wasted in Sussex. In the first place, 'every geologist' did not then 'declare it to be completely abortive'. Farey was the only one who did. Second, it is now known that four separate attempts were made here at four different places and that at least two shafts were sunk. But the cost, although enormous, was not £80 000 but about £30 000, as Townsend stated. Did Herschel's book simply reproduce a printer's error?

The first historian of the Geological Society, William Fitton, was equally misleading, when he confirmed Herschel's opinion that much money was wasted here, adding:

---

[45] The stratigraphic distribution of coal in Sussex and elsewhere in Britain was already complicated by other phenomena. The first is that 'coal' of varying quality was found at other levels than that of the Coal Measures. Smith was well aware of this at least from his work in Yorkshire from 1813 (Hemingway & Owen 1975). But as Buckland (1817, p. 289) emphasized 'no good coal has ... been yet found in England in any stratum more recent than that the new red sandstone, or red rock marl. That of ... Yorkshire, being above the lias and in the oolite formation, is of so bad a quality as scarcely to form an exception to this position.' Its low quality barred such other coals from being worth the great expense of prospecting for them.

> the assemblage [here] is very nearly the same in mineral composition with that of the coal measures...: differing from it only in geological place and the character of its fossils. It is not surprising therefore at a time when the geological relations of the groups in England were less understood than at present, these carboniferous portions of the Wealden group should have excited hopes of discovering coal ... the borings, which some years ago were conducted ... at Bexhill, were much more excusable than has been supposed (Fitton 1833b, p. 49).

But these trials, most of which were not by boring, had proved 'inexcusable' to John Farey over 25 years before. Gideon Mantell (1790–1852) later made a similar claim, when he sold his fossil collection to the British Museum in 1838, writing 'we venture to add in confirmation of its importance that various abortive attempts at mining etc which have been made in Sussex would never have been undertaken if such a collection had previously been formed'.[46]

His claim again ignores John Farey, who had fought on scientific grounds for the abandonment of the Bexhill, and all other Sussex, attempts from 1806 onwards, while they were being carried on, before the Geological Society of London was established. Farey made his own large collections of fossils and rocks in Sussex in demonstration of his stratigraphic work there, but these, unlike Mantell's, the British Museum refused to acquire in 1828.[47] They are now lost. William Smith had also added Sussex fossils to his fine collection while working in Sussex as a canal engineer from 1809. These fossil collections were one of the main grounds on which they could so confidently advise where a particular coal search might succeed and another would stand no chance.

The mineral prospector and the art of stratigraphy

Smith's frustrations as a mineral prospector are clear. In the unpublished Preface to his intended 1806 book (Cox 1942) he wrote:

> I have long left off puzzling about the Origin of the Strata, and content myself with knowing that it is so formed and arranged. 'The

---

[46] Mantell MSS, British Museum archives CE4/18, appeal dated February 1838.
[47] British Museum archives, Sophia Farey letter to Charles Koenig, 4 June 1828, Original Letters and Papers VI, and Officers Reports, C 2327, 10 July 1828.

whys and wherefores' and their attendant train of Chemical Properties, cannot come within the province of a Mineralogical Surveyor. I must therefore content myself with the simple occupation of a Fossil finder, to collect the different Specimens of the Principal Strata, and arrange them as they are in the Earth.[48]

Smith was already concerned at being asked 'how' stratigraphy had originated, rather than 'what' that stratigraphy was, and what its value might be. It was the 'what' which most interested Smith, because he knew his stratigraphic work had real economic significance. This is most clearly expressed in his draft letter, dated 6 July 1817, undoubtedly to John Ingram Lockhart (1765–1835), MP for Oxford 1807–1818, who had raised Smith's financial plight before a Committee of the House of Commons on 23 May 1817. This read:

> You like many others seemed more desirous of knowing the Theory of the Principles I have proceeded upon rather than the useful purposes to which the discoveries may be applied ... the latter is by far the most valuable part of my Work – I have indeed considered all Theories of so little use that I have not yet reduced my own to writing.[49]

The English landed classes were slow to realize the great economic significance of Smith's discoveries. This was not helped by the deep recession which coincided with the publication of Smith's map at the end of the Napoleonic Wars in 1815. In September 1811, Sir Joseph Banks, who had astutely realized the great significance of Smith's results from 1802, wrote to Barthelemy Faujas de St Fond (1741–1819) in France:

> Geology becomes more & more a fashion. I hope we shall before long advance the Limits of that Science. We have now some practical men [Smith, Farey and others] well versd in stratification who undertake to examine the subterranean Geography of Gentlemens Estates in order to discover the Fossils likely to be useful for Manure, for Fuel etc ... If employment begins to be given to these people the Consequence must be a Rapid improvement (de Beer 1960, p. 191).

If Banks was early to realize the great importance of Smith's discoveries, others were not.

William Maclure (1763–1840), the 'father of American Geology', met Smith in 1815 (Cox 1942, p. 47) and purchased his 1815 *Geological Map*, which survives in Philadelphia. Yet in 1819 Maclure could write 'French and English geologists ... do not yet know the relative position of the chalk and coals, because coals have not been found in the same basin with chalk: coals occupy basins filled with different kinds of rocks, and have no resemblance to the rocks found covering the chalk' (Maclure 1819, p. 213). This was exactly what Smith *had* been able to establish, from 1797 onwards. But neither Smith nor Farey were sufficiently encouraged as mineral prospectors. This lack of support was another cause of Smith's imprisonment for debt in June 1819 for nearly ten weeks in the King's Bench Prison, London (Eyles 1969, p. 157). Smith's frustrations became final when G. B. Greenough's rival *Geological Map* was published by the Geological Society of London in May 1820 (Greenough 1820), while Smith was exiled in Yorkshire (Knell 2000). Smith's bitter reaction to this rival publication, 'the ghost of my old map ... mocking me in the disappointments of a science', was recorded by Cox (1942, p. 42). Fuller is right to claim that Greenough's 'action cut the sales of [Smith's] own map to nothing, and ruined his other geological ventures. In short the gentlemen scientists of London bankrupted him' (Fuller 1995, p. 8). William Phillips, already quoted, confirmed that, despite Greenough's map being 'a vast improvement on Smith's, to [Smith] nevertheless belongs the great merit of originality, but I fear his pocket is not so much the heavier for that, as it ought to be'.[50]

Another of Smith's ruined 'other geological ventures' was his series of new *County Geological Maps* issued from January 1819. These had carefully recorded on their covers the differing potential availabilities of coal in English counties (Fig. 4).

But Smith had at least announced his real vocation. On the *Prospectus* to his great *Geological Map* (1815) he first called himself a 'mineral surveyor' (Fig. 5). He repeated this in his most influential, but still never completed, book, *Strata Identified* (Smith 1816–1819). When the *Memoir* to his *Geological Map* appeared, this described him as 'Engineer and Mineral Surveyor' (Smith 1815), while in his *Stratigraphical System of Organised Fossils* he called himself 'civil engineer and mineral surveyor' (Smith 1817). The term, 'mineral surveyor', as a member of the Geological Society had already called

---

[48] OUMSM, Box 36, folder 8.
[49] OUMSM, letter, Smith to Lockhart, February 1821, OUMSM, Box 40, folder 7.

[50] Linnean Society London, Winch MSS, letter, Phillips to N. J. Winch, 30 December 1820.

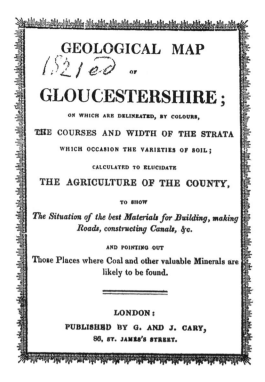

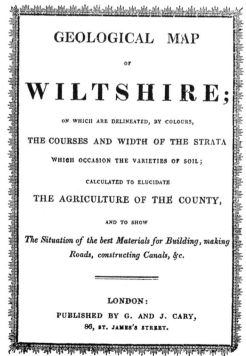

Fig. 4. The two differing styles of cover for Smith's *County Geological Maps*, only one of which 'points out those places where coal and other valuable minerals are likely to be found'.

William James in 1809, had been invented by Farey, by at least May 1808, and was used on the title page of his *magnum opus* in 1811 (Torrens 1993, p. 64). The specific claim that Smith had 'introduced this new art' was made both for Smith in 1817 (Farey 1818, pp. 173, 177, 179) and by Smith in 1818 (Sheppard 1920, p. 219).

Smith had earlier, in unpublished work, as we see above, instead used the phrase 'mineralogical surveyor'. But, as Farey was quick to point out, 'mineral surveying' had nothing to do with mineralogy 'differing most essentially from the microscopic poreings over the individual stones or substances met with while examining the surface of the Earth to which mere Mineralogists are too prone. [It was] a British Art originating with Mr William Smith' (Farey 1816, pp. 355–356). Farey was equally right to claim in 1823 that 'already the strata of England and Wales ... are *become the standards* for comparing and classifying the strata of other countries; even those, from whence we were ... told that all geological knowledge must emanate' ([Farey] 1823). Such surveying for minerals using stratigraphic knowledge is an aspect of the history of geology, and of Smith's work, which has received too little attention, despite its importance today.

Smithian stratigraphy and geological time

Shells in plenty mark the strata,
And though we know not yet awhile
What made them range, what made them pile,
Yet this one thing full well we know–
How to find them ordered so.
  (William Smith 1829: Cox 1942, p. 70)

In the context of this volume, we should try to clarify the complex situation regarding Smith and geological time. The problem is that we still know little of what Smith actually believed. He certainly always thought that the rocks he had been so busy studying had been created, and ordered, by the Creator. In about 1801–1803 (Cox 1942, p. 84), Smith wrote of these strata being both regular and having a natural arrangement produced by centrifugal forces. In this he was following a long tradition (Webby 1969). Fuller, relying on Smith's unpublished writings of 1801, slightly corrected in 1803, has since commented that 'Smith's strata originated by unique fiat, a single act of simultaneous creation that took place before the earth had begin to revolve. Strata so conceived carried no implied age, nor any relative age differences among them' (Fuller 1995, p. 4). Yet Phillips had

This Day is published, by JOHN CARY, *No.* 181, *Strand;*

(Dedicated, by Permission, to the Right Honourable Sir JOSEPH BANKS, Bart. President of the Royal Society, and sanctioned by the BOARD OF AGRICULTURE, the ROYAL INSTITUTION, and the SOCIETY OF ARTS, MANUFACTURES, AND COMMERCE;)

A MAP
OF THE
STRATA
OF
ENGLAND AND WALES,
WITH PART OF
SCOTLAND;
EXHIBITING THE
COLLIERIES, MINES, AND CANALS,
THE
MARSHES AND FEN-LANDS ORIGINALLY OVERFLOWED BY THE SEA,
AND THE
𝔙𝔞𝔯𝔦𝔢𝔱𝔦𝔢𝔰 𝔬𝔣 𝔖𝔬𝔦𝔩,
According to the VARIATIONS in the SUBSTRATA:
Illustrated by the most descriptive Names of Places and of local Districts;
SHOWING, ALSO,
THE RIVERS, SITES OF PARKS, AND PRINCIPAL SEATS OF THE NOBILITY AND GENTRY,
THE OPPOSITE COAST OF FRANCE,
And the LINES of STRATA neatly coloured.

By WILLIAM SMITH, MINERAL SURVEYOR.

MR. J. CARY *takes the liberty of informing the Patrons and Subscribers to the Map of the Strata of England and Wales, by Mr. Smith, Mineral Surveyor, that the Copies are now ready for delivery. On the other side is annexed the Price of the Maps in Sheets, and in the various ways of Mounting; and Mr. C. entreats that the Subscribers, on ordering their Copies, will have the goodness to inform him in what way they would prefer to have them sent.*

London, 181, Strand.

Fig. 5. The prospectus for Smith's *Geological Map* of 1815 (National Museum of Wales).

quoted Smith to have regarded, by about 1795, 'each stratum as [having been] successively the bed of the sea, and contained in it the mineralised monuments of the races of organic beings then in existence' (Phillips 1844, p. 14). Phillips then had his own nepotic reasons for being less than precise, as he was building his own academic reputation (Knell 2000), in which any truth about Smith's, by then, outdated geological beliefs, could have proved harmful to him. So we do face real ambiguity here.

The problem is that Smith's ideas clearly changed as his geological enquiries took him from a provincial to a metropolitan situation after 1803. This removed him from dealings with Somerset colliers and farmers into new circles involving at their very centre the president of the Royal Society, Banks. Certainly in 1797 Smith had written that 'of mineral bodies there is not a doubt but that they were all formed at the same time, as the great Mass of which coal is no insignificant part not in heaps but regularly dispersed' (Douglas & Cox 1949, p. 185). But by 1806, probably as a result of discussions with John Farey, who was remarkably original in his geo-theological thoughts (Torrens 1998a), Smith could write:

The great variety of Fossils, Shells and Plants which are so finely preserved in the solid strata, have handed down to us, the most indubitable history of their own habitations, companions, and manner of living, and have also given us the clearest and best account of the formation of the Earth ... [and] in every part of the Creation there seems to have been one grand line of succession – a wonderful scale of organisation leading on in the same train, towards perfection [so that] it appears that animation and destruction has been going on in regular succession – and we must hence conclude that the earth had been an infinite time in arriving at perfection. [Smith concluded] Fossils will ... throw much light upon the mysterious pursuit of the Geologist. The first [of three] separate histories of fossils are those lodged in Strata, which have never been disturbed, since the creation – they are clearly co-eval with the first formation of their surrounding substances and have handed down to us the clearest, and most unequivocal history of the most ancient Inhabitants of the Earth, and the periods and progress of its formation. [Smith added finally that] if a calculation was formed, on this immense quantity of animal and vegetable matter – the time required for the Perfection, and Decay, and subsequent formation, into Strata which have evidently been formed in deep and quiet water – would stagger the faith of Many.[51]

For Smith to talk here as he does, of 'histories, succession and time staggering the faith of many', must imply that he had undergone major changes of mind, since his earlier provincial writings in 1801–1803, although he always maintained his belief that all such rocks recorded acts of creation rather than depositions. We clearly need to learn more of Smith's confused, but changing, views on geological theory as we do of Smith himself.

S. Brecknell (Oxford) gave vital help, both in allowing access to the Smith archive at Oxford and in providing detailed guidance to it and photocopies from it. Innumerable other archivists, curators and librarians have also helped me. S. Knell (Leicester) and M. Taylor (Edinburgh) kindly read and commented on a first draft of this paper. BP Exploration Ltd provided financial support. S. Baldwin (Witham), J. Fuller (Tunbridge Wells), J. Martin (Roudham), J. Morrell (Bradford), S. Pierce (Wincanton) and the late J. Thackray gave specific help and support. This paper could never have been attempted without them.

References

ANON. 1804. Batheaston Mining Company. *Monthly Magazine*, **17**, 507–508.

---

[51] OUMSM, Box 38, folder 2, 'Material transcribed in about 1806 for Smith's never published work on the Strata then in preparation'.

BAKER, T. H. 1911. *Records of the Seasons, Prices of Agricultural Produce and Phenomena Observed in the British Isles*. Simpkin Marshall, London.

BUCHANAN, C. A. & BUCHANAN, R. A. 1980. *The Batsford Guide to the Industrial Archaeology of Central Southern England*. Batsford, London.

BUCKLAND, W. 1817. Description of a series of specimens from the Plastic Clay near Reading, Berks. *Transactions of the Geological Society of London*, **4**, 277–304.

BUCKLAND, W. 1837. *The Bridgewater Treatises VI: Geology and Mineralogy Considered with Reference to Natural Theology*. Pickering, London.

BURKHARDT, F. 1991. *The Correspondence of Charles Darwin*, Vol. 7, 1858–1859. Cambridge University, Cambridge.

CARDWELL, H. 1987. Capability Brown at Radley: Part 2, The influence of his work on the subsequent history of the estate. *The Radleian 1987*, 80–88.

CONYBEARE, W. D. & PHILLIPS, W. 1822. *Outlines of the Geology of England and Wales*. Phillips, London.

COX, L. R. 1942. New light on William Smith and his work. *Proceedings of the Yorkshire Geological Society*, **25**, 1–99.

CUNNINGTON, R. H. 1975, *From Antiquary to Archaeologist*. Shire Publications, Aylesbury.

DE BEER, G. 1960. *The Sciences were Never at War*. Nelson, London.

DONOVAN, D. T. 1966. *Stratigraphy: An Introduction to Principles*. Murby & Co., London.

DOUGHTY, M. W. 1978, Samborne Palmer's diary: technological innovation by a coal mine owner. *Industrial Archaeology Review*, **3**, 17–28.

DOUGLAS, J. A. & COX, L. R. 1949, An early list of strata by William Smith. *Geological Magazine*, **86**, 180–188.

DUFF, P. M. D. & SMITH, A. J. (eds). 1992. *Geology of England and Wales*. Geological Society, London.

ELLENBERGER, F. 1988. *Histoire de la Géologie*, Vol. 1. Lavoisier, Paris.

EYLES, J. M. 1967. William Smith: the sale of his geological collection to the British Museum. *Annals of Science*, **23**, 177–212.

EYLES, J. M. 1969. William Smith: some aspects of his life and work. *In*: SCHNEER, C. J. (ed.) *Toward a History of Geology*. MIT, Cambridge (Mass.) and London, 142–158.

EYLES, J. M. 1974. William Smith's home near Bath: the real Tucking Mill. *Journal of the Society for the Bibliography of Natural History*, **7**, 29–34.

EYLES, J. M. 1985. William Smith, Sir Joseph Banks and the French geologists. *In*: WHEELER, A. & PRICE, J. H. (eds) *From Linnaeus to Darwin, Commentaries on the History of Biology and Geology*. Society for the History of Natural History, London, 37–50.

FAREY, J. 1806. Coal. *In*: REES, A. (ed.) *The Cyclopaedia*, Vol. 8. Longman, London.

FAREY, J. 1808. [Announcement]. *Sussex Weekly Advertiser*, 18 January 1808.

FAREY, J. 1813. Notes and Observations on ... Mr Robert Bakewell's *Introduction to Geology*. *Philosophical Magazine*, **42**, 356–367.

FAREY, J. 1814. Notes and Observations on ... Mr Robert Bakewell's *Introduction to Geology*. *Philosophical Magazine*, **43**, 27–34.

FAREY, J. 1816. A letter from Dr William Richardson to the Countess of Gosford ... *Philosophical Magazine*, **47**, 354–364.

FAREY, J. 1818. Mr Smith's geological claims stated. *Philosophical Magazine*, **51**, 173–180.

[FAREY, J.] 1823. Notice of Sowerby's *Mineral Conchology* volumes 1–4. *Monthly Magazine*, **55**, 543.

[FITTON, W. H.] 1818. Geology of England [review of Smith's publications 1815–1817]. *Edinburgh Review*, **29**, 310–337.

FITTON, W. H. 1833a. *Notes on the Progress of Geology in England*. Richard Taylor, London.

FITTON, W. H. 1833b. *A Geological Sketch of the Vicinity of Hastings*. Longman, London.

FULLER, J. G. C. M. 1969. The industrial basis of stratigraphy. *Bulletin of the American Association of Petroleum Geologists*, **53**, 2256–2273.

FULLER, J. G. C. M. 1992. The invention and first use of stratigraphic cross-sections by John Strachey, F. R. S., (1671–1743). *Archives of Natural History*, **19**, 69–90.

FULLER, J. G. C. M. 1995. *'Strata Smith' and his Stratigraphic Cross Sections, 1819*. Tulsa and London: AAPG, Tulsa, and Geological Society, London.

FULLER, J. G. C. M. 2000. *The Church of King Charles the Martyr, Tunbridge Wells: A New History*. Friends of the Parish Church, Tunbridge Wells.

GREENOUGH, G. B. 1819. *A Critical Examination of the First Principles of Geology*. Longman, London.

GREENOUGH, G. B. 1820. *A Geological Map of England and Wales*. Longmans, London.

HANCOCK, J. M. 1977. The historical development of concepts of biostratigraphic correlation. *In*: KAUFFMAN, E. G. & HAZEL, J. E. (eds) *Concepts and Methods of Biostratigraphy*. Dowden, Hutchinson and Ross, Stroudsburg, 43–22.

HEMINGWAY, J. E. & OWEN, J. S. 1975. William Smith and the Jurassic coals of Yorkshire. *Proceedings of the Yorkshire Geological Society*, **40**, 297–308.

HERSCHEL, J. F. W. 1831. *The Preliminary Discourse on the Study of Natural Philosophy*. Longman, London.

JONES, S. 1999. *Almost Like A Whale*. Doubleday, London.

KELLAWAYS, G. A. (ed.). 1991. *Hot Springs of Bath*. City Council, Bath.

KIDD, J. 1815. *A Geological Essay on the Imperfect Evidence in Support of a Theory of the Earth*. Oxford University, Oxford.

KNELL, S. J. 2000. *The Culture of English Geology 1815–1851*. Ashgate, Aldershot.

LAUDAN, R. 1987. *From Mineralogy to Geology*. University of Chicago, Chicago.

MACLURE, W. 1819. Hints on some of the outlines of geological arrangement. *American Journal of Science*, **1**, 209–213.

MITCHELL, W. S. 1869. The centenary of William Smith's birth. *Geological Magazine*, **6**, 356–359.

MITCHELL, W. S. 1872. Somersetshire Coal Canal. *Bath Chronicle*, 8 August 1872, 8.

MITCHELL, W. S. 1874, How the study of stratigraphical geology commenced near Bath. *Bath Chronicle*, 26 March 1874, 7.

MOORE, C. 1867. On abnormal conditions of Secondary deposits. *Quarterly Journal of the Geological Society of London*, **23**, 449–568.

[MUIRHEAD, L.] 1812. Review of Farey's *Derbyshire*. *Monthly Review*, **69**, 153.

MURCHISON, R. I. 1832. Announcement of the second award of the Wollaston Fund. *Proceedings of the Geological Society of London*, **1**, 362.

PHILLIPS, J. 1844. *Memoirs of William Smith*. Murray, London.

PHILLIPS, J. 1871. Evidence of Professor J. Phillips. *In*: *Report of the Commissioners appointed to inquire into the several matters relating to Coal in the United Kingdom*, Vol. 2. HMSO, London, 490–501.

PHILLIPS, W. 1816. *Outlines of Mineralogy and Geology* (second edition). Phillips, London.

POLLARD, D. 1983. Bath stone quarry railways 1795–1830. *Bristol Industrial Archaeological Society Journal*, **15**, 13–19.

ROGERS, K. H. 1976. *Wiltshire and Somerset Woollen Mills*. Pasold Research Fund, Edington.

RUPKE, N. A. 1983. *The Great Chain of History*. Clarendon, Oxford.

SEDGWICK, A. 1832. Announcement of the first award of the Wollaston Prize. *Proceedings of the Geological Society of London*, **1**, 270–279.

SHEPPARD, T. 1920. *William Smith: His Maps and Memoirs*. Brown, Hull.

SMITH, W. 1806. *Observations on the Utility, Form and Management of Water Meadows*. Bacon, Norwich.

SMITH, W. 1815. *A Memoir to the Map and Delineation of the Strata of England and Wales*. Cary, London.

SMITH, W. 1816–1819. *Strata Identified by Organised Fossils*. Smith, London.

SMITH, W. 1817. *Stratigraphical System of Organised Fossils*. Williams, London.

SOWERBY, J. 1812. *The Mineral Conchology of Great Britain*, Vol. 1. Sowerby, London.

STIRLING, A. H. W. 1908. *Coke of Norfolk and his Friends*. John Lane, London.

STRICKLAND, H. E. 1835. Memoir on the geology of the Vale of Evesham. *The Analyst*, **2**, 1–10.

TAYLOR, P. M. 1838. A short account of Lead Mines in Co. Clare. *Journal of the Geological Society of Dublin*, **1**, 385–387.

TAYLOR, R. C. 1848. *Statistics of Coal*. Moore, Philadelphia.

THORNE, R. G. 1986. *The House of Commons 1790–1820, Members*. Secker & Warburg, London.

TORRENS, H. S. 1975. The Somerset coal canal caisson lock. *Bristol Industrial Archaeology Journal*, **8**, 4–10.

TORRENS, H. S. 1979. Richard Warner. *Newsletter of the Geological Curators Group*, **2**(4), 188.

TORRENS, H. S. 1983. A family's origins and an incredible coincidence, the Cruse family of Bath, Avon and Warminster, Wiltshire. *Journal of the Bristol and Avon Family History Society*, **34**, 14–15.

TORRENS, H. S. 1990. The four Bath philosophical societies 1779–1959. *In*: ROLLS, R. & GUY, J. R.

(eds) *A Pox on the Provinces*. Bath University, Bath, 181–188.

TORRENS, H. S. 1993. Patronage and problems: Banks and the Earth sciences. *In*: BANKS, R. E. R. *et al.* (eds) *Sir Joseph Banks: A Global Perspective*. Royal Botanic Gardens, Kew, 49–75.

TORRENS, H. S. 1997a. Geological communication in the Bath area in the last half of the eighteenth century. *In*: JORDANOVA, L. J. & PORTER, R. (eds) *Images of the Earth* (second edition). BSHS, Oxford, 217–246.

TORRENS, H. S. 1997b. Some thoughts on the complex and forgotten history of mineral exploration. *Journal of the Open University Geological Society*, **17**(2), 1–12.

TORRENS, H. S. 1998a. Geology and the natural sciences: some contributions to archaeology in Britain 1780–1850. *In*: BRAND, V. (ed.) *The Study of the Past in the Victorian Age*. Owbow Monograph 73, Oxford, 35–59.

TORRENS, H. S. 1998b. Le 'Nouvel art de prospection minière' de William Smith et le 'Projet de Houillre de Brewham': un essai malencontreux de recherche de charbon dans le sud-ouest de l'Angleterre, entre 1803 et 1810. *In*: GAUDANT, J. (ed.) *De la Geologie a son Histoire (Livre Jubilaire pour Franois Ellenberger)*. CTHS, Paris, 101–118.

TORRENS, H. S. 1998c. Coal hunting at Bexhill 1805–1811: how the new science of stratigraphy was ignored. *Sussex Archaeological Collections*, **136**, 177–193.

TOWNSEND, J. 1813. *The Character of Moses Established for Veracity as an Historian*. Longman, London.

TOWNSEND, J. 1914. *News of a Country Town*. Oxford University, Oxford.

VINTNER, D. 1964. The archaeology of the Bristol coalfield. *Journal of Industrial Archaeology*, **1**, 36–47.

WADE MARTINS, S. & WILLIAMSON, T. 1994. Floated water meadows in Norfolk: A misplaced innovation. *Agricultural History Review*, **42**, 20–37.

WATERSTON, C. D. 1982. John Farey's mineral survey of south-east Sutherland and the age of the Brora coalfield. *Annals of Science*, **39**, 173–185.

WEBBY, B. D. 1969. Some early ideas attributing easterly dipping strata to the rotation of the Earth. *Proceedings of the Geologists' Association*, **80**, 91–97.

WEBSTER, T. 1826. Observations on the Strata at Hastings in Sussex. *Transactions of the Geological Society of London* (series 2), **2**, 31–36.

WOODWARD, H. B. 1895. *The Jurassic Rocks of Great Britain, Vol. 5*, The Middle and Upper Oolitic Rocks of England. HMSO, London.

WOODWARD, H. B. 1907. *The History of the Geological Society of London*. Geological Society, London.

WRIGLEY, E. A. 1962. The supply of raw materials in the Industrial Revolution. *Economic History Review*, **15**(2), 1–16.

YOUNG, A. 1798. [Account of his visit to Crook's farm]. *Annals of Agriculture*, **31**, 81–83.

Genesis and geochronology: the case of John Phillips (1800–1874)

JACK MORRELL

8 Randall Place, Bradford, West Yorkshire BD9 4AE, UK

Abstract: In 1841 John Phillips proposed that there were three great periods of past life on the Earth, namely the Palaeozoic, the Mesozoic and the Cainozoic, terms which are still used today. This was by no means Phillips' sole contribution to geochronology and this paper examines his evolving views on it over a span of forty years. In the 1820s he adopted the Deluge as a notion which reconciled Genesis and geology. From the 1830s he adopted a liberal Christian position, which saw attempts at such reconciliation as futile and dangerous, and incurred the wrath of so-called scriptural geologists. From 1853 to his death, Phillips was a public figure as successively deputy reader, reader, and professor of geology in the University of Oxford. He was also president of the Geological Society from 1858 to 1860. The publication of Darwin's *Origin of Species* (1859) not only provoked him to reaffirm his liberal Christian beliefs but also induced him to give greater attention to geochronology as a weapon to be used against Darwinian evolution.

In 1841 John Phillips (Fig. 1) proposed that there were three great periods of past life on the Earth, namely, the Palaeozoic, the Mesozoic and the Cainozoic, terms which are still used today. His definition of Palaeozoic was controversial because it referred to all forms of life before the epoch of the New Red Sandstone. It subsumed Murchison's Devonian and Silurian systems, which were named after areas where they were best represented. Murchison was outraged by Phillips' notion of the Palaeozoic because it obliterated his own zealously promoted geological systems on which he had built his reputation. Peace was restored in 1843 when Murchison adopted Phillips' wide definition of Palaeozoic in exchange for Phillips' public acceptance of the Devonian system as a specific period in the history of the Earth and of life on it, but without reference to a particular locality (Rudwick 1985, pp. 381–386, 392–393). Phillips' triple terminology is his most enduring contribution to geochronology but it was by no means a one-off effort. For almost fifty years Phillips pondered the problems of geochronology. In order to reveal the variety of approaches he took, I propose to examine the three views he held about Genesis and geochronology. Firstly, I shall look at the 1820s when Phillips was not unhappy with the Deluge as a notion which reconciled Genesis and geology. Secondly, I shall examine the 1830s during which he adopted a liberal Christian position, which saw attempts at such reconciliation as futile and dangerous, while maintaining that estimating the great antiquity of the Earth was impossible. Thirdly, I shall discuss the 1860s

Fig. 1. John Phillips (1800–1874).

when his opposition to Darwinian evolution induced him to reaffirm his liberal Christianity and to publish what Joe Burchfield has called 'the first important quantitative determination of the Earth's age based exclusively on geological data'

From: LEWIS, C. L. E. & KNELL, S. J. (eds). *The Age of the Earth: from 4004* BC *to* AD *2002*. Geological Society, London, Special Publications, **190**, 85–90. 0305-8719/01/$15.00 © The Geological Society of London 2001.

(Burchfield 1974, p. 59). Having forced Darwin to withdraw from the third and subsequent editions of the *Origin of Species* his estimate of 300 million years as the age of the Weald, Phillips then advised William Thomson (later Lord Kelvin) in his assault on Darwin's view that 'past periods of time' had been 'incomprehensively vast', a view he had derived from Lyell (Darwin 1859, p. 293). Phillips was, therefore, a major figure in the debates on the age of the Earth in the 1860s.

The Deluge, Genesis and geology

Phillips was the orphaned nephew and chief pupil of William Smith from whom he learned palaeontological stratigraphy. His first big work, his book of 1829 on the Yorkshire coast, was mainly an exercise in Smithian stratigraphy which correlated Yorkshire strata with those of southwest England (Phillips 1829). But Phillips had also to explain the huge erratic blocks which litter east Yorkshire and the vast deposits of diluvium (clay and gravel) which make up much of the large area of Holderness in the southeast of the county. His answer was that there had been an irresistible and huge flood, which he called the deluge. It had carried over hills big erratic blocks of granite 100 miles from Shap to Flamborough and it had created Holderness by depositing diluvium there. This flood had happened after the stratification of Yorkshire had been completed because in Holderness there were lake deposits, containing skeletons of deer and elk, which rested on the diluvium. From such phenomena Phillips concluded that in the Earth's history there had been Antediluvian, Diluvian and Postdiluvian periods. He also inferred that the layering of gravel deposits indicated that, besides a mighty flood, there had been quiet waters and local currents. Unlike William Buckland, the leading English diluvialist, Phillips made no attempt in his Yorkshire coast book to relate his deluge to the historical event recorded by Moses. On the contrary he averred that it was ill-advised to try to compare a 'chronology of nature' with the records of history (Phillips 1829, pp. 16–19). At this time Phillips clearly believed that geology should be pursued freely and without reference to religious doctrine. In these ways he was not a scriptural geologist: he did not try to make geological results conform to literal readings of scripture. On the other hand, up to 1832 he was a fellow traveller of the reconcilers in that he was not unhappy when geologists produced results which broadly and loosely were not incompatible with scripture. Indeed, as he told the good people of York in 1831, geology could be the handmaid of religion because it provided evidence of physical changes of the same kind as those recorded in the Bible. Thus the scriptural reference to the appearance of dry land corresponded roughly with the great mountain elevations which he believed had occurred after the Tertiary period. Phillips reserved his analysis of what he called 'the narratives' of the Creation and the Deluge for the learned citizens of Halifax, also in 1831. The Genesis story showed the eternal providence of God, the creator of all, and was written to be intelligible to Jews and to induce religious contemplation. The point of the narrative of the Flood was to show that God the creator continued to watch over His world, with power to destroy and save. The Flood, controlled by God, was a moral warning.[1]

The liberal Christian

After 1831 Phillips divorced himself entirely from scriptural geology and adopted a liberal Christian position. Phillips was a practising Anglican who was professor of geology at King's College, London, from 1834, whose motto was *sancte et sapienter*. He had learned from experience that Catholics and Dissenters, especially Unitarians, could be good Christians and he deplored doctrinal sectarianism. He wanted a broad, accommodating and intellectually tolerant Church. He supported the way in which liberal Christians asserted the mutual autonomy of science and theology, yet accepted or insisted on the ultimate congruence of natural with revealed religion. They believed that the books of nature and of scripture were in ultimate accord because their Author was the same. This view gave geologists the liberty to pursue their science with total independence while encouraging them to reveal evidence of design and a designer in the natural world. In the hands of liberal Christians, geology led to natural theology which, it was hoped, would strengthen Christian belief. These views were denounced by scriptural geologists who enjoyed a vigorous phase in the 1830s and 1840s. Generally they believed that the Mosaic record irrefutably affirmed the creation of the universe in six days about six thousand years ago, and they viewed any separation of religion from science as heresy (Morrell & Thackray 1981, pp. 224–245).

---

[1] Phillips, lecture notes, York, 23 March 1831, box 103, f 1; on scripture and geology, Halifax, 26 October 1831, box 111, f 4; Phillips Papers, Oxford University Museum.

In the 1830s Phillips seems to have been indebted to two friends, both clergymen, for his liberal Christian view about the Genesis and geology question. Firstly, in his presidential address to the Geological Society in 1831, Adam Sedgwick, professor of geology at Cambridge, recanted his diluvial heresy, i.e. he renounced his former belief that gravel had been produced by one violent deluge which he had equated with the Mosaic flood. Sedgwick went on to stress that there was general accord between historical traditions and geological results, though the foundations of each were independent. Secondly, Phillips respected the views of John Kenrick, a Unitarian and classical scholar in York. Drawing on his study of Greek traditions of the deluge, Kenrick viewed Genesis as a mythic fiction which was deemed literally true in popular belief. He therefore thought it futile to try to fit the stones to the inspired words (Sedgwick 1831, pp. 312–314; Kenrick 1834).

Phillips' own fieldwork led him to the same conclusion. For example, further research on gravel and marl deposits in southeast Yorkshire indicated to him that there had been several inundations of which the latest was anterior to the movement of waters responsible for erratic blocks. In the second edition of 1835 of his Yorkshire coast book, Phillips replaced the term deluge by 'fundamental diluvial formation' and 'all that violence of water' in order to avoid confounding a geological deluge with the Mosaic or historical catastrophe. On the question of dating the various inundations, Phillips refused to give an estimate: 'How long anterior is not the question – the order of succession of the events is all that we are now concerned with' (Phillips 1835, pp. 25, 41, 150).

In the 1830s Phillips was a prolific writer of treatises on geology and could not avoid the topic of the antiquity of the globe. He was in no doubt about the vast antiquity of the Earth. The total depth of sedimentary strata (about ten miles maximum) and the nature of conglomerate rocks convinced him that it was folly to contract 'the long periods of geology into the compass of a few thousand years'. On the question of geological time he realized in general that the series of stratified rocks provided a scale. But he judged the task of determining the age in years of the crust of the Earth as then 'entirely hopeless'. However, he did suggest in the late 1830s two possible modes of attack. The first involved determining 'the rate of secular refrigeration of the globe'. It may seem surprising that Phillips, a geologist, recommended this physical method two decades before William Thomson employed it in the early 1860s. But by the late 1830s Phillips had become a devotee of physical geology. In 1834 he measured the increase of temperature with depth in Britain's deepest coal mine and convinced himself, *contra* Charles Lyell, that the theory of the long-term cooling of the Earth was valid. In 1836 Phillips was quick to welcome what he called 'the deductions from mechanical principles' and the 'demonstrations in geological dynamics' of William Hopkins, the Cambridge mathematical physicist who tutored Thomson in the early 1840s. The second approach suggested by Phillips in the late 1830s involved measuring 'the mean or extreme rate of production of stratified deposits at the present day'. Knowing the total thickness of such strata, and making various hazardous assumptions, he thought it possible to calculate the amount of time taken for their deposition (Phillips 1834, 1836, p. xix, 1838, pp. 293–294). This approach was used in 1860 in the calculation Phillips made of the age of the Earth's crust.

The views of Sedgwick, Buckland, Murchison and Phillips about the great antiquity of the globe were attacked by several scriptural geologists, including one who was particularly troublesome for Phillips. He was William Cockburn, Dean of York Minster, a member of the Yorkshire Philosophical Society of which Phillips was Secretary, a generous contributor (£50) to the Yorkshire Museum (in York) of which Phillips was Keeper, and a facilitator of Phillips' meteorological observations made at the top of Minster. Cockburn was therefore not opposed to science; but he thought that Phillips and company undermined revelation and encouraged infidelity. From 1838 to 1850 Cockburn issued a stream of letters and pamphlets which reasserted the age of the Earth as about 6000 years. He attacked the peripatetic British Association for the Advancement of Science, of which Phillips was Assistant Secretary, as a mouthpiece for irreligious geology.

Usually Phillips did not respond directly to Cockburn's taunts. He did not wish to aggravate Cockburn's scratches into a sore and he wanted to avoid playing Cockburn's game of giving estimates of the Earth's age. Cockburn took advantage of Phillips' silence: Phillips, he claimed, was an intellectual snob so occupied with chronicling infinite time that he did not deign to reply to 'a humble mortal of a pliocene period' (Cockburn 1838a, b, 1840, p. 38, 1845, 1849). By this time Phillips' own views on Genesis and geology had assumed the form they retained for the rest of his life. With specific reference to Cockburn's attack on Buckland, Phillips told a friend: 'the Mosaic cosmogony is one thing, and ... Christ's sermon on the mount never alludes to it; geology is another thing equally independent of

Christian faith, practice and evidence; I will not say that Moses was a geologist; perhaps that whole account of the creation of the world is a fine poem rather than history'.[2]

The age of the Earth

From the late 1830s to 1859 Phillips continued to view the age of the Earth as a magnificent but insoluble problem. In his 1855 *Manual of Geology* he reviewed no less than nine possible ways of establishing geological chronology, all of which used stratified deposits and not one of which was viable (Phillips 1855, p. 618). But the publication in November 1859 of Darwin's *Origin* focused Phillips' mind. Darwin had devoted a chapter to the imperfection of the geological record and another to the geological succession of organic beings. In Phillips' second address as President of the Geological Society, given in February 1860, he made it clear that for him the record of marine deposits was reasonably complete and showed no evidence of evolution having occurred. Phillips thought that evolution from common ancestors was a desperate view because the structures and adaptations of living things were special, various and 'determinate', and it denied the providential plan of creation. On the question of geological time, Phillips attacked Darwin's 'abuse of arithmetic' in his 'geological calculus' of 300 million years for the denudation of the Weald by marine action. For Phillips this was an inconceivable number for three reasons: Darwin's assumption of one inch per century as the rate of marine erosion was too low and sat uncomfortably with the maximum rate of erosion on the Yorkshire coast of eighty inches per year; he ignored any denudation by the atmosphere and rivers; and he ignored the effect of undercutting on the rate of erosion of cliffs (Phillips 1860a, pp. xxxii, xxxvii, lii).

Later in 1860 Phillips delivered at Cambridge and published his Rede lecture on *Life on the Earth: Its Origin and Succession*. In a long discussion of the antiquity of the Earth he used sedimentation data to calculate that sedimentary rocks were about 96 million years old. He was aware that he had made several questionable assumptions and that his calculation would be totally vitiated if the cooling of the Earth, which he accepted, had deranged the steady rate of sedimentation. He therefore tried to take account of this derangement in a second calculation which gave a result of between 38 and 64 million years. He again attacked Darwin's estimate of 300 million years for the denudation of the Weald (Phillips 1860b, pp. 122–137).

Darwin and his friends had a low opinion of Phillips. Huxley was irritated by Phillips' calm judiciousness and thought he would go to that part of Hell which accommodated, according to Dante, 'those who are neither on God's side nor on that of the Devil's'. Darwin judged Phillips' *Life on Earth* to be 'weak, washy, stilted stuff', 'unreadably dull', and he deplored 'the namby-pamby, old-woman style of the cautious Oxford professor'. Lyell thought it 'fearfully retrograde'.[3] Even so, Darwin reluctantly regarded Phillips as an important judge so from 1861 he withdrew from the third and subsequent editions of the *Origin* his Wealden calculation.

In June 1861 Phillips was consulted by William Thomson about whether he or geologists in general subscribed to Darwin's 'prodigious durations for geological epochs'. He told Phillips that preliminary calculations gave the age of the Sun as 20 million years and that of the Earth at the widest stretch as 200 to 1000 million years. In his prompt reply Phillips informed Thomson, whom he had met regularly at British Association meetings, that Darwin's computations were absurd and that Phillips' own calculations, given in his *Life on Earth*, gave 96 million years as the age of stratified rocks. Phillips welcomed Thomson's solar calculation as very interesting and advised Thomson to look at his address of 1859 to the Geological Society.[4] In 1862 Thomson published his estimate of the age of the Sun's heat as falling between 100 and 500 million years. The following year, in the full version of his paper on the secular cooling of the Earth, he estimated the age of the Earth's crust as 98 million years on the basis of certain assumptions. If they did not hold, then the range became 20 to 400 million years. Though Thomson did cite Phillips' published views about the disturbing influences in subterranean temperature measurements, he said nothing in public about Phillips' estimate of the age of the Earth until in 1869 he castigated 'orthodox geologists' for ignoring it (Thomson 1862, reprinted in Thomson 1891, pp. 349–368; Thomson 1863, reprinted in Thomson 1890, pp. 295–311; Thomson

---

[2] Phillips to Henry Robinson, wrongly dated 1837, [1838], Robinson papers, Humberside County Archives, DD x/65/27.

[3] Darwin to Lyell, 2 December 1859, in Burkhardt *et al.*, 1991, p. 409; Darwin to Hooker, 17 December 1860, in Burkhardt *et al.* 1993, pp. 531–532; Darwin to Hooker, 15 January 1861, Darwin to Horner, 20 March 1861, Darwin to Gray, 5 June 1861, in Burkhardt *et al.* 1994, pp. 8–9, 62–63, 162–163.
[4] Thomson to Phillips, 7 June 1861, Phillips papers, Oxford University Museum; Phillips to Thomson, 12 June 1861, Kelvin papers, Cambridge University Library.

1871 (read 1869) reprinted in Thomson 1894, pp. 73–124; Phillips 1859, pp. xlvi–l). For his part Phillips was the first leading British geologist to welcome Thomson's physical approach to geology. As president of the geology section of the British Association in 1864, Phillips recommended to his brethren of the hammer the computations of Thomson and Samuel Haughton on the age of the Earth, 98 million and 2300 million years respectively. Simultaneously Lyell, as president of the Association, reiterated his opposition to the cooling Earth theory. Next year Phillips, this time as president, spoke favourably about Thomson's work on the Earth's age (Phillips 1864, 1865, p. liv; Lyell 1864, p. lxix). Thomson was glad that Phillips was on his side and that they were remarkably *ad idem* about the age of the Earth, in opposition to what Thomson called 'the Hutton, Lyell and Huxley banking and discount company (unlimited)'. He could have added J. D. Hooker who was also unconvinced by Thomson's physical estimates.[5]

Phillips' support of Thomson was based on his belief, which he practised, that geology needed the aid of the collateral sciences of not just zoology, botany and chemistry, but also natural philosophy and astronomy. For him the study of the Earth's past overlapped with terrestrial and cosmic physics, a view not held in England by any other of that small band called heroic geologists. In his 1865 address to the British Association, Phillips revealed his view of nature as a unified totality: 'the greater our progress in the study of the economy of nature, the more she unveils herself as one vast whole; one comprehensive plan; one universal rule, in a yet unexhausted series of individual peculiarities. Such is the aspect of this moving, working, living system of force and law' (Phillips 1865, pp. lviii). Sentiments such as these show why Thomson regarded Phillips as a true geologist (Smith & Wise 1989). Moreover he shared with Thomson a longstanding commitment to the theory of the cooling Earth *and* an enduring enthusiasm for measuring subterranean temperatures. From 1855 at the latest Thomson knew that by the late 1830s Phillips had not only espoused the theory of the refrigeration of the Earth but had also himself measured the rise in temperature with depth in the crust of the Earth. In 1855 Thomson cited as reliable Phillips' figure of 1°F per 45–60 feet (Thomson 1855). It is not widely known that in the late 1860s, when Thomson and Huxley were debating whether geology needed the assistance of physics, Thomson and Phillips came together in instrumental alliance. From 1868 Thomson and his former pupil, J. D. Everett, began to supervise a long stint of underground temperature measurements. For these Thomson designed a hermetically sealed glass case which protected what was then known as a Phillips' self-registering maximum thermometer. This instrument, made commercially by Casella in the 1860s for research in physics, was based on the first ever maximum thermometer, which had been designed and made as long ago as 1832 by John Phillips (Phillips 1832; Everett 1869, pp. 176–177, 1870, p. 37, 1871, p. 25; Middleton 1966, p. 155).

For permission to cite or refer to manuscripts, I am grateful to the librarian, University Museum, Oxford (Phillips papers), the keeper of manuscripts, Cambridge University Library (Kelvin papers) and the archivist, Humberside County Archives (Robinson papers).

References

BURCHFIELD, J. D. 1974. *Lord Kelvin and the Age of the Earth*. Macmillan, London.
BURKHARDT, F. *ET AL.* (eds). 1991. *The Correspondence of Charles Darwin. Volume 7: 1858–1859.* Cambridge University Press, Cambridge.
BURKHARDT, F. *ET AL.* (eds). 1993. *The Correspondence of Charles Darwin. Volume 8: 1860.* Cambridge University, Cambridge.
BURKHARDT, F. *ET AL.* (eds). 1994. *The Correspondence of Charles Darwin. Volume 9: 1861.* Cambridge University, Cambridge.
COCKBURN, W. 1838a. *A Letter to Professor Buckland concerning the Origin of the World.* Hatchard, London.
COCKBURN, W. 1838b. *A Remonstrance, Addressed to his Grace the Duke of Northumberland, upon the Dangers of Peripatetic Philosophy.* Hatchard, London.
COCKBURN, W. 1840. *The Creation of the World. Addressed to R. I. Murchison and dedicated to the Geological Society.* Hatchard, London.
COCKBURN, W. 1845. *The Bible defended against the British Association: being the Substance of a Paper read in the Geological Section, at York, on the 27th September, 1844. To which is added a Correspondence between the Dean and some Members of the Association.* Whittaker, London and York.
COCKBURN, W. 1849. *A New System of Geology; Dedicated to Professor Sedgwick.* Colburn, London.
DARWIN, C. R. 1859. *The Origin of Species.* Murray, London. Reprinted 1968, Penguin, Harmondsworth.
EVERETT, J. D. 1869. [Second] Report of the Committee for the Purpose of investigating the Rate of Increase of Underground Temperature downwards in various Localities of Dry Land and under Water. *Report of the British Association for the Advancement of Science 1869*, 176–189.

---

[5] Thomson to Phillips, 22 December 1868, 1 June 1869, Phillips papers, Oxford University Museum.

EVERETT, J. D. 1870. Third Report. *Report of the British Association for the Advancement of Science 1870*, 29–37.

EVERETT, J. D. 1871. *Report of the British Association for the Advancement of Science 1871*, 14–25.

KENRICK, J. 1834. On the alleged Greek traditions of the Deluge. *Philosophical Magazine*, **5**, 25–33.

LYELL, C. 1864. Presidential Address. *Report of the British Association for the Advancement of Science 1864*, lx–lxxv.

MIDDLETON, W. E. K. 1966. *A History of the Thermometer and Its Use in Meteorology*. Johns Hopkins University, Baltimore.

MORRELL, J. B. & THACKRAY, A. W. 1981. *Gentlemen of Science. Early Years of the British Association for the Advancement of Science*. Clarendon, Oxford.

PHILLIPS, J. 1829. *Illustrations of the Geology of Yorkshire; or, a Description of the Strata and Organic Remains of the Yorkshire coast*. Wilson, York.

PHILLIPS, J. 1832. Description of a new self-registering maximum thermometer. *Report of the British Association for the Advancement of Science 1831 and 1832*, 574–575.

PHILLIPS, J. 1834. On subterranean temperature observed at a depth of 500 yards at Monk Wearmouth. *Philosophical Magazine*, **5**, 446–451.

PHILLIPS, J. 1835. *Illustrations of the Geology of Yorkshire. Part 1. The Yorkshire Coast*. Murray, London.

PHILLIPS, J. 1836. *Illustrations of the Geology of Yorkshire. Part II. The Mountain Limestone District*. Murray, London.

PHILLIPS, J. 1838. *A Treatise on Geology forming the Article under that Head in the Seventh Edition of the Encyclopaedia Britannica*. Black, Edinburgh.

PHILLIPS, J. 1855. *Manual of Geology: Practical and Theoretical*. Griffin, London and Glasgow.

PHILLIPS, J. 1859. Presidential Address. *Quarterly Journal of the Geological Society of London*. **15**, xxv–lxi.

PHILLIPS, J. 1860a. Presidential address. *Quarterly Journal of the Geological Society of London*. **16**, xxvii–lv.

PHILLIPS, J. 1860b. *Life on the Earth: Its Origin and Succession*. Macmillan, Cambridge and London.

PHILLIPS, J. 1864. Address to Geology Section. *Report of the British Association for the Advancement of Science 1864*, 45–49.

PHILLIPS, J. 1865. Presidential Address. *Report of the British Association for the Advancement of Science 1865*, li–lxvii.

RUDWICK, M. J. S. 1985. *The Great Devonian Controversy. The Shaping of Scientific Knowledge among Gentlemanly Specialists*. Chicago University, Chicago and London.

SEDGWICK, A. 1831. Presidential Address. *Proceedings of the Geological Society of London*, 1834, **1**, 270–316.

SMITH, C. W. & WISE, M. N. 1989. *Energy and Empire. A Biographical Study of Lord Kelvin*. Cambridge University, Cambridge.

THOMSON, W. 1855. On the use of observations of terrestrial temperature for the investigation of absolute dates in geology. *Report of the British Association for the Advancement of Science 1855*, 18–19.

THOMSON, W. 1862. On the age of the Sun's heat. *Macmillan's Magazine*, **5**, 288–293.

THOMSON, W. 1863. On the secular cooling of the Earth. *Philosophical Magazine*, **25**, 1–14.

THOMSON, W. 1871. Of geological dynamics. *Transactions of the Glasgow Geological Society*, **3**, 215–240.

THOMSON, W. 1890. *Mathematical and Physical Papers. Volume 3*. Cambridge University, Cambridge.

THOMSON, W. 1891. *Popular Lectures and Addresses. Volume 1*. Macmillan, London.

THOMSON, W. 1894. *Popular Lectures and Addresses. Volume 2*. Macmillan, London.

'Had Lord Kelvin a right?': John Perry, natural selection and the age of the Earth, 1894–1895

BRIAN C. SHIPLEY

Department of History, Dalhousie University, Halifax, Nova Scotia, Canada, B3H 3J5
(*email*: bshipley@is2.dal.ca)

Abstract: The Marquis of Salisbury's 1894 address to the British Association for the Advancement of Science sparked an important development in the debate on the age of the Earth. It led John Perry, a physicist, to produce the first mathematical rebuttal of Lord Kelvin's calculations, which had since 1862 functioned as an argument against the theory of evolution by natural selection. Perry wished to affirm the independence of geology from physics, keeping each branch of science to its proper domain. With the support of his mathematical friends, Perry tried privately to induce Kelvin to modify his views. This effort failed, however, and the discussion became public in *Nature*. Perry supported his calculations with Heaviside's new mathematical methods, and also with empirical data, though these were later undermined by Kelvin's experiments. Perry was uncomfortable with his position as Kelvin's critic, however, because he held his old teacher in great esteem. Although Kelvin never stopped believing that the Earth was too young for natural selection to have taken place, geologists and biologists responded very positively to Perry's results, and no longer felt they had to justify their conclusions to physicists. The answer to 'Had Lord Kelvin a right?', ultimately depended on one's scientific politics.

In August 1894, the Third Marquis of Salisbury (Robert Cecil, 1830–1903) delivered the presidential address to the meeting of the British Association for the Advancement of Science (BAAS) at Oxford (Salisbury 1894). His theme was the greatest unsolved scientific problems of the time; in particular, he descended with some force upon Darwin's theory of evolution by natural selection, which Salisbury found unconvincing, and believed was weakening under new doubts. In making his argument against Darwinism, Salisbury, who was a politician rather than a scientist, relied upon the work of Lord Kelvin (Fig. 1) (William Thomson, 1824–1907; knighted 1863, created Baron Kelvin 1892) in two important respects. First, most famously, he presented Kelvin's 100 Ma (million years) calculation of the age of the Earth, as evidence that there had not been enough time for natural selection to turn jellyfish into humans (Burchfield 1975; Smith & Wise 1989). Then, on the subject of an alternative to natural selection as the cause of evolution, Salisbury invoked 'the judgment of the greatest living master of natural science among us, Lord Kelvin,' concluding his remarks with a direct quotation from Kelvin's own presidential address at the British Association in 1871, twenty-three years before:

I have always felt ... that the hypothesis of natural selection does not contain the true theory of evolution, if evolution there has been in biology. ... I feel profoundly convinced that the argument of design has been greatly too much lost sight of in recent zoological speculations. Overpoweringly strong proofs of intelligent and benevolent design lie around us ... showing to us through nature the influence of a free will, and teaching us that all living things depend on one everlasting Creator and Ruler (Salisbury 1894, p. 15; from Thomson 1872).

With this address, Salisbury did more than just fuel the simmering tensions between geologists (who felt that they had already made adequate revisions to their estimates of the Earth's age) and physicists (who demanded even shorter limits). He also made it perfectly clear that the real stake in the debate was not the quantitative age of the Earth *per se*, but the validity of the theory of evolution by natural selection. Salisbury's call for a return (with Kelvin) to a belief in divine design as a guiding principle for biological studies could be taken as a repudiation of decades of work conducted within an evolutionary framework. Thus, although the debate on the age of the Earth was conducted between

From: LEWIS, C. L. E. & KNELL, S. J. (eds). *The Age of the Earth: from 4004 BC to AD 2002*. Geological Society, London, Special Publications, **190**, 91–105. 0305-8719/01/$15.00 © The Geological Society of London 2001.

Fig. 1. Lord Kelvin (William Thomson, 1824–1907). Science Museum/Science and Society Picture Library.

Fig. 2. John Perry (1850–1920). An unorthodox but esteemed teacher, he is also remembered as a reformer of mathematical education. Science Museum/Science and Society Picture Library.

physicists and geologists, the science of biology was just as deeply involved.

By the time of Salisbury's comments in 1894, geologists had had a full generation to come to terms with Kelvin's limitations (Burchfield 1975; Smith & Wise 1989). However, there was a feeling of increasing discontent among geologists that the physicists were not making any concessions to geological evidence in return. The best-known example of this sentiment is to be found in Sir Archibald Geikie's own presidential address to the British Association, two years earlier in Edinburgh (Geikie 1893). It is worth noting the role which such speeches played in this debate, probably because they gave a prime opportunity to address scientists in many fields simultaneously, and thus ensure that one's message was received by the widest possible audience. Geikie (1835–1924) was the director-general of the Geological Survey of the United Kingdom, the most eminent living British geologist, and had first-hand experience of the modifications which had been made to uniformitarianism over the previous decades, in response to Kelvin's criticisms.

In his 1892 address, Geikie expressed confidence that a flaw must exist in Kelvin's mathematics, which after all were based on a highly idealized model of the Earth. He was very conscious that the physicists' calculations involved many assumptions, any change to which might completely alter the result. While the most extreme physicists now argued for an age as short as 10–20 Ma, Geikie felt sure that geologists' extensive evidence for the Earth being much older would eventually be vindicated when the oversight, wherever it lay, was discovered. Nevertheless, neither Geikie nor anyone else in the geological or biological camps was in a position to respond to Kelvin's argument in mathematical terms. Although they could appreciate the logic of the Earth's gradual cooling, and hence its formation at some finite time in the past, and although many criticized Kelvin's assumptions that the Earth's composition, density, and thermal properties were uniform throughout, no one was able to work out mathematically an alternative model. However, with influential figures like the Marquis of Salisbury keeping the issue in front of the general scientific public, it was only a matter of time before a physicist with a taste for geology, a permissive attitude towards mathematics, and a sympathy for the underdog took up the problem. This is where John Perry (Fig. 2) came in.

John Perry: 'Few really understood him'

John Perry (1850–1920) is not a very well-known figure in the history of science. Born to a Protestant family in Ulster, in the north of Ireland, he was educated first as an apprentice in industrial engineering (Armstrong 1920; Turner 1926). In 1868, he entered Queen's College, Belfast, where he studied engineering under James Thomson, Kelvin's older brother. Ironically, Perry's early interest in geology was frustrated at this time: he narrowly missed getting a prize in the subject and later 'quarreled with the University viva voce examiner' about a controversial fossil.[1] Nevertheless, Perry graduated in 1870, taking a bachelor's degree in engineering with high honours, and subsequently becoming a lecturer in mathematics and physics at Clifton College in Bristol. In 1874, however, Perry's career really began to take off when he was appointed to an honorary assistantship in the Glasgow laboratory of Sir William Thomson (later Lord Kelvin). The year that he spent there proved to be very influential, making him an active researcher, and putting him on the path to greater things. For the rest of his life, Perry looked back fondly on his early association with Kelvin, and generally credited his mentor with much of his later success.

In 1875, Perry became a professor at the newly established Imperial College of Engineering in Tokyo, along with W. E. Ayrton (1847–1908), who was to become his long-term collaborator in research. They returned from Japan after three years, but continued to work together, first in industry, and then at the Finsbury Technical College, London, which Ayrton co-founded in 1879 and Perry joined in 1882, as professor of mechanical engineering and applied mathematics.[2] Perry remained at Finsbury until 1896, when he advanced to a professorship at the Royal College of Science, in South Kensington, a post which carried many more responsibilities. After he retired from teaching in 1913, Perry continued his long-standing involvement in the activities of the British Association, until his death in 1920.

The bulk of Perry's scientific output was made up of over fifty papers that he co-authored with Ayrton, in general physics and especially in electrical engineering. Some of the earliest of these, written while the two were busy in Japan, reveal an interest in the very subject that would figure so prominently in Perry's discussion with Lord Kelvin many years later: heat conduction in stone (Ayrton & Perry 1878). Perry also had a strong inclination towards the application of science to technology: he patented many inventions, and he and Ayrton are particularly remembered for their work on early ammeters (Gooday 1995). Despite his many publications, however, Perry did not leave a deep mark on the history of science. Much of his energy went into teaching, and his engineering students generally worked in industry rather than in scientific research. Perry put much effort into pedagogical reform, especially with regard to the practical use of mathematics (Perry 1900a, b; Brock 1996). He had a reputation for unpredictability, a flair for whimsy, an anti-authoritarian streak, a passion for defending the underdog, and a strong proclivity to follow his own path. These characteristics endeared him to his friends, even if they made him hard to fathom at times. According to H. E. Armstrong, Perry's chemistry colleague at Finsbury, Perry 'more than once remarked to me, [that] few really understood him' (Armstrong 1920, p. 752). Each of these dimensions of his character came out in the discussion on the age of the Earth.

Perry's motivation: 'The logic was irresistible'

It took John Perry barely two months from the time he read Salisbury's address until he reached his solution. Like Salisbury, Perry was convinced that the debate had reached an impasse, with neither the physicists nor the geologists willing to alter their position. But whereas Salisbury had taken this stalemate to indicate that the doctrine of natural selection was on shaky ground, Perry believed that the opposite was true: that error instead must lie somewhere in Lord Kelvin's calculations. In part, Perry's conviction stemmed from the vast amount of data which geologists and palaeontologists had accumulated, supporting the idea that the Earth was much older than the ever-shorter time limits imposed by physicists like Peter G. Tait (1831–1901). In this respect, Perry allied himself with the geologists' camp, breaking ranks with the physicists, who had always denied that Kelvin's calculations were open to question. However, Perry had another reason for believing that Kelvin was wrong, one which showed that he was no less conscious of the need to remain faithful to the precepts of physics. Perry felt that

---
[1] Cambridge University Library, Kelvin papers, MS. Add. 7342 (hereafter CUL/KP), P62, Perry to Kelvin, 26 December 1894.
[2] Guildhall Library London, City and Guilds of London Institute Archives, Minute-book of Subcommittee C: Finsbury Technical College, MS. 21821, p. 110.

there was a fundamental contradiction inherent in Kelvin's own principles, which Salisbury had brought to renewed public attention by quoting Kelvin's 1871 text in favour of divine design, or at least a guiding providence. 'The logic was irresistible,' Perry told his friend, the physicist Oliver Lodge (1851–1940), explaining that:

> If Lord Kelvin was right it was absolutely necessary that a providence should be discovered which had been doing for organic beings what Darwin had done with his pigeons. Now such a providence is destructive of Kelvin's theory & of all principles of Natural Philosophy & there must evidently be a flaw in Kelvin's reasoning.[3]

Kelvin's argument against an indefinitely long age of the Earth, Perry pointed out, was based on the constancy of physical laws. Strict uniformitarianism, as it had once been propounded by Lyell and others, was in Kelvin's view completely untenable, because it blatantly violated the laws of thermodynamics (Thomson 1864, 1871a, b). Yet the solution to the problem of evolution proposed by Kelvin and his followers was equally invalid, Perry now argued. For if natural selection was rejected as the mechanism by which life had developed, then there was no alternative but to appeal to some intelligent and controlling force which had acted instead. This was exactly the point that the Marquis of Salisbury had made in his address. And this suggestion, Perry avowed, was far more of an abomination to good scientific reasoning than the mere idea that Kelvin might have overlooked some factor in his calculations. Consequently, Perry felt confident in undertaking a review of Kelvin's work, with every prospect of success:

> Once it became clear to my mind that there was necessarily such a flaw, its discovery was no mere question of chance & I confidently announced to several persons my belief in a flaw before I had tried to discover it.[4]

For Perry, then, it was the mass of geological and biological evidence in favour of evolution that made him sure that Kelvin was wrong, but it was his principled belief that Kelvin's error (as he saw it) ought not to be allowed to stand, that ultimately induced him to take up the question.

---

[3] University College London, Lodge Papers MS. Add. 89 (hereafter UCL/LP), Perry to Lodge, 31 October 1894.
[4] UCL/LP, Perry to Lodge, 31 October 1894.

'Professor Perry's conclusion is very far from obvious'

When Kelvin originally made his calculations of the time that it had taken for the Earth to cool to its present surface temperature, he assumed that the Earth was of homogeneous composition (Thomson 1864). In addition to being an easily workable case, he thought that it would yield the longest potential age, and thus be as generous as possible to geologists. The only piece of field data employed was the rate at which temperature increased immediately below the Earth's surface, roughly 1°F for every 50 feet of depth. Given that the Earth's surface had once been at the same temperature as the core, when the whole Earth was a molten mass, it was merely a question of computing how long it had taken for the system to cool from its origins to its present conditions. Although Kelvin's simplistic model had its critics, it was generally believed that any factor which caused the Earth to cool more quickly would mean that Earth was even younger (Perry 1895a). So while geologists like the Reverend Osmond Fisher (1817–1914) proposed more complex interpretations of the Earth's internal structure, they did not attempt to calculate the resultant impact on the age of the Earth (Fisher 1889; Burchfield 1975).

John Perry's contribution, worked out in October 1894, was two-fold. First, he recognized that the faster that heat was conducted outwards from the Earth's core, the *longer* it would take to obtain the present observed temperature gradient at the surface. Perry suggested that the hot, dense, semi-fluid material of the Earth's interior would almost certainly conduct heat much better than the surface rocks whose properties were the basis of Kelvin's calculations (Perry 1895a). Instead of only one kind of rock making up the Earth, there were now two: a thin outer shell (with observed thermal behaviour), and an interior through which heat flowed much more readily. If the Earth had cooled faster than Kelvin estimated, then it had lost more of its original heat, but this same heat would have kept the surface warmer for longer, meaning that more time had elapsed before the present state was reached. Kelvin's assumption of a homogeneous Earth turned out *not* to be the most favourable case. This realization was important because it was highly counter-intuitive. Edward Poulton (1856–1943), professor of zoology at Oxford, emphasized that:

> Professor Perry's conclusion is very far from obvious, and without the mathematical reasoning would not be arrived at by the vast

majority of thinking men. The 'natural man' without mathematics would say ... it is quite clear that increased conductivity, favouring escape of heat, would lead to more rapid cooling, and would make Lord Kelvin's age even shorter (Poulton 1896, p. 813).

Poulton's statement also drew attention to the second aspect of Perry's contribution: he provided a way of treating mathematically this new, two-layer model of the cooling Earth. This was harder than it might seem: even Perry did not attempt to work out the general case. Instead, he ingeniously defined his variables so as to greatly simplify the calculation, using the same multiple n to represent both the conductivity and thickness of the interior, relative to the surface layer. If the ratio of conductivity to heat capacity in the interior was increased by the factor m, then Kelvin's estimate of 100 Ma would be multiplied n^2/m times (meaning that a higher heat capacity of internal material would also increase the age) (Perry 1895a; Smith & Wise 1989). Taking a few examples, Perry could easily show that Kelvin's age might be made tens or hundreds of times longer. Whatever the true age was, there was no reason to believe that Kelvin had accurately determined the upper limit. This seemed to be the flaw that Geikie, Perry and others had intuitively believed must exist. All that remained was for Perry to check his work with a few friends, put the matter before his mentor Kelvin, and the wrong could be corrected.

'Your affectionate pupil': the campaign to convince Kelvin

Lord Kelvin was essentially the only person whose approval John Perry had to win for his new calculations, since it was on Kelvin's authority that the age of the Earth had been limited in the first place. Even geologists who accepted a relatively young age had no compelling independent reason to do so, and if Kelvin changed his mind then Salisbury's objection to natural selection would be removed. Perry naturally wanted to convince Kelvin on his own mathematical and physical terms, which Perry shared. He had no intention of challenging Kelvin publicly, as some earlier participants in the debate, like Huxley, had done. Perry was devoted to Kelvin, and had no interest in creating an awkward situation for both of them. As he explained to Lodge:

> I think I speak truly when I say that because I can only obtain credit in this matter & my old teacher can only lose credit, any pleasure that I have is a purely scientific pleasure – I am glad for the Biologists & Geologists but I am as much without personal pleasure in my share of the business as it is possible to be.[5]

Thus, Perry sought to sway Kelvin privately, perhaps aided by the opinions of a few other scientists. He wanted to act as a 'collaborator' rather than as a 'contestant.' If Kelvin willingly announced a revision to his own estimates, there would be no need for a confrontation, allowing Perry to fulfill his duties both to science and to his scientific father figure. First, however, Kelvin would have to agree to consider the matter.

After sending his argument and calculations in mid-October to his close friends George Fitzgerald (1851–1901) and Joseph Larmor (1857–1942), who were both well-known, respected physicists, Perry wrote to Kelvin a few days later, enclosing a copy of the same material and explaining its significance, emphasizing the increase in age by the factor of n^2/m. By appealing to other mathematical physicists, Perry made it clear that he considered this a problem within physics, rather than between physics and geology:

> My friends Prof. Fitzgerald & Dr Larmor say that the artifice by which I have evaded difficult mathematics ... is legitimate; without their approval I dare not even as a privileged old pupil have taken the liberty of putting this before you. Fitzgerald says that I am certainly right in my argument & he praises my method of attacking the problem.[6]

After sending two more letters, with new calculations, over the following week, however, Perry still had not managed to get a response from Kelvin.[7] By now, Perry had gathered the support of further physicists, including Osborne Reynolds, Olaus Henrici and Arthur Rücker, which he hoped would encourage Kelvin to take the matter seriously. Perry told Kelvin that 'we all agree in thinking that your argument must be modified. I know & told them that my great difficulty was in getting you to reconsider the question.'[8] He typically signed his letters 'your affectionate pupil,' reminding the septagenarian Kelvin that they were on the same side, and reassuring his old master that Perry meant him no harm or disrespect.

---

[5] UCL/LP, Perry to Lodge, 31 October 1894.
[6] CUL/KP, P56, Perry to Kelvin, 17 October 1894.
[7] CUL/KP, P57, Perry to Kelvin, 22 October 1894; P58, Perry to Kelvin, 23 October 1894; P55, Perry, 'The Age of the Earth,' MS. dated 23 October 1893 [sic].
[8] CUL/KP, P58, Perry to Kelvin, 23 October 1894.

Lacking results, however, Perry's next line of attack was to confront Kelvin personally. His opportunity came on 28 October 1894, less than two weeks after he had first written to Kelvin, at a dinner at Trinity College, Cambridge. Perry related the encounter with relish to his friend Lodge the next day:

> I sat beside him [Kelvin] last night at Trinity & he had to listen. I knew beforehand that he would not read my documents, & he hadn't but I gave him a lot to think of & his pitying smile at my ignorance died away in about 15 minutes. I think he will now really begin to consider the matter. Geikie was opposite, his eyes gleaming with delight. He wont [sic] publish the thing but means to apply gentle conversational pressure to Kelvin. The face of McK. Hughes was also a sight – a thing of joy. Glazebrook, Forsyth & many others welcomed the excitement of a new idea.[9]

Perry was evidently pleased by the reception which the broader scientific community had given to his work, including Geikie and McKenny Hughes (1831–1917), professor of geology at Cambridge. But his relief at finally getting Kelvin to listen, although he 'did not seem to think me right,' was tempered by the knowledge that the battle was still far from won (Perry 1895a, p. 227). Geikie's recollection of the same event, written two weeks later to Raphael Meldola (1849–1915), agrees:

> I met him [Perry] at Cambridge a fortnight ago and had a talk with him on the subject. . . . The great thing now is to get Lord Kelvin to reconsider the matter. Prof. Perry and he were sitting opposite to me at the after-dinner gathering in the Trinity combination room and I was amused with the persistent way he pressed the subject on Lord Kelvin. He got a little way on, but confessed that it would probably take some time to get the veteran to review his age and universally accepted opinions. I have long felt that there must be a flaw somewhere in the physical argument, for the conclusion it leads to is, from the geological & biological side, impossible.[10]

Despite Perry's initial triumphant reaction, this incident would come back to haunt him once the momentum that he had initially built up began to slip away.

Several weeks later, in mid-November, and with still no definite sign of movement from Kelvin, Perry resorted to a more desperate tactic. At this point, he was reluctant to give up on the effort he had made to solve what he saw as an important problem; no less did he wish to abandon the geologists and biologists in whom he had already begun to inspire a new hope. Perry was thus willing to take a chance, and put his pride on the line. He wrote to the one person who he knew could influence Kelvin: P. G. Tait, Kelvin's long-time collaborator. The only problem was that Tait was a deep opponent of natural selection and took an even harder line on the age of the Earth than did Kelvin, advocating a maximum of 10 Ma, a mere tenth of Kelvin's most widely accepted estimate of 100 Ma (Thomson 1871a). Perhaps not surprisingly, Tait responded less than warmly to Perry's request for advice on Kelvin, claiming to have 'entirely missed your point' (Perry 1895a, p. 226). Tait's objection was not to Perry's mathematics, but to his logic. He claimed that Perry's result was 'absolutely obvious', but at the same time unconvincing, because there was simply no way of knowing what the Earth's interior was really like. Kelvin had chosen 'the simple and apparently possible case of uniform conductivity all through having no data whatever'. According to Tait, one could not invalidate Kelvin's work simply by showing that other cases were possible.

Needless to say, Perry was not satisfied. He sensed a logical flaw in Tait's argument, hinging on the distinction between possible and probable ages of the Earth. In order to prevent biologists from appealing to the extreme age of the Earth as part of the case for natural selection, anti-evolutionary physicists had to claim that their calculated ages were not just *possible*, but actually *likely*. Perry, to refute them, had only to show that other ages were *equally* likely, and thus reduce *all* calculations to the status of mere possibility. Tait, however, tried to turn the tables on Perry, by implying that Perry was seeking a *more* probable age than Kelvin's, while Kelvin had not intended his case to be more likely than any others. Perry was not fooled however; and, unable to resist, he responded vigorously:

> I should have been on the whole better satisfied if you had opposed my conclusions. You say I am right, and you ask my object. Surely Lord Kelvin's case is lost, as soon as one shows that there are *possible* conditions as to the internal state of the Earth which will give many times the age which is your and his limit. However I ought not to bore you, and I am much obliged to you for answering me. I cannot help it! I must put the matter before you. What troubles me is that I cannot see one

---

[9] UCL/LP, Perry to Lodge, 29 October 1894.
[10] Imperial College, Archives, Meldola papers [hereafter IC/MP], letter-book, p. 149, Geikie to Meldola, 15 November 1894.

bit that you have reason on your side, and yet I have been so accustomed to look up to you and Lord Kelvin, that I think I must be more or less of an idiot to doubt when you and he were so 'cock-sure.' ... If I were alone in my opinion, I should still have the courage, I think, to write as I do; but as I have already told you, I did not venture to write and speak to Lord Kelvin, or write to you until I found that so many of my friends agreed with me.[11]

Perry thought that Tait was trying to trick him into believing that his new calculations were acceptable, in an effort to trap a more eminent scientist (perhaps Huxley or Geikie) in the same error:

> One who did not know you or Sir William [Lord Kelvin] would think that the admission in your letter was a surrender, but I know you better. You see something probably that I have neglected. You won't oppose me, but you will wait for some strong opponent. ... Well I dont [sic] mind saying that I am very unwilling to rise and meet the winds and if you now ask me my *object* I would say that I should have liked your help in getting Lord Kelvin to go into the matter, but having now said what I had to say, I want to let the subject drop. It would not be becoming in me who owe [sic] so much to Lord Kelvin, to attempt to force his hand more than I have already tried.[12]

Frustrated by Tait's obstinacy, Perry showed for the first time that he was uneasy about taking the debate too far. Despite his strong beliefs, he was unwilling to damage his relationship with Kelvin. But Tait's reply the next day spurred him to one final effort.

Tait's response to Perry was just a brief note. Apart from a scornful comment about the insatiable demands of '*advanced* geologists' for time, it contained but a single statement of substance: 'What grounds have you for supposing the inner materials of the Earth to be better conductors than the skin?' (Perry 1895a, p. 226). This simple question proved to be a turning point. Perry replied quickly but cautiously, reasserting that 'It is for Lord Kelvin to prove that there is *not* greater conductivity inside. Nevertheless I will state my grounds' (Perry 1895a, p. 226). These were two-fold: first, there was 'no doubt of a certain amount of fluidity inside' the Earth, which would lead to 'very much greater quasi-conductivity inside than of true conductivity in the surface rocks'. This referred to the transfer of heat in the interior, not only by flow through rock, but by the flow of the rock material itself. Second, and crucially, Perry said that 'surely there can be no doubt of the conductivity of rock *increasing* with the temperature'. To support this claim, he put forward some experimental data from the Swiss physicist Robert Weber, which had been quoted by Jospeh D. Everett in a work on physical constants (Everett 1886). These data were the first new empirical evidence in the discussion on the cooling of the Earth since Kelvin's use of the observed thermal properties and temperature gradient of surface rocks in his initial calculations. Finally, Perry had found an approach that would convince Kelvin to enter the discussion, even though it meant that he had been forced to make his argument specific when he had always preferred to emphasize its generality. Regardless of how unhelpful Tait's attitude had been, the senior physicist had succeeded in moving the debate to the next stage, by forcing Perry to provide the kind of information to which Kelvin would respond.

One week later, Perry wrote to Tait again, with further information from Everett about the reliability of Weber's results. Perry was happy to tell Tait that Everett had judged him to 'have established a strong presumption in favour of the increase of rock conductivity with temperature' (Perry 1895a, p. 227). Around the same time, in the first week of December, Kelvin finally contacted Perry and asked to see the relevant documents. Perry eagerly replied:

> I need hardly say how glad I am that you have again spoken about the Earth business, and as you have requested, I enclose copies of my correspondence with Prof Tait and also a printed copy of the original documents which I sent you before. ... I would have sent you a printed copy sooner but I know your much employment [sic] & did not wish to seem too persistent.[13]

This printed pamphlet contained Perry's original calculation, and a second, more sophisticated model of the cooling Earth, both of which had been produced in October, circulated first in manuscript (including to Kelvin, as described above), and printed for private distribution at the beginning of November 1894.[14] No effort had yet been made to publish Perry's results: the

---

[11] CUL/KP, P59c, Perry to Tait, 26 November 1894.
[12] CUL/KP, P59c, Perry to Tait, 26 November 1894.
[13] CUL/KP, P59, Perry to Kelvin, 7 December 1894.
[14] CUL/KP, P59a, printed pamphlet, 'On the Age of the Earth,' 1 November 1894.

intention still was to generate an informal consensus first, and thus avoid embarrassment on all sides. Whether this hope would be fulfilled now hinged on Kelvin's response.

'I think I must try': Kelvin's response

One week after he had received John Perry's documents, Lord Kelvin made his first written statement on the validity of the new calculations. It was by now mid-December, almost exactly two months since Perry had first announced his discovery, and just over four months since Salisbury's address. Kelvin began his reply on a positive note, telling Perry 'your n^2/m theory is clearly right', referring to Perry's equation for the factor by which Kelvin's age should be multiplied (Perry 1895a, p. 227). The senior physicist had two reservations, however: first, that the mathematics of the 'true case' (as opposed to an idealization of the Earth's properties) would be even more complex than Perry had accounted for; and second, 'that we must try and find how far Robert Weber's results can be accepted as trustworthy'. For Kelvin, the second problem was much more interesting than the first. As far as the mathematics of the possible age was concerned, he was willing to admit that 'it is quite possible that I should have put the superior limit a good deal higher, perhaps 4000 [Ma] instead of 400'. The real question, however, was physical and empirical, and concerned not the possible but the probable age of the Earth. Here, Kelvin's curiosity was aroused by the prospect of finding out exactly how conductivity did change with temperature: 'it will be worth while to make further experiments on the subject, and I see quite a simple way, which I think I must try, to find what deviation from uniformity of conductivity there is in slate, or granite, or marble, between ordinary temperatures and a red heat'.

But Kelvin warned Perry that the argument against natural selection ultimately did *not* depend on experimental data. However long the age of the Earth turned out to be, 'it can bring no comfort in respect to demand for time in Palaeontological Geology. Helmholtz, Newcomb, and another [Kelvin himself], are inexorable in refusing sunlight for more than a score or a very few scores of million years of past time' (Perry 1895a, p. 227). The duration of the Sun's heat was closely linked to the debate on the cooling Earth, since both limited the possible time in which natural selection could have acted. These three concerns (difficult mathematics, the need for empirical data, and the age of the Sun) formed the basis of Perry and Kelvin's correspondence in the last weeks of 1894 and early 1895. With the discussion now split into three parts, each took on a life of its own, some more fruitful than others.

'The want of rigour is of no importance'

On the subject of mathematics, Perry had already been in contact with his friend Oliver Heaviside (1850–1925), a reclusive and eccentric mathematician who had developed powerful methods of analysis that were well-suited to problems such as a multi-layer model of the cooling Earth (Nahin 1985). Perry was very excited about the method of 'operators', which allowed Heaviside to replace Perry's calculations, valid only for special cases, with a general solution. One of Heaviside's unexpected findings was that the cooling time in Perry's model was so sensitive to the values chosen for internal conductivity and surface thickness, that, in some cases, it was possible to obtain a younger age (the intuitive guess) rather than an older one. This showed, Perry told Kelvin, that 'Tait was not quite right in saying that it needed no mathematics to show that greater conductivity inside gives greater age'.[15] Heaviside also found that Perry's chosen example gave very close to the maximum age, although the assumption of a thicker crust would give a similar effect.[16] These calculations, first made in November, were later published in the journal *The Electrician*, where they provided a significant example of the utility of Heaviside's new approach (Heaviside 1899).

As keen as Perry was on Heaviside's novel methods, though, they were unlikely to convince Kelvin, a committed traditionalist who was noted for his refusal even to accept the modern, Maxwellian version of electromagnetic theory (Knudsen 1985). As early as 1885, Perry had pointed out to Lodge that Kelvin was slipping behind in this respect: 'He doesn't know Maxwell's theory. If only he knew how beautiful it was & how it holds together'.[17] Perry's mathematical and physical allies, on the other hand, including Fitzgerald and Larmor, were from a younger generation, and were committed 'Maxwellians' (Hunt 1991). Although Kelvin had supported Heaviside in the past, on this occasion he remained unmoved by Heaviside's radical approach, which was widely criticized for its disregard of conventional mathematical

---

[15] CUL/KP, P61, Perry to Kelvin, 17 December 1894.
[16] UCL/LP, Heaviside to Lodge, 19 December 1894.
[17] UCL/LP, Perry to Lodge, 14 June 1885.

standards (Nahin 1985). Perry tried to convince Kelvin that, as long as Heaviside's results were physically testable: 'the want of rigour is of no importance'.[18] In this case, however, Kelvin was eager to start with the testing itself. It was characteristic of his scientific method to 'look and see', to investigate empirically first, to formulate a mechanical model, and only then to apply mathematics (Smith & Wise 1989).

'Observation; observation only'

As early as 1868, Kelvin had made it clear that further empirical study was necessary to understand properly the current rate of the Earth's cooling. In his address 'On Geologic Time', which attracted Huxley's angry attention, he had asked:

> as regards underground heat, where must we apply to get evidence? Observation; observation only. We must go and look. We must bore the Earth here in the neighbourhood. We must examine underground temperature in other places. We must send out and bore under the African deserts, where water has not reached for hundreds of years. The whole Earth must be made subject to a geothermic survey (Thomson 1871a, p. 21).

In developing his new model, Perry had not questioned Kelvin's assumptions about surface temperature, but rather about the properties of internal material, which could not be observed directly. Even before he received a reply from Kelvin, Perry claimed he had 'been consulting with one of our Finsbury Assistants as to the best way of arriving at experimental results on Temperature & Conductivity of rock'.[19] Perry was no less eager than Kelvin to obtain the best possible experimental results, if such data were to play a part in the discussion. Although Perry sided with the geologists, he continued to behave as a physicist in this respect. It was to be several months, however, before results of the experiments were available. In the meantime, Perry pressed Kelvin on the age of the Sun, in an effort to prevent the debate from slipping away.

'The argument from the sun's heat seems to me quite weak'

Since the 1860s, Kelvin had argued that the Sun had only been shining on the Earth for a relatively short period of time, less than 100 Ma. Along with some more minor considerations about geodynamics, this was part of his overall argument that life on Earth could not have existed long enough for evolution to have occurred by natural selection (Burchfield 1975; Smith & Wise 1989). Perry knew that in order to vindicate geologists and evolutionary biologists, it would be necessary to show that Kelvin's assumptions in making this calculation could also be modified to permit a much older age. He had already told Tait that 'the argument from the sun's heat seems to me quite weak' (Perry 1895a, p. 226). But Tait was not interested in debating this point, and other than Kelvin, Perry was not able to find a receptive audience for this argument. Astronomers had no vested interest in the age of the Sun, and geologists and biologists had positive evidence relating only to the Earth. Perry and Kelvin exchanged several letters on solar heat, but Kelvin refused to alter his assumptions (Smith & Wise 1989, pp. 544–548).[20] This stalemate exemplified the point the discussion had reached. Perry had managed to get Kelvin to listen to his arguments, but, against Perry's hopes and expectations, his old master was unwilling to revise his pronouncements. For Perry, and more particularly for his friends and admirers who were less hesitant about possibly offending Lord Kelvin, it looked as if the time had come to publish after all.

'The thirty pieces of silver'

On 3 January 1895, the fruits of three months of behind-the-scenes labour appeared in the pages of *Nature*. John Perry's paper 'On the age of the Earth' consisted of his original manuscripts, as they had been privately printed and circulated, along with excerpts from his correspondence with Tait, and Kelvin's first reply of 13 December 1894 (Perry 1895a). There was also a brief preface that explained how this had all come about, which began, with careful neutrality: 'It has been thought advisable to publish the following documents'. No apology was offered for any embarrassment to Tait, who was cast in an unfavourable light by the exposure of his short-tempered responses. Although Perry was willing to acknowledge his authorship of the paper, rather than submitting it anonymously, he portrayed himself (as always) as acting with the support of a larger community of scientists. Perry

---

[18] CUL/KP, P61, Perry to Kelvin, 17 December 1894.
[19] CUL/KP, P61, Perry to Kelvin, 17 December 1894.

[20] CUL/KP, P62, Perry to Kelvin, 26 December 1894; P63, Perry to Kelvin, 8 January 1895.

took the opportunity of praising Heaviside's contribution, and further claimed that, 'Only for Prof. Fitzgerald's encouragement and sympathy, it is very probable that this document would never have been published' (Perry 1895a, p. 224).

Perry's true feelings about publication were expressed more freely, however, buried at the bottom of a footnote in the middle of the paper:

> Some of my friends have blamed me severely for not publishing the above document sooner. I was Lord Kelvin's pupil, and am still his affectionate pupil. ... He has been uniformly kind to me, and there have been times when he must have found this difficult. One thing has not yet happened: I have not received the thirty pieces of silver (Perry 1895a, p. 226).

Two of the strongest forces in Perry's professional life were his deep personal principles and his admiration for Kelvin: here, exactly at this point, the two clashed inescapably. As Perry said in his original manuscript sent to Larmor and Fitzgerald, now reprinted in *Nature*: 'his calculation is just now being used to discredit the direct evidence of geologists and biologists, and it is on this account that I have considered it my duty to question Lord Kelvin's conditions' (Perry 1895a, p. 224). It was bad enough that Perry had been forced by his own values to attack the man who had been the greatest influence on him: now the whole world was to know about it, too. Perry tried to make light of this predicament by jokingly casting himself in the role of Judas Iscariot, being paid thirty pieces of silver by the priests in the temple to betray his master (Matthew 26: 14–16). The magnitude of comparing Kelvin to Christ indicates that Perry did not take the matter so lightly, however, and that a mere joke, however odd, would not resolve the tension.

There is little trace of Perry's anxiety in his remaining letters to Kelvin during the first two weeks of January. In fact, there is no mention at all of the fact that the debate had now gone public. Perry continued to expound both the value of Heaviside's mathematics, and the impossibility of estimating the age of the Sun with any degree of precision, but Kelvin maintained an unconcerned, teacherly air of superiority. He referred Perry to a 60-year-old text for mathematical help, chiding him gently: 'You have not noticed how vastly it simplifies and shortens the numerical work from that of Heaviside's'.[21]

When Perry described a simple experiment he had devised to measure quickly and easily the variation of heat conductivity in rock with temperature, Kelvin wrote back immediately: 'I am sure you will get no satisfactory result from the experiment you propose', but offering to share details of his own set-up 'if we meet in the tea-room [at the Royal Society] on Thursday'.[22] This was the end of the (surviving) correspondence between Perry and Kelvin on the age of the Earth.

Neither Perry nor Kelvin was quite ready to abandon the topic, though. Perry sent a second letter to *Nature* in early February, chiefly to describe the use of Heaviside's operators to solve a more general version of the problem, in which conductivity could vary widely with temperature (Perry 1895b). Perry also made sure to mention publicly that he had been corresponding with Kelvin about experiments to verify Weber's data. More importantly, Perry stressed that his conclusions were 'really independent of whether R. Weber's results are correct or not. Lord Kelvin has to prove the impossibility of the rocks inside the Earth being better conductors' (Perry 1895b, p. 342). With this paper, Perry hoped to meet Kelvin's mathematical objections, and to establish the validity of his challenge to Kelvin's age, regardless of the experimental outcome. He was eager to have the matter resolved as soon as possible, now that it had come so far. Even if Kelvin was oblivious to Perry's uncomfortable situation, it was ominously present in other people's minds.

Shortly after publishing his second paper in *Nature*, Perry became severely ill with influenza, and spent some time in a feverish, delirious state. Late in February, still in this condition, he wrote to his friend Oliver Lodge, to whom he had previously confided his personal views on the discussion with Kelvin. On this occasion, though, he wrote to defend himself against a perceived attack in a letter that Lodge had recently sent him:

> All these ten days I have slept little, but had uneasy dreams all the time & the central *motif* of many of my dreams has been a letter received from you saying that I had behaved unfairly to Lord Kelvin. And what was wrong in my reference to the 30 pieces of silver? What an unspiritual, unimaginative man you must be. ... I wish I knew exactly how to pitch into you so you would really feel it – but you are so pachydermatous! . ... I had one

---

[21] CUL/KP, P66, Kelvin to Perry, 10 January 1895.

[22] CUL/KP, P67, Perry to Kelvin, 14 January 1895; P68, Kelvin to Perry, 15 January 1895.

conversation with Kelvin at Cambridge; Geikie *did* chance to overhear part of it – On that you have a charge of 'baiting Kelvin among the Geologists at Cambridge' as if he had been St Paul 'warring with beasts at Ephesus'.[23]

Here followed a long (if not totally coherent) defence of Geikie's character, and assurances of the absence of any ill-intent on Perry's part. After drifting further off track, the exhausted Perry managed to marshal his faculties and finish the letter:

> What I meant to write about was your letter about my treatment of Lord Kelvin. But you don't understand! You might as well talk about a man's ill treatment of his wife or his mother. How can one ill treat a person that one loves as I do Lord Kelvin? O my poor dear Lodge. What an unspiritual man you must be! And no doubt you know that if I committed any sin whatsoever, Kelvin would forgive me because he knows how much I love him?

From this candid expression of his personal dilemma, it seems clear that Perry had little personal interest in pursuing the debate any farther. Troublingly, Lodge appeared to have taken Perry's reference to the thirty pieces of silver as a boast, rather than as an admission of shame. Now that the situation was literally giving Perry nightmares, it could only be hoped that geologists at least would benefit from his sacrifice.

Unfortunately for Perry, Kelvin's own paper in *Nature*, which appeared the following month, March, reported empirical data from various sources (including new results from Weber), all of which indicated that Perry was wrong: that the thermal conductivity of rocks *decreased* as their temperature went up (Thomson 1895). There was certainly now no chance of convincing Kelvin to accept a higher figure for the age of the Earth. In fact, having been led to re-examine the question, he now concluded that the true age was probably much less than 100 Ma. Far from helping the geologists, Perry seemed to have made their situation worse. In an attempt to limit the damage, Perry composed a final article for *Nature*, making (with Fitzgerald's assistance) a lengthy and detailed review of all the evidence that bore on the age of the Earth, from both physics and geology (Perry 1895*c*). He now fully realized that Tait had done him no favour in getting him to back up his abstract model with specific data, since Kelvin had exploited this opening to discredit Perry's approach altogether.

In this final essay, Perry stressed the simple logic of his argument: all he wanted to do was show that higher ages were possible, for whatever reason, and thus that physics could provide no meaningful upper limit to the age of the Earth. Since the interior of the Earth was unknown, no amount of laboratory work could determine what its conductivity really was; Kelvin might have disproved one way in which heat could flow faster on the inside than at the surface, but there were any number of other possibilities available. The age of the sun was even more arbitrary: there was no reason that existing physical models could not be modified to provide a much longer time for the existence of life. Perry sought to give geologists all the grounds they needed to make their own estimates with confidence. Having abandoned the project of converting Kelvin, it no longer mattered what Kelvin believed, as long as geologists realized that they were not bound by his decrees. As soon as Perry could step out of the discussion with a clear conscience, he could begin to return to normal relations with Kelvin, an equally important goal.

'It is impossible to exaggerate the importance'

Neither Perry nor Kelvin can be seen as victors in this controversy. The most important outcome, however, was its impact on geologists and evolutionary biologists, who were the real subjects of Salisbury's attack and Perry's defence. If it is not surprising that their overall response was positive, it is nevertheless informative to see how news of Perry's findings spread, and how its reception varied. The chief architect of the dissemination of Perry's work to biologists was his colleague at Finsbury Technical College, Raphael Meldola. An ardent Darwinian evolutionist, Meldola was probably one of Perry's main inspirations in tackling the age of the Earth, and he was one of Perry's first confidants in the matter. Even before Perry sent his work to Fitzgerald and Larmor, he mentioned it in a note to Meldola as a topic they had discussed before.[24] When Perry's pamphlet was printed privately in early November, Meldola took it upon himself to distribute copies to interested parties.

---

[23] UCL/LP, Perry to Lodge, 25 February 1895.

[24] IC/MP, letter-book, p. 148, Perry to Meldola, 8 October 1894.

Among the recipients of this document were the two most eminent living evolutionists, Thomas Huxley (1825–1895) and Alfred Wallace (1823–1913). Meldola informed Huxley that Perry was 'an old pupil of Lord Kelvin's' who could 'allow us 100 times the amount of time allotted by Lord Kelvin for geological & biological evolution without violating any physical principle'.[25] Although 'most of the leading mathematical physicists in the country' had seen it and could 'find no material flaw in Perry's method', Meldola cautioned that it was still 'to be considered as private till Lord K. has considered it & given his opinion'. Huxley, who was near the end of his life, made only a modest response, thanking Meldola, but admitting that he could 'do no more than catch the general drift of [Perry's] argument'.[26] Wallace, on the other hand, took up the matter enthusiastically, despite being at an equally serious mathematical disadvantage: 'I was very glad to hear of Prof. Perry's discovery of Lord Kelvin's error & to have the paper though *that* is very unintelligible to me'.[27] He did, however, grasp the importance of increased internal conductivity, which he took as a vindication of Fisher's earlier arguments about the structure of the Earth (Fisher 1889). The senior biologists thus received Perry's result warmly, but they were not overwhelmed or shocked by it. This suggests that the age of the Earth was not felt so strongly, after all, as a limit to the validity of evolution by natural selection, at least for the older generation who had not grown up under Kelvin's restrictions.

Some of the most appreciative responses came from geologists. John Wesley Judd (1840–1916), professor at the Royal College of Science and past-president of the Geological Society, declared that 'It is impossible to exaggerate the importance of this work or its probable influence on both physicists & biologists, as well as geologists'.[28] Judd was a keen supporter of Darwinian evolution, and knew something of the history of the debate on the age of the Earth through his interest in Charles Lyell. The geologist with perhaps the most at stake, however, was Archibald Geikie, who as director-general of the Geological Survey was the leader of British geology, and who was already involved in the debate. Geikie told Meldola that: 'As you may believe, [Perry's] results very greatly interest me'.[29] Of all the respondents, Geikie was the most aware at this point (partly because he had witnessed Kelvin's reaction at Cambridge) that the battle was far from over. He realized that Perry's argument would require a substantial amount of promotion, to receive general acceptance. It was convenient, then, for Geikie to draw attention to Perry's work in concluding his timely article on 'Twenty-five years of geological progress in Britain', published in *Nature* in mid-February, within weeks of papers by both Perry and Kelvin (Geikie 1895). Geikie simply took this opportunity to remind readers that geologists had made many concessions, and physicists none, despite all of the evidence being on the geological side. But with the sudden appearance of Perry on the scene, Geikie reported, there was now 'every prospect that the physicists will concede ... a very much greater time' (Geikie 1895, p. 370).

Geikie's hopes were, in the event, too sanguine: the physicists (meaning Kelvin and Tait) conceded nothing. If anything, the eager expectations of geologists and biologists could only have reaffirmed Kelvin's commitment to a short age. When Kelvin published his final experimental results in June 1895, he received some criticism from Weber (whose experimental results Perry had first quoted), but there was no reply from Perry himself (Thomson & Murray 1895; Weber 1895). Nevertheless, Perry's contribution was remembered appreciatively. At the BAAS meeting the following summer, evolutionary biologist Poulton told members of the Zoological Section that Perry had indeed removed the 'barriers across our path' by showing that physicists' arguments were illusory rather than effective (Poulton 1896, p. 815). With the Earth's age not limited to any specific duration, Poulton, a dedicated Darwinian selectionist, encouragingly instructed his fellow biologists that 'we are free to proceed, and to look for the conclusions warranted by our own evidence'.

Geikie's attitude was similar, that Perry provided a sort of liberation. He was aware, though, that Kelvin had not changed his mind, and thus still presented an obstacle to complete geological independence. In the following years, Geikie criticized Kelvin both publicly and privately for his narrow-mindedness, writing to the physicist in 1898: 'I must postpone the pleasure of

---

[25] Imperial College, Archives, Huxley papers, 22.210, Meldola to Huxley, 14 November 1894.
[26] IC/MP, letter-book, p. 149, Huxley to Meldola, 16 November 1894.
[27] IC/MP, letter-book, p. 149, Wallace to Meldola, 18 November 1894.
[28] IC/MP, letter-book, p. 149, Judd to Meldola, 15 November 1894.

[29] IC/MP, letter-book, p. 149, Geikie to Meldola, 15 November 1894.

listening to your latest blast of the anti-geological trumpet. ... [T]he geological & biological arguments for a longer period than you would allow seem to me so strong that I do not see how they are to be reconciled with physical demands.'[30] A year later, he told a BAAS audience of geologists that:

> Lord Kelvin has never taken any notice of the strong body of evidence adduced by geologists and palaeontologists in favour of a much longer antiquity. ... [I]n none of his papers is there an admission that geology and palaeontology, though they have again and again raised their voices in protest, have anything to say in the matter that is worthy of consideration. It is difficult satisfactorily to carry on a discussion in which your opponent entirely ignores your arguments, while you have given the fullest attention to his (Geikie 1900, p. 724).

Although Geikie put the most emphasis on the positive power of geological evidence (much of it derived from rates of erosion and sedimentation), he cheered Perry's bold insurrection against his disciplinary colleagues as a 'remarkable admission from a recognised authority on the physical side' (Geikie 1895, p. 723).

Conclusion: 'Had Lord Kelvin a right?'

Although the scientific exchange on the age of the Earth is generally referred to as a 'debate' or 'controversy', the communications between John Perry and Lord Kelvin do not have to be seen in such a light. To the extent that there was common ground between them, in mathematical and experimental physics, their opinions did not differ very much. Perry saw himself as working very much within a framework established by Kelvin, and he hoped to collaborate with his old teacher, rather than challenging him openly. Indeed, he originally took up the question of the age of the Earth in part to save the principles of physics from Salisbury's invocation (using Kelvin's own text) of divine intervention. On the other hand, to the extent that Perry and Kelvin did not see eye to eye, the issues were not technical details, but broader philosophical questions of probability versus possibility, and of the authority of one science over another.

Previous commentators have noted that these philosophical differences 'meant that no ultimate reconciliations could, or indeed need,

---

[30] CUL/KP, G48, Geikie to Kelvin, 25 February 1898.

occur between pupil and master' (Smith & Wise 1989, p. 607). Rather than seeking closure, Perry and Kelvin aimed simply to win as many adherents as possible to their divergent positions. The issue of who 'won' was a matter of perspective: to physicists and mathematicians who followed Tait's reasoning, Kelvin's arguments had not been overturned, since he had been able to defend his assumptions with experimental data. To geologists like Geikie and Judd, though, and to proponents of natural selection like Poulton and Meldola, Perry had shown that much longer ages could be equally calculated, and so these scientists could make their own estimates based on their own evidence. Ultimately, the side one took on the age of the Earth was a function of whether one thought that Kelvin's physics should have authority over the science of geology.

In understanding this question, however, it is important to remember that the boundaries of scientific disciplines evolve over time, and that 'physics' and 'geology' are negotiated rather than natural categories. For example, when Huxley admonished Kelvin in 1869 for interfering in geological matters which were outside his domain, Kelvin replied with astonishment:

> Who are the occupants of 'our house,' and who is the 'passer-by'? Is geology not a branch of physical science? Are investigations experimental and mathematical, of underground temperature, not to be regarded as an integral part of geology? .... For myself, I am anxious to be regarded by geologists, not as a mere passer-by, but as one constantly interested in their grand subject, and anxious, in any way, however slight, to assist them in their search for truth (Thomson 1871b, pp. 232–3; re: Huxley 1869).

Not everyone agreed with Kelvin's assessment of the subordinate status of geology to physics, however. Perry repeatedly stated his belief in the independence of geological knowledge, saying that it was '*impossible for a physicist*' to obtain a correct estimate of the age of the Earth (Perry 1895c, p. 583). For him, 'the real question' was 'Had Lord Kelvin a right to fix 10^8 years, or even 4×10^8 years, as the greatest possible age of the Earth?' (Perry 1895a, p. 227). Ironically, in thus defending the rights of geologists, Perry further contributed to the reification of the distinction between 'physicist' and 'geologist.' This was despite the fact that Perry and his allies show that not all 'physicists' supported Kelvin, just as there were geologists who did endorse shorter ages of the Earth. The division of scientists into one camp or another was a matter of politics, as the confrontational language often employed by Geikie and others indicates.

From the beginning, as in Salisbury's address, the main scientific question was political: the validity of natural selection as the mechanism of biological evolution. Participants' pre-existing commitments to views on this matter were arguably the main reason why the exchange between Perry and Kelvin did not seem to change very many minds. Perry based his personal position on a strong belief in the independence of geological evidence from theoretical conjectures, and an even firmer conviction that there was no place in any kind of science for metaphysical appeals to divine action. However, it should not be overlooked that the political dimension of this affair very nearly extended to the national stage as well, on which Kelvin had recently played an important role, much to Perry's consternation.

In a letter to Lodge, written in June 1894, just before Salisbury's address, Perry complained privately about scientists who abused their cultural authority by interfering in extra-scientific affairs: 'Who would pay the slightest attention to Lord Kelvin on politics if he were not what he is in Math. Physics. He commits the unpardonable sin when he talks of politics'.[31] What was this political role of Kelvin, to which Perry referred? It concerned the debate on 'Home Rule' for Ireland, which was the most important issue in mid-1890s Britain. Governments stood and fell on their handling of this question, in which Kelvin, Irish-born himself, took a keen and active interest. Smith and Wise have shown convincingly that William Thomson played a 'powerful role' in promoting Liberal Unionist opposition to Home Rule, concluding that '[w]ithout such an active political interest, Sir William Thomson would not have become Lord Kelvin' (Smith & Wise 1989, p. 803). Interestingly, what has not been emphasized before, is that the Conservative prime minister who raised Kelvin to the peerage in 1892 was none other than the Marquis of Salisbury. Two years later, Salisbury was quoting Kelvin approvingly in his address to the BAAS, thus completing the exchange between science and politics.

It seems clear that this political link was obvious at least to Lodge, who was a resident of Liverpool, a city with a keen awareness of Irish issues. Presumably recalling Perry's complaint about Kelvin's political activities a few months earlier, Lodge teased the Irish-born Perry about having a political motivation of his own in challenging Kelvin's calculation of the age of the Earth. Perry strongly denied any such bias, replying indignantly: 'As for 'political interest'. [Lodge's phrase] However one may joke ... I should hold myself really disgraced if I let my political feelings influence me, even a very little, in bringing forward a scientific matter'.[32] Although it is further interesting to note that the first recipients of Perry's calculations, Larmor and Fitzgerald, were also Irish, the situation is complicated by the fact that both physicists shared Kelvin's opposition to Home Rule. Larmor eventually became a Conservative member of parliament, while Fitzgerald, Perry's greatest supporter, was the son of a bishop in the Church of Ireland (Hunt 1991).

So while it cannot be suggested that the Home Rule question influenced which side participants took in the debate on the age of the Earth, the fact remains that it did play a role in *precipitating* Perry's contribution, which itself was an important nucleus for discussion among the generation of scientists who succeeded Darwin, Lyell and Huxley. Because of the connection between Salisbury and Kelvin, and specifically because the politician quoted the scientist not only on the age of the Earth, but also on the plausibility of divine providence in biological development, Perry was induced to enter the debate for the first time in 1894, overcoming a previous reluctance (Poulton 1896). Because of Perry's refusal to let Salisbury's characterization of Kelvin as unscientific (in his disciple's eyes, at least) stand, the age of the Earth became an explicit topic of debate once again, as long-simmering tensions broke through. Although this episode may seem minor in the overall story of the age of the Earth, it is valuable because it provides an interesting glimpse of what the problem had come to mean at the end of the nineteenth century, in the final years before the discovery of radioactivity redefined the age of the Earth and thus (along with the rediscovery of Mendelian genetics) the operation of natural selection. The question that Perry felt most deeply, and which inspired him to act, was essentially political: 'Had Lord Kelvin a right?' It was an equally pressing question for geologists and evolutionary biologists of the 1890s, in whom Perry found a grateful and enthusiastic audience.

S. Hong read the first version of this paper in 1996, and made many helpful observations. A later version was

---

[31] UCL/LP, Perry to Lodge, 25 June 1894.

[32] UCL/LP, Perry to Lodge, 31 October 1894.

presented to the 1998 Joint Atlantic Seminar in the History of Biology, at Johns Hopkins University. I am grateful to H. Gay for informing me of the documents relating to Perry and the age of the Earth in the Meldola papers. G. Gooday and B. Hunt also contributed timely and valuable suggestions. And I am pleased to thank J. Morrell and P. W. Jackson, who reviewed the manuscript for the Geological Society, for their supportive guidance. All errors are of course my own. Unpublished archival material is quoted with the kind permission of the following institutions: the Syndics of Cambridge University Library (Kelvin papers); University College, London, Library Services (Lodge papers); and the Archives of Imperial College of Science and Technology, London (Meldola papers and Huxley papers). My research has been generously supported at the University of Toronto by a Connaught Scholarship and an Ontario Graduate Scholarship, and at Dalhousie University by the Killam Trust, the Social Sciences and Humanities Research Council of Canada, and the University's Research Development Fund.

References

ARMSTRONG, H. E. 1920. Prof. John Perry, F.R.S. *Nature*, **105**, 751–753.

AYRTON, W. E. & PERRY, J. 1878. Experiments on the heat-conductivity of stone, based on Fourier's 'Théorie de la Chaleur'. *Philosophical Magazine*, Series 5 **5**, 241–267.

BROCK, W. H. 1996. *Science for All: Studies in the History of Victorian Science and Education*. Variorum, Aldershot.

BURCHFIELD, J. 1975. *Lord Kelvin and the Age of the Earth*. Science History Publications, New York.

EVERETT, J. D. 1886. *Units and Physical Constants*. Macmillan, London.

FISHER, O. 1889. *Physics of the Earth's Crust* (second edition). Macmillan, London.

GEIKIE, A. 1893. President's Address. *Report of the British Association for the Advancement of Science, Edinburgh 1892*, 3–26.

GEIKIE, A. 1895. Twenty-five years of geological progress in Britain. *Nature*, **51**, 367–370.

GEIKIE, A. 1900. Presidential Address, Section C – Geology. *Report of the British Association for the Advancement of Science, Dover 1899*, 718–730.

GOODAY, G. J. N. 1995. The morals of energy metering: constructing and deconstructing the precision of the Victorian electrical engineer's ammeter and voltmeter. *In*: WISE, M. N. (ed.) *The Values of Precision*. Princeton University Press, Princeton, 239–282.

HEAVISIDE, O. 1899. *Electromagnetic Theory*, vol. 2. The Electrician, London.

HUNT, B. J. 1991. *The Maxwellians*. Cornell University Press, Ithaca.

HUXLEY, T. H. 1869. President's Annual Address. *Quarterly Journal of the Geological Society of London*, **25**, xxxviii–lii [Reprinted in: HUXLEY, T. H. 1898. *Discourses Biological and Geological* (Collected Essays, vol. 8). Appleton, New York, 'Geological Reform', 305–339].

KNUDSEN, O. 1985. Mathematics and physical reality in William Thomson's electromagnetic theory. *In*: HARMAN, P. M. (ed.) *Wranglers and Physicists: Studies on Cambridge Physics in the Nineteenth Century*. Manchester University Press, Manchester, 149–179.

NAHIN, P. J. 1985. Oliver Heaviside, fractional operators, and the age of the Earth. *Institute of Electrical and Electronics Engineers Transactions on Education*, **E-28**, no. 2 (May), 94–104.

PERRY, J. 1895*a*. On the age of the Earth. *Nature*, **51**, 224–227.

PERRY, J. 1895*b*. On the age of the Earth. *Nature*, **51**, 341–342.

PERRY, J. 1895*c*. The age of the Earth. *Nature*, **51**, 582–585.

PERRY, J. 1900*a*. The teaching of mathematics. *Nature*, **62**, 317–320.

PERRY, J. 1900*b*. *England's Neglect of Science*. Unwin, London.

POULTON, E. B. 1896. Presidential Address, Section D – Zoology: A naturalist's contribution to the discussion upon the age of the Earth. *Report of the British Association for the Advancement of Science, Liverpool 1896*, 808–828.

SALISBURY, R. C. 1894. President's Address. *Report of the British Association for the Advancement of Science, Oxford 1894*, 3–15.

SMITH, C. & WISE, M. N. 1989. *Energy and Empire: A Biographical Study of Lord Kelvin*. Cambridge University Press, Cambridge.

THOMSON, W. (LORD KELVIN). 1864. On the Secular Cooling of the Earth. *Transactions of the Royal Society of Edinburgh*, **23**, 157–170 [Reprinted in: KELVIN, 1890. *Mathematical and Physical Papers*, vol. 3. Cambridge University Press, London, 295–311].

THOMSON, W. (LORD KELVIN). 1871*a*. On Geological Time. *Transactions of the Geological Society of Glasgow*, **3**, 1–28 [Reprinted in: KELVIN, 1894. *Popular Lectures and Addresses*, vol. 2. Macmillan, London, 10–72].

THOMSON, W. (LORD KELVIN). 1871*b*. Of Geological Dynamics, *Transactions of the Geological Society of Glasgow*, **3**, 215–240 [Reprinted in: KELVIN, 1894. *Popular Lectures and Addresses*, vol. 2. Macmillan, London, 73–131].

THOMSON, W. (LORD KELVIN). 1872. President's Address. *Report of the British Association for the Advancement of Science, Edinburgh 1871*, lxxiv–cv [Reprinted in: KELVIN, 1894, *Popular Lectures and Addresses*, vol. 2. Macmillan, London, 132–205].

THOMSON, W. (LORD KELVIN). 1895. The age of the Earth. *Nature*, **51**, 438–440.

THOMSON, W. (LORD KELVIN) & MURRAY, J. R. 1895. On the temperature variation of the thermal conductivity of rocks. *Nature*, **52**, 182–184.

TURNER, H. H. 1926. John Perry – 1850–1920. *Proceedings of the Royal Society of London*, **A111**, i–vii.

WEBER, R. 1895. Letter to the editor. *Nature*, **52**, 458–459.

John Joly (1857–1933) and his determinations of the age of the Earth

PATRICK N. WYSE JACKSON

Department of Geology, Trinity College, Dublin 2, Ireland (e-mail: wysjcknp@tcd.ie)

Abstract: John Joly (1857–1933) was one of Ireland's most eminent scientists of the late nineteenth and early twentieth centuries who made important discoveries in physics, geology and photography. He was also a respected and influential diplomat for Trinity College, Dublin, and various Irish organizations, including the Royal Dublin Society. Measuring the age of the Earth occupied his mind for some considerable time – a problem he was to address using a diverse range of methods. His sodium method of 1899, for which he is best known, was hailed by many as revolutionary, but it was later superseded by other techniques, including the utilization of radiometric dating methodologies. Although Joly himself carried out much research in this area, he never fully accepted the large age estimates that radioactivity yielded. Nevertheless, Joly's work in geochronology was innovative and important, for it challenged earlier methods of arriving at the Earth's age, particularly those of Lord Kelvin. Although his findings and conclusions were later discredited, he should be remembered for his valuable contribution to this important and fundamental debate in the geological sciences.

John Joly was born on 1 November 1857 in Hollywood House (the Rectory), Bracknagh, County Offaly, the third and youngest son of the Reverend John Plunket Joly (1826–1858) and Julia Anna Maria Georgina née Comtesse de Lusi. The Joly family originated from France, but came to Ireland from Belgium in the 1760s. Joly's great-grandfather served as butler to the Duke of Leinster who gave the living of Clonsast parish to the family in the early 1800s (Dixon 1934; Nudds 1986).

After his father's sudden death at a young age, the Joly family moved to Dublin where John Joly received his secondary education at the celebrated Rathmines School in which he was enrolled from 1872 to 1875. Although he did not excel in the classroom, he was popular, nevertheless, and became known as 'The Professor' on account of his tinkering with chemical apparatus and other gadgets. In 1875 after a bout of poor health Joly spent some time in the south of France, before returning to Dublin in 1876. He then entered Trinity College, Dublin, where he remained for the rest of his life. He followed courses in classics and modern literature, but later concentrated on engineering. In 1882 he sat for the degree of Bachelor of Engineering and gained first place and certificates in several subjects including engineering, experimental physics, mineralogy, geology and chemistry.

In adulthood Joly was a distinctive and unforgettable man. Tall, with hair swept off his forehead, a bushy moustache, and pince-nez perched on his nose (Fig. 1), he spoke with what was considered to be a foreign accent, but in reality the rolled r's were simply used to conceal a slight lisp (Dixon 1941). He was a keen traveller and made many trips, especially to continental Europe and the Alps, with his life-long friend Henry Horatio Dixon (1869–1953), one-time professor of botany at Trinity College, Dublin. Joly loved the sea and was a notable yachtsman who made frequent voyages along the western seaboard of the British Isles. He also served as a Commissioner of Irish Lights and carried out some work for the Admiralty on signalling and safety at sea (Nudds 1986, 1988).

Academic career

Although Joly considered a career abroad, he acceded to his widowed mother's wishes and remained in Ireland, at Trinity College, Dublin. There he was employed to carry out some teaching and to assist in the research of the professor of civil engineering. During this time (1882–1891) he was able to engage in a great deal of research, although this was largely in aspects of physics and mineralogy, rather than engineering. As with most experimental research of this period, apparatus did not exist, or was not available, and so had to be invented and built by the researcher himself. Among the first pieces invented by Joly were a new photometer and a hydrostatic balance.

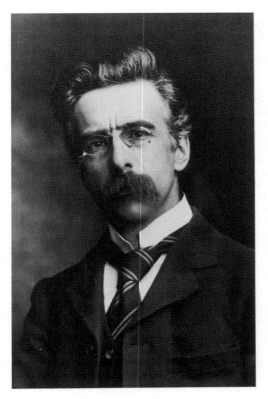

Fig. 1. John Joly (1857–1933), professor of geology and mineralogy, Trinity College, Dublin. Photograph taken May 1901 (courtesy of the Department of Geology, Trinity College, Dublin).

It was during this time that he developed an interest in mineralogy and began to accumulate a large collection of very fine Irish, continental and American mineral specimens (Wyse Jackson 1992).[1] He invented the steam calorimeter for measuring the specific heat of minerals. At the same time he worked on the density of gases and the detection of small pressures, and the steam calorimeter later played an important role in the kinetic theory of gases (Somerfield 1985).

In 1891 Joly was appointed assistant to George Francis Fitzgerald, professor of natural and experimental philosophy, and the following year he was elected a Fellow of the Royal Society (of London), an indication of how important his contemporaries considered his work to be. Surprisingly, he was not admitted as a member of the Royal Irish Academy for another five years. Over the ensuing years he addressed a tremendous range of scientific subjects. In 1894,

together with Dixon, he published an important paper that explained for the first time the mechanism of the ascent of sap in trees and plants, and in 1895 Joly exhibited colour slides to the Royal Society. His method, the 'Joly process of colour photography', was the first successful process of producing colour photographic images from a single plate. However, his assertion that he had invented the technique was challenged by an American. Legal battles ensued but although Joly's priority was finally established, the process was soon superseded by other methods. Nevertheless, the 'Joly process' is important as it is essentially the method used in colour photography today.

In 1897 Joly succeeded William Sollas in the chair of geology and mineralogy at Trinity College, a position he retained until his death in 1933. Unlike his predecessors Joly was essentially a physicist, not a geologist, and consequently he published little on the geology of Ireland. He carried out much research on minerals, but it was his work on radioactivity and radium that was his most important. This led to the establishment of the Irish Radium Institute in 1914, which exploited the medical advantages of the radioactive element.[2] But Joly also worked on determining the age of the Earth, and it is for his contributions to this debate that he is now best remembered.

Joly's sodium method

Joly's contribution to the question of the age of the Earth was one of a number made by Irish-born or Irish-based academics (O'Donnell 1984; Wyse Jackson 1992) (Table 1). The first was that of Archbishop James Ussher, who in the seventeenth century arrived at a date of 4004 BC from a reading of biblical events (Reese et al. 1981; Brice 1982; Fuller 2001). Others included Belfast-born William Thomson (later Lord Kelvin) who argued, based on the cooling rate of an initially molten planet, that the Earth was between 20 and 400 million years old (Thomson 1864). However, his last paper on the subject, published in 1899, firmly placed the limits between 20 and 40 million

---

[1] The collection is still on display in the Geological Museum, Trinity College, Dublin.

[2] In Ireland Joly pioneered the medical benefits of radioactivity. With a Dublin doctor, Water Clegg Stevenson (1876–1931), Joly established the Radium Institute at the Royal Dublin Society in 1914 whose work was in treating cancer patients (Joly 1931). Their 'Dublin method' was the first to utilize radium emanation (radon) enclosed in hollow needles in the treatment of tumours (Murnaghan 1985). Joly was very proud of this work.

Table 1. *Estimates of the age of the Earth made by various Irish-born or Irish-based* scientists*

Author	Age determination	Date	Method
James Ussher	4004 BC	1650	Biblical chronology
William Thomson (Lord Kelvin)	20–400 Ma	1864	Cooling Earth
Samuel Haughton	2298 Ma	1865	Cooling Earth
Samuel Haughton	1526 Ma	1871	Sediment accumulation
Samuel Haughton	200 Ma	1878	Sediment accumulation
William Sollas*	17 Ma	1895	Sediment accumulation
William Thomson (Lord Kelvin)	20–40 Ma	1899	Cooling Earth
John Joly	90–100 Ma	1899	Sodium accumulation
John Joly and Ernest Rutherford	20–470 Ma (Devonian)	1913	Pleochroic halos
John Joly	47–188 Ma	1915	Sediment mass
John Joly	300 Ma	1923	Thermal cycles
John Joly	300 Ma	1930	Sodium accumulation
Edward J. Conway	700–2350 Ma	1943	Ocean chemistry

years (Thomson 1899), a value that antagonized many geologists (Dalrymple 2001; Shipley 2001). Samuel Haughton and William Sollas, Joly's predecessors at Trinity College, Dublin, also entered the geochronological debate, and their contributions are discussed below.

Joly's first documented thoughts on the antiquity of the Earth were in written verses penned on 28 August 1886 in the Wicklow Mountains south of Dublin (Nudds 1983). One sonnet considered the age of the enigmatic trace fossil *Oldhamia antiqua* Forbes from the Cambrian slates of Bray Head, which Joly suggested was a witness to the long, slow changes that had affected the Earth:

> Is nothing left? Have all things passed thee by?
>   The stars are not thy stars! The aged hills
>   Are changed and bowed beneath repeated ills
> Of ice and snow, of river and of sky.
> The sea that raiseth now in agony
>   Is not thy sea. The stormy voice that fills
>   This gloom with man's remotest sorrow shrills
> The memory of the futurity!
>   We – promise of the ages! – Lift thine eyes,
>   And gazing on these tendrils intertwined
> For Aeons in the shadows, recognize
>   In Hope and Joy, in heaven-seeking Mind,
>   In Faith, in Love, in Reason's potent spell
> The visitants that bid a world farewell!

Joly's first scientific foray into the matter of geochronology came thirteen years later with the publication of his first and probably most celebrated, if somewhat controversial, paper on the subject (Joly 1899). Simply put, Joly examined the rate of sodium input into the oceans and by simple mathematics arrived at an estimate for the age of the Earth. The idea came to him whilst sailing off the east coast of Ireland with Henry Dixon in 1897, collecting coccoliths and plankton from the Irish Sea (Dixon & Joly 1897, 1898) and observing the feeding behaviour of seabirds (Joly 1898).[3]

At that time Joly was unaware, as were his contemporaries, of the pioneering work of the English astronomer Edmond Halley (Cook 1998). In 1715 Halley had proposed to the Royal Society that salt concentrations in lakes that had no discharge rivers, should be measured every 100 years, as he considered that from the incremental increase of the salt, the age of the lake could eventually be deduced. Once enough data had been collected over time, inferences about the age of the ocean, and therefore of the Earth, could be drawn from the results. While Halley accepted that mankind had dwelt upon the Earth for about 6000 years, as stated in the Scriptures, he also considered that it was 'no where revealed in Scripture how long the Earth had existed before this last Creation' (Halley 1715, p. 296). He thus regretted that the ancient Greek and Latin authors had not recorded the saltiness of the sea 2000 years ago, since 'the World may be found much older than many have hitherto imagined' (Halley 1715, p. 299). He therefore recommended to the Society that experiments be started 'for the benefit of future Ages'. But the Society does not seem to have heeded his advice and Halley's idea was only rediscovered in 1910 (Becker, 1910a).

Although aware of Mellard Reade's important book of 1879, Joly was unaware of his valuable paper published in the *Proceedings of the Liverpool Geological Society* in 1877, which examined the volume of sulphates, carbonates and chlorides in the oceans, and their rate of

---

[3]Joly had bought a yacht, *Gweneth*, soon after his appointment to the chair of geology and mineralogy, which carried a salary of £500 per annum.

accumulation. Using this information Reade calculated the time taken for the oceans to reach their present concentration of these substances: 25 million years, nearly half a million years, and 200 million years respectively for sulphates, carbonates and chlorides. These figures gave an estimate of the minimum age of the Earth (Mellard Reade 1877, p. 229). Although Joly later acknowledged these pioneering publications of Halley and Mellard Reade he noted, without explanation, that their schemes, unlike his, would not have produced reliable results (Joly 1915).

Brilliant in its simplicity, Joly's paper of 1899 fired the imagination of both scientific and general audiences, and for perhaps a decade his 'sodium method' held sway amongst geochronologists. It relied upon a number of assumptions, and on data regarding ocean and river characteristics published by John Murray in the 1880s. The fundamental tenet was the uniformitarian stance that the rates of denudation of sodium-bearing rocks, and the discharge of the rivers into the oceans, had remained uniformly constant over geological time. The age of the Earth was thus derived by the simple formula:

$$\text{Earth's age} = \frac{\text{Volume of sodium in the ocean}}{\text{Rate of annual sodium input}}$$

which yielded an age of 90–100 million years.

Joly read his paper to the Royal Dublin Society at a meeting in Leinster House on 17 May 1899, and it was published several months later in September in the *Scientific Transactions*, the premier journal of the Society. Reactions to his ideas began to appear in the scientific press within six months. Review articles were published in several journals, including the *American Journal of Science* (Anon. 1899) and *Geological Magazine* (Fisher 1900), but the Reverend Osmund Fisher's review was by far the most testing, because, he argued, the processes invoked by Joly were *not* uniform throughout geological time. Additionally, Fisher suggested that Joly's figures for the volume of sodium delivered into the oceans by rivers might be at fault, and that Joly did not take into account the effect of 'fossil sea water' which, Fisher noted, was trapped in sediments and elsewhere. Fisher also suggested that Joly's estimate, that 10% of the sodium chloride came from rainwater, was too high. Finally he made a little swipe at Joly's written style which he contended was rather convoluted and not very clear. He has a point. These minor criticisms aside, Fisher recognized that Joly's essay had 'opened up an entirely new line for the investigation of geological time' (Fisher 1900, p. 132).

William Ackroyd, Public Analyst for Halifax, was of the opinion that a great deal of the oceanic salt was transported back onto land; he put this figure at 99%, and on the basis of this recalculated the age of the Earth and derived an age of 8000 million years (Ackroyd 1901*a*). This was the first offering in a public debate on Joly's theory played out in the pages of *Chemical News* and *Geological Magazine* (Ackroyd 1901*a, b, c, d*; Joly 1901*b, c*). Ackroyd argued that it was important to know the ratio in river water of transported (or recycled) sea salt and that derived from rocks through solvent denudation. He included data that demonstrated that only 0.02% of the chlorides in the water of Malham Tarn near Craven were derived from the surrounding limestones. From this one must assume he considered that most were derived from atmospheric water. In his papers in *Geological Magazine* (Ackroyd 1901*c, d*) he criticized Joly for not appreciating the significance of this work or for ignoring it altogether.

At the British Association for the Advancement of Science meeting held in 1900 at Bradford, Joly's findings must have created quite a stir as his report was ordered by the general committee to be published *in extenso* (Joly 1900*c*). William Sollas, in his presidential address to Section C – Geology, sided with Joly's finding, stating that 'there is no serious flaw in the method, and Professor Joly's treatment of the subject is admirable in every way' (Sollas 1900). Sollas, however, did question the reliability of the data concerning the river discharge of sodium (Sollas 1900). Later papers by Sollas (1909), Becker (1910*b*) and Clarke (1910) laid minor criticisms at Joly's door. Sollas suggested that some modification could be made for the fact that volcanic activity was at certain times in the past more pronounced that at present times, and that this would have had some effect on the supply of sodium to the oceans. In addition he argued that modern ocean water, at normal temperature, had only a slight corrosive effect on salt contained in rocks. At higher temperatures of between 180°C and 370°C far higher volumes of sodium would have been dissolved. From this we surmise that Sollas believed that the early oceans were considerably hotter than those of today. Sollas recalculated the annual discharge of the rivers from which he derived a date of 78 million years for the age of Earth, but suggested that it lay within the range 80–150 million years. Clarke (1910) examined the rate of removal of sodium from the landmass and arrived at a figure of 80 million years, while Becker (1910c) suggested that Joly's figure of 10% for the contribution of sodium recycled from the atmosphere was closer to 6%. Subsequently, a considerable number of papers also discussed Joly's sodium method (for example, Rudzki 1901) which were either in broad agreement with his ideas (e.g. McNairn 1919) or raised a number of objections (see for example Shelton 1910, 1911; Holmes 1913). Shelton said that while Joly's scheme was instructive, some of the underlying foundations on which it was based needed careful consideration. Joly, Shelton argued, did not make adequate allowance for salt recycled via the atmosphere, nor for fossil salt contained in sedimentary rocks. Shelton also noted that Joly's age determination was based on analyses of sodium content, which given the accuracy of instrumentation available could not be done with any great accuracy. In particular, Arthur Holmes reasoned that the rocks would have had to lose more sodium into the oceans than they had ever contained, for Joly's figures to add up.

Joly's responses to these criticisms (Joly 1900b, 1901b, c) strongly reinforced his uniformitarian principle; however, he did accept that his estimate of the role of rainwater in providing sodium chloride might have been underestimated and required further experimental work. With respect to fossil seawater Joly stated that it could only have contributed 0.9% of oceanic sodium chloride and, as such, was negligible. His response to Sollas was that 'there is much reason to believe that the nineteen rivers ... afford an approximation as to what the world's rivers yield' (Joly 1900c). Indeed, he stated in 1911 that the findings of Sollas, Becker and Clarke, together with his own, gave concurrent results of circa 100 million years, and proudly anticipated that this determination would not be 'seriously challenged in the future' (Joly 1911a).

Joly went further in defence of his ideas, in that he devised various experiments which he hoped would generate acceptance of some of his theoretical assumptions made in 1899. One of these was to invent a fractionating rain-gauge (Joly 1900a), which he hoped would allow him to collect rainwater over incremental time periods. Subsequent analysis of the amount of dissolved sodium chloride in this rainwater would enable him to quantify the volume of sodium in the oceans from this pluvial source. While demonstrating how his rain-gauge would operate, Joly fails to record whether it was ever put to use or was effective!

Joly also examined the rate of solution of various igneous materials in fresh and salt water (Joly 1900d) which showed that of the four tested (basalt, orthoclase, obsidian and hornblende) the basalt from the Giant's Causeway in County Antrim dissolved more readily than the others, and that salt water was a more effective solvent than fresh water. Not surprisingly, the obsidian proved the most resistant to solution. Joly noted that his results for the rates of denudation were far lower than those demonstrated by field study and he argued that additional factors such as organic acids, wetting and drying, and other erosive processes were more important than solution of rocks by water. Nevertheless, he made an allowance for the solvent action of the ocean, by reducing his age estimate by a few million years to 96 million (Joly 1911a).

After the initial peak of interest that closely followed on from his 1899 paper, many of Joly's subsequent papers on the subject were simply reports of lectures, or reiterations of the original theory. In 1915 with the publication of *Birthtime of the World* (Joly 1915) there followed renewed interest in the sodium method – but it did not last. The theory was finally consigned to the scientific scrap-heap by several geologists (Harker 1914; Barrell 1917; Gregory 1921; Chamberlin 1922) who, in the damning words of Arthur Holmes, 'rejected it as worthless' (Holmes 1926, p. 1056). By the mid-1920s the scientific community was focused on the new theories based around radioactivity (Lewis 2001). Paradoxically Joly also carried out much useful research in this developing area, but he himself could never consign his sodium method to the waste basket, although by 1930 he had accepted some major modifications suggested by A. C. Lane in 1929, which pointed at a figure of 300 million years for the method (Joly 1930).

Unusually, and perhaps uniquely for publications of the Royal Dublin Society, a second impression of the original paper had to be produced in November 1899 as all the stocks had been distributed and demand continued. This allowed Joly to rectify a number of small errors that had appeared in the appendix of the first impression. The paper was also printed in North America in its entirety in the *Annual Report of the Smithsonian Institution* for 1899, and so Joly's ideas and methodologies were rapidly transmitted throughout the scientific community on either side of the Atlantic (Joly 1901a). It was an important contribution to the growing body of scientific opinion that refuted the low estimate of the age of the Earth of 20–40 million years propounded by Kelvin (see, for example, Shipley 2001), which was eventually dispelled with the advent of radioactive dating methods.

It is interesting to note that it is now thought that on the basis of the volume of chloride contained in brines found in deep-seated groundwater, the salinity of the earliest oceans is considered to be 1.5 to 2 times saltier than that of today (Knauth 1998a, b). Thus the oceans are not becoming progressively saltier through the release of sodium and chloride during denudation, as was Joly's contention. Quite the contrary, Knauth (1998a) argued that if all the present deposits of subsurface salt were returned to the oceans, they would be 30% saltier than at present.

Sediment accumulation

At the same time as the 'sodium method' was gaining acceptance, others continued to estimate the age of the Earth using sediment accumulation as their gauge. This method owed its origins to the work of John Phillips in 1860 (Phillips 1860; Morrell 2001) who determined the thickness of the global sedimentary pile and, using a figure for the rate of sedimentation, arrived at 96 million years for the age of the Earth. Many other calculations using this method followed (see Lewis 2001), perhaps the most celebrated of which were those of the Reverend Samuel Haughton, third professor of geology at Trinity College, Dublin. Haughton, who was a supporter of Kelvin's methods for estimating the age of the Earth and an opponent of Charles Darwin's theory of evolution, initially achieved an astonishing estimate of 2298 million years, based on the same principles as those used by Kelvin (Haughton 1865). In the context of sediment accumulation, however, Haughton is remembered for his principle that 'the proper relative measure of geological periods is the maximum thickness of the strata formed during these periods' (Haughton 1878), which of course necessitated a global estimate of the maximum thickness of each sedimentary sequence, and the determination of the rate at which those sediments accumulated. Using this principle he first published an age for the Earth of 1526 million years (Haughton 1871), but uncomfortable with the vast timescales he was deriving, which were in stark contrast to those deduced by Kelvin, seven years later he attempted the calculation again – and arrived at much the same answer. This time, however, he conceded: 'If we admit (which I am by no means willing to do) that the manufacture of strata in geological times proceeded at *ten times* this rate, or at the rate of one foot for every 861.6 years ... This gives for the whole duration of geological time a *minimum of two hundred million years*' (Haughton 1878, p. 268), a value much more acceptable to the wider geological community. Much later Arthur Holmes argued that the uniformity of sedimentation rates assumed by Haughton was incorrect and that his principle would be better stated as 'the time elapsed since the end of any geological period is a function of the sum of the maximum thicknesses accumulated during all the subsequent periods.' (Holmes 1947, p. 119).

By the 1890s the methodology had become the standard means of measuring the age of the Earth and a great many authors attempted it, including Charles Doolittle Walcott in the United States who derived a date of 35–80 million years based on measured sections in North American sedimentary basins (Walcott 1893), and Mellard Reade in England who estimated that 'The time that has elapsed since the commencement of the Cambrian is therefore in round figures 95 millions of years' (Mellard Reade 1893, p. 100). In 1895 William Johnston Sollas, fifth professor of geology at Trinity College, following Haughton's principle, calculated the Earth's age to be 17 million years (Sollas 1895), one of the lowest figures ever established. However, a later calculation, based on a total sediment thickness of 335 000 feet or 63 miles, and sedimentation rates of 3 and 4 inches per century resulted, respectively, in ages of 148 and 103 million years (Sollas 1909). Sollas noted though, that it was difficult to determine accurately the rate of sediment accumulation, which he acknowledged could be anything between 2 and 12 inches per century.

Joly was delighted since these dates largely concurred with his and thus confirmed the strength of his sodium method determination. In 1909 he examined Sollas' figures for himself and agreed with his results (Joly 1911a). However, in 1914, in a lecture to the Royal Dublin Society, he argued that sediment mass, not thickness, was a more accurate measure, which widened Sollas' results to a minimum of 47 million years and a maximum of 188 million years. Then, on the basis that he believed sedimentation rates were not uniform through geological time (Joly 1915), he reduced the mean of these limits, 117 million years, to a figure of 87 million years which concurred well with his sodium method results. This marked a change in the uniformitarian stance that he had adopted in 1899 for the rate of sodium accumulation in the oceans.

By 1910 these methods were being supplanted by the age determinations generated by radioactive decay methods. Although in its infancy, the study of radioactivity was beginning to yield ages for the Earth that were considerably older than 100 million years.

Radioactivity and pleochroic halos

Joly was just one of many scientists drawn into the field of radioactivity following its discovery by Henri Becquerel in 1896 and the subsequent discovery of radium by Marie and Pierre Curie in 1898. By 1904 Ernest Rutherford had already suggested that the presence of radioactive minerals might provide a measure of the age of those minerals (Rutherford 1905), and by 1905 Robert Strutt (later Lord Rayleigh) reported widespread presence of radioactive elements in rocks. Over the next five years he went on to examine helium as a means of determining ages. In the United States in 1907 Bertram Boltwood was the first researcher to examine the uranium–lead series (Badash 1968; Dalrymple 2001) and, using this method, in 1911 Arthur Holmes published a date of 1640 million years for the Archaean of Ceylon (Holmes 1911; Lewis 2001). Immediately Joly was on the back-foot defending his geochronology from these new data. In his book *Radioactivity and Geology* (Joly 1909c) and in an important paper entitled 'The Age of Earth' Joly (1911a) questioned whether the decay rate of uranium had been constant throughout geological history, as suggested by others. He said that this assumption was without strong basis and that the calculations of age limits were based on derived radioactive products, rather than radioactive parent elements. Turning to his own previous methods for support of his unease, Joly noted that the extremely slow rates of sediment accumulation implied by these vast ages were difficult to credit:

> If the recorded depths of sediments have taken 1400 million years to collect, the average rate has been no more than one foot in 4000 years! This seems incredible: and if we double the depth of maximum sedimentation it still remains incredible. But, if possible, still more incredible is the conclusion respecting solvent denudation to which radioactivity drives us. If the sodium in the ocean has taken 1400 million years to accumulate, the rivers are now bearing to the sea about 14 times the average percentage of the past ... It seems quite impossible to find any explanation of such an increase (Joly 1911a, p. 379).

Joly was not alone in his concerns – the American geochronologist George Becker also voiced unease with radiometric dates, as did the American Committee on the Measurement of Geological Time by Atomic Disintegration, which reported that uranium as a whole decayed more rapidly in the past, and therefore the dates derived from it could be overestimated by 25% (Lane 1925). Lane was in fact ahead of his time in recognizing that uranium might have several isotopes (not confirmed for another four years, and U^{235} was not discovered until 1936) and thus 'as a whole' (averaging all isotopes) uranium appeared to have decayed more rapidly in the past. U^{235} decays six times as fast as U^{238}. Holmes (1913), though, considered it highly unlikely that radioactive decay rates had varied through geological time: as unlikely as finding that the laws of physics and chemistry had changed over time!

Joly's first paper on radioactivity and geology was published in *Nature* in 1903 where he discussed the potential of using radium to date the age of the Earth (Joly 1903). In a number of later papers published in 1908 and 1909 he calculated the volume of radioactive elements, including radium and thorium, in terrestrial and oceanic rocks of various ages (Joly 1908b, 1909a),[4] and from seawater (Joly 1908a, 1909b). Much of this work received widespread release in his book *Radioactivity and Geology* (Joly 1909c), and the outcome of this study was formulation of his ideas pertaining to internal heat sources in the Earth, which had implications for a later geochronological method.

In 1907 Joly realized that small dark rings, or pleochroic halos as they came to be known, which he had observed in biotite in some granites, were the products of radioactive decay in zircons enclosed within the biotite crystals (Joly 1907a) (Fig. 2). Previously it had been suggested that these were due to the presence of organic pigments in the minerals, but Strutt had earlier demonstrated the radioactive properties of zircon, to which Joly attributed the halos. The size of the halo was related to the type of radioactive decay product and the range of the rays produced (Joly 1911b), while the intensity, he argued, was due to the duration of radioactive decay. He observed complex halos with distinctive inner and outer rings, or corona, in a greisen from Saxony in Germany (Joly 1910), and soon afterwards he and his research assistant, Arnold

---

[4] It would be interesting to re-examine Joly's rocks from the Simplon and St Gothard tunnels through the Alps (Joly 1907b, 1912), now housed in the Geological Museum in Trinity College, Dublin, to determine how accurate his results were. Do they match with modern calculations of the radioactive elements contained in these rocks? If not, what does this say about Joly's methodologies or the reliability of his equipment? Perhaps, as Léo Laporte has pointed out, any differences may result from the different half-lives then in use.

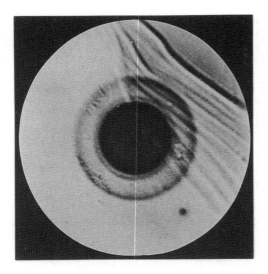

Fig. 2. A single pleochroic halo developed in biotite in the Leinster granite from Garryellen, County Carlow, Ireland. The inner dark disc is due to radon (a gas derivative of radium); the succeeding inner ring is due to radium A alpha decay while the outer darker ring was produced by radium C alpha decay (from Joly 1911b, plate III, fig. 4).

Lockhart Fletcher, attributed the development of the outer rings to the alpha decay of radium C (now known as the isotope bismuth^{214}) while the inner rings were produced by radium A (now known as the isotope polonium^{218}) (Joly & Fletcher 1910). Joly was also able to distinguish halos produced by other radioactive sources including thorium and uranium, and the radii of the halos produced by the thorium and uranium decay series were given in the Huxley Lecture that Joly delivered in Birmingham in October 1912 (Joly 1913).

In 1913 Joly and Ernest Rutherford developed a unique methodology to date a rock based on its pleochroic halos, which required knowledge of the mass of the nucleus of the halo and the number of alpha rays required to produce a certain intensity of halo. Using specimens from the Leinster Granite,[5] Rutherford, working in Manchester, produced artificial halos in mica and measured the number of alpha rays required to produce them. Meanwhile, in Dublin, Joly measured the mass of the nuclei. Between them they tabulated the ages of 30 halos which ranged from 20 to 470 million years. They concluded that the age of the Devonian was not less than 400 million years, which concurred with results found by Holmes two years earlier; however, a proviso

---

[5] County Carlow in Ireland.

read: 'that if the higher values of geological time are so found to be reliable, the discrepancy with estimates of the age of the ocean, based on the now well-ascertained facts of solvent denudation, raises difficulties which at present seem inexplicable' (Joly & Rutherford, 1913, p. 657).

Three years later Joly measured halos in younger rocks from the Vosges (Joly 1917), and later still he measured the radii of rare halos in the Tertiary granites from the Mourne Mountains. The radii of the latter were 7% smaller than those of the Leinster Granite, which themselves were 10% smaller than those he had recognized in Archaean rocks (Joly 1923b). Joly concluded in triumph: 'It would seem as if we might determine a geological chronology on the dimensions of these halo-rings!' (Joly 1922a).

No sooner had he come to this exciting conclusion than he discovered small halos in Archaean rocks from Norway. His first specimens of this material had been lost when his laboratory at Trinity College was occupied by troops during the Easter Rebellion in 1916 (Joly 1920). Six years later he received fresh material and discovered the small halos which he concluded represented full decay of a radioactive element that was no longer present in the Earth's crust. He called this new radioactive element 'hibernium' after his homeland (Joly 1922b) it was later found to be samarium.

In 1922, Joly reiterated his contention that radioactive decay rates were not constant throughout geological history. This was partially based on his observations of halos whose characteristics were not consistent. In particular, the innermost rings produced by uranium in some halos were not consistent with the known ionization curves of uranium alpha particles. He suggested that these rings were caused by the faster decay of uranium in the past, or by the decay of a uranium isotope that was no longer present. The fact that thorium halos of all ages were constant in size was also difficult to explain (Joly 1922a). Holmes (1926) argued that the inconsistencies of the uranium halos were due not to time but to other factors, including the presence of the recently discovered rare isotope, actinium. Nevertheless, he accepted that Joly's scheme of correlating halo radii with time would eventually give a scale against which the ages of other halos could be determined.

Subsequently, however, Joly's halo data were examined, and the accuracy of his measurements was questioned by Kerr-Lawson (1927) who was unable to detect the differences in radii that Joly had claimed. Nevertheless, his imaginative attempt to employ pleochroic halos as a measure for geological time was noteworthy.

Joly's objections to dates derived by radiometric methods were also rejected by the late 1920s when further information about isotopes, atomic weights and other radioactive elements became available (e.g. Holmes & Lawson 1927). Holmes, who was at the forefront of radiometric research (Lewis 2001), argued that, at worst, errors of a few percent might be attributed to the unranium–lead methodology (Holmes 1926).

Earth's surface history and thermal cycles

Joly's final contribution to the geochronological debate came about from his interest in isostasy, the internal heat of the Earth, and its signature on the surface of the Earth. Kelvin's ideas on a cooling Earth were long dismissed, and it had been recognized by Strutt and Joly that the Earth had an internal heat source, resulting from decay of the radioactive elements it contained. Joly's pioneering suggestion was that the internal heat of the Earth built up over a period of time (Joly 1909c).

By 1923 Joly had developed a theory in which he explained that the surface features seen on the Earth were formed as a direct result of its own internal heat source. This theory subsequently became known as the theory of 'thermal cycles' (Joly 1928). The heat source was responsible for the melting of the basaltic crustal horizons (termed the basaltic magmatic layer by Joly) that lay beneath the continental crust. The continents, which he described as 'granitic scum' (Joly 1925, p. 176), rested directly in isostatic balance on a basaltic magmatic layer that was up to 70 miles in thickness, and which enclosed the whole globe. The water that made up the oceans simply filled the voids between the continents. In a thermal cycle, he explained, this basaltic magmatic layer melted periodically due to the build-up of heat generated by radioactive decay. Melting led to changes in the volume of this horizon beneath the continents, which in turn generated tidal currents in the basaltic magma. Joly argued that such a change in volume could have resulted in an increase in the diameter of the Earth, and that the continents would have been isostatically buoyed up. This effect would have been negated by the resultant sinking of the continental masses due to the decreased density of the molten layers, and consequently there would have been widespread transgressive events.

What is interesting is what Joly speculated happened on cooling of the molten basaltic magmatic layer. Heat was lost as it migrated from beneath the continental areas to oceanic regions, from where it was largely lost into the oceans (Joly 1923a). Joly argued that the cooling of the melted level caused it to recrystallize and shrink, and that continents would return through a series of vertical movements to their former isostatic levels. He pointed out, quite correctly, that vertical movements were often associated with lateral or horizontal Earth movements. Both, he argued, were generated by the same thermal processes outlined above. But how did he explain the orogenesis that was also a feature of these complex processes? He postulated that the most lateral movement occurred when the basaltic level beneath the continents was in a molten state, at which time it would have had a similar density to the granitic continental masses above. Beneath the oceans thermal currents significantly reduced the thickness of the basaltic magmatic layer, which he deduced would have cracked, and which would have seen the injection of new basalt. The area of the ocean floor thus increased significantly, and this increase in area resulted in the generation of compressive forces which forced the oceanic floor to press against the margins of the continental crust. This initiated orogenesis or mountain building. Joly estimated that these orogenic events or 'revolutions', as they were called, took place once every 50 million years or so. He tabulated five or six revolutions of different ages which he correlated with various mountain orogenic belts. These included the Caledonian, Appalachian and the Alpine 'revolutions'. From the evidence of global tectonics, Joly considered that the Earth was 300 million years old.

Many of his ideas on global tectonics were articulated in his book *The Surface History of the Earth* (Joly 1925, 1930) and in his papers on thermal cycles (Joly 1928) and the Earth's surface structure (Joly & Poole 1927). Naturally he entered the debate on continental drift and argued that there was evidence that the continents drifted westwards. This delighted some Irish wags, who highlighted in the popular press this scientific proof that Ireland was moving further away from England.

Joly's ideas on global tectonics were typically complex and imaginative, and certainly deserve fuller examination and assessment – this is beyond the scope of this present paper – Greene (1982) has discussed the evolution of tectonic theory at length, from forerunners such as Elie de Beaumont and Eduard Suess, to those later geologists that included Joly, who worked on what Greene termed 'the fourth global tectonics'. Oreskes (1999) also analyses in depth Joly's contribution to the theory of continental drift.

Conclusion

As early as 1922 Joly noted that the expression the 'age of the Earth' was ambiguous (Joly 1922a), and it can now be seen that Joly's own position on this matter was also ambiguous. In 1930, in the second edition of his *Surface History of the Earth*, published three years before his death, he wrote: 'the age of the Earth is not the same as geological time' (Joly 1930). It is my contention that Joly understood 'age of the Earth' to mean the time since the onset of denudation and biological processes, while 'geological time' included older Archaean times. In essence, Joly was taking a philosophical stance which allowed his low age estimates to stand as correct estimates for the age of the Earth, yet at the same time he did not reject out of hand the accepted estimates of Holmes and others, derived by radiometric means, which gave a longer measure of geological time. Nevertheless, he remained sceptical of dates determined by radiometric methods.

In the light of his work on radioactivity it is somewhat surprising that Joly did not accept that his sodium method yielded erroneously low age estimates, or acknowledge that what it measured was not the age of the Earth, but an estimate of the age of the oceans. Although it is clear today that the oceans are older than Joly's method might suggest, his sodium method actually gives a good measure of the residence time of sodium in the oceans, which is approximately 260 million years (Mittlefehldt 1999). Up until his death Joly continued to hold the view that 300 million years was a good estimate of the age of the Earth.

Joly's age methodologies and breadth of research in geochemistry, mineralogy and radioactivity made a major contribution to the debate on the age of the Earth played out at the end of the nineteenth and beginning of the twentieth centuries. His sodium method held centre-stage among geologists and geochemists for some years, and stimulated much debate on the subject of the age of the Earth. Additionally, Joly's unique attempts to employ pleochroic halo radii as measures for geological time were extremely valuable and, in the light of the exciting avenues opening up in the field of radioactive research in the first decades of the 1900s, may well have appeared to have the distinct potential to estimate the actual age of the Earth. Even though it was subsequently demonstrated that the ratio between original radioactive elements and their decay derivatives yielded better estimates of the age of the Earth, Joly's many faceted investigations on the subject helped to stimulate research that ultimately led others to the geochronological methodologies in vogue today.

Joly spent all of his professional life working in Trinity College, Dublin, during which time he wrote 269 scientific papers and several books, including an autobiography. He was the recipient of many awards and had a high international reputation. He was a Fellow of the Royal Society which awarded him the Royal Medal in 1910, and he received the Boyle Medal of the Royal Dublin Society in 1911 and the Murchison Medal (of the Geological Society of London) in 1923. He became a Fellow of Trinity College in 1919 (the first person in 250 years who did not have to pass a rigorous examination) and was president of the Royal Dublin Society (1929–1932). He was conferred with the honorary degrees of Doctorate of Literature from the University of Michigan, and Doctorate of Science from the University of Cambridge and the National University of Ireland. Forty years after his death he was honoured by having a crater on Mars named after him (Batson & Russell 1995), which is appropriate given Joly's work on the nature and origin of Martian 'canals' (Joly 1897).

He was by all accounts a very popular man, loved and respected by many. On his death many friends contributed to a memorial fund that is still used to promote an annual lecture on a geological theme, which is usually given by a leading foreign authority. In addition his name and memory are perpetuated by the Joly Geological Society, the student geological association of Trinity College, Dublin, founded in 1960.

I thank C. Lewis for her kind invitation to present this paper at the meeting *Celebrating the Age of the Earth*, which formed the William Smith meeting at the Geological Society in London in June 2000. My research on John Joly has benefited from numerous discussions with colleagues; in particular I thank J. R. Nudds and G. L. Herries Davies. This paper has been improved by the comments of C. Lewis, S. Knell and the reviewers L. Badash and L. Laporte, for which I am most grateful.

References

ACKROYD, W. 1901a. On the circulation of salt and its bearing on chemico-geological problems, more particularly that of the age of the Earth. *Chemical News*, **83**, 265–268.

ACKROYD, W. 1901b. On the circulation of salt and its bearing on chemico-geological problems, more particularly that of the age of the Earth. *Chemical News*, **83**, 56–57.

ACKROYD, W. 1901c. On the circulation of salt in its relations to geology. *Geological Magazine*, New Series, Decade 4, **8**, 445–449.

ACKROYD, W. 1901d. On the circulation of salt in its relations to geology. *Geological Magazine*, New Series, Decade 4, **8**, 558–559.

ANON. 1899. Review of 'An estimate of the Geological Age of the Earth', by J. Joly, University of Dublin. *American Journal of Science*, **158**, 390–392.

BADASH, L. 1968. Rutherford, Boltwood, and the age of the earth; the origin of radioactive dating techniques, *Proceedings of the American Philosophical Society*, **112**(3), 157–169.

BARRELL, J. 1917. Rhythms and the measurements of geologic time. *Geological Society of America Bulletin*, **28**, 745–904.

BATSON, R. M. & RUSSELL, J. F. (eds) 1995. *Gazetteer of Planetary Nomenclature 1994*. US Geological Survey Bulletin **2129**.

BECKER, G. F. 1910a. Halley on the age of the ocean. *Science*, **31**, 459–461.

BECKER, G. F. 1910b. Reflections on J. Joly's method of determining the ocean's age. *Science*, **31**, 509–512.

BECKER, G. F. 1910c. The age of the Earth. *Smithsonian Miscellaneous Collections*, **56**(6), 1–28.

BRICE, W. R. 1982. Bishop Ussher, John Lightfoot and the age of creation. *Journal of Geological Education*, **30**(1), 18–24.

CHAMBERLIN, T. C. 1922. The age of the earth from the geological viewpoint. *Proceedings of the American Philosophical Society*, **61**(4), 247–271.

CLARKE, F. W. 1910. A preliminary study of chemical denudation. *Smithsonian Miscellaneous Collections*, **56**(5), 1–19.

COOK, A. 1998. *Edmond Halley: Charting the Heavens and the Seas*. Clarendon, Oxford.

DALRYMPLE, G. B. 2001. The age of the Earth in the twentieth century: a problem (mostly) solved. *In*: LEWIS, C. L. E. & KNELL, S. J. (eds) *The Age of the Earth: from 4004 BC to AD 2002*. Geological Society, London, Special Publications, **190**, 205–221.

DIXON, H. H. 1934. John Joly 1857–1933. *Obituary Notices of the Royal Society*, **3**, 259–286.

DIXON, H. H. 1941. *John Joly: Presidential Address to the Dublin University Experimental Science Association 1940*. Dublin University, Dublin.

DIXON, H. H. & JOLY, J. 1897. Coccoliths in our coastal waters. *Nature*, **56**, 468.

DIXON, H. H. & JOLY, J. 1898. On some minute organisms found in the surface waters of Dublin and Killiney Bays. *Scientific Proceedings of the Royal Dublin Society*, **8**, 741–752.

FISHER, O. 1900. Review of 'An estimate of the Geological Age of the Earth', by John Joly. *Geological Magazine*, New Series, Decade 4, **7**, 124–132.

FULLER, J. G. C. M. 2001. Before the hills in order stood: the beginning of the geology of time in England. *In*: LEWIS, C. L. E. & KNELL, S. J. (eds) *The Age of the Earth: from 4004 BC to AD 2002*. Geological Society, London, Special Publications, **190**, 15–23.

GREENE, M. T. 1982. *Geology in the Nineteenth Century: Changing Views of a Changing World*. Cornell University, Ithaca, New York and London.

GREGORY, J. W. 1921. The age of the Earth. *Nature*, **108**, 283–284.

HALLEY, E. 1715. A short account of the cause of saltiness of the oceans, and of the several lakes that emit no rivers; with a proposal by help thereof, to discover the age of the world. *Philosophical Transactions*, **29**, Number 344, 296–300.

HARKER, A. 1914. Some remarks on geology in relation to the exact sciences, with an excursus on geological time. *Proceedings of the Yorkshire Geological Society*, **19**, 1–13.

HAUGHTON, S. 1865. *Manual of Geology* (first edition). Longman, Green, Longman, Roberts and Green, London.

HAUGHTON, S. 1871. *Manual of Geology* (third edition). Longman, Green, Reader and Dyer, London.

HAUGHTON, S. 1878. A geological proof that changes in climate in past times were not due to changes in position of the Pole; with an attempt to assign a minor limit to the duration of geological time. *Nature*, **18**, 266–268.

HOLMES, A. 1911. The association of lead with uranium in rock-minerals and its application to the measurement of geological time. *Proceedings of the Royal Society*, A**85**, 248–256.

HOLMES, A. 1913. *The Age of the Earth*. Harper & Brothers, London and New York.

HOLMES, A. 1926. Estimates of geological time, with special reference to thorium minerals and uranium haloes. *Philosophical Magazine*, Series 7, **1**, 1055–1074.

HOLMES, A. 1947. The construction of a geological time-scale. *Transactions of the Geological Society of Glasgow*, **21**, 117–152.

HOLMES, A. & LAWSON, R. W. 1927. Factors involved in the calculation of radioactive minerals. *American Journal of Science*, **13**, 327–244.

JOLY, J. 1897. On the origin of the canals of Mars. *Scientific Transactions of the Royal Dublin Society*, **6**, 249–268.

JOLY, J. 1898. A shag's meal. *Nature*, **59**, 125.

JOLY, J. 1899. An estimate of the geological age of the Earth. *Scientific Transactions of the Royal Dublin Society*, **7**, 23–66.

JOLY, J. 1900a. A fractionating rain-gauge. *Scientific Proceedings of the Royal Dublin Society*, **9**, 283–288.

JOLY, J. 1900b. Geological Age of the Earth. *Geological Magazine*, New Series, Decade 4, **7**, 220–225 [reprinted in *Nature*, 1900, **62**, 235–237].

JOLY, J. 1900c. On Geological Age of the Earth. *Report of the British Association for the Advancement of Science, Bradford 1899*, Section C3, 369–379.

JOLY, J. 1900d. Some experiments on denudation in fresh and salt water. *Report of the British Association for the Advancement of Science, Bradford 1899*, Section C3, 731–732. [also published in the proceedings of the 8th International Geological Congress, Paris 1901, and enlarged in 1902 in the *Proceedings of the Royal Irish Academy*, **24A**, 21–32].

JOLY, J. 1901a. An estimate of the geological age of the Earth. *Annual Report of the Smithsonian Institution* (for 1899), 247–288.

JOLY, J. 1901b. The circulation of salt and geological time. *Chemical News*, **83**, 301–303
JOLY, J. 1901c. The circulation of salt and geological time. *Geological Magazine*, New Series, Decade 4, **8**, 354–350.
JOLY, J. 1903. Radium and the geological age of the Earth. *Nature*, **68**, 526.
JOLY, J. 1907a. Pleochroic halos. *Philosophical Magazine*, Series 6, **13**, 381–383.
JOLY, J. 1907b. Distribution of radium in the rocks of the Simplon Tunnel. *Nature*, **76**, 485.
JOLY, J. 1908a. The radioactivity of sea water. *Scientific Proceedings of the Royal Dublin Society*, **11**, 253–256 [reprinted in part, and expanded, in *Philosophical Magazine*, Series 6, **15**, 385–393].
JOLY, J. 1908b. On the radium-content of deep-sea sediments. *Scientific Proceedings of the Royal Dublin Society*, **11**, 288–294 [reprinted, with additions, in *Philosophical Magazine*, Series 6, **16** 190–197].
JOLY, J. 1909a. On the distribution of thorium in the Earth's surface materials. *Philosophical Magazine*, Series 6, **18**, 140–145.
JOLY, J. 1909b. On the radium-content of sea water. *Philosophical Magazine*, Series 6, **18**, 396–407.
JOLY, J. 1909c. *Radioactivity and Geology*. Archibald Constable and Co., London.
JOLY, J. 1910. Pleochroic halos. *Philosophical Magazine*, Series 6, **19**, 327–330.
JOLY, J. 1911a. The age of the Earth. *Philosophical Magazine*, Series 6, **22**, 357–380 [reprinted in the *Annual Report of the Smithsonian Institution* (for 1911), 271–293 (1912)].
JOLY, J. 1911b. Radiant Matter. *Scientific Proceedings of the Royal Dublin Society*, **13**, 73–87.
JOLY, J. 1912. The radioactivity of the rocks of the St Gothard Tunnel. *Philosophical Magazine*, Series 6, **23**, 201–211.
JOLY, J. 1913. Pleochroic haloes. *Bedrock* [reprinted in the *Annual Report of the Smithsonian Institution* for 1914, 313–327].
JOLY, J. 1915. *Birth-time of the World and other scientific essays*. T. Fisher Unwin, London [lecture delivered 1914 to Royal Dublin Society].
JOLY, J. 1917. The genesis of pleochroic haloes. *Philosophical Transactions of the Royal Society*, **217**, 51–79.
JOLY, J. 1920. *Reminiscences and Anticipations*. Fisher Unwin, London.
JOLY, J. 1922a. The age of the Earth. *Nature*, **109**, 480–485 [reprinted: The Age of the Earth, Royal Institution Library of Science, *Earth Science*, 1971, **3**(3), 140–151. Friday Evening Discourses in Physical Sciences held at the Royal Institution 1922].
JOLY, J. 1922b. Haloes and Earth-history: a new radioactive element. *Nature*, **109**, 517–518.
JOLY, J. 1923a. The movements of the Earth's surface crust. *Philosophical Magazine*, Series 6, **45**, 1167–1188.
JOLY, J. 1923b. Pleochroic haloes of various geological ages. *Proceedings of the Royal Society*, Series A, **102**, 682–705.
JOLY, J. 1925. *Surface History of the Earth* (first edition). Oxford University, Oxford.
JOLY, J. 1928. The theory of thermal cycles. *Gerlands Beiträge zur Geophysik*, **19**, 415–441.
JOLY, J. 1930. *Surface History of the Earth* (second edition). Oxford University, Oxford.
JOLY, J. 1931. History of the Irish Radium Institute (1914–1930). *Royal Dublin Society Bicentenary Souvenir 1731–1931*. Royal Dublin Society, Dublin, 23–32.
JOLY, J. & FLETCHER, A. L. 1910. Pleochroic halos. *Philosophical Magazine*, Series 6 **19**, 631–648.
JOLY, J. & POOLE, J. H. J. 1927. On the nature and origin of the Earth's surface structure. *Philosophical Magazine*, Series 7, **3**, 327–330.
JOLY, J. & RUTHERFORD, E. 1913. The age of pleochroic haloes. *Philosophical Magazine*, Series 6, **25**, 644–657.
KERR-LAWSON, D. E. 1927. Pleochroic haloes in biotite from near Murray Bay, Province of Quebec. *University of Toronto Studies, Geological Series*, **24**, 54–70.
KNAUTH, L. P. 1998a. Salinity history of the Earth's early ocean. *Nature*, **395**, 554–555.
KNAUTH, L. P. 1998b. Salinity, temperature, and oxygen history of the Precambrian ocean. *Abstracts with Programs – Geological Society of America*, **30**(7), 272.
LANE, A. C. 1925. Measurement of geological age by atomic disintegration. *Proceedings – Lake Superior Mining Institute, Annual Meeting*, **24**(45), 106–116.
LANE, A. C. 1929. The Earth's age by sodium accumulation. *American Journal of Science*, Series 5, **17**, 342–346.
LEWIS, C. L. E. 2001. Arthur Holmes' vision of a geological timescale. *In*: LEWIS, C. L. E. & KNELL, S. J. (eds) *The Age of the Earth: from 4004 BC to AD 2002*. Geological Society, London, Special Publications, **190**, 121–138.
MCNAIRN, W. H. 1919. How old is the world? *Popular Astronomy*, **27**(2), 1–8.
MELLARD READE, T. 1877. President's Address. *Proceedings of the Liverpool Geological Society*, **3**(3), 211–235.
MELLARD READE, T. 1879. *Chemical Denudation in Relation to Geological Time*. Daniel Dogue, London.
MELLARD READE, T. 1893. Measurement of geological time. *Geological Magazine*, New Series, Decade 3, **10**, 97–100.
MITTLEFEHLDT, D. W. 1999. Sodium. *In*: MARSHALL, C. P. & FAIRBRIDGE, R. W. (eds) *Encyclopedia of Geochemistry*. Kluwer, Dordrecht, 577–578.
MORRELL, J. 2001. Genesis and geochronology: the case of John Phillips (1800–1874). *In*: LEWIS, C. L. E. & KNELL, S. J. (eds) *The Age of the Earth: from 4004 BC to AD 2002*. Geological Society, London, Special Publications, **190**, 85–90.
MURNAGHAN, D. 1985. The Irish Radium Institute, Dublin. *In*: MOLLAN, R. C., DAVIS, W. & FINUCANE, B. (eds). *Sòme People and Places in Irish Science and Technology*. Royal Irish Academy, Dublin, 102–103.

NUDDS, J. R. (ed.) 1983. *Upon Sweet Mountains: a Selection of Poetry by John Joly, F.R.S.* Trinity Closet, Dublin.

NUDDS, J. R. 1986. The life and work of John Joly (1857–1933). *Irish Journal of Earth Sciences*, **8**, 81–94.

NUDDS, J. R. 1988. John Joly; brilliant polymath. *In*: NUDDS, J. R., MCMILLAN, N., WEAIRE, D. & MCKENNA LAWLOR, S. (eds) *Science in Ireland 1800–1930: Tradition and Reform*. Trinity College, Dublin, 163–178.

O'DONNELL, S. 1984. Irish Time. *Eire-Ireland*, **19**(2), 131–134.

ORESKES, N. 1999. *The Rejection of Continental Drift*. Oxford University, Oxford.

PHILLIPS, J. 1860. *Life on the Earth: its Origin and Succession*. Macmillan, Cambridge.

REESE, R. L., EVERETT, S. M. & CRAUN, E. D. 1981. The chronology of Archbishop James Ussher. *Sky and Telescope*, **62**(5), 404–405.

RUTHERFORD, E. 1905. Present Problems in Radioactivity. *Popular Science Monthly*, **67**, 5–34.

RUDZKI, M. P. 1901. Sur l'Age de la Terre. *Bulletin de l'Académie des Sciences de Cracovie*, **1901**, 72–94.

SHELTON, H. S. 1910. The age of the earth and the saltness of the sea. *Journal of Geology*, **18**, 190–193.

SHELTON, H. S. 1911. Modern theories of geologic time. *The Contemporary Review*, 195–209.

SHIPLEY, B. C. 2001. 'Had Lord Kelvin a right?': John Perry, natural selection and the age of the Earth, 1894–1895. *In*: LEWIS, C. L. E. & KNELL, S. J. (eds) *The Age of the Earth: from 4004 BC to AD 2002*. Geological Society, London, Special Publications, **190**, 91–105.

SOLLAS, W. J. 1895. The age of the Earth. *Nature*, **51**, 533.

SOLLAS, W. J. 1900. Evolutionary Geology. *Report of the British Association for the Advancement of Science, Bradford 1899*, 449–463 [reprinted 1915 in *The Age of the Earth*. T. Fisher Unwin, London].

SOLLAS, W. J. 1909. Anniversary address of the President: position of geology among the sciences; on time considered in relation to geological events and to the development of the organic world; the rigidity of the Earth and the age of the oceans. *Proceedings of the Geological Society*, **65**, lxxxi–cxxiv.

SOMERFIELD, A. E. 1985. John Joly Geologist, Physicist and Mineralogist 1857–1933. *In*: MOLLAN, R. C., DAVIS, W. & FINUCANE, B. *Some People and Places in Irish Science and Technology*. Royal Irish Academy, Dublin, 64–65.

THOMSON, W. (LORD KELVIN). 1864. On the Secular Cooling of the Earth. *Transactions of the Royal Society of Edinburgh*, **23**, 157–170.

THOMSON, W. (LORD KELVIN). 1899. The age of the Earth as an abode fitted for life. *Philosophical Magazine*, Series 5, **47**, 66–90.

WALCOTT, C. D. 1893. Geologic time, as indicated by the sedimentary rocks of North America. *Journal of Geology*, **1**, 639–676.

WYSE JACKSON, P. N. 1992. A Man of Invention: John Joly (1857–1933), engineer, physicist, and geologist. *In*: SCOTT, D. S. (ed.) *Treasures of the Mind: A Trinity College Dublin Quatercentenary Exhibition*. Sothebys, London, 89–96, 158–160.

Arthur Holmes' vision of a geological timescale

CHERRY L. E. LEWIS

History of Geology Group, 21 Fowler Street, Macclesfield, Cheshire, SK10 2AN, UK
(*email*: clelewis@aol.com)

Abstract: Arthur Holmes (1890–1965) was a British geoscientist who devoted much of his academic life to trying to further the understanding of geology by developing a radiometric timescale. From an early age he held in his mind a clear vision of how such a timescale would correlate and unify all geological events and processes. He pioneered the uranium–lead dating technique before the discovery of isotopes; he developed the principle of 'initial ratios' thirty years before it became recognized as the key to petrogenesis, and he wrote the most widely read and influential geology book of the twentieth century. But despite all this, much of his contribution to geology has gone unrecognized in the historical literature. This paper attempts to redress this omission, to dispel some of the myths about Holmes' life, and to trace his contribution to the development of the geological timescale.

Of all subjects of speculative geology, few are more attractive or more uncertain in positive results than geological time (Charles Walcott, 1893, p. 676).

Arthur Holmes (1890–1965) is arguably the greatest British Earth scientist of the twentieth century. Amongst many other accolades he was awarded both the Wollaston and Penrose Medals, the highest awards given, respectively, by the Geological Societies of London and America, and in 1964, the year before he died, he was honoured with the most prestigious award that a geologist can receive – the Vetlesen Prize, which is equivalent in stature to the Nobel Prize. Only two other British scientists have subsequently received it.

Today there is a passing generation of geologists who grew up with Arthur Holmes' famous book *Principles of Physical Geology* (Holmes 1944) and although the fourth edition, written by one of his students, Donald Duff, is still in use today, none of the later editions compare with the first one written for RAF cadets during the Second World War, whilst Holmes was on fire-watching duty. One of those rare scientists able to bridge the gap between science and the layman, a love of literature and an interest in philosophy lay at the root of Holmes' easy mastery of the language and a desire to communicate his subject (Reynolds 1968).

He dedicated *Principles of Physical Geology* to 'the reader who wishes to see something of the "wild miracle" of the world we live in through the eyes of those who have tried to resolve its ancient mysteries'. His use of the phrase 'wild miracle' aptly illustrates the undiminished sense of wonder he found in his science, which comes across in much of his work, but particularly in this book. Dynamically written and lavishly illustrated, it soon became an international bestseller, despite the rather high price of thirty shillings. The somewhat cumbersome title, apparently chosen as a tribute to Charles Lyell's *Principles of Geology*, was soon dropped by those who bought the book, and ever since it has been known more simply and fondly as 'Holmes'. Published in 1944, the first print run of 3000 copies sold out almost immediately and, despite a paper shortage both during and after the war, the first edition was reprinted no fewer than eighteen times. But for someone who wrote such an extraordinarily influential book that dominated geological teaching for half a century, and who contributed so much to our understanding of the evolution and age of the Earth, remarkably little is known about the man (Fig. 1).

Biographical background

Contrary to a popular myth that seems to have originated in the obituary written by Holmes' second PhD student, Kingsley Dunham, Arthur Holmes was *not* from 'Northumbrian farming stock' (Dunham 1966, p. 291), a phrase that lends a sense of grandeur to a man whose background was far from that. He was in fact born on 14 January 1890, at 62 Glen Terrace in Hebburn, which borders Jarrow, of 1930s hunger-march

From: LEWIS, C. L. E. & KNELL, S. J. (eds). *The Age of the Earth: from 4004 BC to AD 2002*. Geological Society, London, Special Publications, **190**, 121–138. 0305-8719/01/$15.00 © The Geological Society of London 2001.

Fig. 1. Arthur Holmes with Doris Reynolds, who later became his second wife, circa 1931.

Fig. 2. The house at 19 Primrose Hill, Low Fell, Gateshead, where Arthur Holmes lived as a boy.

fame. Both are districts of Gateshead, the industrial heartland of northeast England, where, from the age of about 10, Arthur Holmes and his parents lived at 19 Primrose Hill, Low Fell (Fig. 2), on a new estate built to house the growing population moving into Gateshead to work on the docks. Arthur's father, David Holmes, was at this time an assistant in an ironmonger's shop and Arthur was proud of the fact that his father was also a skilled carpenter. A cabinet made by him is still in the family's possession. Later on, and until his retirement, David Holmes appears to have been an insurance salesman. Arthur's mother, Emily Dickinson, is believed to have been a teacher and came from Newcastle. One grandfather, Thomas Holmes, was a cordwainer (shoemaker) and the other, John Dickinson, a metal moulder. Neither family was evidently from farming stock.

In 1901, Arthur entered the local Higher Grade School in Gateshead which was renowned as one of the best municipal schools in the country for its teaching of science. A surviving school certificate[1] indicates that he was a very bright and industrious child, and apparently he could be seen walking to school every morning reading a book (Wilson pers. comm., 1999). He obviously had an enquiring mind, for years later he recalled that from an early age he had been fascinated by the date of Creation – 4004 BC – as calculated by Archbishop Ussher, and written in the margin of his parent's Bible. 'I was puzzled by the odd "4"', he wrote, 'Why not a nice round 4000 years? And why such a very recent date? And how could anyone know?' (Holmes 1963, p. xv). But brought up as he was in a religious community – his parents were staunch Methodists – all this young mind was told in answer to his enquiries, was not to question the 'Word of God'. Thus it was with delight he realized a few years later, 'that there were people who audaciously doubted that the world was made in six days'. The man who enlightened him on this matter was his sixth-form master, Mr J. McIntosh, whose brilliant teaching of physics was largely responsible for the direction taken by the rest of Holmes' life. McIntosh introduced his pupils to the *Popular Lectures and Addresses* of Lord Kelvin who, in Holmes' view, had transformed 4004 years into forty million for the age of the Earth. McIntosh also familiarized his pupils with the work of the Swiss geologist Edward Seuss, whose first volume of *The Face of the Earth* had recently been translated into English. These two works stimulated, and now reflect, Holmes' life-long interest in the physics of geology.

---

[1] A certificate from University of Oxford shows that he passed the Oxford Local Examinations as a Junior Candidate in 1905, and was placed in the First Class of the honours list. Royal Holloway University Library Archive, Doris Reynolds, Box 2.

Historical background

The tremendous controversy between geologists, biologists and physicists over the age of the Earth has been thoroughly documented by Dalrymple (2001) and Shipley (2001). At the turn of the twentieth century geological time was a topic of supreme interest and importance, as is evident from William Sollas' Presidential Address to Section C (Geology) of the British Association for the Advancement of Science (BAAS) in 1900, where he likened it to the Djini in Arabian tales because, no matter how many times you disposed of it, geological time would irrepressibly come up for discussion again and again. Sollas also recognized just how crucial 'time' was to the further understanding of geology: 'How immeasurable would be the advance of our science could we but bring the chief events which it records into some relation with a standard of time!' he lamented (Sollas 1900, p. 717). But what he did not realize was that the means by which this advance would be made had already been discovered some four years earlier.

At the British Association Meeting held in 1896 J. J. Thomson, head of the Cavendish Laboratory in Cambridge and President of Section A (Physics) reported how 'The discovery at the end of last year by Professor Röntgen of a new kind of radiation ... has aroused an amount of interest unprecedented in the history of physical science' (Thomson 1896, p. 701). Initially this discovery of X-rays and the exceptional interest that it generated overshadowed the related discovery of radioactivity the following year, due to the enormous medical possibilities promised by X-rays; but following the dedicated work of Marie and Pierre Curie a growing excitement over radioactivity eventually triggered an explosion of activity in physics laboratories around the world. Marie had discovered two new elements. The first she called polonium after her native Poland; the second she called radium from the Latin word *radius* meaning 'ray', because radium appeared to give out 'mysterious rays'. In the same year, Thomson discovered the electron and realized that the mysterious rays emitted by radium were in fact streams of electrons. This then led to the obvious conclusion that the atom could no longer be regarded as the smallest particle of matter, a theory first proposed by Democritus almost two and a half thousand years ago, and little challenged since that time. The nuclear age had dawned.

New discoveries relating to the atom accrued one after the other. In 1902 Ernest Rutherford and Frederick Soddy, working in Montreal, shocked the world when they reported that in the process of radioactive decay, an atom of radium spontaneously turned into an atom of radon, releasing an atom of helium as it did so (Rutherford & Soddy, 1902a, b). From one element two new elements had been born, both of which were in a completely different physical state from their parent – a metal was changing to two gases. It caused a sensation. The following year Pierre Curie, working with his assistant, Albert Laborde, detected that radium was constantly releasing heat. As electrons were emitted from the atom during radioactive decay, energy was given out in the form of heat: 'Every hour radium generates enough heat to melt its own weight in ice', they announced (Curie & Laborde 1903). Here at last was the evidence for which geologists had been waiting. While the physicists might be right in that the Earth was still cooling down from a time when it had been a molten globe (Dalrymple 2001; Shipley 2001), what they had not known was that at the same time as it was cooling, radioactive elements within the Earth were generating enough heat to prolong that cooling for a period of time far greater than that required by the geologists and biologists. Geological time took on new dimensions.

Since 1895 it had been noted that helium occurred in rocks and once Rutherford and Soddy had established that its presence there was due to the decay of radioactive elements (Rutherford & Soddy 1903; Lewis 2000), it was but a short step to realizing that if the rate of helium production could be established, and the amount of helium that had accumulated in a rock was measured, then a relatively simple calculation would show how long it had taken for the helium to accumulate, and the time since the rock's formation could be established. Accordingly, an apparatus was arranged so that the helium atoms, being naturally and spontaneously emitted from a very small but very accurately known quantity of radium, were impelled through a specially designed chamber. The passage of each particle set up a tiny electric current which gave a 'kick' to the needle of an electrometer. By counting the kicks, the production rate of helium was measured.

But the decay rate was not all that Rutherford needed to know in order to determine an age. The law of radioactive decay states that the number of radioactive atoms that decay in a given time is dependent upon the number of radioactive atoms that are present. Put simply, if there are 100 atoms of a radioactive element in a rock which decay at a rate of 10% a year, then after the first year ten will have decayed and 90 will be left, but after the second year only nine will have decayed (10% of 90) and 81 will be left. Thus the number that decay is directly proportional to the number that is present. Conversely of course, the numbers of daughter elements increase in the same way, and provided that none of the parent or daughter atoms escape, then the sum of parent and daughter will always equal the original amount present. Thus when analysing a sample it was necessary to know not only how much of the daughter element was present, but how much of the parent was there too. In due course, Rutherford showed that all radioactive elements decayed at a constant and measurable rate, the value of which was termed the 'half-life', which was the time taken for any element to decay to half its original value. But at that time the half-lives of most elements were extremely difficult to measure and determined ages were frequently revised as more accurate half-lives were obtained.

The first tentative radiometric age determination was completed on a fergusonite mineral by Rutherford in 1904 and yielded an age of 40 Ma (Rutherford 1905), but at that time the production rate of helium was still poorly constrained. Within a year, however, a new and more accurate assessment of the rate became available, and a recalculation of the mineral's age showed it to be closer to 500 Ma. However, Rutherford was already aware that even this was a minimum estimate, 'for some of the helium has probably escaped' (Rutherford 1906, p. 189).

First dates

So it was against this background of dramatic and exciting scientific discoveries that Arthur Holmes completed his schooling and in 1907 won a scholarship to study physics at the Royal College of Science in London, one of only six offered each year. By the end of his second year he had passed the BSc in physics, but in that second year he had also taken a subsidiary course in geology, a subject that not only continued to interest him, but which he considered was more likely to offer him employment. Thus by the end of his second year he had decided to change courses and become a geologist.

Holmes had been studying in the physics department for the Diploma of the Associateship of the Royal College of Science (ARCS). This course consisted of two years of general science studies, after which it was necessary to pass exams (which Holmes appears to have called the BSc) before being allowed to progress to the third year, where one could become more specialized. Few people actually changed departments at this stage, and in moving from physics to geology, Holmes caused some consternation amongst his teachers who advised him against it. On completion of three years' study students would be awarded the Associateship Diploma and, subject to the opinion of the rector and approval of the professor of the department, a fourth and final honours year could be devoted to research work. Although having officially moved to the geology department by this time, Holmes actually carried out his research year back in the physics department, under Strutt.

On 1 January 1911, in the fourth and final year of his course, Holmes started to keep a diary.[2] On the first page he summarized the events of 1910 and described how, in May of that year, he heard of a competition for the position of petrologist at the British Museum (Natural History) which he decided to enter, stopping his college studies altogether in order to work for it. The competition took the form of onerous exams, which required a knowledge of advanced mathematics, optical crystallography and inorganic chemistry, as well as English composition and the translation of three out of four languages – Latin, French, German or Greek. Having taken the exams in October he writes in the diary:

> While waiting for the result of my competition I began some research work (having been awarded a schol.[arship] of £60) for the fellowship of the Imperial College. This I did in the Physical department under Strutt and got on so successfully that I cut short my Xmas vacation at both ends, in order to put in more time.

The recently appointed professor of physics at the Royal College of Science, more familiarly called Imperial College of Science after 1910, was Robert Strutt. He had trained at the Cavendish Laboratory in Cambridge where his father, John Strutt the third Baron Rayleigh, had been head of the laboratory and taught J. J. Thomson. In his turn, Thomson had become head of the laboratory and taught Strutt's son, Robert, who became the fourth Baron Rayleigh on his father's death. Perhaps understandably, these two Strutts and Barons Rayleigh are often confused in the literature. Like everyone else at the Cavendish at that time, Robert Strutt had been swept along with the rising tide of interest in radioactivity, and between 1905 and 1910 he carried out a thorough investigation of the helium contents of a wide variety of minerals. He also investigated the problem of helium leakage, having been astonished to discover the large quantity of helium diffusing from powdered monazite placed

---

[2] This diary, and another from 1922, will form part of the Arthur Holmes Collection being formed at the Geological Society. At present they are held in the author's collection.

under vacuum at room temperature. As a result of these studies Strutt realized that helium ages could never be relied upon to give accurate results. Another method was required, and he set Holmes the task of finding it. This was the research Holmes referred to in his diary.

In 1905 an American chemist, Bertram Boltwood, had observed that radium appeared to be a disintegration product derived from the radioactive decay of uranium (hitherto radium had been thought to be the starting point in the decay chain) and he had set up an experiment with Rutherford to demonstrate this association (Rutherford & Boltwood 1905). At the same time, in fact reported in the same journal volume, Boltwood also noted that: 'In reviewing the various published analyses of minerals containing notable proportions of uranium ... one cannot fail to be impressed by the frequent and almost invariable occurrence of lead as one of the other constituents' (Boltwood 1905, p. 255). He was also impressed by the information supplied to him by W. F. Hillebrand of the US Geological Survey who could not remember ever having found uranium in a mineral without its being accompanied by lead: '... the association has often caused me thought', Hillebrand remarked to Boltwood (Boltwood 1905, p. 256). Thus in 1905 these empirical observations indicated quite convincingly that radium was an intermediate product in the decay chain from uranium, and that *lead* was the final product. But it still had to be proved.

Over the following two years Boltwood attempted to find the necessary proof, but due to the scarcity of suitable minerals containing levels of uranium high enough to be measured, he was unable to obtain new material for analysis and in the end was forced to use published values for uranium and lead, determined by other analysts, many of which he considered were not 'particularly accurate'. Furthermore, he also complained that there was 'an unfortunate tendency on the part of many mineralogists to carry out an analysis merely for the purpose of assigning the mineral some definite chemical formula, which often leads to the overlooking or ignoring of a number of the minor constituents' (Boltwood 1907, p. 23). In order to *prove* that lead was the final inactive decay product of uranium, it was necessary for Boltwood to demonstrate that the ratio of uranium to lead *increased* as the age of the mineral increased, therefore it was essential that he had access to values of both uranium and lead from the same sample. And since the amount of uranium present was derived by measuring radium, it was also necessary to have radium values.

Unfortunately, uranium, lead and radium were often not included in determinations made for evaluating the chemical formula of a mineral, since, quite reasonably, the considerable extra work involved was not deemed necessary for that purpose. This therefore limited Boltwood to a small number of analyses suitable for his requirements. But despite these many difficulties, Boltwood determined the uranium–lead ratios for more than 40 of these published analyses, and calculated a radiometric age for ten of them. Significantly he found that the uranium–lead ratio increased as the age of the mineral increased, from which he concluded that here was 'proof that lead is the final disintegration product of uranium' (Boltwood 1907, p. 88).

Thus in 1910, when Holmes was set the task of finding a more reliable means of determining the age of minerals than the helium method, he hoped to produce additional support for Boltwood's 'strong evidence' (Holmes 1911, p. 248) (note, not the *proof* Boltwood considered he had already found) that lead was the ultimate decay product of uranium. By making some new analyses specifically designed for that purpose he hoped to avoid all the flaws inherent in the published values used by Boltwood.

Holmes was fortunate in having access to a series of uranium-bearing nepheline syenites from the Christiana district of Norway, near Oslo, which were ideal for the purpose. However, in order to obtain enough material to work with, approximately 100 kilograms of rock were required in which zircon represented at least 0.1%,[3] for zircon was known to be the best mineral for harbouring radioactive elements. The rock was crushed and the 'tedium of separation' yielded 0.3–2.0 g from each of no less than seventeen different radioactive minerals. These separations were then powdered for use in the lengthy analysis.

Of the seventeen minerals – almost all of which were repeatedly analysed between two and five times to verify results – only eight were deemed suitable to be included in the final age calculation. The remaining nine were considered to have already contained lead when they first formed, and were thus found unsuitable for age dating purposes. By averaging the U/Pb ratios of the eight acceptable minerals, the first age determination ever made by the uranium–lead method yielded a result of 370 Ma, and since the geological age of the rock was believed to be Devonian, 370 Ma became the first radiometrically defined control point for that geological period.

---

[3] Holmes to Sederholm, 1 February 1912. Courtesy of Åbo Akademis Bibliotek, Åbo, Finland.

Table 1. *Holmes' first geological timescale (1911)*

Geological period 1911	Holmes' U/Pb ages (Ma)	Geological period today	Age range today (Ma)
Carboniferous	340	Lower Carboniferous	362–330
Devonian	370	Upper Devonian	380–362
Silurian or Ordovician	430	Silurian	443–418
Precambrian in:		Precambrian	
–	–	Late Proterozoic	900–544
Sweden	1025	Middle Proterozoic	1600–900
	1270		
United States	1310	Middle Proterozoic	1600–900
	1435		
Ceylon	1640	Early Proterozoic	2500–1600

Comparison of the very first U/Pb age determinations ever made, with the range of present-day values for the relevant geological period. Despite all the difficulties inherent in techniques of the time, and the fact that isotopes had still to be discovered, Holmes' ages nevertheless fall within the present-day range. Note how the Precambrian was, at that time, completely undifferentiated.

In addition to his own new analysis Holmes also recalculated some of the age data published by Boltwood, in line with more recent decay rates, and assigned a geological period to each of those ages (Table 1), something which, in Holmes' opinion, Boltwood had 'unfortunately omitted' (Holmes 1911, p. 249). Perhaps this barb was an attempt to redress Boltwood's comments about mineralogists, but it also highlights the fact that Holmes realized there was no point in simply having an age for its own sake. To him the only thing that mattered was that each radiometric date was a new control point on the geological timescale, and so to be of use each date must be assigned to a geological period. Thus Holmes went to some lengths to obtain this information from the US Geological Survey for Boltwood's data. However, prior to radiometric dating the only way to assign a stratigraphic age, or geological period, to an igneous rock was from its stratigraphic relationship to the nearest fossil-bearing sediments. But because the nature of igneous rocks is that they often cut across sedimentary rocks, this frequently led to the wrong geological period being assigned to igneous rocks which, when they were radiometrically dated, would mean that the relatively accurate control point would date the wrong period. It was a circular problem.

Problems can also arise from the misuse of terminology. In an attempt to avoid confusion when discussing the age of a rock, and in accordance with terminology widely used by Holmes, throughout this paper the term 'geological age' refers to the stratigraphic age of an igneous rock, i.e. the geological period (Devonian, Carboniferous, etc.) to which the rock had been assigned. Thus, the 'geological age' of a rock is to be clearly distinguished from its 'radiometric age', which is measured in years. The term 'absolute age', an expression frequently used in the literature when referring to a radiometric age, is specifically avoided in deference to Holmes' contempt for that 'meaningless term' (Holmes 1962, p. 1238).

Holmes' early research provided further evidence that lead was indeed the stable end product of uranium but, like Boltwood, Holmes too concluded it was highly improbable that thorium also decayed to lead. The oldest age in Holmes' suite was 1640 Ma, which pushed geological time further back than it had ever been before. As it began to dawn on him that 'the time available for the Precambrian was at last beginning to approach its correct proportion' (Holmes 1963, p. xvi) he realized that 'this very method may in turn be applied to help the geologist in his most difficult task, that of unravelling the mystery of the oldest rocks of the earth's crust (Holmes 1911, p. 256). Holmes had embarked on his lifetime's quest 'to graduate the geological column with an ever-increasingly accurate time scale' (Holmes 1911, p. 256). The paper was read by Robert Strutt to members of the Royal Society on 6 April 1911 while Holmes was in Mozambique. Sir Archibald Geikie, president at the time, was in the Chair.

Fortunately for science, Holmes did not win the competition for the post of petrologist at the Natural History Museum referred to in his diary. Although he came top in the mineralogy exam, his proficiency in Latin apparently let him down and overall he came second to William Campbell Smith. But with a scholarship of only £60 a year to live on in London, he was constantly seeking a supplementary income. He occasionally earned ten shillings writing book reviews for *The Times*, but when, early in 1911, he was offered a six-month contract prospecting for minerals in Mozambique for the handsome sum of £35 a month, he had no option but to accept.

Mozambique

All of what we know about Holmes' adventures in Mozambique results from the diary he kept during the trip and from letters he sent home to his parents and to his best friend from school, Bob Lawson. Their friendship seems to have lasted at least forty years until the late 1930s when Holmes remarried and appears to have lost touch with Lawson. But up until that time they jointly collaborated on many projects – 'I want to discuss it with you ... and ask your help in the maths involved',[4] being a typical appeal from Holmes. Lawson, like Holmes, had shone at school and had also taken a physics degree, but at the local university in Newcastle – perhaps because he was not eligible for a scholarship. Nevertheless, Holmes recognized Lawson's superior mathematical ability, as is evidenced in this letter from Mozambique: 'Your correspondence with Dr. Jude is highly complimentary to you and is great evidence of the height to which you must now have climbed mathematically. In that direction I feel absolutely ignorant – even though I know probably far more than the average geological man.'[5]

The problem Holmes needed help on in this instance concerned the still highly speculative ideas about the internal temperature of the Earth. The question was one of cosmic importance since, if it was found that the Earth's internal heat was due entirely to radioactive decay, then it would follow that the sun and stars may be heated by radioactive decay also. However, Strutt's detailed work on the distribution of radioactive elements had shown that if these elements were equally distributed throughout the Earth's interior, then the present-day temperature of the surface should still be too hot to support life. Since this was clearly not the case, Strutt had gone on to calculate that these elements must be largely confined to a surprisingly thin crust, perhaps only 45 miles in depth. What, then, was the composition of the remainder of the Earth, and was it hot or cold?

Holmes had obviously been giving these problems some thought in Mozambique, particularly while suffering from bouts of malaria when he was unable to do much else other than think, as he records in his diary: 'Feeling unwell and on the verge of fever all day. Lay about only and read. Did not attempt any work. Fooled on speculating on origin of Earth. Have concluded that the inner sphere of high density [core] is of different origin to the siliceous acid and uranium-bearing exterior of much less density.'[6] A few days previously he had written to Lawson:

One thing which has struck me in connection with earth cooling and the conditions of the earth's interior is this. You know that the Radium in the earth's crust is easily sufficient to account for the radiation of heat into space. It would be interesting to see what kind of temperature gradient the corresponding amount of Radium of 1,500 million years ago would give. There was sea then and therefore the temperature of the surface must have been under 100°C. From rough calculations I have made out here (necessarily not very reliable!) I estimate the Ra at 20% more [than 1,500 Ma ago] and the temperature at 60°C [at the surface] – assuming there is exactly enough Radium for present radiation. If that be the case under rigid maths, it almost proves that the earth is cold in its interior as otherwise the earth would have notably grown hotter.

These wonderful and extremely important letters, now held by Holmes' daughter-in-law,[7] give an indication of the level of understanding about the Earth's interior prevailing at that time and Holmes determination to develop a geological timescale: 'I am in hopes of gradually building up a geological timescale, and hope it might do for a DSc! There's conceit if you like. Still, I may as well confess to you that a DSc is my present aim and object, and with other published work I think it ought not now to be far away – if only I can avoid having to pass the honours BSc.'[8] Having changed from physics to geology Holmes would have been required to take the geology BSc, since he had missed his third-year

---

[4] Holmes to Lawson, 14 September 1911. Courtesy of Karla Holmes.
[5] Holmes to Lawson, undated, but probably 7 September 1911. Courtesy of Karla Holmes.
[6] Diary entry, 25 September 1911.
[7] Karla Holmes, widow of Geoffrey Holmes (1924–1992), Arthur Holmes' second son by his first wife Maggie. Only the letters from Holmes have survived, and those to Bob Lawson are for some reason incomplete. At the time of writing Karla lives in Vienna. She has four children.
[8] Holmes to Lawson, 14 September 1911. Courtesy of Karla Holmes.

exams while studying for the competition at the British Museum. In fact he never did take a degree in geology, and he did not get his DSc for another six years.

Holmes became extremely ill with malaria in Mozambique, such that a notice of his death was telegraphed to a local paper back home (Hedberg 1957). Fortunately the company had found no minerals of economic worth so that Holmes' six-month contract was not extended as anticipated, for it seems unlikely that he would have survived much longer in that 'underhand sort of country'.[9] But his enthusiasm for his work was undaunted by his illness and whilst still on board the RMS *Edinburgh Castle*, which carried him for the last part of the journey home, he was already writing to established geologists such as Alexander du Toit in South Africa and J. J. Sederholm in Finland, asking their help in obtaining Precambrian samples so that he could compare their ages with those he had found in Mozambique. As he explains to Sederholm in a later letter: 'My two chief objects are the formation of a geological time scale and also the correlation of the various members of the Pre-Cambrian rocks in different parts of the world'.[10] These were objectives he was to pursue for the rest of his life.

On his return from Mozambique, Holmes was given a demonstratorship at Imperial College and immediately started to write his now famous and highly influential little book *The Age of the Earth* (Holmes 1913). In it he endeavoured to give as full an account as possible of the prevailing methods for measuring geological time, in an attempt to explain the 'extraordinary discrepancy' of the conclusions drawn from the two most prominent methods of dealing with it, namely the age of the Earth derived from geological methods, as opposed to the age determined by radioactivity, for 'The other arguments may be dismissed without further discussion'. He eulogized:

> From the mists of controversy which for half a century have hung over the subject, the two hour-glass methods alone emerge, and the final issue must be fought out between them. In the one the world itself is the hour-glass, the accumulating materials are salt, the sedimentary rocks and calcium carbonate.... Each set of data is intimately related to the others and all stand or fall together. In the other case the accumulating materials are helium and lead, and the hour-glass is constituted by the minerals in which they collect (Holmes 1913, p. 166).

By clearly setting out the arguments for each of the two methods he hoped that his treatment of the subject would 'stimulate a greater interest in the time-problem, or provide material for further discussion' and thereby move the debate forward. Already he was erecting 'wickets to be bowled at' (Holmes 1932, p. 556), a cricketing term he used to illustrate the need for discussion of speculative ideas or tentative conclusions – a theme that runs throughout his work. He was never afraid to erect these wickets even when, as happened on occasions, he was shown to be completely on the wrong track. He believed that science moved forward through discussion and speculation, which were justified 'if they do no more than stimulate a search for the more elusive pieces of the jigsaw we are doing our best to put together' (Holmes 1931, p. 451).

In his 'Review of the evidence', the final chapter in *The Age of the Earth*, Holmes regretted that confidence in the pioneer work on radioactivity had been shaken by advocates of the geological methods who, having previously been deprived of sufficient time for geological processes, now felt that the pendulum had swung too far, such that they found it impossible to accept what they considered was an excessive period of time offered by radioactivity. 'Evidently we are at a parting of the ways', he wrote, 'The fundamental assumptions on which the arguments are based cannot both be right. One of them must be rejected. Which is it to be?' (Holmes 1913, p. 167). This almost 'journalistic' style of writing became his hallmark, and the way it injects a feeling of excitement into the text is undoubtedly what endeared him to his readers both inside and outside the profession.

The real problem, he believed, like Kelvin had before him, lay in the geologists' stubborn adherence to the uniformity of geological processes: 'Uniformity proved a great advance, but in detail is apt to lead us astray if applied too dogmatically' (Holmes 1913, p. 174). For to accept geological uniformity dogmatically and the short timescales which that implied would be to deny the uniformity of atomic disintegration over time, and that would be like saying that the laws of physics and chemistry had varied over time. Not surprisingly he came down heavily on the side of radioactivity as the only way forward, and concluded with his prophetic vision for the future of geology: 'With the acceptance of a reliable time-scale, geology will have gained an invaluable key to further discovery. In every branch of the science its mission will be to unify and correlate, and with its help a fresh light will be thrown on the more fascinating problems of the Earth and its Past' (Holmes 1913, p. 176).

---

[9] Diary entry 3 October 1911.
[10] Holmes to Sederholm, 1 February 1912. Courtesy of Åbo Akademis Bibliotek, Åbo, Finland.

Discovery of isotopes

In the same year that Holmes' book *The Age of the Earth* was published, Frederick Soddy published his ideas on isotopes (Soddy 1913), and Holmes was the first to recognize the enormous impact this was to have on age dating techniques. It was no longer acceptable simply to measure the total amount of uranium and lead present in a sample, because it was now apparent that lead was derived from three different sources. Uranium decayed to ^{206}Pb, thorium decayed to ^{208}Pb, and a third lead, ^{207}Pb, was at that time believed to be the non-radiogenic primeval lead that had been around since formation of the Universe. So now when determining an age, as well as measuring the parent element – uranium or thorium – it was first necessary to identify in what proportions these three leads were present, and the only way to do that, in the days before mass spectrometers, was to measure their atomic weights.

Established in 1910, the Radium Institute in Vienna immediately attracted some of the best scientists in Europe, such as Viktor Hess and Georg Von Hevesy, both of whom went on to win Nobel prizes. So it is an indication of Bob Lawson's ability that he too was invited to work at the Institute in 1913. The Vienna team became experts at measuring atomic weights of isotopes, and with Lawson's help Holmes was able to determine with far greater accuracy the age of the minerals he dated. But with the outbreak of the First World War a year later, Lawson found himself unable to get home and was thus detained in Vienna for the next four years. This was no bad thing since it was a highly productive time for his collaboration with Holmes who was also not required for active service on account of recurring attacks of malaria, although Holmes was forced to publish the first part of their joint paper independently since 'all correspondence [with Vienna] is temporarily at an end owing to the war' (Holmes & Lawson 1914, p. 840).

But despite the greater accuracy with which ages could now be measured, all radioactive procedures were then unhappily suffering from the damage inflicted on them by the helium method, which Holmes had shown gave ages almost half those measured by the uranium method. Consequently many geologists considered *all* radiometric ages to be completely unreliable. But it must have come as a considerable disappointment when, at a BAAS meeting in 1915 chaired by Rutherford, Holmes once again put forward his arguments for an Earth of great antiquity (Holmes 1915), only to discover that it was Frederick Soddy of all people who came to the geologists' defence. Soddy hoped that geologists would not be in any immediate hurry to decide between the geological and radioactive estimates of the age of the Earth, for he considered that 'owing to the element of uncertainty about the initial stages of the disintegration and the long periods involved, there was a great *terra incognita*, and the new theory of isotopes made it necessary to take into account many possibilities, some of them referred to by Professor Joly ... there might well be unknown factors still to be discovered, sufficiently important to bring the two methods into closer agreement' (Soddy 1915, p. 434). Joly's argument was that radioactive decay may have been more rapid in the distant past (Wyse Jackson 2001) and if so, then the Earth need not be as old as it appeared from radioactive methods. It is therefore quite astonishing to find Soddy the physicist being prepared to consider Joly's views, ideas that Holmes the geologist had dismissed three years earlier when arguing in his book *The Age of the Earth* for the uniformity and constancy of the laws of physics and chemistry.

To compound these problems, the extremely time-consuming chemical methods of analysing uranium and lead were made even more so by the need to determine the atomic weight of each analysis, and once Lawson returned home from Vienna at the end of the war, access to this facility was closed. Thus for the latter part of the war and until he left for Burma in 1920, Holmes, almost the only person in the world then working on age dating techniques, concentrated exclusively on his petrographic work (Lewis 2000).

New key to petrogenesis

In 1923 the National Research Council in America had established a committee for the 'Measurement of Geologic Time by Atomic Disintegration', its objectives being to collate and monitor all the dating of rocks being done around the world. In 1926 a sister committee on the 'Physics of the Earth' spawned a subcommittee for 'The Age of the Earth' which was commissioned to specifically address the wide variance in geological time readings derived from the geological methods, as opposed to the radiometric methods, a problem that was still considered a serious geological conundrum. Holmes was on the first committee, and a founder member of the subcommittee that was chaired by the Yale mineralogist Adolph Knopf. Other members of the subcommittee were the astronomer Ernest Brown, the physicist Alois Kovarik, the palaeontologist Charles Schuchert and a geologist, Alfred Lane, who was also chairman of the committee for the 'Measurement of Geologic Time by Atomic Disintegration'. At the time all of these individuals worked at either Yale or Harvard, placing Holmes in England at a considerable disadvantage.

Holmes had been given, or set himself, the task of evaluating every single radiometric analysis of uranium, lead, helium and thorium ever published. Rejecting the vast majority of them, now amounting to many hundreds, because they did not meet his high standards, he found only thirty U/Pb ratios, worldwide, to be acceptable. This scrutiny entailed a huge amount of work at a time when Holmes was deeply engrossed in evolving his ideas on continental drift (Lewis in press), but five years later the committee was ready to present their results and Knopf wrote to Schuchert about progress on the book in which their report would be published (Knopf et al. 1931):

> The last batch of galley[s] from Holmes has just come in and I am at work getting it in shape. He has kindly left fifty-nine references to be compiled on this side of the Atlantic, has added an extraordinary amount of new material, and slashed up the galleys; I think that these galleys will not meet with a warm welcome when they arrive in Washington. I was told by the editor the other day, while I was in Washington, that the printer said that Holmes' manuscript was the worst he had ever received.[11]

While this may be the case, the fact that Holmes had written a total of 336 pages, which included 639 references, while the other four contributors to the report managed 131 pages between them, is an indication perhaps, of how important the radiometric method for dating rocks and determining the age of the Earth had become. It was now twenty years since Holmes had started his work on radiometric dating and, although there was still much to be done in that arena, there was little left to say about the other methods except to historically review them. Holmes' rejection of the vast majority of analyses brought home the fact that there was still an urgent requirement for a much simpler dating procedure that could be applied to common igneous rocks such as granite or basalt. Analyses needed to be done much more quickly and routinely, thereby building up the database much more rapidly and accurately. In 1932 Holmes thought he had found this new method when he published a seminal paper entitled 'The Origin of Igneous Rocks' (Holmes 1932).

A great deal of Holmes' work involved investigation of the evolution of the Earth's crust, which, when combined with his petrographic studies, had naturally led him to question the genesis of many rock types. Whereas it was beginning to be considered that basalts might be derived from melting in the substratum (mantle), the origin of granites was still a hotly debated topic, and Holmes was looking for a method that would determine whether the pre-magmatic source of granites was granitic or basaltic. He believed that a new decay scheme, that of ^{41}K to ^{41}Ca, would be able to distinguish between the two, since the K/Ca ratio was much higher in granites than in basalts and therefore the proportion of Ca represented by ^{41}Ca would, over geological time, be increased in the granites something like twenty-seven times as fast as that in the basalts.

Using Figure 3 to illustrate his ideas, Holmes showed how OE represented the ^{41}Ca/Ca ratio at the time when he considered that the granitic layer (crust) had separated from the basaltic layer (mantle) in the newly formed Earth, and when the ratio would have been the same in both kinds of material. But over geological time, that ratio would greatly increase in the granite (EF), while only slightly increasing in the basalt (ED) due to the K/Ca ratio being almost three times higher in granites than in basalts. If a new granite was then initiated in the Tertiary (T) and the magma was produced by 'refusion' of granitic crustal material (melting of the crust) at F, then the present ^{41}Ca/Ca ratio would fall somewhere near G, whereas if the magma was differentiated from basaltic material at D, the present ratio of the new granite would be close to C. The large difference between the two ratios would be sufficient to identify their pre-magmatic sources.

The K/Ca decay scheme, however, was still theoretical and had not been applied to real rocks, although a suite had been chosen and was awaiting analysis. But Holmes was confident: 'If the method proves to be practicable', he wrote, 'and I can see no reason why it should fail, it will supply a master key for unlocking many of the pre-magmatic secrets of petrogenesis.' And to this he added a footnote:

> It is worthy of note that when the rate of disintegration of K^{41} is accurately known, such lines as EG and EB, and therefore their intersection at point E will be determinable. E corresponds as exactly as one could wish with the date of the earth's origin, and its determination will therefore provide for the first time a direct estimate of the age of the earth. When suitably corrected, the values of Ca^{41}/K for potash felspars [sic] will give a measure of

---

[11] Knopf to Schuchert, 11 February 1913. Courtesy of Yale University Library.

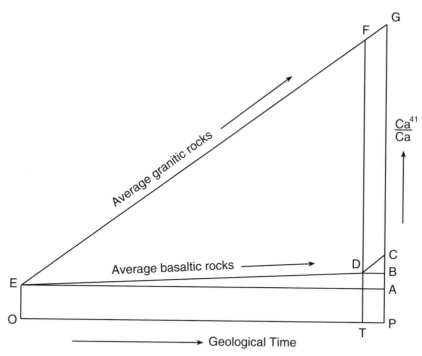

Fig. 3. Holmes' diagram to explain the principles of his new key to petrogenesis, using the apparent decay scheme of ^{41}K to ^{41}Ca, which in 1932 he believed would provide a new method for dating rocks. OE represents the ^{41}Ca/Ca ratio 'at the time when the separation of the granitic and basaltic layers took place in the newly-formed Earth', and when the ratio would have been the same in both kinds of material. But over geological time, that ratio would greatly increase in the granite (EF), while only slightly increasing in the basalt (ED), due to the larger amount of K contained in granites. If a new granite was then initiated in the Tertiary (T) and the magma produced by 'refusion' of granitic crustal material at F then its present-day ratio would fall somewhere near G. On the other hand, if the magma was derived from basaltic material at D the present-day ratio of the new granite would be close to C. The large difference between the two present-day ratios would be sufficient to distinguish between their pre-magmatic sources. This description of the modern-day concept of 'initial ratios' was 30 years ahead of its time. (Redrawn from Holmes 1932).

the age of the granites containing them. It will then be possible to build up a geological timescale, to fix the ages of granites of unknown geological position and to effect a detailed correlation of Pre-Cambrian rocks from continent to continent in far more detail than is possible at present.

It is difficult to emphasize adequately the enormous significance of this paper by Holmes. The 'master key' he is describing here is what geochemists today call 'initial ratios', and since the 1960s initial ratios have been the bedrock upon which modern geochemistry was, and still is, being built. The ability to determine the original isotopic ratio of almost any igneous rock opened the door to our understanding of the evolution of the Earth's crust and mantle, and now allows us to date the vast majority of igneous rocks. In addition, Holmes saw that the age of the Earth could ultimately be determined using these principles. But because the decay scheme he was using in this paper turned out to be erroneous, the method did fail, for it had not yet been established which of the potassium isotopes was radioactive. It is in fact ^{40}K that decays to both ^{40}Ca and ^{40}Ar, but ^{40}K was still to be discovered by Alfred Nier some three years later (Harper 1973). Nevertheless, the principles of Holmes' new key to petrogenesis are precisely those in use today, but it was another thirty years before initial ratios were 'discovered' with the recognition that the ideas described here by Holmes, but never ascribed to him, could be applied to the rubidium–strontium method (Compston & Pidgeon 1962).

Only Kalvero Rankama in his excellent book *Isotope Geology* (Rankama 1954, p. 314) seems to have recognized in print the potential of Holmes' method, but it was Norman Snelling who pointed out to me the full significance of the paper, having come across it himself in 1963, following his contribution to a paper on the Ntungamo gneiss dome in Uganda (Snelling 1965):

> I sent a typescript to Holmes in 1963 (because I thought he would be interested in the age of the gneisses, little knowing that it would be the genetic significance of the initial ratio that would excite his attention) and he sent me a reprint of his 1932 paper with an exclamation mark against the date. Until then I just did not know that everything I said in my contribution regarding the significance of initial Sr isotope ratios had been worked out by Holmes some thirty years before (Snelling pers. comm., 1998).

When presented with the Penrose medal in 1956, Holmes described how his greatest satisfaction arose from this work on the K/Ca method, although ultimately it resulted in the greatest disappointment of his career (Hedberg 1957).

Helium again?

In 1928, Fritz Paneth, who had worked with Bob Lawson at the Vienna Institute of Radium during the First World War and was now professor of chemistry in Berlin, had developed very precise techniques for measuring minute amounts of helium. The problem with the uranium–lead technique was that in the common rocks such as basalt, uranium and lead were present in such small quantities as to then be immeasurable. Helium, on the other hand, was ubiquitous – for each uranium atom that decayed to lead, eight helium atoms were produced, so large concentrations of helium were generated over geological time, even in the common rocks. The problem here was not the shortage of uranium, but the surplus of helium which either could not be contained in the mineral over long periods of geological time, or was lost in the preparation process. But Paneth considered that precisely because the concentrations of uranium in common minerals (those not normally selected for age dating) were so low, the much smaller volumes of helium produced would have a good chance of being retained in the mineral. Holmes seized the initiative and arranged to have some rocks analysed in Paneth's laboratory in Berlin.

Samples were taken from two famous dolerites from northern England, known to be of very different geological (stratigraphic) ages: the Whin Sill was believed to have a very late Carboniferous age, while the Cleveland Dyke was considered to be middle to early Tertiary. When Paneth sent the helium results back, the Whin Sill gave a radiometric age of 182 Ma while the Cleveland Dyke gave 26 Ma, values Holmes considered 'to be in excellent agreement with the geological evidence' (Holmes & Dubey 1929, p. 794). The ideal test of these results would have been to compare the helium ages with the more reliable lead ages analysed on the same rocks. But it was precisely because these two rocks contained so little lead that they had to be dated by the helium method, thus controls had to be found in lead ages from other rocks *believed* to be of the same geological age. The problem was, of course, that very few reliable lead ages were available anywhere in the world, so Holmes had to make do with what was available.

Today we know that the Whin Sill was indeed intruded during the late Carboniferous, but its radiometric age is around 295 Ma. Similarly, the Cleveland Dyke is known to be early Tertiary, but its radiometric age is about 60 Ma. The huge discrepancy between today's radiometric ages and the helium results found by Paneth for these rocks well illustrates not only how inaccurate the helium method was, but, more importantly, how badly constrained the geological ages of most igneous rocks were at that time. Although in this case the geological ages of the Whin Sill and Cleveland Dyke were quite well constrained, the so-called 'Late Carboniferous' control gave a lead age of 192 Ma, while the 'Miocene' control gave a lead age of 36 Ma. It was not the lead ages that were wrong, but the geological ages of these rocks. The result was that the wrongly assigned geological ages of the control rocks facilitated the acceptance of wildly inaccurate helium ages for the Whin Sill and Cleveland Dyke and led Holmes to believe that the helium method would solve all his dating problems.

In 1929 Holmes was not aware of the poor geological constraints on the control rocks. As far as he was concerned, both of Paneth's helium results were only 10 Ma younger than the uranium–lead determined ages for those geological periods. And although he recognized the helium results to be 'slightly low', he considered they concurred 'quite satisfactorily with the scanty results based on lead ages'. Despite the historical problems with helium, Holmes could not suppress his excitement when reporting these results: '... there is now available a practical means ... of building up a geological time-scale which, checked by a few reliable lead-ratios here and there, should become far more detailed than could ever be realised by means of the lead method alone' (Holmes & Dubey 1929, p. 795).

In March 1932 Holmes went to the United States to give the prestigious Lowell Lecture Series at Harvard and other universities, at the invitation of Reginald Daly. During his trip he discussed with his colleagues there the age dating results he had obtained using Paneth's new helium method that appeared to be so successful. On an earlier occasion he had put out a plea to American universities to help him with his dream of building a geological timescale – their wealthy research establishments being the only ones that could really afford to consider this 'Herculean task' (Holmes 1926, p. 482) – so while in America, and with the positive helium results from Paneth to support his case, he raised the subject again. His reasoning was persuasive and it was agreed that William Urry, then working with Paneth in Berlin, would come to the United States and finally make Holmes' dream come true.

The idea was to use basalts because it was easier to assign a geological age to basalts than to any other common igneous rock type. In addition, basalts were widely distributed in space and time and so representatives of all ages were readily available. But the crucial factor determining a sample's suitability was how well its position in the geological column could be determined and rocks would only be chosen if their geological age could be really well defined. The samples, their geological ages ranging from the Cambrian to the top of the Tertiary, were carefully selected from sites all over the USA, Canada and Europe.

Working at the Massachusetts Institute for Technology (MIT), Urry took nearly four years to analyse the thirty-nine samples initially selected, but when Holmes saw the results he was overjoyed. He was just writing the third edition of his book *The Age of the Earth* and in it he compared the helium ages with the established uranium–lead ages, believed to be from the same geological horizons (Holmes 1937a, fig. 3). The correspondence was astonishing and Holmes wrote euphorically:

> ... it is remarkable how consistently the age estimates fall into appropriate positions. That this stringent test of internal consistency is satisfactorily met must be regarded as the final proof that the ages calculated from lead and helium ratios are at least of the right order and that no serious error is anywhere involved (Holmes 1937a, p. 179).

But as the third edition was going to press a problem was noticed by an assistant professor of physics, Robley Evans, who had recently come to MIT to develop yet another new method for determining helium ages. When Evans' new technique was sufficiently well-established and giving consistent results, ages were determined on some of the same samples that Urry had used and the two sets were compared. Urry's ages were found to be significantly older than those determined by Evans. After exhaustive investigations the problem was traced to a faulty calibration of the ionization chamber used by Urry (Harper 1973) which resulted in all his ages being consistently too old, and fortuitously in line with the uranium–lead ages. It was a devastating moment as they realized that five years of hard work would have to be discarded. Helium was still leaking, even from these barely radioactive samples. Goodman & Evans (1941, p. 527) summed up the position: 'These general inconsistencies in helium age ratios indicated quite clearly that there was some fundamental failure in the helium method'. But it was too late for Holmes' book that was already in print.

The missing isotopes

Until 1929 it was still thought that the only isotope of lead not to be derived from radioactive decay was ^{207}Pb. Furthermore, no isotopes of uranium had been found. It therefore came as a considerable surprise when Frederick Aston, working at the Cavendish Laboratory in Cambridge on an early mass spectrograph, produced some new results suggesting that ^{207}Pb could not, after all, be the non-radiogenic lead isotope, but that it must result from the decay of a hitherto unknown isotope of uranium. Aston discussed the problem with Rutherford who not only agreed with his deductions but went on to calculate that the new uranium isotope must have an isotopic number of 235. Indeed, it was uranium 235 which, because it represented less than 1% of total uranium, had so far gone unnoticed. But if ^{235}U decayed to ^{207}Pb, what then was the non-radiogenic isotope of lead? Did it exist at all? Well Holmes had already identified samples that contained far too much lead to have been simply derived from radioactive decay, given the amount of uranium or thorium present, so there must have been some lead there at the same time as the parent elements were incorporated into the mineral. But what was its isotope number? The question remained unanswered for several more years.

On completing his PhD in 1936, Alfred Nier had been invited to work at Harvard University by K. T. Bainbridge, who had established an enviable reputation for his spectrographic studies of the masses and relative abundances of isotopes. The invitation had been accompanied by a large grant for the construction of a bigger and better mass spectrometer, and Nier was happy to accept (Nier 1982). He soon became interested in the problems of measuring geological time, and recognized that the possibility of being able to measure accurately the abundances of individual lead isotopes, hitherto only determined by their atomic weights, would add a new dimension to the field. However, the key to the successful exploitation of the lead method lay in knowing the isotopic ratio of the two *uranium* isotopes, a value that had not yet been determined. Having obtained, with some difficulty, samples of rare and highly volatile uranium compounds, in 1939 Nier was able to measure the ^{238}U/^{235}U ratio. He determined it to be 139 (Nier 1939), very close to the value of 139.6 accepted today.

Following this success, a study of the isotopic composition of twelve common lead ores was commenced, of which nine had previously had chemical atomic weight determinations made on them, all yielding a value close to 207.21, in spite of their wide variation in geological age and geographical location (Nier 1938). This led Nier to assume that the isotopic abundances would be equally consistent, but results from the initial six samples were so surprising that Nier was made to repeat them. Despite having a constant atomic weight, the individual isotopic abundances were shown to vary considerably from sample to sample, but results on all twelve samples eventually confirmed this wide variation in the lead isotopes and also illustrated that it was not a random variation since those which contained more ^{206}Pb also had more ^{207}Pb and ^{208}Pb. Nier interpreted these results by proposing that all the lead ores must consist of a mixture of *primeval* leads of fixed composition, which had existed at the time of the formation of the Earth, to which had been added radiogenic leads generated by the subsequent decay of uranium and thorium, and that these mixtures fortuitously resulted in a constant atomic weight.

Before commencement of the study, great efforts had been made to ensure that all the equipment, built by hand, was free from contamination before the new analyses were made, so that in addition to seeing the previously known isotopes of Pb, for the first time ^{204}Pb, the true non-radiogenic lead isotope, was finally identified as a measurable peak and recognized as the missing lead isotope. Its peak had previously been obscured by ^{204}Hg (mercury), which was used for calibration purposes.

In June 1938 Nier wrote to Holmes informing him of his results,[12] knowing that they would be of considerable interest to him, and offering to collaborate with Holmes if he would provide some samples. Given that Nier only required ten milligrams of lead for an analysis, which could be done in two days, it is extremely surprising that Holmes took three months to reply and when he did so he said that he had no samples suitable for analysis.[13] One interpretation of Holmes' uncharacteristic response was that his first wife, Maggie, was extremely ill with cancer – she died three weeks after Holmes wrote his letter (Lewis 2000) – and that Holmes was too distraught to concentrate on his work. But in fact all that year Holmes had been embroiled in a very public argument through the pages of *Economic Geology* over his ideas about the origin of primary lead ores (Holmes 1937*b*, 1938; Knopf 1937; Wells 1938; Graton 1938; Keevil 1938). Nier's data seemed to clinch the arguments against Holmes' case for in his reply to Nier, Holmes writes: 'I was surprised to find that both Graton and Keevil used your results to discredit my deductions'. Consequently Holmes might well have not felt inclined to collaborate with Nier at this time. Given his estrangement from his wife (Lewis 2000) this seems a likely explanation, since his reply to Nier certainly indicates that he recognized the value of Nier's 'most important work'. With the intervention of the Second World War it was to be another six years before Holmes contacted Nier again, whereupon a remarkable correspondence commenced between them and continued over the following sixteen years, both sides of which are still preserved at the University of Minnesota where Nier worked when he left Harvard.

In the early years of the Second World War, Nier continued to publish more results on the variations in lead isotopes on a range of rocks, but then he was required to work on development of the atomic bomb and was unable to continue his research on lead. It was thus April 1945 before Holmes' second letter, written nearly five months previously, finally found Nier in New York on 'leave of absence' from Minnesota University. Holmes was requesting copies of all Nier's papers written on the subject of lead isotopes, since he had been cherishing the idea, ever since the first one had been published, that the data would shed light on the time when the Earth's primeval lead had begun to be contaminated by radiogenic lead (Lewis in press). But in the course of this work, Holmes also recognized the value of using some of Nier's data as control points in the construction of a geological timescale.

---

[12] Nier to Holmes, 22 June 1938. Courtesy of Minnesota University Archives.

[13] Holmes to Nier, 2 September 1938. Courtesy of Minnesota University Archives.

Construction of a geological timescale

Since 1911 when Holmes constructed his first geological timescale based on only a very few data points (Table 1) he had tried on many occasions to improve the resolution of the timescale as new data became available, and notable attempts before the Second World War can be found in Holmes (1920), Holmes & Lawson (1927) and Holmes (1933). The latter resulted from a lecture delivered as part of the Lowell Lecture Series to the Washington Academy of Sciences in 1932. In it Holmes showed a diagram of maximum sediment thickness plotted against geological time derived from lead ratios, and for the first time he ventured to place boundaries at the top and bottom of each geological period, from the Tertiary to the Cambrian (Holmes 1933, fig. 10), but no explanation is given as to how the sediment thicknesses were determined, or how accurate the lead ages were. However in 1947, following Nier's work on lead isotope ratios, Holmes made a similar, but much more rigorous attempt, to construct a geological timescale (Holmes 1947). Recognizing that although the 'maximum thickness' of sediments was indeed some measure of geological time, it could however, only represent a balance between sediment income and sediment loss, thus it was clear that some independent measure of time was essential, other than the rates of sediment deposition and erosion.

Unfortunately, only five of Nier's samples were deemed suitable, since the majority were Precambrian rocks, which the timescale would not cover due to the impossibility of getting estimates for the thickness of Precambrian sediments. Each of the five lead ratios was rigorously examined by Holmes. The age was corrected for the loss of lead due to alteration of the minerals and a 'most probable' radiometric age for the sample was duly arrived at. These were then placed in their appropriate positions within the sedimentary pile at the points considered most likely to represent the geological age of the sample, which had also been meticulously analysed. An interpolated curve joined up the dots (Fig. 4) and although Holmes was fully aware that this method of interpolation had obvious weaknesses, it suited his purpose at the time as he was unable to find a better alternative. Two curves, A and B were constructed which, because of uncertainties placed on the geological ages of the

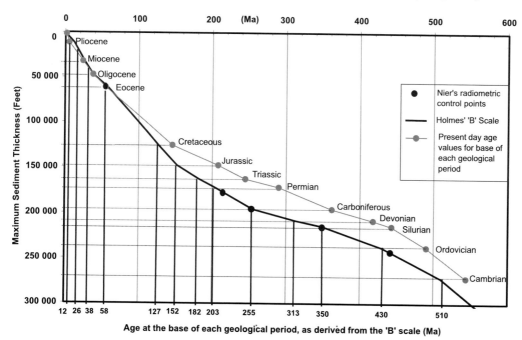

Fig. 4. Holmes' 1947 geological timescale. Using five well-constrained lead ages determined by Alfred Nier (black dots), Holmes positioned these according to their geological age (which had been defined as accurately as possible) at their correct locations within the sedimentary pile, to act as control points in the timescale. By joining up the dots he defined the 'B scale' (black line) from which could be read the date of the base of each geological period, or the date of any rock that had a known geological age. For comparison, present-day radiometric ages for the base of each geological period (grey dots) are plotted against Holmes' estimate for the thickness of each geological period. (Adapted from Holmes 1947).

Table 2. *Holmes' 1947 'B scale' age determinations at base of geological periods*

Holmes' 'B scale' (Ma)		Today's values (Ma)
1	Pleistocene	1.6
12	Pliocene	5.1
26	Miocene	24
38	Oligocene	38
58	Eocene	55
–	Palaeocene	65
127	Cretaceous	146
152	Jurassic	208
182	Triassic	245
203	Permian	290
255	Carboniferous	362
313	Devonian	418
350	Silurian	443
430	Ordovician	490
510	Cambrian	544

Comparison of the age values determined by Holmes (as read from the 'B scale' – originally drawn accurately on graph paper) with present-day values.

samples, defined upper and lower geological age limits. These uncertainties however, favoured the B scale, and this was the one that became widely used.

Although now we can see the discrepancies in Holmes' results with today's values (Fig. 4 and Table 2), the B scale enjoyed an unexpected success for more than ten years in a wide variety of scientific applications. When he came to revise it in 1960, Holmes said he had come to 'bury the B scale, not to praise it' (Holmes 1960, p. 184), for he recognized that so much work had been done in the interim that the B scale should be viewed as 'no more than historical interest' and that in turn, any new scale would need revising in a few years' time. By the 1960s many new dating techniques were available, mass spectrometers were relatively sophisticated machines following their rapid development during research on the atom bomb, and there was a hunger for data driven by the exciting potential of proving, or otherwise, the theory of continental drift.

'Father' of geological timescales

From the early 1950s until his death, Holmes advised the British Museum (Natural History) on the evolving geological timescale, and the preserved correspondence with Helen Muir-Wood provides a fascinating record of the changing details over that period. In 1965 a large timescale was erected at the entrance of the Museum for all

Fig. 5. Arthur Holmes, circa 1956, 'Father' of the geological timescales.

to see,[14] copied from that published in the second edition of Holmes' *Principles of Physical Geology* (Holmes 1965), then just published.[15] As it turned out, it was a fitting memorial to his life-long vision. Arthur Holmes died of bronchial pneumonia a few months later, at Bolingbroke hospital in London, on 20 September 1965. It was Alfred Nier who christened Holmes the 'Father' of geological timescales (Fig. 5),[16] seeing him as the leading pioneer in developing the geological timescale, which he had spent a lifetime pursuing.

Today 'time' is the framework onto which we hang all geological events and from which we developed a unifying theory that explains all geological processes, for it was development of the K/Ar timescale of magnetic reversals which finally proved to be the key that unlocked the theory of seafloor spreading, and all that followed from that. The great 'unifying' theory

---

[14] Muir-Wood to Holmes, 12 May 1965. By permission of the Trustees of the Natural History Museum.
[15] Holmes to Muir-Wood, 13 May 1965. By permission of the Trustees of the Natural History Museum.
[16] Nier to Holmes, 3 May 1960. Courtesy of Minnesota University Archives.

of plate tectonics has largely been unified by understanding the *rates* of geological processes, which could only be achieved by measuring time – exactly what Holmes predicted back in 1913 (Holmes 1913, p. 176).

I gratefully acknowledge Research Grant No. 20235 from the Royal Society which assisted with travel to visit Holmes' family in Vienna, and to research archives in the United States. I would also like to thank H. Torrens, I. Fairchild and other members of the Geology Department at Keele University for their sustained support of this work.

References

BOLTWOOD, B. B. 1905. On the ultimate disintegration products of the radio-active elements. *American Journal of Science*, **20**, 253–267.

BOLTWOOD, B. B. 1907. On the ultimate disintegration products of the radio-active elements. Part II. The disintegration products of uranium. *American Journal of Science*, **23**, 77–88.

COMPSTON, W. & PIDGEON, R. T. 1962. Rubidium–strontium dating of shales by the total-rock method. *Journal of Geophysical Research*, **67**, 3493–3502.

CURIE, P. & LABORDE, A. 1903. Sur la chaleur degagee spontanement par les sels de radium. *Comptes Rendus*, **136**, 673–675.

DALRYMPLE, G. B. 2001. The age of the Earth in the twentieth century: a problem (mostly) solved. *In*: LEWIS, C. L. E. & KNELL, S. J. (eds) *The Age of the Earth: from 4004 BC to AD 2002*. Geological Society, London, Special Publications, **190**, 205–222.

DUNHAM, K. C. 1966. Arthur Holmes. *Biographical Memoirs of Fellows of the Royal Society*, **12**, 291–310.

GOODMAN, D. & EVANS, R. D. 1941. Age measurements by radioactivity. *Geological Society of America Bulletin*, **52**, 521–529.

GRATON, L. C. 1938. Ores: from magmas, or deeper? A reply to Arthur Holmes. *Economic Geology*, **33**, 251–286.

HARPER, C. T. (ed.) 1973. *Benchmark Papers in Geology. Geochronology: Radiometric Dating of Rocks and Minerals*. Dowden, Hutchinson & Ross, Stroudsberg, PA.

HEDBERG, HOLLIS, D. 1957. Presentation of Penrose Medal to Arthur Holmes. *Proceedings of the Geological Society of America for 1956*, 69–74.

HOLMES, A. 1911. The association of lead with uranium in rock-minerals, and its application to the measurement of geological time. *Proceedings of the Royal Society (A)*, **85**, 248–256.

HOLMES, A. 1913. *The Age of the Earth*. Harper, London and New York.

HOLMES, A. 1916. Contribution to the discussion on radioactive evidence of the age of the earth, in Section C – Geology. *Report of the British Association for the Advancement of Science, Section C – Geology*, 432–434.

HOLMES, A. 1920. The measurement of geological time. *Discovery*, **1**, 108–114.

HOLMES, A. 1926. Rock-lead, ore-lead, and the age of the Earth. *Nature*, **117**, 482.

HOLMES, A. 1931. Problems of the Earth's Crust: Discussion in Section E (Geography) of the British Association for the Advancement of Science. *Geographical Journal*, **78**, 433–455.

HOLMES, A. 1932. The origin of igneous rocks. *Geological Magazine*, **69**, 543–558.

HOLMES, A. 1933. The thermal history of the Earth. *Journal of the Washington Academy of Sciences*, **23**, 169–195.

HOLMES, A. 1937a. *The Age of the Earth* (Third Edition). Nelson, London.

HOLMES, A. 1937b. The origin of primary lead ores. *Economic Geology*, **32**, 763–782.

HOLMES, A. 1938. The origin of primary lead ores: Paper II. *Economic Geology*, **33**, 829–867.

HOLMES, A. 1944. *Principles of Physical Geology*. Nelson, London.

HOLMES, A. 1947. The construction of a geological time-scale. *Transactions of the Geological Society of Glasgow*, **21**, 117–152.

HOLMES, A. 1960. A revised geological time-scale. *Transactions of the Edinburgh Geological Society*, **17**, 183–216.

HOLMES, A. 1962. 'Absolute' age; a meaningless term. *Nature*, **196**, 1238.

HOLMES, A. 1963. Introduction. *In*: Rankama, K. (ed.) *The Geologic Systems: The Precambrian, Volume 1*. Interscience, New York, xi–xxiv.

HOLMES, A. 1965. *Principles of Physical Geology* (second edition). Nelson, London.

HOLMES, A. & DUBEY, V. S. 1929. Estimates of the ages of the Whin Sill and the Cleveland Dyke by the helium method. *Nature*, **123**, 794–795.

HOLMES, A. & LAWSON, R. W. 1914. Lead and the end product of thorium (Part 1). *Philosophical Magazine*, **28**, 824–840.

HOLMES, A. & LAWSON, R. W. 1927. Factors involved in the calculation of the ages of radioactive minerals. *American Journal of Science*, **13**, 327–344.

KEEVIL, N. 1938. Thorium–uranium ratios of rocks and their relation to lead ore genesis. *Economic Geology*, **33**, 685–696.

KNOPF, A. 1937. The origin of primary lead ores. *Economic Geology*, **32**, 1061–1064.

KNOPF, A., SCHUCHERT, C., KOVARIK, A. F., HOLMES, A. & BROWN, E. W. 1931. Physics of the Earth – IV: The Age of the Earth. *Bulletin of the National Research Council, Washington*, **80**.

LEWIS, C. L. E. 2000. *The Dating Game*. Cambridge University, Cambridge.

LEWIS, C. L. E. (in press.) Arthur Holmes' unifying theory: from radioactivity to continental drift. *In*: OLDROYD, D. (ed.) *The Earth Inside and Out: Some Major Developments in Geology in the Twentieth Century*. Geological Society, London, Special Publications.

NIER, A. O. 1938. Variations in the relative abundances of the isotopes of common lead from various sources. *Journal of the American Chemical Society*, **60**, 1571–1576.

NIER, A. O. 1939. The isotopic constitution of uranium and the half-lives of the uranium isotopes. *Physical Review*, **55**, 150–153.

NIER, A. O. 1982. Some reminiscences of isotopes, geochronology and mass spectrometry. *Annual Review of Earth and Planetary Science*, **9**, 1–17.

RANKAMA, K. 1954. Possible application of radiometric ^{40}Ca to petrogenetic problems. *In*: RANKAMA, K. (ed.) *Isotope Geology*, Pergamon Press, London, 314–315.

REYNOLDS, D. L. 1968. Memorial of Arthur Holmes. *American Mineralogist*, **53**, 560–566.

RUTHERFORD, E. 1905. Present problems in radioactivity. *Popular Science Monthly*, **67**, 5–34.

RUTHERFORD, E. 1906. The production of helium from radium and the transformation of matter. *In: Radioactive Transformations*. Yale University, New Haven, 187–193.

RUTHERFORD, E. & BOLTWOOD, B. 1905. The relative proportion of radium and uranium in radioactive minerals. *American Journal of Science*, **20**, 55–56.

RUTHERFORD, E. & SODDY, F. 1902a. The cause and nature of radioactivity – Part 1. *The London, Edinburgh and Dublin Philosophical Magazine and Journal of Science*, **6**, 370–396.

RUTHERFORD, E. & SODDY, F. 1902b. The cause and nature of radioactivity – Part II. *The London, Edinburgh and Dublin Philosophical Magazine and Journal of Science*, **6**, 569–585.

RUTHERFORD, E. & SODDY, F. 1903. Radioactive change. *Philosophical Magazine*, **5**, 576–591.

SHIPLEY, B. C. 2001. 'Had Lord Kelvin a right?': John Perry, natural selection, and the age of the Earth, 1894–1895. *In*: LEWIS, C. L. E. & KNELL, S. J. (eds) *The Age of the Earth: from 4004 BC to AD 2002*. Geological Society, London, Special Publications, **190**, 91–105.

SNELLING, N. 1965. Written contribution to: NICHOLSON, R. The structure and metamorphism of the Ntungamo Gneiss Dome and their time relation to the development of the dome. *Quarterly Journal of the Geological Society*, **121**, 159–161.

SODDY, F. 1913. Intra-atomic charge. *Nature*, **92**, 399–400.

SODDY, F. 1915. Contribution to the discussion on radioactive evidence of the age of the Earth, in Section C – Geology. *Report of the British Association for the Advancement of Science, Manchester*, 434.

SOLLAS, W. J. 1900. President's Address, Section C – Geology: Evolutionary Geology. *Report of the British Association for the Advancement of Science, Bradford*, 711–730.

THOMPSON, J. J. 1896. President's Address to Section A – Mathematical and Physical Science. *Report of the British Association for the Advancement of Science, Liverpool*, 699–707.

WALCOTT, C. D. 1893. Geological time as indicated by the sedimentary rocks of North America. *Journal of Geology*, **1**, 639–676.

WELLS, R. C. 1938. The origin of primary lead ores. *Economic Geology*, **33**, 216–217.

WYSE JACKSON, P. N. 2001. John Joly (1857–1933) and his determinations of the age of the Earth. *In*: LEWIS, C. L. E. & KNELL, S. J. (eds) *The Age of the Earth: from 4004 BC to AD 2002*. Geological Society, London, Special Publications, **190**, 107–119.

The age of the Earth in the United States (1891–1931): from the geological viewpoint

ELLIS L. YOCHELSON[1] & CHERRY L. E. LEWIS[2]

[1] *Department of Palaeobiology, National Museum of Natural History, Washington, DC 20560-0121, USA (email: yochelson.ellis@nmnh.si.edu)*
[2] *History of Geology Group, 21 Fowler Street, Macclesfield, Cheshire SK10 2AN, UK (email: clelewis@aol.com)*

Abstract: In North America, prior to the Second World War, discussions on the age of the Earth were a minuscule part of the geological literature, as demonstrated by the small number of papers indexed to the subject in bibliographies. Indeed, during the first quarter of the twentieth century, there were few general papers on this topic circulating among those geologists who dealt with sedimentary rocks and fossils; nevertheless, evidence is provided that many geologists were aware of the 'debate' going on in Britain.

As the methodology for determining the length of geological time dramatically changed during the four decades represented here, so too did the evolution of ideas about the age of the Earth. These can conveniently be divided into three time periods: before, during and after the discovery that radioactivity could be applied to the dating of rocks. The first section reviews the attitudes of geologists in America to the age of the Earth in the 1890s. It is followed by their reactions to the discovery of radioactivity. The third part discusses two major publications on the age of the Earth which reflect the ultimate acceptance, by geologists, of the long timescale revealed by radioactivity. Because much of the early work on radioactivity was being done in Europe, American geologists were marginally later than their British counterparts in accepting the concept of radiometric dating, but by the end of the period under consideration they led the field in geochronology.

Although there is mention in the Bible of the 'everlasting hills', in a more modern context, Charles Darwin's estimate of 300 million years (Ma) for the erosion of the Weald in southern England, given in the first edition of *On the Origin of Species*, marked the beginning of serious discussion and argument amongst geologists and physicists as to the numerical age of the Earth. The enormous timespan required by Darwin for evolution to produce the predicted changes in organisms incensed William Thomson (later Lord Kelvin), a British physicist of international standing, and led him to calculate the age of the Earth for himself using the laws of thermodynamics. The story of the diminution of the age of the Earth on the basis of new calculations and new assumptions by Lord Kelvin is now well known (Burchfield 1975; Albritton 1980; Knell & Lewis 2001) and needs no further elaboration here.

Even though Darwin said that his calculation was only an attempt to give a crude idea of geological time, a year later the American edition of his book carried a footnote revising some of the assumptions made and a disclaimer that he had been 'rash and unguarded in the calculation'. By the third edition, printed in 1861, it had disappeared altogether. However, a letter from Darwin to James Croll, dated 31 January 1869, some eight years after the calculation had been removed, indicates the difficulty still faced by Darwin: 'I am greatly troubled by the short duration of the world according to Sir W. Thomson for I require for my theoretical views a long period before the Cambrian formation'.[1]
In America, several editions of this book were reprinted but printings were also pirated, so it is difficult to determine what impact, if any, Darwin's number had on the American geological community. Certainly the biologists in America who supported Darwinian evolution also supported a long timespan. But overall, whereas this subject generated much argument and research in Britain, in America there was little published comment on the topic at this

---

[1] Held by the American Philosophical Society, Philadelphia, Pennsylvania, USA.

time. The 'age' debate involving biologists and geologists versus the physicists was primarily a British affair.

Survey of literature

The bibliographies of North American geology, compiled by the United States Geological Survey (USGS), began only a few years after the organization was founded in 1879, and continued for about three-quarters of a century. The indexes to these volumes provide a thumbnail survey of what was published on a plethora of geological subjects and, as such, can indicate the extent to which any one topic was of interest to the larger geological community. It is possible that a few papers are inappropriately indexed, but because the compilation of bibliographies and the indexing of the papers listed was an activity which lasted for so long, there must have been a continuity of internal consistency in the indexing effort. By their very nature, these bibliographies included almost no literature from the field of physics or chemistry where one might find more papers on the subject under consideration. Their aim was to document literature in the science of *geology*, and in that sense they are the appropriate source for data regarding this particular work since they reflect the interest shown in the subject by *geologists*.

The oldest paper indexed on the age of the Earth in the first comprehensive volume, covering the period 1785–1918, is from 1885 by a non-geologist named Templin who was much concerned with the age of the world (Nickles[2] 1924, p. 152). Four papers are listed from 1893 and these are considered below. Five years after that, the views of Lord Kelvin were made well known in North America with publication of his paper 'The age of the earth as an abode fitted for life', in both the *Annual Report of the Smithsonian Institution* (Thomson 1898) which, interestingly, was before that paper appeared in Britain (although its subject matter had been discussed since Kelvin addressed the Victoria Institute on it in 1897), and in *Science* (Thomson 1899). Nine more titles are indexed from the first part of the 1900s, including a publication by Joly (1901), and two additional titles are subindexed to 'radioactivity, evidence of'. Under the headings 'Uranium' and 'Radioactivity' we find two other publications by Joly (1908, 1909), the latter being his book *Radioactivity and Geology*, which suggests that American geologists were becoming aware of developments in Europe by this time.

However, only a single paper by Boltwood (1905) is listed in the compilations, and it is not apparent where, or even if, it was ever indexed. As is well known, this Yale professor is generally credited with having proposed that lead was derived from the decay of uranium, and he made the first uranium–lead age estimates based on published U/Pb data (Dalrymple 2001; Lewis 2001). He wrote a series of important papers on these subjects, but the fact that he was a chemist probably explains their omission from this index, despite their relevance to geology.

In the following volume of bibliographies (1919–1928), eleven titles are given under the general heading of 'Earth, Age' (Nickles 1921, p. 765), suggesting a growing interest in the problem. However, among the general papers, all save one can be considered 'popular' accounts, textbooks or abstracts, which in essence simply mention the age of the Earth and do not discuss it at any length. During the final compilation considered here (1929–1939), the bibliography has 14 titles indexed under 'Earth, age, general' (Thom 1944, p. 1163), but an additional 68 titles are under more specific subheadings, about two-thirds of these being concerned, in one way or another, with radioactivity. This reflects a gradual movement away from a variety of 'geological' methods of calculating time to concentration on the concept of radioactive decay as a means of determining geological time, which echoed, and eventually overtook, events occurring in Britain.

Notwithstanding the apparent lack of geological literature concerning the age of the Earth during the four decades under consideration, interpretation of American attitudes to the age of the Earth, from a perspective other than simply counting titles, may be clarified by detailing several events in America. Accordingly, the concentration herein is on a few significant events and publications that appeared within three sequences of time which are before, during and after the discovery that radioactivity could be applied to the determination of geological time.

1891–1900: countdown

The few 'age of the Earth' estimates made in North America before 1900 largely used the traditional methods established by geologists in Britain; these were based either on rates of erosion or deposition of sedimentary strata, and their results produced figures similar to those derived by British workers who generally arrived at an answer of about 100 Ma. From the lack of discussion of the subject in the few American textbooks on geology from this era, it appears that

---

[2] See Bassler (1947) on Nickles.

Fig. 1. Three directors photographed during April 1896, near Harpers Ferry, West Virginia, with the Potomac River in the background. Charles Doolittle Walcott, USGS director (1894–1907), is holding a small hammer in his left hand; John Wesley Powell, director of the Bureau of Ethnology (1879–1902) and of the USGS (1880–1894), is holding a hat in his only hand; Sir Archibald Geikie, director-general of the Geological Survey of Great Britain (1882–1901), is holding a cap in his left hand. Photo by J. P. Iddings, USGS photographic library collection, Denver, Colorado, USA.

this 100 Ma time interval was tacitly accepted in North American professional circles, although digressions from this value can also be found.[3]

One need only cite the activities of Wilber John McGee. That gentleman, who styled himself, without periods, as W J McGee ('no stop' McGee to friends and 'full stop' to detractors), was an important member of the USGS and a close advisor to its director John Wesley Powell (Fig. 1). In 1892, the USGS was in grave political difficulty with Congress concerning irrigation in the western United States (Stegner 1953; Dupree 1957). The USGS had been given the responsibility of determining the appropriate sites for irrigation dams, which led to claims on Federal land being staked as the USGS surveyors were working. But shortly after the work began, a bureaucratic decision was made that no public land claims anywhere in the west would be valid until after all the irrigation surveys were completed. This put the USGS, through no fault of its own, in direct conflict with settlers in the west. It would have been almost impossible to make the situation of the USGS much worse than it was, yet McGee managed it by starting a controversy concerning the age of the Earth, while at the same time making scathing remarks about certain individuals of high rank.

At the annual meeting of the American Association for the Advancement of Science meeting in 1892, McGee presented results he had obtained for the age of the Earth: '... he used erosion as a measure of time to fix the length of recent geological epochs and then extrapolated using Dana's ratios of the lengths of the different eras,[4] to fix the age of the Earth at 15,000 million years, of which 7,000 million had elapsed

---

[3] The number of 100 Ma deserves some consideration. It is easy to roll off the tongue in discussion, and with 'about' added, it conveys a great deal yet does not fall into the trap of spurious precision. 'One hundred' is also a 'good' number. When one examines the history of institutions, centennial celebrations tend to attract more attention than bicentennials; there are few celebrations of sesquicentennials and almost no one pays attention to a 125th anniversary.

[4] Dana considered the ratio of the Palaeozoic:Mesozoic:Cenozoic to be 12:3:1 (Dana 1895).

since the beginning of the Palaeozoic' (Rabbitt 1980, p. 216). McGee's (1892a) estimates still stand as the largest assessments for the age of the Earth ever determined. The contrast with Kelvin's meagre 100 million years could not have been more marked, but the following year Clarence King, the first director of the USGS, was to provide Kelvin with ammunition to reduce even further his age for the Earth.

Only five years after graduating from Yale in 1862, King was in charge of the Geological Exploration of the Fortieth Parallel, an enormous tract of land about 100 miles wide, covering the Cordilleran ranges from eastern Colorado to the California boundary line. The seven-volume report that resulted reached perhaps the highest standard yet obtained by government publications, and within it King introduced into American mapping the system of denoting topography by contour lines. By 1879 he was appointed as the first director of the USGS, but after less than two years in that office, King left to seek fame and fortune in the mining business, and ultimately found neither.

King had instituted, and partly financed, laboratory work by the physicist Carl Barus on several physical problems in geology. Amongst other investigations, Barus had determined the melting point of an igneous rock, diabase, under various conditions of temperature and pressure. The melting temperature of 1400°C was found to be much lower than the assumed value (7000°C) used by Kelvin for his calculations of the age of the Earth. King, making assumptions about a rigid Earth lacking any liquid layer and steadily cooling, but with temperature increasing towards the core, combined inferences on temperature gradients in the Earth with the new data of Barus on the fusion of diabase to obtain a number for the age of the Earth of 24 million years. He then compared his age to that determined by Kelvin and others for the age of the sun (Knell and Lewis 2001) and concluded: 'The earth's age, about twenty-four millions of years, accords with the fifteen or twenty millions found for the sun' (King 1893, p. 20). His paper ended with a challenge to geologists to demonstrate that things were otherwise:

> ... the concordance of results between the ages of sun and earth, certainly strengthens the physical case and throws the burden of proof upon those who hold to the vaguely vast age, derived from sedimentary geology.

For practical purposes, this 1893 publication was King's last scientific work. As noted, during 1892 John Wesley Powell, who replaced King as director, ran afoul of Congress and as a result the USGS budget was slashed. There were rumours that King might resume directorship of the USGS, so it is plausible to suggest that he may have written this paper as part of a campaign to refurbish his scientific reputation and thereby gain support for reappointment to his former position. This did not happen.

King's paper was favourably received overseas by those more oriented towards physics in the arguments concerning the age of the Earth. Osmond Fisher, the British physicist, said it was 'impossible not to admire King's "ingenious argument"' (Fisher 1893), although he questioned the result as he did not believe that the rigidity of the Earth had been clearly established. Kelvin was so impressed with King's interpretation of Barus's temperature data that he extensively quoted from King's paper in a letter to *Nature* which he felt would 'be interesting to readers of Nature' (Thomson 1895, p. 440). He freely admitted that the new data on the melting temperature of rocks would 'reduce my estimate of 100 million to ... a little less than ten million years' but concurred with King, that once the effects of pressure had been taken into account, 'we have no warrant for extending the earth's age beyond 24 million years'.

King's paper was published in January 1893. In February, following a pause in the hammering of the USGS by Congress, the Geological Society of Washington was founded by Charles Doolittle Walcott (Fig. 1) (Yochelson 1993). The second and third meetings of the new society were devoted to vigorous discussion of King's work, to which considerable exception was taken by almost all of the participants. McGee was in the forefront of critics and rushed into print with more scathing comments (McGee 1893a), alienating both academic friends of the USGS and their remaining political allies in Congress. At the same time: 'Apparently he [McGee] saw nothing incongruous in adding a footnote to correct his own estimate of the age of the Earth from 15,000 million to 6,000 million years, and the time since the Palaeozoic from 7,000 million years to 2,400 million years. He had found a mistake in his arithmetic' Rabbitt reports (1980, pp. 227–228). All the furore apparently stemmed from arguments as to whether man had been in America during the Pleistocene (McGee 1892b, 1893b).

Barus gave a short rejoinder to McGee's publication, and included several perceptive comments in defence of King:

> What we did was an endeavor to remove the preponderating element, and I must reiterate that if our respite had not been cut short by

recent unfavorable legislation, other things would have been brought out in their turn and in due time. Perhaps it is heresy to state that an immense future awaits laboratory research in physical geology; but stating it, one would like to refer not so much to the punching of clay or the pulling of taffy candy, as to legitimate physical measurement (Barus 1893, p. 23).

These are restrained remarks from a scientist who had been elected to the National Academy of Sciences in April 1892, but discharged from the USGS in August because of budgetary cuts, while McGee still kept his government position.

That same year of 1893 when the Geological Society of Washington was founded, Walcott was a vice-president of Section E (Geology and Geography) of the American Association for the Advancement of Science. One of the duties of office was an annual address, to be given at the meeting in Madison, Wisconsin. His subject was geological time, and although there is no documentation on why he expounded on this topic, almost certainly it stemmed from the uproar created by King's calculations.

The resulting paper (Walcott 1893) may be the most thorough study ever made of stratigraphic thickness and sedimentation rates as a method for the estimation of geological time. Without attempting to be chauvinistic, the semi-arid western United States has superior outcrops to those of the British Isles where calculations of the rates of deposition and erosion of strata had previously been made. The sequences in America were generally longer and less tectonically disturbed than those of Europe. The United States has the Grand Canyon! Thus Walcott based his calculations on the thickest rock sections throughout the west, which, he considered, represented 'almost uninterrupted sedimentation during Palaeozoic time'. From his examination he deduced:

> It is easy to vary these results by assuming different values for area and rate of denudation, the rate of deposition of carbonate of lime, etc.; but there remains, after each attempt I have made that was based on any reliable facts of thickness, extent and character of strata, a result that does not pass below 25,000,000 to 30,000,000 years as a minimum and 60,000,000 to 70,000,000 years as a maximum for post-Archean Geologic time (Walcott 1893, pp. 675–676).

Walcott's other significant contribution in this paper was in proposing a new time-ratio for the Cenozoic, Mesozoic and Palaeozoic. In devising a standard time-unit, termed a 'geochrone' from which the word 'geochronology' as a measure of geological time evolved, Henry Shaler Williams, a part-time associate of the USGS since 1883, had proposed a ratio of $1:3:15$. This compared with Dana's similar estimate of $1:3:12$, but Walcott felt from his knowledge of the Mesozoic section in southwest Utah, combined with publications and discussions with those who had worked on the thick coal beds of the Cretaceous Laramie and the 20 000 feet of limestone in the Cretaceous around the Gulf of Mexico, that the post-Palaeozoic was under-represented. He therefore determined the ratio to be $2:5:12$, which is in accord with present-day data derived from radiometric dating, although the time durations he proposed for those era were 2.9 Ma, 7.24 Ma, and 17.5 Ma respectively.

The greatest weakness in Walcott's assumptions for his age calculations was, understandably, in his estimation of the thickness of Precambrian strata. He had measured more than two miles of strata below the Cambrian in the Grand Canyon from which he calculated that the Algonkian had a duration of 17.5 Ma, and that the Archaean had endured at least 10 Ma. Nevertheless, he made an important observation which supported arguments put forward by Sir Archibald Geikie (Fig. 1), then director-general of the Geological Survey of the United Kingdom. Some geologists had suggested that sedimentation and erosion rates had been more rapid in the geological past in an attempt to 'squeeze' the evidence observed in the rock record, which appeared to require inordinate amounts of time, into the ever-diminishing timescale provided by Kelvin and the physicists. However, Walcott, like Geikie, found that there was no proof whatsoever of this. He observed how Algonkian sands, shales and limestones were identical in appearance and characteristics to those from later epochs, and how the thin, even layers, sun cracks and ripple marks were evidence for *slow* deposition. 'There is absolutely nothing to indicate more rapid denudation and corresponding deposition in this early pre-Cambrian series than we find in the Palaeozoic, Mesozoic or Cenozoic formations' he stated emphatically (Walcott 1893, p. 660). He concluded: 'geologic time is of great but not of indefinite duration. I believe that it can be measured by tens of millions of years, but not by single millions or hundreds of millions of years' (Walcott 1893, p. 676). Walcott's study was printed four times in three journals and in the Smithsonian Annual Report. This may have been the most widely distributed paper in North American geology during the nineteenth, and perhaps even the twentieth, century (Yochelson 1989).

As well as the formation of the Geological Society of Washington, another significant event concerning geology to occur in 1893 was publication of the first volume of the *Journal of Geology*. Founded by Thomas Chrowder Chamberlin, later considered to be America's top-ranking geologist,[5] a prestigious list of editors included Chamberlin and Walcott, as well as associate editors from Britain (Sir Archibald Geikie), Germany, France, Austria, Norway, Sweden, Brazil and Canada. Chamberlin was deeply interested in the subject of geological time, thus it is not surprising that this first volume contained Walcott's paper 'Geologic Time', as well as an article by Henry Williams (an assistant editor) entitled 'The Making of the Geological Time-Scale', and the abstracts of three other papers on geological time. One of these abstracts was of King's paper discussed above, but the other two reported attempts to estimate geological time from the conventional methods of sediment deposition and erosion.

The first of these was by Warren Upham, a geologist with the USGS, who recognized that 'the diversified types of animal life in the earliest Cambrian faunas surely imply a long antecedent time for their development ... therefore, the time needed for the deposition of the earth's stratified rocks and the unfolding of its plant and animal life must be about 100 million years' (Bownocker 1893). A similar result was also found by T. Mellard Reade, in Britain, who concluded that the time since the beginning of the Cambrian was around 95 million years, although, he noted: 'When the enormous length of pre-Cambrian time is added to the above, the estimate is found to agree very closely with that of Sir Archibald Geikie, i.e. 100 to 600 millions years' (Kummel 1893).

As the century closed, North American geologists had an increasing opportunity to follow European developments in their own national literature. As mentioned earlier, Kelvin's final paper on the age of the Earth, which applauded the work of King, was reprinted in *Science* (Thomson 1899). This provoked a strong rejoinder from Chamberlin who attacked almost all of Kelvin's 'assumptions' on which he had based his estimates for the age of the Earth for more than 30 years, with an authority that few geologists had previously dared to do (Chamberlin 1899). Irritated by the tone of Kelvin's address which Chamberlin considered was 'permeated with an air of retrospective triumph and a tone of prophetic assurance', Chamberlin listed numerous instances of Kelvin's 'sure assumption', 'certain truth' and 'no possible alternative', and enquired whether '... these definite statements, bearing so much the air of irrefutable truth, be allowed to pass without challenge?' The most fundamental tenet that Chamberlin challenged was that regarding formation of the Earth itself. Questioning the rate at which meteorites had aggregated to form the Earth, Chamberlin argued that if they had accreted slowly then it did not necessarily follow that a molten Earth had ever existed. 'Is not the assumption of a white-hot liquid earth still quite as much on trial as any chronological inferences of the biologist or geologist?' he asked. If this fundamental assumption proved to be wrong then the whole of Kelvin's argument was undermined. Chamberlin spent many subsequent years developing his 'Planetesimal Hypothesis' which demonstrated that the early Earth was initially a cold body (Brush 2001).

As in Britain, geologists in North America took both sides of the argument, and, as in Britain, the majority like Chamberlin, McGee, Walcott and Upham who voiced an opinion supported a long timescale for the Earth's evolution, while a few, like King, sided with the physicists who largely favoured a short timescale. Thus there is evidence that the age of the Earth was of interest to geologists in the United States during the last decade of the nineteenth century. However, for the majority of those working on the ground, with enormous areas to map, its significance is placed in context by James Dwight Dana, one of the patriarchs of American geology. Writing in the fourth and final edition of his classic *Manual of Geology*, Dana devoted less than four pages to summarizing the then current literature and the various methods proposed for estimating geological time. Unconcerned by the debate, he wisely concluded that 'Time is long, very long. ... Geologists have no reason to feel hampered in their speculations, while the extreme results of calculations are between 10,000,000 and 6000,000,000' (Dana 1895, p. 1026).

1901–1917: ignition

The Smithsonian Institution continued its practice of reprinting selected articles from various

---

[5] In the first edition of American Men of Science (1903), the one hundred outstanding scientists in each field were ranked in importance. Thomas Chamberlain was judged the most important geologist in America, with Grove K. Gilbert second and Charles Walcott number three. Nearly a century later, this still seems an accurate evaluation. In 1934, Chamberlain's son, R. T. Chamberlain, wrote a memorial published in volume 15 of the Biographical Memoirs, National Academy of Sciences.

journals in their Annual Report, though just how the selection was made, or by whom, is not always known. It is not clear how well-developed the libraries in most American universities were at this time, but the tradition which lasted well beyond the Second World War was that the geology department was located in the oldest building on campus. Equipment, and presumably library holdings, reflected their status and direct access to the foreign literature may have been limited. However, because of the widespread distribution of the Annual Report which American geologists could readily obtain, reproduction of papers in the volume certainly added not only to their availability, but, more importantly, to their weight. As three of John Joly's papers were printed in this period (Joly 1901, 1908, 1912), the age of the Earth appears to have continued to be a topic of interest, with Joly's ideas (Wyse Jackson 2001) being given particular prominence.

In the United States during this interval, George Ferdinand Becker (1847–1919)[6] was a key person. With a background in geology, mathematics and physics, Becker, like Walcott, was one of the original members of the USGS and, like Walcott, he was a prolific writer. A reasonable assessment is that Becker was an original thinker and during his lifetime was probably the premier geophysicist in North America. Dalrymple (2001) has discussed Becker's first pronouncements on the age of the Earth which he made in 1904 when, as head of the division of chemical and physical research of the USGS he addressed the International Congress of Arts and Sciences (ICAS) held that year in St. Louis (Becker 1904). The discovery of radioactivity had been made less than a decade previously and geologists, both in Britain and America, were still unaware of the impact it was to have on their science. Thus Becker reviewed the prevailing arguments regarding the age of the Earth, siding, on the whole, with Kelvin and the physicists.

Interestingly, Ernest Rutherford, then professor of physics at McGill University in Montreal, was also present at the ICAS meeting. Rutherford, a physicist trained at the Cavendish Laboratory in Cambridge, England, had already made an international reputation for himself with his discovery two years earlier that one element could spontaneously decay to another (Lewis 2000, 2001). So when it came to his turn to talk on 'Present Problems in Radioactivity', he gave his audience a thorough review 'of the more important problems that have arisen during the development of the subject' (Rutherford 1905, p. 5). In particular, he explained how radioactivity was responsible for the internal heat of the Earth, and how helium could be used to date minerals: 'I think there is little doubt that ... this method can be applied with considerable confidence to determine the age of the radioactive minerals'. This is believed to be the first time that the concept of dating rocks by radioactivity was publicly announced.

Americans had been quick to dabble in radioactivity when in 1901 it was discovered that air, and even the mists of Niagara Falls, contained radioactive matter (Badash 1969). Then in 1903 Joseph John Thomson, head of the Cavendish Laboratory in Cambridge, where most of the work on radioactivity originated, inaugurated the Silliman Lectures at Yale and urged his American hosts to examine the waters around New Haven. Henry Bumstead and Lynde Phelps Wheeler, both friends of Bertram Boltwood, subsequently made the first serious study of radioactivity in America, following up Thomson's suggestion. Boltwood had in fact begun investigations into the phenomenon as early as 1899, only three years after radioactivity had been reported, but no publication ensued. It was not until Boltwood heard that Rutherford would be coming to lecture at Yale in the spring of 1904, that he began a serious study of the uranium decay series.

This work by Boltwood (1905, 1907) was seminal in revealing the relationship between uranium and lead and it has received much critical discussion and evaluation (e.g. Badash 1968, 1969; Dalrymple 2001; Lewis 2001). However, it has only a retrospective claim to fame,[7] for it made no immediate impact upon American scientists, a view that is borne out by an article written for *Scientific American* in December 1907, on 'The Age of the Earth's Consolidation: How Scientists Estimate Periods of Time'. In a two-page summary of all the arguments regarding the age of the Earth, a single paragraph is reserved for radioactivity: 'Several years ago, when the enthusiasm over the radium discoveries was at its height, there were those who admitted a terrestrial history of a thousand million years. But mysterious as radium still remains, it is doubtful if such a view is generally held to-day'

---

[6] In 1920 Arthur L. Day published a memorial to Becker in volume 31 of the *Geological Society of America Bulletin*.

[7] For another example of retrospective fame, recall that the president of the Linnaean Society in 1859 pronounced that little of major interest had transpired in the year Darwin and Wallace first announced their ideas on evolution; it was subsequent events which gave the 1859 announcement ins significance.

(See 1907, p. 366). Thomas See, who was actually an astronomer, preferred an age estimate for consolidation of the globe not exceeding eight or ten million years.

In 1908, Becker touched again on the age of the Earth, with two papers. In the first, he considered the issue of a cooling globe, following in the footsteps of Kelvin, to arrive at his own age estimate: 'Notwithstanding the inadequacy of the data, I can not but believe the 60-million-year earth here discussed is a fair approximation ... It is in good accord with geological estimates from denudation and sedimentation, with the age of the ocean as inferred from sodium content, and with the age of the moon as computed by Sir George Darwin' (Becker 1908a, p. 233).

The second paper of that year went more directly to the significance of radioactivity for geology. With a good grasp of the subject, Becker summarized radioactivity for geologists who 'can not be expected to be very familiar with radioactivity'. Nevertheless, when it came to estimating the age of the Earth, he summarized the traditional methods used to calculate it, emphasizing how results largely concurred with his own estimate of 60 million years, and highlighted only the problems that had been encountered with radioactivity. For example, Becker calculated ages for some mineral analyses from their published uranium–lead ratios, despite the fact that Boltwood had written extensively on problems related to the use of published analyses (see Lewis (2001) for more on these problems), and he had specifically told Becker that the analyses were unsuitable because the minerals showed varying stages of alteration. It was not surprising, therefore, that minerals from sources that appeared to Becker to be geologically coeval, gave ages ranging from 1.67 to 11.47 billion years. But Becker concluded from this discrepancy that helium would be a better basis than lead for estimating the age of minerals, despite the problems that that technique was already known to have. Furthermore, he considered that there was 'no convincing evidence that the law of decay is so simple as is assumed' (Becker 1908b, p. 135), proposing that it might fail under various conditions and thus could not be relied upon to give consistent results. Like many of his contemporaries who were on the border between geology and physics, John Joly for example (Wyse Jackson 2001), the notion of constant rates of decay in radioactivity was a great uncertainty.

This paper received a rousing two-page comment in *Science*: 'The whole paper is of extraordinarily suggestive character, not only in the direction of pure speculation, where scanty data can ever be gathered, but in offering several points of contact with direct laboratory measurement' (Day 1908, p. 527). Day, appointed as the first director of the newly organized geophysical laboratory of the Carnegie Institution of Washington in 1906, was anxious to show the importance of his new organization to these types of problems. In a broader sense, however, Day's remarks were actually in tacit support of Becker's view of the fallibility of radiometric dating.

Two years later, still unable to accept the potential of radioactivity, Becker (1910) reiterated his thoughts on the subject in a paper entitled 'The age of the Earth':

These three methods [stratigraphy, sodium solution, and cooling of the Earth] seem to be mutually confirmatory and to give results which converge towards some value near 60 million or perhaps 65 million years. This being granted, it follows that radioactive minerals cannot have the great age attributed to them (Becker 1910, p. 28).

Like many of their counterparts in Europe, many eminent geologists in North America took a dim view of radioactive age determinations at this time.

If there were any strong geological champions of radioactive methods in America at this time, they were not at all obvious. Alfred Church Lane, who was later most prominent in this field, produced only a short paper in which he proposed yet another methodology for calculating the Earth's age (Lane 1913), suggesting that deep-sea red clays or snow in Arctic and Antarctic regions be examined for 'cosmic dust' from meteorites as another way of deriving an approximate age.[8] It was not until 1916 that another academic, Joseph Barrell,[9] argued for the importance of radioactive age determinations in interpreting the record of sedimentation.

Barrell was a respected professor of geology at Yale University. He was of a younger generation of geologists than Becker, and was generally considered to be a rising star in the profession; unfortunately, he died young. 'Rhythms and the measurement of geologic time' (Barrell 1917) was his *magnum opus*. Charles Schuchert,

---

[8] It is interesting to note that 'cometary dust' is now being looked for in polar ice cores, not to date the age of the Earth, but to detect whether it should be considered a potential hazard to life (Samuel 2001).
[9] In 1929 Charles Schuchert authored a memorial to Barrell in volume 12 of Biographical Memoirs, National Academy of Sciences.

a palaeontologist who had never completed grammar school but rose to be a professor at Yale University and was one of the most influential teachers of his generation, was unstinting in his praise of this publication, as was Arthur Holmes, a geologist on the other side of the Atlantic who was by then the world's leading authority on radiometric dating (Lewis 2000, 2001, in press). Even though Schuchert was a colleague of Barrell's, his judgement of the importance of Barrell's paper went far beyond mere loyalty.

In addition to providing a great deal of information on sedimentation, Barrell, like Becker before him, again explained radioactive decay in terms more comprehensible to geologists than that given by physicists. It may be that because he was at Yale University where the first work on uranium–lead radiometric dates had been deployed, he had a better understanding and appreciation of the technique than his colleagues elsewhere in the country. It cannot be proven, but it seems probable that he and Boltwood associated informally, for the Yale faculty was not large at that time and there was more interaction between members of the various departments. On the other hand, it is interesting to note that many of Barrell's ideas reflect those of Arthur Holmes written a few years earlier, in the first edition of his book *The Age of the Earth*, and that the format of Barrell's extensive paper (159 pages)[10] is very similar to that of Holmes' small book, which Barrell references (Holmes 1913). In fact, in the section 'Measurements Based on Radioactivity' Barrell decided, presumably with Holmes' consent, that:

> ... the best presentation will be to quote freely from a recent article by Arthur Holmes [1915], who by his research has contributed much to this subject. The following three topics [Accumulation of Helium; Pleochroic Halos; Accumulation of Lead] are quoted entire, as they give in brief space the essentials of the methods and the original article will be seen by but few American geologists. The importance of the whole subject from a geological standpoint is such that this presentation should be given in an American geological publication (Barrell 1917, p. 845).

Although slightly ambiguous, it seems that Barrell was advocating the publication in full of Holmes' paper in an American journal, an event that did not occur. Further evidence that Holmes' work influenced Barrell can be seen in a letter written to Holmes in 1924 by Schuchert: 'Many years ago I read your splendid ... little book The Age of the Earth. I soon directed Barrell to it and I loaned him my copy. So you see you have had influence over both of us ...'.[11]

Whatever the background, Barrell was able to provide a coherent view that the dates from this new method of radiometric dating were not disharmonious with more traditional geological thinking. In essence, Barrell noted the many uncertainties in calculations based on rock thickness and rates of deposition, but, perhaps more importantly, it was Barrell's sympathetic views to the position that geologists had been placed in that may have won them over. His extensive paper cannot be properly summarized here, but one quote may illustrate his approach. After briefly mentioning earlier estimates of time, Barrell came to Kelvin and then the dramatic impact of the discovery of radioactivity:

> Not only did physicists destroy the conclusions previously built by physicists, but based on radioactivity, methods were found of measuring the life of uranium minerals and consequently of the rocks which envelop them. Instead of limiting earth history to less than 40,000,000 years, they now granted upwards of 1500,000,000. Many geologists, adjusted to the previous limitations, shook their heads in sorrow and indignation at the new promulgations of this dictatorial hierarchy of exact scientists. In a way, this scepticism of geologists was a correct mental attitude. The exact formulations of a mathematical science often conceal the uncertain foundations of assumptions on which the reasoning rests and may give a false appearance of precise demonstration to highly erroneous results. No better illustration could be given than that of the case in point. This scepticism was incorrect, however, unless it led to a careful and unprejudiced re-examination of the postulates on which rested the geologic measurements of time (Barrell 1917, p. 749).

---

[10] A footnote tells the reader that the paper was second in a series of papers composing a 'Symposium on the Interpretation of Sedimentary Rocks' presented at the Albany meeting of the Geological Society of America on 29 December 1916. Barrell was given a special dispensation by the Publication Committee of the Society to publish it in full, since he felt that condensation would greatly diminish the value of the paper. Interestingly, it was first given in two parts to the University of Illinois Public Science Lectures on 16 and 18 January, back in 1912.

[11] Schuchert to Holmes, 8 October 1924. Courtesy of Yale University Library.

It is interesting to compare that quote with a remarkably similar one by Arthur Holmes from 1913 given in Knell & Lewis (2001). It may well be that this was the publication in the United States which tipped the scales against Becker's view and towards the beginning of acceptance of dates for a very old Earth. At this point, however, North America became involved in the First World War. That event, as a minimum, began the shattering of the fabric of life in America, though not to the extent that it did in Europe, and forced people into new modes of thinking. Both Becker and Barrell died in 1919, just after the war, before they had any further opportunity to publish more and clarify their respective positions on the best methodology for determining the age of the Earth.

1918–1931: liftoff

During this third interval, two different organizations were involved in disparate publications on the age of the Earth. Because each had different objectives and audiences, they are best treated as distinct entities. The earlier is the American Philosophical Society (APS), headquartered in Philadelphia. The APS is the oldest scientific society in America, having been founded before the United States was formed, for the purpose of 'promoting useful knowledge'. In some ways, membership of the organization represents the pinnacle of academic achievement in the United States.

APS council minutes for 25 November 1921, record the following: 'On motion, The Age of the Earth'. The subject had been chosen as the topic for a symposium, with the suggestion that it be considered from varying viewpoints, to be allocated as follows:

Professor Boltwood [Yale University] – Radioactivity
Professor Shapley [Harvard University] – Astronomy
Dr. Walcott [Smithsonian Institution] – Geology
Dr. J. M. Clark [sic] [New York State geologist and palaeontologist] – Palaeontology
Doctors [W. B.] Scott and [H. N.] Russell were appointed as a subcommittee on the Symposium.[12]

There is no indication as to why this particular subject was chosen for the annual meeting.

A reasonable surmise is that the recent publication of the astronomer Henry Norris Russell (1921), which suggested a figure between 1100 Ma and 8000 Ma as maximum limits for the Earth's crust (Dalrymple 2001), could well have been the trigger, particularly since Russell was later appointed to the subcommittee. On the other hand, perhaps they were following the lead set by Britain in the previous year. In 1921 the British Association for the Advancement of Science had held a joint discussion between physicists, geologists, astronomers and biologists on the age of the Earth. It marked a turning point in the acceptance of radiometric dating in that country and there was now a distinct change of opinion in favour of the longer age estimates (Lewis in press). No doubt the Americans wished to review the topic for themselves, and followed the example set by Britain in not restricting the discussion to geologists.

Whatever the initiative, William Scott, then serving as APS president, was a distinguished vertebrate palaeontologist and would certainly have supported the topic, so the subcommittee sent out invitations in a week. Anyone who has been involved in organizing a symposium will empathize with Scott's words in a letter to John Mason Clarke on 6 February 1922: 'Of those originally selected by the Committee to speak in the Symposium, you are the only one who has been able to accept.'[13] Had the original people been available, a very different meeting would have resulted, but a 'second choice' was assembled. The papers were presented on 22 April 1922 and published later that year in the order in which they were given.

Thomas Chamberlin, unable to attend in person because he was ill, produced a publication on the 'Geological Viewpoint' of the age of the Earth, which was as long as the other three papers combined (Chamberlin 1922). Beginning with brief comments on the development of geology as an early science, he touched on Darwin and Wallace, describing how they had need of even more time than the geologists: 'So they took the lead and the team became a tandem, Biology prancing in front, Geology trotting on in the thills'[14] (Chamberlin 1922, p. 247). From his review he had three conclusions: (1) that estimates from sediments would eventually fall into harmony with those from radioactivity and astronomy;

---

[12] Council minutes are in the archives of the American Philosophical Society, Philadelphia, Pennsylvania, USA (X, 1; minutes, committee on General Meeting, 1921–1923).

[13] The quoted letter is in the archives of the American Philosophical Society, Philadelphia, Pennsylvania, USA (V, A, 4; letters sent by Wm. Scott, 1921–1925).
[14] 'Thills' are the poles or shafts on either side of a horse that attach the animal to a wagon or cart. Even in Chamberlin's day, the term was uncommon.

(2) that sediments can give no verdict as to the *total* age of the Earth, except to show that it is very old, since their starting point cannot be established; (3) that salinity of the oceans was not ready to render a verdict at all, since 'it has more need of a court of inquiry than a place on the witness stand'.

John Clarke gave the 'Palaeontological Viewpoint' and admitted right at the start 'that there can be little hope of arriving either at a reliable or an approximate conclusion as to the age of the earth through this palaeontological channel' (Clarke 1922, p. 272). The 'Astronomical Viewpoint' was summarized in a scant three pages by Ernest Brown (Yale University), which provided no figures on age (Brown 1922), while William Duane (Harvard Medical School) gave an equally scant three pages on radioactive calculations, noting the difficulties which were involved in differentiating isotopes of lead (Lewis 2001), but also giving no numbers for the ages to which he alluded. He concluded in vague terms:

> The ages calculated from radio-active data represent the length of time during which we may suppose the chemical elements to have been in more or less mechanical contact with each other. They do not represent the time that has elapsed since the earth may have reached a state capable of supporting organic life as we now know it (Duane 1922, p. 288).

The following year Holmes reviewed the published proceedings of the American symposium for *Nature*. He considered the chief feature of interest to be Chamberlin's 'spirited attempt to show how the geological estimates may be brought into harmony with the revised deductions from radioactivity and astronomy' (Holmes 1923, p. 302). Holmes had long been an admirer of Chamberlin. Writing to him in 1912, while still a student, Holmes had said: 'You have done more probably than any other living geologist to clothe the dry bones of geological fact with the fascination of co-ordinating theories'.[15] But despite his admiration for Chamberlin, Holmes clearly felt that the Americans had not done the subject justice, and dismissed the other three papers in a single sentence: 'The remaining papers call for little comment.'

The symposium added nothing new to the debate, except the hope for reconciliation of different approaches. The age of Earth was tangential to the interests of most American geologists, and it seems unlikely that publication of the symposium had much impact on the geological community, particularly the younger members. The *Proceedings of the American Philosophical Society* was simply not a journal that one would expect a new professor of geology to examine on a regular basis. In the 1920s the literature was burgeoning and it was hard enough to keep up with the journals within the field of geology. The APS symposium is of historical interest however, for it catalogues the last remnants of the nineteenth century approach to dating the Earth.

The second publication on the age of the Earth, mentioned at the start of this section, was published nine years later, in 1931, and particularly documents the escalation of radiometric dating that occurred during this period.

The National Academy of Sciences (NAS) was founded in 1863. During most of its first half century it was fairly inactive but the 1913 semicentennial, sparked by the visionary astronomer George Ellery Hale, at last breathed some life into the organization. Formation of the National Research Council (NRC) in 1916, under the NAS, then added a more active component to the group, as many of the members of NRC committees were not the older scientists who formed the bulk of the NAS membership. During the First World War, the NRC performed valuable service in adding a scientific component to the American war effort, and after that conflict ended, the NRC scrambled to find new outlets for the momentum that had been created. It soon surpassed the NAS in size and found its place at the centre of scientific policy for about two decades. If the relationship between NAS and NRC led to confusion, those within the NRC sometimes led to bewilderment.

The various NRC components were organized along both political and scientific lines. The division of geology and geography (G&G) was under physical sciences. In 1923 this Division (G&G), under the chairmanship of Andrew Cowper Lawson, a well-known California geologist, created a Committee on the Determination of Geologic Time by Atomic Disintegration. For years, the sparkplug of that committee, which commonly had eleven or so members, was its chairman Alfred Lane.[16] The annual reports of the committee were technical and added rigour to this growing field of knowledge.

In 1925, Joseph Ames, then head of physical sciences, suggested a joint committee on the relations of physics to geology. The NRC was careful

---

[15] Holmes to Chamberlin, 4 June 1912. Courtesy of the Chicago University Library.

[16] Lane was chairman of this NRC committee for 24 years, yet never was elected to the National Academy of Sciences. In 1953, Robert L. Nicols wrote his memorial in *Proceedings of the Geological Society of America*, Annual Report for 1952.

to point out that there would be no overlap between this new initiative and Lane's committee. This gave rise to the Committee on Physics of the Earth which existed from 1926 until 1936, and which generated four subcommittees whose brief was to each produce a bulletin 'to give the reader, presumably a scientist but not a specialist in the subject, an idea of its present status together with a forward-looking summary of its outstanding problems'. Bulletins entitled *Volcanology*, *The Figure of the Earth* and *Meteorology* had all been published by the time the fourth volume in this series on *The Age of the Earth* (Knopf 1931a) was completed. By the end of January 1929, almost all manuscript copy was in hand and it went to the printer on 1 April 1929. However, Arthur Holmes, whose onerous task was to evaluate every single radiometric analysis ever published, held up the printers for almost two years (Lewis 2001) and it was not actually published until June 1931. For those deeply interested in trivia, the printing cost was $4489.36. After publication the subcommittee which had prepared the book was dissolved.

Even though this bulletin is a reasonably thick tome, an accurate summary is that it is a nine-chapter book by Holmes (1931) on the general subject of geological time and radioactivity, with emphasis on the latter. Holmes reviewed the acceptability, or otherwise, of all radiometric dates, continent by continent, preparing a summary geological timescale, based on the results for each area. But problems lay not only with dating a mineral but also in determining which geological period it was from. Nevertheless, the first radiometric timescale for North America was determined, and is given in Table 1.

There are another fifty pages by Alois Kovarik that might be considered a 'how to do it' manual for the Holmes contribution (Kovarik 1931). The committee's chairman, Adolph Knopf, wrote a summary of the volume (Knopf 1931a) and a five-page article which determined, as Chamberlin had done nine years earlier, that the number of assumptions in Joly's methodology of calculating the Earth's age from sodium in the sea had too many uncertainties to be useful (Knopf 1931b). Ernest Brown, who had also participated in the APS symposium, this time took seven pages rather than three, to indicate that there were no hard figures on the age of the Earth from the field of astronomy (Brown 1931).

That left the sixty-page report by the palaeontologist Charles Schuchert, who opened with a quotation from Holmes (Schuchert 1931). This paper was in part a refinement of earlier data gathered by Walcott (1893), Clarke (1910) and Chamberlin (1922) on sedimentation and denudation, along with figures from the work of foreign authors. The data on total thickness of the world's stratigraphic column and presumed rates of deposition are interesting, though hardly exciting. A small part of the essay is concerned with Precambrian time, then the greatest area of uncertainty. Schuchert's final few pages summarize the work of Matthew on horse evolution, plus a few other points to indicate that rates of evolution may vary through time. His conclusions differed little from those by Barrell written twenty-four years earlier.

This bulletin is significant, for the publication had the authority of the NRC behind it. As uncertain as radiometric determinations still were, in part because of sample contamination, laboratory procedures, geological uncertainties and poorly constrained decay rates, the results were now far more impressive than any other method of trying to date the ancient past. Radiometric dating had broken the mental time barrier of 100 million years and expanded it into the billions. But despite the enormous amount of work that had gone into the bulletin over the five years since its inception, Holmes' conclusion for the age of the Earth was that it still sat at 1600 million years, the same value he had determined for it back in 1913.

Table 1. *Ages of North American minerals in 1931*

Mineral	Probable geological ages	Millions of years
Uraninite, Mexico	Post-Middle Miocene?	35
Brannerite, Idaho	Miocene?	38
Pitchblende, Colo. and Wyo.	Late Cretaceous–early Tertiary	60
Uraninite, N. Carolina	Late Carboniferous or early Permian?	250
Uraninite, Glastonbury, Conn.	Late Palaeozoic: possibly mid-Devonian	290
Uraninite, Branchville, Conn.	Late Ordovician?	390
Uraninite, Ontario and Quebec	Middle Precambrian	1050
Uraninite, Llano Co., Tex.	Precambrian	1100
Uraninite, Black Hills, S.D.	Precambrian	1460

After Holmes (1931).

From 1931 onward, in the broadest sense both sedimentologists and palaeontologists accepted the numbers for ages they were given by the geochronologists and did not attempt to check or corroborate them by other methods. In effect, the paper by Schuchert in this bulletin was the last major effort in America to discuss alternatives to dating by radioactivity, and even here they were treated more as historical curiosities than opportunities for further research.

Discussion

John Phillips, the remarkably able British palaeontologist, was considering the problem of the age of the Earth even before Darwin's calculation on erosion of the Weald was first published in 1859. But despite some of the earlier publications by Kelvin in the 1860s and 1870s, the subject did not begin to have a serious impact on geologists in North America until the 1890s when Kelvin's persistent questioning of the uniformitarian principles, closely adhered to by some geologists, required that they think again.

In the 1890s, bibliographies of the USGS indicate that as a percentage of the published literature, articles on the age of the Earth were extremely small in number. While this was certainly the case, it is too strong to state that American geologists were simply not interested in the subject. The debate triggered by WJ McGee in 1892 reflects an awareness of the issues that had been under discussion since the 1860s in Britain. Furthermore, the relatively prompt publication of works written on the subject near the turn of the century, such as those by Kelvin and Joly, indicates that those in charge of deciding what foreign literature was reprinted in American journals were also aware of the significance of the debate. On the other hand, it may be fair to note that there were far more pressing problems in mapping and stratigraphy for geologists to pursue, which would provide answers that were far less speculative than the new avenues of research in geochemistry and geophysics. Geology had had a far longer history in the British Isles than in North America and, although the number of geologists in the United States and Britain during the 1890s is not precisely known, it seems they were quite similar. Therefore, if one divides areas by the number of geologists and the duration of studies, it is obvious that Americans had far more of the basic geology to cover with far fewer people. Accordingly, speculations with no immediate application could wait.

Counteracting that effect was the fact that because the scientific community of that era was tiny by today's standards, a handful of people held a great deal of influence, which had bearing on what was published. In 1907 Walcott left the directorship of the USGS where, amongst others, he had been the boss of George Becker, to became the fourth secretary of the Smithsonian Institution (Yochelson 1998, 2001). That organization was small and for some years Walcott read all manuscripts and had ultimate control over what was printed, thus Joly's two later papers would not have appeared without his tacit authorization. Furthermore, under Walcott as secretary, the Annual Report of the Smithsonian began to publish original papers along with the reprintings, and he modified the Smithsonian Miscellaneous Collections from a quarterly, designed for long monographs, to a shorter paper format with quick publication. So, although papers such as Becker's on 'The age of the Earth' (1910) and Clarke's (1910), were formally authorized by the director of the USGS, it would not have appeared in a Smithsonian format without first going through the hands of secretary Walcott, who, as we have seen, was himself particularly interested in that topic.

Walcott was president of the NAS from 1917 to 1923, and was one of the founders of the NRC in 1916. It was he who had crafted the post-war structure of the NRC, which propelled it to the forefront of scientific policy-making, thus few scientific sparrows fell in Washington without Walcott hearing of their fall. Even though he was just out of office in the NAS when the Committee on the Measurement of Geologic Time by Atomic Disintegration was formed in 1923, he was still actively involved with the NRC. If still more evidence of interlocking directorates is needed, Ames, who laid the foundation for the Committee on Physics of the Earth, was also a member of the National Advisory Committee for Aeronautics; the founder and chairman of that organization was Walcott! In fact, despite appearances, Walcott was not manipulating strings. It was precisely because the scientific community was so small, that ideas circulated more freely and, accordingly, worthwhile projects were more readily implemented.

The concept of radioactive decay as a dating mechanism appears to have had little immediate impact on geologists, or even palaeontologists whose concern it was to relatively date the rocks, despite the profound paradigm shift in thinking it was to become:

> Few scientists directly and explicitly engaged in disputes and debates over programs and

theories, gave public appraisals of them or actively participated in their development. Most geologists seem to have been content to leave aside such activities and get on with their own work (LeGrand 1988, p. 272).

This comment, although referring to development of the theory of continental drift, is equally applicable to an earlier generation of geologists who were mostly unconcerned about the age of the Earth. But these attitudes were not unique to North America whose geologists absorbed, or resisted, the new ideas about radioactivity and the age of the Earth at much the same rate as their colleagues in Britain.

In the early part of the twentieth century, most of the work on the application of radioactivity to dating was being done in Britain, and even there only a handful of people such as Rutherford and Holmes, and Joly in Ireland were involved. Despite his prodigious output, Holmes' work does not seem to have been reprinted in the widely read American journals. It appears that his ideas circulated amongst a small group of like-minded American geologists, such as Chamberlin, Barrell and Schuchert, to whom Holmes sent his reprints and received theirs in return. Nevertheless, Holmes' influence was felt through Barrell's work, who in 1917 illustrated to geologists that the ages being obtained by radiometric dating were not entirely out of line with rates of sedimentation and erosion. This seems to have been the turning point at which geologists in North America began to take serious notice of radiometric dating.

Following the First World War, interest in the physics of the Earth rapidly escalated in America, encouraged by the NRC, and by the early 1920s America seems to have been only a year behind Britain in largely accepting the long timescales offered by radioactivity, although in both countries there were dissenters for some years to come. The wealth of laboratories in America, compared to Britain, was appealed to by Holmes in 1926 to help him build a geological timescale (Lewis 2001). By the late 1920s a fruitful alliance had developed between the Geophysical Laboratory in Washington and the Cavendish Laboratory in Cambridge, which led to one of the most important advances in the field of geochronology – the first mass spectrographic analysis of lead isotopes by Frederick Aston (Harper 1973).

By the start of the Second World War, America had become the undisputed leader in the field of geochronology, as Alfred Nier's mass spectrometry instrumentation and skill dramatically improved radiometric dating, and the Earth's age rapidly increased from 1.6 to 3.5 billion years. During the war Nier, Patterson and many others involved in geochronology worked on the Manhattan Project that gave the world the atom bomb, a consequence of which was highly sophisticated mass spectrometers available for geochronology, once the world returned to normality. Nier's work, carried out before the war in the Massachusetts Institute of Technology and after it at the University of Minnesota, paved the way for Claire Patterson. In 1956, Patterson, in his ultra-clean laboratory at the California Institute of Technology, determined the still-accepted age for the Earth of 4.5 billion years, from the study of lead in meteorites (but see Hofmann 2001).

From the geologist's point of view, in both Britain and America, the issue of the age of the Earth was essentially resolved by 1931, for once radioactivity was finally accepted as a dating technique, all subsequent efforts were elaborations on the original theme. Kelvin made the world increasingly young; Holmes, Nier, Patterson, and a host of other investigators made it steadily older. But it made little difference to the geologist on the ground whether the Earth was one or five billion years old.

However, things have not remained static since the 1950s and only a few years ago, based on new radiometric interpretations, a dramatic shortening in the length of the Cambrian Period was reported, when an 'absolute age' of $544\,\text{Ma} \pm 5\,\text{Ma}$ was assigned to the end of that period. Unsettling as this was, the wholehearted and rapid unquestioning acceptance of the new dates may turn out to be more upsetting. The four decades covered by this paper, during which the age of the Earth was much debated, provided time for discussion, argument and, most importantly, development of new approaches. It also allowed time for older opponents of a new idea to die off and new ideas to gain converts, more commonly among the younger workers in a field. One cannot help but wonder if the present rapid transmission of bold new ideas in many fields of geology, their hasty acceptance and ofttimes their subsequent rejection, is actually an improvement on scientific methodology.

For generations, many palaeontologists earned their bread and butter by practising the craft of relative dating from a study of the fossils enclosed in rocks. They have been superseded to a large extent by those who employ radiometric methods of dating. These new dates are useful and have opened many new vistas on how to approach fundamental problems in palaeontology. At the same time, echoes of the old debate can still be found in their reactions to the term 'absolute age', for such a slogan pre-empts any discussion, and to palaeontologists it suggests the hubris of physicists who know their answer is correct simply because it is cast in numbers. Radiometric determinations may give a date of 544 Ma ± 5 Ma, but any figure with a plus or minus cannot be 'absolute'. Forty years ago Holmes made the same complaint: 'I wish to appeal to my fellow geologists and workers in the rapidly growing subject of geochronology to discontinue a habit or fashion which is both unnecessary and misleading. I refer to the habit of calling radiometric age "absolute" age' (Holmes 1962, p. 665). Perhaps we should have heeded him then.

Another recent development in palaeontology has come from biology, with the ability to calibrate changes from the earliest known life to the present, based on the assumption that molecular and genetic change are as constant as the assumptions about radioactive decay. With this in mind, one might reflect on a comment by Clarke: 'There are species that have held their own without change throughout the ages... and there are others which have yielded so rapidly to change that their evolution is explosive. The same facts are true for *groups* of animals; and for the entire organic world there have been earth-wide periods of long stagnation as well as of rapid intensive change' (Clarke 1922, p. 282). These observations would still be relevant today, but now the palaeontologists are told by geneticists to accept that even though changes on the outside of organisms clearly occur at variable rates, changes on their insides are at a constant rate. Whereas this paper is not the place to discuss these arguments it is, nevertheless, our view that one can state the need for caution without elaboration.

The strongly felt need in evolutionary biology for a long timescale has been met, and early life is now recognized at about 3500 Ma (Schopf 1999). So it is interesting to consider yet another Clarke quote from 1922, which has bearing on the next great quest – the search for life, or former life, on Mars and, indirectly therefore, on the age of the origin of life:

The inception of life was the most solemn moment in the history of the Universe. We invite certain astronomers to refrain from further speculations and presumptions as to life in other worlds, and followers of Arrhenius from pursuing life spores through interplanetary space. These notions seem to be very exciting to the emotional public and there is indeed no shred of evidence of these things, no matter what conditions may be predicted of other worlds than this (Clarke 1922, p. 277).

Seventy years later the emotional public is still excited by this concept and the possibility that there may be other life 'out there' attracts massive funding, but there is still no shred of evidence.

The age of the Earth is of great intellectual interest and has been for many centuries. It is of prime concern in arguments for and against the concept of evolution (Strahler 1987), the age of the Universe, and those concerned with a fundamentalist interpretation of the six days of creation. In an attempt to counter that latter view, one geologist boldly asserted: 'I am confident that someday the concept of geologic time will be acclaimed as one of the more wonderful contributions from natural science to general thought' (Albritton 1980, p. 11). We find we must agree.

Ms J. Goldblum (National Academy of Sciences) was remarkable in producing old records of several National Research Council committees from the NAS archives. Mr J. James Ahern (American Philosophical Society) was kind enough to search out material relating to the 1922 APS symposium, and arranging for permission to quote. Friends R. Thomas (Franklin and Marshall College), H. Yoder (Geophysical Laboratory, Carnegie Institution of Washington) and R. Clarke (National Museum of Natural History) removed so many blunders that they cannot possibly be held responsible for any that remain. Removal of additional blemishes, directing attention to overlooked references, and clarification of turgid writing were provided by reviews from L. Badash and P. Wyse Jackson. To all we are indebted.

References

ALBRITTON, C. C. JR. 1980. *The Abyss of Time: Changing Concepts of the Earth's Antiquity after the Sixteenth Century*. Freeman, Cooper, San Francisco.

BADASH, L. 1968. Rutherford, Boltwood, and the age of the Earth; the origin of radioactive dating techniques. *Proceedings of the American Philosophical Society*, **112**, 157–169.

BADASH, L. (ed.). 1969. *Rutherford and Boltwood: Letters on Radioactivity*. Yale University, New Haven.

BARRELL, J. 1917. Rhythms and the measurement of geologic time. *Geological Society of America Bulletin*, **28**, 745–904.

BARUS, C. 1893. Mr McGee and the Washington Symposium. *Science*, **22**, 22–23.

BASSLER, R. S. 1947, In Memoriam [John Milton Nickles], *Bibliography and Index of Geology exclusive of North America*, Vol. 11 (1945–1946). Geological Society of America.

BECKER, G. F. 1904. Present problems of geophysics. *Science*, **20**, 545–556.

BECKER, G. F. 1908*a*. Age of a cooling globe in which the initial temperature increases directly as the distance from the surface. *Science*, **27**, 227–233, 392.

BECKER, G. F. 1908b. Relations of radioactivity to cosmology and geology. *Geological Society of America Bulletin*, **19**, 113–146.

BECKER, G. F. 1910. The age of the Earth. *Smithsonian Miscellaneous Collections*, **56**(6), 1–28.

BOLTWOOD, B. B. 1905. On the ultimate disintegration products of the radioactive elements. *American Journal of Science*, series 4, **20**, 253–267.

BOLTWOOD, B. B. 1907. On the ultimate disintegration products of the radioactive elements, Part 2. The disintegration products of uranium. *American Journal of Science*, series 4 **23**, 77–88.

BOWNOCKER, J. A. 1893. The Age of the Earth. By Warren Upham. *Journal of Geology*, **1**, 203–204.

BROWN, E. W. 1922. The age of the Earth from the point of view of astronomy. *Proceedings of the American Philosophical Society*, **61**(4), 283–285.

BROWN, E. W. 1931. The age of the Earth from astronomical data. *In*: KNOPF, A. (ed.) *Physics of the Earth – IV: The Age of the Earth. Bulletin of the National Research Council, Washington*, **80**, 460–466.

BRUSH, S. G. 2001. Is the Earth too old? The impact of geochronology on cosmology, 1929–1952. *In*: LEWIS, C. L. E. & KNELL, S. J. (eds) *The Age of the Earth: from 4004 BC to AD 2002*. Geological Society, London, Special Publications, **190**, 157–175.

BURCHFIELD, J. D. 1975. *Lord Kelvin and the Age of the Earth*. Science History, New York.

CHAMBERLIN, T. C. 1899. Remarks on Lord Kelvin's address on the age of the earth as an abode fitted for life. *Science*, **9**, 889–901.

CHAMBERLIN, T. C. 1922. The age of the Earth from the geological viewpoint. *Proceedings of the American Philosophical Society*, **61**(4), 247–271.

CLARKE, F. W. 1910. A preliminary study of chemical denudation. *Smithsonian Miscellaneous Collections*, **56**(5), 1–19.

CLARKE, J. M. 1922. The age of the Earth from the paleontological viewpoint. *Proceedings of the American Philosophical Society*, **61**(4), 272–282.

DALRYMPLE, G. B. 2001. The age of the Earth in the twentieth century: a problem (mostly) solved. *In*: LEWIS, C. L. E. & KNELL, S. J. (eds) *The Age of the Earth: from 4004 BC to AD 2002*. Geological Society, London, Special Publications, **190**, 205–221.

DANA, J. D. 1895. *Manual of Geology* (fourth edition). American Book Company, Cincinnati.

DAY, A. L. 1908. Geology and radioactive substances. *Science* (new series), **28**, 526–527.

DUANE, W. 1922. The radio-active point of view. *Proceedings of the American Philosophical Society*, **61**(4), 286–288.

DUPREE, A. H. 1957. *Science in the Federal Government: a History of Policies and Activities to 1940*. Belknap, Cambridge (Mass.).

FISHER, O. 1893. Rigidity not to be relied upon in estimating the Earth's age. *American Journal of Science*, **45**, 464–66.

HARPER, C. T. (ed.) 1973. *Benchmark Papers in Geology. Geochronology: Radiometric Dating of Rocks and Minerals*. Dowden, Hutchinson & Ross, Stroudsberg, PA.

HOFMANN, A. W. 2001. Lead isotopes and the age of the Earth – a geochemical accident. *In*: LEWIS, C. L. E. & KNELL, S. J. (eds) *The Age of the Earth: from 4004 BC to AD 2002*. Geological Society, London, Special Publications, **190**, 223–236.

HOLMES, A. 1913. *The Age of the Earth*. Harper Brothers, London and new York.

HOLMES, A. 1915. Radioactivity and the measurement of geological time. *Proceedings of the Geologists' Association*, **26**, 289–309.

HOLMES, A. 1923. The Age of the Earth. *Nature*, **112**, 302–303.

HOLMES, A. 1931. Radioactivity and geologic time. *In*: KNOPF, A. (ed.) *Physics of the Earth – IV: The Age of the Earth. Bulletin of the National Research Council, Washington*, **80**, 124–459.

HOLMES, A. 1962. 'Absolute' age: a meaningless term. *Nature*, **196**, 665.

JOLY, J. 1901. An estimate of the geologic age of the Earth. *Annual Report of the Smithsonian Institution for 1899*, 247–288.

JOLY, J. 1908. Uranium and geology. *Science*, **28**, 697–713 [reprinted in 1909: *Annual Report of the Smithsonian Institution for 1908*, 355–384].

JOLY, J. 1909. *Radioactivity and Geology*. Archibald, Constable, New York

JOLY, J. 1912. The age of the Earth. *Annual Report of the Smithsonian Institution for 1911*, 271–293.

KING, C. 1893. The age of the earth. *American Journal of Science*, series 3, **45**, 1–20.

KNELL, S. J. & LEWIS, C. L. E. 2001. Celebrating the age of the Earth. *In*: LEWIS, C. L. E. & KNELL, S. J. (eds) *The Age of the Earth: from 4004 BC to AD 2002*. Geological Society, London, Special Publications, **190**, 1–14.

KNOPF, A. (ed.). 1931a. *Physics of the Earth – IV: The Age of the Earth. Bulletin of the National Research Council, Washington*, **80**.

KNOPF, A. 1931b. Age of the ocean. *In*: KNOPF, A. (ed.) *Physics of the Earth – IV: The Age of the Earth. Bulletin of the National Research Council, Washington*, **80**, 66–72.

KOVARIK, A. F. 1931. Calculating the age of minerals from radioactive data and principles. *In*: KNOPF, A. (ed.) *Physics of the Earth – IV: The Age of the Earth. Bulletin of the National Research Council, Washington*, **80**, 73–123.

KUMMEL, H. B. 1893. Measurement of geological time. By T. Mellard Reade, C.E., F.G.S. *Journal of Geology*, **1**, 205.

LANE, A. C. 1913. Meteor dust as a measure of geologic time. *Science*, **37**, 673–674.

LEGRAND, H. E. 1988. *Drifting Continents and Shifting Theories*. Cambridge University, Cambridge.

LEWIS, C. L. E. 2000. *The Dating Game*. Cambridge University, Cambridge.

LEWIS, C. L. E. 2001. Arthur Holmes' vision of a geological timescale. *In*: LEWIS, C. L. E. & KNELL, S. J. (eds) *The Age of the Earth: from 4004 BC to AD 2002*. Geological Society, London, Special Publications, **190**, 121–138.

LEWIS, C. L. E. (in press.) Arthur Holmes' unifying theory: from radioactivity to continental drift.

In: OLDROYD, D. (ed.) *The Earth Inside and Out: Some Major Developments in Geology in the Twentieth Century*. Geological Society, London, Special Publications.

MCGEE, W. J. 1892*a*. Comparative chronology. *American Anthropologist*, **5**, 327–344.

MCGEE, W. J. 1892*b*. Man and the glacial period. *Science*, **20**, 317.

MCGEE, W. J. 1893*a*. Note on the 'age of the earth'. *Science*, **21**, 309–310.

MCGEE, W. J. 1893*b*. Man and the glacial period. *American Anthropologist*, **6**, 85–95.

NICKLES, J. M. 1921. Bibliography of North American Geology 1919–1928. *US Geological Survey Bulletin* **823**.

NICKLES, J. M. 1924. Geologic literature on North America 1785–1918, Part II, Index. *US Geological Survey Bulletin* **747**. [Bulletin 746 covering the bibliography was issued in 1923 and replaced a number of annual volumes and earlier compilations of the literature.]

RABBITT, M. C. 1980. *Minerals, Lands, and Geology for the Common Defense and the General Welfare: Volume 2 1879–1904*. United States Government Printing Office, Washington, DC.

RUSSELL, H. N. 1921. A superior limit to the age of the Earth's crust. *Proceedings of the Royal Society of London*, **A99**, 84–86.

RUTHERFORD, E. 1905. Present problems in radioactivity. *Popular Science Monthly*, **67**, 5–34

SAMUEL, E. 2001. Sting in the tail. *New Scientist*, **2283**, 5.

SEE, T. J. J. 1907. The age of the Earth's consolidation. *Scientific American Supplement*, **64**, 366.

SCHOPF, J. W. 1999. *Cradle of Life: The Discovery of the Earth's Oldest Fossils*. Princeton University, Princeton.

SCHUCHERT, C. 1931. Geochronology, or the age of the Earth on the basis of sediments and life. *In*: KNOPF, A. (ed.) *Physics of the Earth – IV: The Age of the Earth. Bulletin of the National Research Council, Washington*, **80**, 10–54.

STEGNER, W. 1953. *Beyond the Hundredth Meridian*. Houton Miflin, Boston.

STRAHLER, A. N. 1987. *Science and Earth History – the Evolution/Creation Controversy*. Prometheus Books, Buffalo.

THOM, E. M. 1944. Bibliography of North American Geology 1929–1939. *US Geological Survey Bulletin* **937**.

THOMSON, W. (LORD KELVIN). 1895. The age of the Earth. *Nature*, **51**, 438–440.

THOMSON, W. (LORD KELVIN). 1898. The age of the earth as an abode fitted for life. *Annual Report of the Smithsonian Institution for 1897*, 337–357.

THOMSON, W. (LORD KELVIN). 1899. The age of the earth as an abode fitted for life. *Science*, **9**, 665–674, 704–711.

WALCOTT, C. D. 1893. Geologic time, as indicated by the sedimentary rocks of North America. *Journal of Geology*, **1**, 639–676 [also published in: 1893, *American Geologist*, **12**, 343–368; 1894, *Proceeding of the American Association for the Advancement of Science*, **42**, 129–160; *Annual Report of the Smithsonian Institution for 1893*, 301–334].

WYSE JACKSON, P. N. 2001. John Joly (1857–1933) and his determinations of the age of the Earth. *In*: LEWIS, C. L. E. & KNELL, S. J. (eds) *The Age of the Earth: from 4004 BC to AD 2002*. Geological Society, London, Special Publications, **190**, 107–119.

YOCHELSON, E. L. 1989. 'Geologic time' as calculated by C. D. Walcott. *Earth Sciences History*, **8**, 150–158.

YOCHELSON, E. L. 1993. The founding. *In*: ROBERTSON, E. C. (ed.) *Centennial History of the Geological Society of Washington 1893–1993*. Geological Society of Washington, Washington, DC, 3–14.

YOCHELSON, E. L. 1998. *Charles Doolittle Walcott, Paleontologist*. Kent State University, Kent, Ohio.

YOCHELSON, E. L. 2001. *Smithsonian Institution Secretary, Charles Doolittle Walcott*. Kent State University, Kent, Ohio.

Is the Earth too old? The impact of geochronology on cosmology, 1929–1952

STEPHEN G. BRUSH

University of Maryland, College Park, Maryland 20742, USA (email: brush@ipst.umd.edu)

Abstract: Estimates of the Earth's age have had significant impacts, not only on geology but also on biology, astronomy and biblical creationism. In the 1930s and 1940s, the age of the universe as estimated from the expanding universe was less than 2000 million years, but the age of the Earth as estimated from radiometric dating was perhaps as great as 3000 million years. Astronomers responded to this contradiction in at least three different ways. Some cosmologists favoured Georges Lemaître's relativistic model, in which the universe remains about the same size for an indefinite period of time before starting its present stage of expansion. Since theories of the origin of the solar system that were popular in the early 1930s assumed an encounter between the Sun and another star, it seemed plausible that the Earth could have been formed around this epoch of 'cosmic congestion'. Edwin P. Hubble, generally regarded as the founder of the expanding-universe theory because of his discovery of the redshift-distance law, doubted that redshifts are actually due to velocities, and seemed to prefer a non-expanding model, though he emphasized that the correct interpretation of the redshifts of distant galaxies was still an open question up until the time of his death in 1953. Fred Hoyle, Hermann Bondi and Thomas Gold proposed a 'steady-state' cosmology: the universe has always existed, so there is no conflict between its (infinite) age and that of the Earth. The discrepancy was finally resolved in the 1950s when astronomers revised their distance scale and boosted the age of the universe to 10 000 million years or more. The current agreement between geologists and astronomers again leaves creationists with no scientific support at all for their claim that both the Earth and universe were created only about 10 000 years ago.

What's in a number? What difference does it make if the age of the Earth is 6000 years, or 24 million years, or 3000 million years?

Estimates of the Earth's age can affect the relations between disciplines, undermine orthodox ideas and encourage the development of alternative theories. For young-Earth creationists, any number significantly greater than Bishop Ussher's 6000 years is a threat to the credibility of the Bible (Morris 1972, p. 80) and invites the dangerous speculation that a 'day' of the creation week is a metaphor for a geological age. For Darwinian evolutionists, who wanted several hundred million years of geological time, Lord Kelvin's 24 million years was a threat to the hypothesis that the slow process of natural selection could have produced present-day plants, animals and humans; it encouraged 'revised' evolutionary theories that could do the job more efficiently and complete it in the time allowed (Burchfield 1990, pp. 157–158; Bowler 1983, pp. 24, 202; 1986, p. 24). And for cosmologists in the 1930s and 1940s, whose models assumed an expanding universe less than 2000 million years old, the radiometric dating of terrestrial rocks by Arthur Holmes and others (Lewis 2001), showing that the age of the Earth was at least 3000 million years, was an anomaly so serious that it helped to inspire the competing steady-state theory.

In this paper I focus on the last of these controversies, asking: how did the age of the Earth influence the history of cosmology between 1929 and 1952? The first of these dates marks the announcement of the linear relation between spectral redshifts and distances of nebulae by the American astronomer Edwin Hubble (1889–1953), and the last is the announcement of the doubling of the distance scale by Walter Baade (1893–1960) and others, the first step towards making the age of the universe definitely greater than the age of the Earth.

The problem faced by cosmologists came from their axiom that 'The Earth cannot be older than the universe'. This axiom might seem obvious to scientists, but it is not accepted by those who believe in the literal truth of the Bible, e.g. the young-Earth creationists. If 'the universe' means not just 'the firmament' but includes 'the lights in the firmament of the heaven' then Genesis

From: LEWIS, C. L. E. & KNELL, S. J. (eds). *The Age of the Earth: from 4004 BC to AD 2002.* Geological Society, London, Special Publications, **190**, 157–175. 0305-8719/01/$15.00 © The Geological Society of London 2001.

tells us that the Earth was created one day *before* the entire universe. In the United States, where creationists control public education in many communities and influence the curriculum in entire states such as California, Texas, and (most recently) Kansas, scientists must keep this fact in mind when speaking to the public. The creationists reject not only biological evolution but also the Big Bang cosmology, plate tectonics and any other scientific theory that assumes a timescale longer than a few thousand years.

A more subtle point must also be made about the axiom. It actually does *not* imply that the universe is older than the Earth, for there is a third possibility: they are the *same* age. That possibility seems to be of interest only to a mathematician, but remember in science 'the same' usually means 'approximately the same'; in geology or astronomy, approximately could mean 'within a few million years'. As we shall see, the possibility that the Earth might have been formed within a few million years after the beginning of the universe was taken seriously in the 1930s and could be used to keep some cosmological theories alive.

The context for the debate: four 'new sciences' and one shared memory

Our story involves the interactions of four scientific fields that arose, or were revived and reorganized, at the beginning of the twentieth century. Tables 1–4 list scientists in those fields whose publications played some role in this particular controversy.

Stellar astronomy, based on observations with large telescopes

These telescopes were mostly in the United States, built as a result of the efforts of George Ellery Hale, Edward Pickering and Percival Lowell. Theories of the evolution of stars before 1925 assumed a gradual transformation of mass into energy in accordance with Albert Einstein's equation $E = mc^2$; after 1930, as a result of the discovery by Cecilia Payne, confirmed by Henry Russell, that the Sun and stars are mostly hydrogen, a theory based on the synthesis of elements by fusion of hydrogen was developed by Hans Bethe and others. Important results were obtained by Americans or by Europeans who came to the USA, listed in Table 1 (Struve & Zebergs 1962; Brush 1979; Lang & Gingerich 1979; DeVorkin 1982).

Theoretical physics/cosmology

Theoretical physics is an old field but in the late nineteenth century it became possible to specialize in theoretical physics without having to be a competent experimenter as well; with the development of quantum theory and relativity, theoretical physicists acquired considerable prestige and resources in Europe, and later in the USA (Table 2); it also attracted a number of mathematicians (Jungnickel & McCormmach 1986; Garber 1999). Cosmology (including cosmogony) had been a sideline for some astronomers who had to earn their living doing observations and calculating celestial mechanics,

Table 1. *Selected leaders in stellar astronomy in the first half of the twentieth century*

Name	Nationality (year of birth–death)	Research
George Ellery Hale	American (1868–1938)	Large telescopes and observatories
Henrietta Leavitt	American (1868–1921)	Period-luminosity relation for Cepheid variables
Ejnar Hertzsprung	Danish (1873–1967)	Distances of galaxies, colour-luminosity diagram
Vesto M. Slipher	American (1875–1969)	Redshifts of galaxies
Henry Norris Russell	American (1877–1957)	Colour-luminosity diagram, hydrogen in sun and stars; refuted encounter theory of origin of solar system
Harlow Shapley	American (1885–1972)	Distances of galaxies
Edwin Hubble	American (1889–1953)	Distances of galaxies, redshift–distance relation
Milton Humason	American (1891–1972)	Relation between redshifts and distances
Walter Baade	German–American (1893–1960)	Revision of distance scale for galaxies
Cecilia Payne [Gaposchkin]	British–American (1900–1979)	Hydrogen in sun and stars

This table (and Tables 2–4) list only those scientists whose work was relevant to the age problem. Nationality does not necessarily imply citizenship, e.g. Baade did most of his research in the USA but never became a US citizen.

Table 2. *Selected leaders in theoretical physics/cosmology*

Name	Nationality (year of birth–death)	Research
Willem de Sitter	Dutch (1872–1934)	Cosmology
James Jeans	British (1877–1946)	Cosmology, origin of solar system
Albert Einstein	German (1879–1955)	Relativity, quantum theory
Arthur S. Eddington	British (1882–1944)	Stellar structure and evolution, relativity, cosmology
Georges Lemaître	Belgian (1894–1966)	Cosmology
George Gamow	Russian–American (1904–1968)	Nuclear physics, cosmology
Hans Bethe	German–American (1906–)	Nuclear physics, synthesis of elements in stars
Fred Hoyle	British (1915–2001)	Cosmology, nucleosynthesis in stars
Hermann Bondi	Austrian–British (1919–)	Cosmology
Thomas Gold	Austrian–American (1920–)	Cosmology

or a dilettante pursuit for amateurs; in the twentieth century it became a professionalized specialty (North 1965; Lang & Gingerich 1979; Kragh 1996).

Atomic physics

Atomic theories go back to antiquity, but only at the end of the nineteenth century did it become possible to study atoms and their components experimentally, with the discovery of X-rays, radioactivity and the electron. The laboratory of Marie and Pierre Curie in Paris, and the Cavendish Laboratory at Cambridge under J. J. Thomson, became the major centres for atomic research (Table 3). Ernest Rutherford quickly became the leader in research on radioactivity, transmutation and the nuclear atom, starting at the Cavendish and returning as its director after making major discoveries at McGill and Manchester (Romer 1964; Kragh 1999; Nye 1999; Dalrymple 2001).

Planetary geology

Here I mean the kind of geology that tries to understand the structure and evolution of the Earth as a whole, and as part of the solar system (Table 4). It was unfashionable in Anglo-American geology during most of the nineteenth century but was revived in the twentieth century and made possible the 'revolution in the Earth sciences' of the 1960s (Glen 1982; Wood 1985).

The co-operation of theoretical physics/cosmology and stellar astronomy led to the result that the universe was about 1800 million years old or less, while the combination of atomic physics and planetary geology led to the result that the Earth was about 2000 million years old or more. When there is a conflict between the conclusions of different sciences, which prevails? A sociologist might expect that astronomy and theoretical physics, having higher prestige than atomic physics and planetary geology in the twentieth century, would win out or at least be able to ignore the contrary results. But here we have to recognize a shared memory: all these

Table 3. *Selected leaders in atomic physics*

Name	Nationality (year of birth–death)	Research
Marie Curie	Polish–French (1867–1934)	Radioactivity
Ernest Rutherford	New Zealander–British (1871–1937)	Radioactivity, nuclear physics
Alfred Nier	American (1911–1994)	Mass spectrometry, abundances of lead isotopes

Table 4. *Selected leaders in planetary geology*

Name	Nationality (year of birth–death)	Research
Thomas Chamberlin	American (1843–1928)	Glacial geology, origin of solar system
Arthur Holmes	British (1890–1965)	Physical geology, geochronology
Harold Jeffreys	British (1891–1989)	Geophysics, seismology, origin of solar system
Claire Patterson	American (1922–1995)	Geochronology, lead pollution

scientists were probably aware of the great controversy between Kelvin and the geologists in the nineteenth century. They knew that Kelvin's estimate of the ages of the Earth and Sun based on the application of a simple physical model – a cooling fluid sphere – and ignoring geological evidence, had led to spectacularly wrong results (Burchfield 1990; Brush 1967, 1996b). In 1923 Arthur Eddington, in a lecture to the Geological Society of London on 'The Borderland of Astronomy and Geology', stated:

> I am sure it will not be supposed that, in presenting the astronomical side of these questions which belong both to geology and astronomy, I have any intention of laying down the law. The time has gone by when the physicist prescribed dictatorially what theories the geologist might be permitted to consider. You have your own clues to follow out to elucidate these problems, and your clues may be better than ours for leading towards the truth. ... Where, as in the new views of the age of the Earth, physics, biology, geology, astronomy, all seem to be leading in the same direction, and producing evidence for a greatly extended time-scale, we may feel more confidence that a permanent advance is being made. Where our clues seem to be opposed, it is not for one of us to dictate to the other, but to accept with thankfulness the warning from a neighbouring science that all may not be so certain and straightforward as our own one-sided view seemed to indicate (Eddington 1923, p. 21).

At that time there seemed to be no danger that the Earth could be older than the universe. Around 1915, ten years after the publication of the first radioactivity estimates of geological time by Rutherford and Robert Strutt, the figure of 1600 million years (Ma) was frequently proposed as the best estimate for the age of the oldest minerals and the Earth's crust (Holmes 1913). Thomas Chamberlin, combining his planetesimal hypothesis (accretion of small solid bodies) for the formation of the Earth with radioactivity and theoretical biology, found about 4260 Ma for its age (Chamberlin 1920; Brush 1996b, p. 73). Russell arrived at a similar estimate for the age of the Earth by a different method, relying solely on the radioactive decay of thorium and uranium – an upper limit of 8000 Ma and a lower limit of 1100 Ma (Russell 1921); the average of those limits would be 4550 Ma. Harold Jeffreys estimated that the shape (eccentricity) of Mercury's orbit indicated an evolution over a period from 1000 to 10 000 Ma, the most probable figure being about 2500 Ma (Jeffreys 1921, 1929, pp. 58–59). During the 1920s radioactivity estimates of the age of the crust based on lead/uranium ratios started to creep up to values around 3000 Ma but this increase was counteracted by the recognition that a significant part of the lead actually is either non-radiogenic or comes from the decay of thorium. Thus values for the age of the Earth in the range of 1600 to 2000 Ma continued to be accepted by geochronologists as late as 1935. Ernest Rutherford (1929) concluded that the best value is about 3400 Ma, although Holmes (1927) thought 3200 Ma was too high (Brush 1996b, p. 74).

It is somewhat puzzling that cosmologists in the 1930s worked with estimates of 2000 to 3000 Ma for the age of the Earth, when Holmes, the acknowledged expert on that subject, did not accept the higher value until the 1940s. The reason is probably that the estimates of Russell (a well-known astronomer), Rutherford (a well-known physicist) and Jeffreys (a well-known geophysicist using astronomical evidence) were the most familiar to them. Those who read the authoritative compendium on *The Age of the Earth* issued by the United States National Research Council (Knopf et al. 1931) would have learned from the short summary at the beginning that 'Holmes ... reduces Russell's estimate ... to 3000 million years as a possible upper limit' and concludes with E. W. Brown's estimate 'from the point of view of the astronomer' of 2000 Ma (Knopf 1931). Holmes, in his long article in that book, quotes (with no reservations) Rutherford's estimate of 3400 Ma (Holmes 1931, p. 153). He then reviews his own correction of Russell's estimate down to 3000 Ma (Holmes 1931, pp. 219–220). The fact that he actually favoured a significantly lower value (1600 to 2000 Ma) elsewhere in this article may have been overlooked. Even his lower limit of 1600 Ma would be difficult to reconcile with the estimate of 1800 Ma for the expanding universe (even less than 1800 Ma if the expansion is slowed by gravity).

In the meantime astronomers had started to talk about much longer timescales for the stellar universe. If the Sun is a typical star, its lifetime might be measured in *millions of millions* of years, so the Earth would be a relative newcomer to the solar system (Holmes 1926). That idea would be inconsistent with the nineteenth-century nebular hypothesis in which the formation of planets was roughly contemporary with the formation of the Sun (Brush 1996a), but could easily be accommodated by the new encounter theories of Chamberlin and Moulton working together in America, and Jeans and

Jeffreys (separately) in Britain (Brush 1996c). These theories postulated a sun existing for an indefinite period before the event – its close encounter with another star – that gave birth to the planets.

James Jeans invented what came to be called the 'long' timescale in astronomy on the basis of his hypothesis that stars derive their energy from the mutual annihilation of positive and negative particles. Although Jeans had toyed with such ideas earlier, it was Eddington's theoretical relation between the mass and luminosity (brightness as seen by an observer at a standard distance) of a star that gave Jeans the opportunity to make a quantitative calculation of stellar ages, on the order of 6 or 7×10^{12} years (6 or 7 Ta in SI units). According to Jeans' hypothesis, all stars are formed with about the same mass and after reaching a maximum brightness gradually become dimmer as they radiate their mass away through particle annihilation. This hypothesis seemed to explain why hotter stars generally have greater luminosity (an empirical correlation that astronomers call the 'main sequence of the Hertzsprung–Russell diagram'; see Brush 1996b, pp. 87–90). A good contemporary discussion of this subject, fascinating to read, is Chapter XI of Eddington's classic monograph on *The Internal Constitution of the Stars* (1926), still easily available.

Jeans found further evidence for his long timescale in statistical studies of the velocities of stars and the shapes of orbits of binary stars. The basis for his arguments here was an analogy between stars and molecules in gases (he had written extensively on the kinetic theory of gases in addition to his work in astronomy). He calculated, on his hypotheses, that the present distribution of stellar velocities is so close to the Maxwellian distribution[1] that the system must be 10^{12} to 10^{13} years old. He also derived, from his hypotheses, the expected statistical distribution of orbital eccentricities for binary stars and found that the actual distribution agrees with it fairly well. Another statistical process was the loss of less massive stars from moving clusters, a process that theoretically should take 10^{12} to 10^{13} years and seemed to have already occurred (Jeans 1929).

The expanding universe

By a remarkable coincidence, research in both theoretical physics/cosmology and in stellar astronomy in the 1920s led to the conclusion that the stellar universe – including perhaps space itself – is expanding. Galaxies are moving away from us; the more distant galaxies are moving faster. This implies that at some time in the past – the beginning of the expansion – they were all crowded together in a small space. Theory and observation left open questions such as: (a) Is the expansion proceeding at a constant rate, speeding up, or slowing down? (b) Was the universe actually created at the beginning of the expansion, or did it exist before that, and if so was it contracting or simply static for an indefinite time period? (c) Will it continue to expand forever, or will gravitational attraction eventually slow it down and turn the expansion into a contraction? (Theoretically it could reach a maximum size and remain there forever.) (d) Is the universe oscillating between periods of expansion and contraction? (e) If there was a universe before the origin of the present expansion period, could stars, planets and life have survived the passage through that origin?

Einstein's general theory of relativity, as worked out in detail by Aleksandr Friedmann, Willem de Sitter, Georges Lemaître and others, allowed several answers to these questions (see the surveys by Robertson (1933), McVittie (1937), Whitrow (1959), North (1965) and the reprints of original papers in Bernstein & Feinberg (1986)). Einstein himself had assumed that the universe is stationary, and would eventually collapse as a result of the mutual gravitation of all its components unless there was a repulsive force that acted to counteract gravitational attraction. This hypothetical repulsive force is represented in his equations by the notorious 'cosmological constant' λ (in the more recent literature a capital lambda, Λ, is used). But if the universe is actually expanding, at a rate continually diminished by gravity, then the constant is not needed. Nevertheless many cosmologists preferred to retain it as an adjustable parameter in the equations of the theory, in case models with $\lambda = 0$ turned out to be unsatisfactory.

Early in the twentieth century, Vesto Melvin Slipher observed the spectral lines of light coming from a number of galaxies outside our own Milky Way galaxy, and found that most of them were shifted toward the red end of the spectrum. At the same time other astronomers were developing methods for estimating the distances of these galaxies, using Henrietta Leavitt's discovery of an empirical correlation between the

---

[1] In physics, a statistical equation describing the distribution of velocities among the molecules of a gas. This equation is related to the well-known 'normal distribution' or 'bell-shaped curve' in statistics. Jeans, who had written a treatise on the 'dynamical' (kinetic) theory of gases, could estimate the 'relaxation time' – how long it takes for a system with a non-Maxwellian distribution to go over to the Maxwellian state.

period and luminosity of a special kind of variable star, called the Cepheid variables.

Although some astronomers had proposed that the redshifts of galaxies increase with distance, their evidence was not very convincing, and it is generally agreed that Edwin Hubble (Fig. 1) was the first to establish the relationship in 1929 (Hetherington 1971a; Smith 1982). If the redshifts are due to the Doppler effect, they can be interpreted as motion away from us. Inspired by suggestions at the 1928 International Astronomical Union meeting in Holland,[2] Hubble enlisted Milton Humason in a project that established a 'roughly linear relation between velocities and distances among nebulae'. He proposed that this relation '... may represent the de Sitter effect, and hence that numerical data may be introduced into discussions of the general curvature of space' (Hubble 1929).

The linear relation was later expressed by the equation

$$v = Hd$$

where v is the average velocity with which distant galaxies move away from us, d is their distance, and H is called the Hubble constant. Since velocity has the dimension length/time, the inverse of the Hubble constant, $1/H$, will be a characteristic *time* for the expansion, called the 'Hubble time'.[3] If the rate of expansion is constant (individual galaxies neither speed up nor slow down), then $1/H$ can be called the 'age of the universe', i.e. the time elapsed since all the galaxies were crowded together in a very small volume. (Since a galaxy moving away from us at constant speed v travels a distance $d = vt$ in time t, its distance was $d = 0$ when $t = 0$.) But Hubble (1929) did not mention any inferences from his velocity–distance law concerning the timescale for expansion; in fact the phrase 'expanding universe' (or any equivalent) does not appear in this paper.

Hubble pursued his research with Humason, and soon provided more extensive confirmations of the redshift–distance relations. But they quickly abandoned the original assumption that the redshifts were caused by *motion* of the galaxies. Humason wrote two years later: 'It is

Fig. 1. Edwin Hubble, American astronomer whose research led to the establishment of the expanding-universe theory. Credit: Hale Observatories, courtesy American Institute of Physics, Emilio Segrè Visual Archives.

not at all certain that the large red-shifts observed in the spectra are to be interpreted as a Doppler effect, but for convenience they are expressed in terms of velocity and referred to as apparent velocities', as in the title of his paper (Humason 1931, p. 35). As we shall see below, in later years Hubble consistently avoided any definite statement that the redshifts of distant nebulae are due to motion (though he admitted that this interpretation did apply to nearby ones), or that the universe is actually expanding, and seemed to doubt the validity of this conclusion. Thus one may say that Hubble 'discovered the expanding universe' in the same sense that Max Planck 'discovered the quantum': he established an empirical formula that seemed to imply the theory and indeed led others to adopt it (and later to assume that he must have adopted it himself) – yet he drew back from explicitly advocating it as a true statement about the world, and on some occasions even suggested that it was false (cf. for the Planck analogy, Kuhn 1978; Brush 2000).

The Hubble–Humason results were generally taken to indicate an expansion time of about 1800 Ma (the reciprocal of the Hubble constant H), while Jan Hendrik Oort's (1931) estimate of 3400 Ma was in general ignored, though Gamow

---

[2] M. L. Humason, transcript of an interview taken on a tape recorder by Bert Shapiro. Niels Bohr Library, American Institute of Physics, College Park, MD. See also Smith (1982, p. 180).

[3] Astronomers usually measure H in units of velocity divided by distance, e.g. kilometres per second divided by megaparsecs. One megaparsec (Mpc) is equal to about 3×10^{19} km, so $H = (1 \text{ km/s})/\text{Mpc}$ corresponds to $1/H \sim 1000$ Ga.

cited it 20 years later (Gamow 1952, p. 34; Kragh 1996, p. 77). In cosmology (as in other sciences) there are always a few people who dissent from the consensus at any given time, and occasionally they turn out to have been 'right'.

Cosmology constrained by terrestrial time

In November 1931 Arthur Eddington, at a meeting of the Royal Astronomical Society, pointed out that the systematic recession of distant galaxies (at that time called extra-galactic nebulae meaning that they are outside our galaxy, the Milky Way) observed by Slipher and Hubble, 'if accepted as genuine, is alarmingly rapid; the universe must have doubled its radius within geological times, and the consequences in regard to the time-scale and the problem of stellar evolution are most startling' (Eddington 1932a, p. 3). Although some (unnamed) scientists have doubted that the redshifts are due to motion, Eddington himself believed they are Doppler shifts and that the universe is indeed expanding at something like the rate implied by Hubble's relation – because he had calculated this rate from first principles and found a similar rate: $H = 528$ km/s/megaparsec (Eddington 1931a). At that time the observational data gave the rate of expansion as being between 430 and 550 ($H = 500$ corresponds to $1/H \sim 2$ Ga). 'Naturally this close accordance of theory and observation has made me believe that both are right and that the observed motions of the nebulae are genuine' (Eddington 1931b, p. 709; see also Eddington 1932b).

But, although he was aware of the difficulty of reconciling this rate, corresponding to a Hubble time of less than 2000 Ma, with the age of the Earth, he was more concerned about its inconsistency with the long timescale for stellar evolution and velocity distribution. At a meeting of the International Astronomical Union in 1932 he argued that this timescale is popular mostly because 'the policy of the evolutionist is to grab as much as possible in order to accomplish something' – not because of 'any striking success of the theory' (Eddington 1933, p. 83). And, indeed, other astronomers were beginning to abandon the long timescale for several reasons, including the appeal of the expanding-universe theory. For example, advances in nuclear physics persuaded astrophysicists that a star does not evolve by simply annihilating mass and losing luminosity; rather, the correlation between mass and luminosity comes from a collection of data points pertaining to different equilibrium states for different stars initially formed with different masses, and the actual evolution of an individual star is more complicated (Sitter 1933; Strömgren 1933; Öpik 1933; Brush 1996b, pp. 92–102; other reasons are reviewed by Bok (1946) and Haar (1950)). The abandonment of the long timescale by 1935 was fortunate for cosmology since *all* models considered at that time seemed to violate it (Robertson 1933, p. 62).

Perhaps the first cosmologist to stress the conflict between the ages of the Earth and the universe was Willem de Sitter. At a discussion on the evolution of the universe at a 1931 meeting of the British Association for the Advancement of Science, he said: '... a thousand million years is a short time in the evolution of the universe. It is only a third or a quarter of the accepted age of the Earth, and I do not think geophysicists will be ready to take off even one single zero'. But most stars were thought to be a least a thousand times older than that (according to the long timescale) while 'the time elapsed since the beginning of the expansion is only a few thousand million years.... I do not think it will ever be found possible to reconcile the two time scales'. Perhaps, he suggested, the contradiction could be subsumed under the general principle of complementarity produced by Niels Bohr in connection with quantum mechanics: 'We must be prepared to allow this "universe", as we have been forced to grant the atom, the freedom to have contradictory properties' (Sitter 1931, pp. 706, 707, 708).

But de Sitter soon found a way to resolve what he now called a 'well known paradox' – the shortness of the Hubble time and the age of the Earth, both 'on the order of' 2000 Ma – as compared with the thousand-times-longer timescale for stars (Sitter 1933). First, in Lemaître's model for the expanding universe, the beginning of the actual expansion from a finite radius is preceded by an indefinitely-long period of static metastability, during which stars might evolve. If there is still a conflict between theories of stellar evolution and relativistic theories of the expanding universe, relativity must win and theories of stellar evolution must be revised (Sitter 1932a, 1933). Second, the encounter theories of the origin of the solar system postulated that the Earth was formed as a result of a stellar collision or close approach; the chances of such an encounter would obviously be much greater at the epoch when the universe had its minimum size, before starting to expand, than now when stars are very far apart. So the fact that the age of the Earth is equal to that of the universe – in the sense that they are of the same order of magnitude – is not surprising (Sitter 1932a, b). In any case the expanding universe is 'an observed fact', not a theory; we just have to find a theory to account for it, based on general relativity (Sitter 1932c).

Jeans, an advocate of the encounter theory, inferred from that theory the consequence that we are most likely the only intelligent life in the universe. The formation of our planetary system, which he dated as about 2000 Ma ago, was a rare event, and because of the expansion of the universe stellar encounters must have become even more rare. But he did not accept the short timescale of the expanding universe derived from Hubble's data, doubting that redshifts are Doppler shifts and invoking as a possible alternative Fritz Zwicky's idea of redshifts as produced by 'the gravitational pull of stars and nebulae on light passing near them'. This would preserve his own long timescale for the universe as a whole (Jeans 1932, pp. 2–5, 79–81).

During the 1930s, the Lemaître model (sometimes called the Lemaître–Eddington model) combined with the Chamberlin–Moulton–Jeffreys–Jeans encounter theory for the origin of the solar system seemed to provide the best account of the co-evolution of the Earth and universe on a 'short' timescale, a few thousand million years since the beginning of the expansion (Kragh 1996). At that time, which was *not* the origin of the universe itself, galaxies were crowded together in a small but finite space; stars collided and formed planetary systems. This compromise was plausible as long as the two crucial dates – the Hubble time and the age of the Earth – were considered to be known only approximately, so that they could be assumed to be about the same.

In these discussions the cosmologists generally respected the accuracy of geochronology as compared with astronomy. For example, de Sitter stated that the determination of the age of the Earth's crust as:

> of the order of a few thousand million years ..., resting on the chemical analysis of minerals and the laws of radioactive processes, leaves only a small margin of uncertainty. The age of the Earth itself is probably not much more, and possibly much less, than twice the age of the crust (Sitter 1933, p. 632).

While Jeans asserted that:

> the two important and best-determined data we have are, first, the age of the Earth, which from consideration of radioactive processes is about 2.10^9 years and, second, the time scale of the recession of the nebulae, also about 2.10^9 years' (Jeans *et al.* 1935, pp. 108–111).

Harold Jeffreys, who was both a geophysicist and a theoretical astronomer, wrote that astronomical evidence relevant to the age of the Earth 'is consistent with geophysics but less definite' (Jeffreys 1931, p. 986). But while the age of the Earth could encourage theorists to favour one version of relativistic cosmology over another, it would certainly not cause them to abandon basic principles of physics such as general relativity. Eddington preferred the value of the Hubble time that he had derived from abstract reasoning about the physical constants and was not very much concerned about whether it was shorter than the age of the Earth.

Another theorist, Paul Dirac, proposed a 'large numbers hypothesis' in which the physical constants changed with time; he estimated that the age of the universe on this hypothesis would be about 700 Ma (Dirac 1937, 1938). He admitted that this 'is rather small, being less than the age of the Earth as usually calculated from data of radioactive decay, but this does not cause an inconsistency, since a thorough application of our present ideas would require us to have the rate of radioactive decay varying with the epoch and greater in the distant past than it is now' (Dirac 1938, p. 204). But he never published an explicit result for the age of the universe taking account of this effect (see, however, Houtermans & Jordan 1946; Hönl 1949.)

Edward Milne suggested that his 'kinematic relativity' might avoid the conflict between terrestrial and cosmic timescales by introducing two different time parameters, one for material particles and one for radiation (Milne 1940, 1952). But although the biologist J. B. S. Haldane (1945) was an enthusiastic supporter of Milne's theory, it gained little acceptance from astronomers (the text by Skilling & Richardson (1947) is an exception). By 1940, the consensus supporting Lemaître's theory had begun to fall apart, and choosing a suitable timescale had become 'the nightmare of the cosmologist' (North 1965, p. 125; see for example Robertson 1940). As the accuracy of determinations of the ages of Earth and universe improved, the discrepancy between them could no longer be attributed to errors in either one (McCrea 1953, p. 355).

Another escape route from 'the nightmare' was also closed off. In 1935, the encounter theory of the origin of the solar system was subjected to devastating criticism by Russell (1935) and could no longer be used to justify a scenario in which the Earth was formed at 'about the same time' as the beginning of the expansion of the universe (Brush 1996c). Harlow Shapley rejected the idea that cosmic congestion 2000 Ma ago, implied by the expanding universe theory, could be causally connected with the origin of the Earth just because the latter 'also' is a few thousand million years old. We are all 'uncomfortable' with the short timescale of the universe, he wrote: 'It does not seem sufficiently dignified that the uncompromisingly majestic universe measure its duration as scarcely greater than the age of the oldest rocks on this small planet's surface' (Shapley 1944, p. 74).

Hubble doubts the expanding universe

Since Hubble is generally regarded as the founder of the expanding-universe theory, one might ask what role he played in the timescale controversy. The answer is: he used it as one of several arguments *against* that theory. In fact Hubble withdrew his public support for the expanding-universe theory in 1935, arguing that the evidence did not favour his original interpretation of redshifts as Doppler shifts and challenging theorists to provide a better interpretation. Even before that time he had rarely discussed the 1800 Ma 'age of the universe' that seemed to be implied by his redshift–distance law. Perhaps the best example of this is in a 1934 article for a popular science journal. He wrote that we should assume redshifts are due to motion 'until evidence to the contrary is forthcoming' but the implication of this assumption is that the nebulae were:

> jammed together in our particular region of space, and at a particular instant, about 2000 million years ago, they started rushing away in all directions at various velocities ... The time scale seems suspiciously short – a small fraction of the estimated age of some stars – and the apparent discrepancy suggests the advisability of further discussion of the interpretation of red-shifts as evidence of motion.

Pending further research, 'the cautious observer refrains from committing himself to the present interpretation and employs the colorless term "apparent velocity".' (Hubble 1934, p. 199).

One could argue that the term 'apparent velocity' is consistent with general relativity: the galaxies do not have a real velocity in the sense that they move *through* space, rather they are carried along by the expansion of space itself (Gribbin 2000, p. 136). But that argument would be misleading since Hubble was also doubting the expansion of the universe. The expanding universe, he wrote, 'is the latest widely accepted development in cosmology' but it depends on assuming redshifts are velocity shifts. 'The 200-inch telescope will definitely answer the question of the interpretation of red-shifts, whether or not they do represent motions' (Hubble 1934, p. 202).

Hubble's work with cosmologist Richard Tolman led to another way to answer the question. Tolman (1934, p. 485) was aware of the timescale problem but that was not the primary motivation for their research. They looked at two simple models based on the assumption that the redshifts (a) are or (b) are not velocity shifts, and applied them to the observational data. For (a) the universe 'is represented by a homogeneous expanding model obeying the relativistic laws of gravitation', while for (b) they used 'a static Einstein model of the universe, combined with the assumption that the photons emitted by a nebula lose energy on their journey to the observer by some unknown effect ...' (Hubble & Tolman 1935, p. 304). This alternative kind of redshift was apparently inspired by the speculations of Zwicky (1929, 1935; see also MacMillan 1932), and was later called the 'tired light' hypothesis. But Hubble did not explicitly advocate Zwicky's idea, though he often seemed to be challenging the theoretical physics/cosmology community to come up with new principles that could provide an alternative interpretation of the redshifts.

The conclusion of the Hubble–Tolman collaboration was that model (a) could be made to fit the data only by assuming a rather high curvature, high density and small size of the universe; it would imply that the 'average' nebulae now observed emitted their light about 3×10^8 years in the past, nearly comparable to the 'time of cosmic expansion – possibly of the order of 10^9 to 10^{10} years (Hubble & Tolman 1935, p. 336). (It is not clear to me why this time was allowed to be as much as five times as long as what was usually estimated to be the Hubble time, 1800 Ma, but it may explain why Hubble did not yet perceive a direct conflict with the age of the Earth.)

According to science writer George Gray, the Hubble–Tolman analysis 'which casts doubt on the reality of the expansion ... has come like a bombshell into the camp of the theorists and is providing a major topic of conversation among astronomers, cosmologists, mathematicians, and other universe explorers' (Gray 1937, pp. 66–67). But it did not disturb for more than a couple of years the widespread acceptance of the expanding-universe theory.

In his later publications, and in earlier correspondence,[4] Hubble took the position that it was not his responsibility as an observer to determine the correct theoretical interpretation

---

[4] E. P. Hubble, letter to W. de Sitter, 23 September 1931. Edwin Hubble Collection, Huntington Library, San Marino, CA, Box 1. Quoted by Hetherington (1982) and by Sharov & Novikov (1993).

of his data. But he went further than that, with rather explicit statements that the 'expanding models are definitely inconsistent with the observations unless a large positive curvature (small, closed universe) is postulated', which gives an unreasonably high density as well as a 'small scale ... both in space and time' (Hubble 1936b, pp. 517, 554). On the other hand the non-expanding model (b) gives a 'rather simple and thoroughly consistent picture'. It is 'more economical and less vulnerable, except for the fact that, at the moment, no other satisfactory explanation [of the red shift] is known'. Moreover, the expanding model (a) assigns a unique location to the observer, which 'is unwelcome and a priori improbable' (Hubble 1936c, pp. 624–626).

In his popular book *The Realm of the Nebulae*, Hubble (1936a) took a completely neutral position in regard to the two models, but in a paper in *Monthly Notices of the Royal Astronomical Society* the next year he asserted that even when observational data are 'weighted in favour of the [expanding-universe] theory as heavily as can reasonably be allowed, they still fall short of expectations': the expanding universe would have to be closed, 'curiously small and dense, and, it may be added, suspiciously young'. If redshifts are *not* primarily velocity shifts, we have a much more reasonable universe in which 'the observable region may extend indefinitely both in space and time' (Hubble 1937b, pp. 509, 513).

In the autumn of 1936, Hubble came to Oxford to give the Rhodes Memorial Lectures and gave perhaps his most explicit statement that the Earth is too old for the expanding-universe theory to be valid. Deviations from the linear redshift–distance relations now corresponded to an expansion (if such it be) being slowed down by gravity, so that the 'true age' of the universe would be a maximum of only 1500 Ma – less than the Hubble time of 1800 Ma: So the

> initial instant ... clearly falls within the life-history of the Earth, probably within the history of life on the Earth. ... Some there are who stoutly maintain that the Earth may well be older than the expansion of the universe. Others suggest that in those crowded, jostling yesterdays, the rhythm of events was faster than the rhythm of the spacious universe today; evolution then proceeded apace, and, into the faint surviving traces, we now misread the evidence of a great antiquity.

But such speculations 'sound like special pleading, like forced solutions of the difficulty' so we must again look for a different explanation of the redshifts: 'some unknown reaction between the light and the medium through which it travels' (Hubble 1937a, pp. 42, 44, 45; see also Hubble 1937c). Other astronomers who were sceptical of the expanding universe were well aware that Hubble shared their doubts, though his reluctance to strongly support any particular alternative meant that he could not be regarded as a leader of an anti-expansion movement.[5]

Hubble revisited the problem in a public lecture in 1940, explaining that by continuing to consider two models he was following the method of multiple hypotheses (usually attributed to T. C. Chamberlin):

> A critical review of previous investigations ... leaves unchanged the conclusion concerning the interpretation of red shifts. No effects of expansion – no recession factor – can be detected. The available data still favor the model of a static universe rather than that of a rapidly expanding universe (Hubble 1940b, p. 407).

Or, as a report in *Publications of the Astronomical Society of the Pacific* summarized the lecture: 'The theory that red shifts measure the recession of the nebulae in a rapidly expanding universe is not borne out by the available observational data' (Hubble 1940a, p. 289).

In his Sigma Xi Lecture in December 1941, published in a number of magazines, Hubble announced that the latest astronomical data implied that a hypothetical expanding universe would be less than 1000 Ma old – 'a fraction of the age of the Earth ... the time scale is probably not acceptable. Either the measures are unreliable or red shifts do not represent expansion of the universe' (Hubble 1942, p. 112). After returning to astronomy following government service in the Second World War, Hubble kept up with Holmes' research on the age of the Earth, including his 1947 value of 3350 Ma (Holmes 1947b), but did not seem to see it as a major argument against expansion.[6] He did not mind being associated with the expanding universe in the public mind, as long as he could stipulate that this was not an established fact but a controversial theory still to be tested by observations with the long-awaited 200-inch telescope.[7]

Shapley, Hubble's long-time rival (Shapley 1969, p. 57; Gingerich 1990), thought one of his

---

[5] N. U. Mayall, letter to E. P. Hubble, 16 March 1937. Hubble Collection, Huntington Library, San Marino, CA.
[6] E. P. Hubble, letter to G. Gamow, 14 April 1948. Hubble Collection, Huntington Library, San Marino, CA.
[7] E. P. Hubble, letter to J. H. Breasted, 17 February 1948. Hubble Collection, Huntington Library, San Marino, CA.

arguments for rejecting the velocity interpretation of redshifts – a radial gradient in density – was weak, but announced that his own work on faint galaxies had uncovered a *transverse* gradient 'considerably greater than the radial gradient suspected by Hubble, which has led to his doubts on the existence of space curvature and on the alleged expansion of the universe. Obviously we are not through with this business' (Shapley 1944, p. 69). Moreover, the apparent high speeds of distant galaxies calculated from their redshifts raised a new problem for the expanding-universe theory: would we have to conclude that these galaxies moved faster than the speed of light? So Shapley agreed with Hubble that a search for alternative explanations of the redshift was appropriate.

Perhaps the last major effort to modify relativistic cosmology in order to comply with the short time period dictated by the age of the Earth was made by Guy C. Omer, Jr, a student of Tolman's at Caltech, who acknowledged extensive discussions with Hubble. As Omer stated the problem, Hubble 'has shown that the observational data which he has obtained do not agree satisfactorily with the homogeneous relativistic cosmological models'. First, the models have too high a density (about 1000 times the current estimates). Second, they are too small – they imply that 'the 100-inch reflector has already surveyed a large fraction of the existing universe. This might be true, but it seems philosophically repugnant.' Third, the maximum age for the homogeneous model is 1200 Ma; 'This is about one third the recent estimation of the age of the Earth as an independent body, made by A. Holmes [1947b]. This is probably the most serious difficulty of the homogeneous model' (Omer 1949, p. 164). Omer constructed a *non*-homogeneous model whose density increases outward from the origin (because the expansion starts at the origin and is propagated outward). Its age was 3640 Ma, slightly greater than Holmes' (1947a, b) estimate of 3350 Ma for the age of the Earth, and within the range of 3000 to 5000 Ma estimated by Bok (1946) from astronomical data. It also agreed with Jeffreys' estimate of about 4000 Ma for the Earth–Moon system from tidal interactions (Jeffreys 1924, p. 229). Tolman called Omer's result a 'satisfactory figure for the time scale' (Tolman 1949, p. 377).

Did Hubble actually reject the expanding universe? One historian who has examined all the available Hubble material, Norriss Hetherington, argues that he continued to favour the relativistic expanding universe for philosophical reasons even though his own research and the age of the Earth refuted it (Hetherington 1971b, 1982, 1989, 1990). For example, unpublished notes for the Hubble–Tolman paper show that the evidence was even more unfavourable to the expanding-universe theory than their published paper admitted, a fact that Hetherington interprets to mean that Hubble let his bias in favour of the expanding model outweigh the empirical facts. But the statements in his publications, while sometimes ambivalent, give me the impression that Hubble did not support the expanding universe after 1935, or at the very least that he wanted his audience to know about the strong empirical evidence against it. Several astronomers who have written about Hubble agree that he rejected the velocity interpretation of redshifts and was sceptical about the reality of expansion (Whitrow 1972, p. 532; Sandage 1989, p. 357; Osterbrock *et al.* 1993, p. 88; Gribbin 2000, p. 106). Perhaps, using Hubble's own term, we should say he *apparently* rejected the theory without explicitly saying so.

A radical solution: steady-state cosmology

By 1948 the timescale problem had become much more serious. Additional astronomical observations reconfirmed the 1800 Ma Hubble time, implying an age of only 1200 Ma for the simplest plausible model with the expansion slowed by gravity. More sophisticated techniques in geochronology, combined with new determinations of lead isotopic abundances (Nier 1938, 1939), made 3350 Ma a firm *lower* limit for the age of the Earth (Holmes 1947a, b) though Jeffreys' critique (1948, 1949) allowed Whitrow (1949, 1954b) and Öpik (1954a, b) to treat it as still controversial. Paul Couderc (1952) insisted that any satisfactory cosmological model must give an age of the universe at least three times as great as the 1200 Ma for the simple expanding model with gravitational effects; only Omer's non-homogeneous model met that criterion.

The age of the Earth may have affected the interpretation of some astronomical observations: it has been suggested that the need to fit the evolution of the universe into a 3 Ga timescale influenced estimates of the age of globular clusters.[8] Even Albert Einstein feared that relativistic cosmology was in danger:

> The age of the universe, in the sense used here, must certainly exceed that of the firm crust of the Earth as found from radioactive minerals. Since determination of age by these minerals

---

[8] A. Sandage, transcript of interview with S. Weart, 22 May 1978, pp. 74–75. Niels Bohr Library, American Institute of Physics, College Park, MD.

is reliable in every respect, the cosmologic theory here presented would be disproved if it were found to contradict any such results. In this case I see no reasonable solution (Einstein 1945, p. 132).

What happened next looks very much like the beginning of a Kuhnian scientific revolution (Kuhn 1962). Since it appeared that the timescale anomaly could not be resolved within the existing paradigm, three cosmologists boldly proposed a new theory that explicitly violated one of the most sacred laws of that paradigm: conservation of mass. Fred Hoyle, Hermann Bondi and Thomas Gold postulated a 'perfect cosmological principle': the universe, in the large, always looks the same to an observer at any time or place. True, distant galaxies are rushing away from us and will eventually become invisible even to our most powerful telescopes, but not to worry! We will not be left alone with only our own galaxy to look at. Other galaxies will appear in their place – formed from matter that is being continually created at a rate just sufficient to keep constant the average density of matter in the visible universe. In this 'steady-state' model the universe has always existed in the past and will always exist in the future. The Hubble time describes the rate of expansion but has nothing to do with the *age* of the universe, which is infinite. By waving their magic wand, Hoyle, Bondi and Gold made the age paradox disappear.

Not so fast, gentlemen! How can you violate the law of conservation of mass (or, relativistically speaking, mass-energy) without throwing away modern physics? The reply, already prepared, comes back: isn't the orthodox cosmology – which Hoyle sarcastically called the 'Big Bang' (Hoyle 1950, p. 113) – also based on a violation of this law, but one that occurred 1800 Ma ago, when all the matter and energy of the universe was supposedly created at one instant? Since, by hypothesis, no one except God was around to observe this illegal event, we have no way to prove scientifically that it really happened – whereas in the steady-state theory, the creation is happening *right now* at a definite rate that can easily be calculated, and even directly observed by a sufficiently delicate measurement. Thus the steady-state theory, unlike the Big Bang, makes a *testable prediction*. Isn't that the essence of science?

At this point Bondi and Gold deviated from the Kuhnian script and started to read from the one written by Karl Popper (1934). Bondi especially emphasized that a theory, if it is to be considered scientific, must make *falsifiable predictions* and the theory must be abandoned if those predictions are refuted (Bondi 1973, p. 11; 1992). The steady-state theory predicted that the universe in the distant past, which we observe by looking far out in space, must be essentially similar to the present (nearby) universe. When that prediction was refuted in the 1960s, Bondi (1990*a*, p. 65) gave up steady-state theory (as did Gold a few years later), while Hoyle, who had never endorsed falsifiability, tried to keep the theory alive with various modifications.

So we learn from this episode that Popper's philosophy is valid only if you believe in it.

There was another contribution from the philosophy of science: in 1954, the American philosopher of science Michael Scriven won a prize offered by the *British Journal for the Philosophy of Science* for an essay on 'What is the logical and scientific status of the concept of the temporal origin and age of the universe?' (Scriven 1954). Öpik (1954*a*, *b*) and Whitrow (1954*b*) also entered the contest but failed to persuade the philosopher-judges to take seriously the *scientific* evidence, though their papers did get published in the journal. In the great tradition of August Comte, who had asserted more than a century earlier that we would never know the chemical composition of the stars, Scriven argued that 'no verifiable claim can be made either that the Universe has a finite age or that it does not. We may still believe that there is a difference between these claims: but the difference is one that is not within the power of science to determine, nor will it ever be' (Scriven 1954, p. 190).

Astronomy blinks

The conflict between astronomical and geological timescales was eliminated in the 1950s when the astronomers (Baade, Thackeray, Sandage and others) revised their distance scale (Baade 1952; Hoyle 1994, p. 263; Osterbrock 1998; Feast 2000; Gribbin 2000). This was in part due to new observations made with the 200-inch telescope which, as Hubble had predicted, resolved the redshift problem and put the expanding-universe theory on a firmer foundation. By doubling and then quadrupling (or more) the distances of most galaxies, these observations increased the age of the universe to more than 10 000 Ma, at least twice the age of the Earth even when the latter was increased to its current value of 4500 Ma.

Since the new astronomical timescale removed a major objection to the Big Bang cosmology without introducing continuous creation, it appeared to many astronomers that the steady-state paradigm switch was no longer necessary. Some supporters of steady-state theory tried to de-emphasize the importance of the timescale problem as a justification for their theory and argued that it was still superior to the Big Bang for other reasons (McCrea 1953, p. 362; Bonnor 1964, p. 164). In effect they invoked another maxim from the philosophy of science: the distinction between the context of discovery and the context of justification (Reichenbach 1938, pp. 6–7, 382–384). It doesn't matter at all where your theory came from (how it was 'discovered'); all that counts is how well it is supported by logical reasoning and empirical evidence (its 'justification'). I do not wish to argue that point but simply want to reinforce one of my major themes by citing a few published statements showing or affirming that the age of the Earth problem *was* an important motivation for proposing the steady-state theory and a reason why it was initially supported by other cosmologists (Bondi 1948, p. 111, 1952, p. 140, 1990b, pp. 189, 191; Bondi & Gold 1948, p. 263; Hoyle 1948, p. 374; Bondi *et al.* 1995, p. 10; McCrea 1968, p. 1296; Whitrow 1954a, p. 85, 1959, p. 175).

The case of the steady-state 'revolution' (which had little support outside Britain and failed in the 1960s, as described in detail by Kragh 1996), has been mentioned here mainly to show how cosmologists acquired so much respect for geochronology that they were sometimes willing to trust its results more than their own theories. Those who did not want to admit that the age of the Earth invalidated their ideas could simply ignore that inconvenient fact. If they wanted to dispute the accuracy of geochronology they had one small straw to grasp at: in 1948 Harold Jeffreys, a respected contrarian whose views on the analysis of data could be disputed but not ignored, criticized the method used by Arthur Holmes to estimate the age of the Earth. Jeffreys argued that there were not enough data to infer a reliable conclusion by the isochron method, and suggested that a value between 2100 and 2600 Ma was more likely to be correct than Holmes' 3350 Ma (Jeffreys 1948, 1949). Holmes (1949) defended his value, but for those who didn't bother to study the technical details, this exchange made it possible to say that Holmes' value was still controversial (Öpik 1954a, b; Whitrow 1949, 1954b).

It is perhaps remarkable that most cosmologists did not look more closely at the methods used by Holmes and others to estimate the age of the Earth. As far as I know the only one who did so was George Gamow, who in his inimitable and entertaining style gave a detailed exposition of the isochron method in one of his popular works, using an analogy with deposits of cattle dung (Gamow 1952).

Another question, which might intrigue those scholars who study the 'social construction of scientific knowledge', is: how did it happen that the astronomers managed to rescue relativistic cosmology and the expanding-universe theory by 'discovering' that they had been using the wrong distance scale, making a major correction just at the time when it was most desperately needed? I know of no evidence that the results of Baade and others were in any way influenced by a desire to expand the distance scale; in fact Baade had started his research on this subject in 1931, before the universe was generally thought to be 'too young'. On the other hand, it seems reasonable that the widely publicized timescale problem would have encouraged everyone to scrutinize as carefully as possible the evidential basis of the astronomical timescale (Treder 1969). Criticizing and attempting to replicate a controversial measurement is just good science.

Finally we come back to Hubble himself: what was his reaction to the rescue of the expanding-universe theory? One biographer has suggested that he 'must have been secretly pleased' when Baade's preliminary results were announced in 1944, indicating that Population II stars must be 'much more distant from one another than Hubble's calculations suggested' since that removed the major objection to Hubble's theory by making the universe as old as the Earth (Christiansen 1995, p. 293). But Osterbrock (1997) notes that Hubble could not keep up with astronomical research during the war because of his own preoccupation with military projects, and concludes that he did not read (or at least did not understand the significance of) Baade's 1944 work. As late as 1951 Hubble doubted that the redshifts of distant nebulae are velocity shifts (Hubble 1951, p. 463). Even after the announcement of Baade's results in 1952, making the Hubble time about 3500 Ma, a journalist reported that Hubble was still not satisfied with the expanding-universe theory. Since the apparent velocities indicated that the expansion (if real) is slowing down, the actual time since the beginning of the expansion would be less than that: 'The time-scale is still an uncertain and disturbing feature ... even the extended "age of the universe" is no greater than current estimates of the age of rocks in the crust of the Earth' (Gray 1953).[9] I do not know if Hubble was aware at that time that the age of the Earth had recently been extended to 4500 Ma (Patterson 1953) – some of the participants involved in

---

[9] See also E. P. Hubble, letter to G. W. Gray, 12 March 1953. Hubble Collection, Huntington Library, San Marino, CA.

that research were at nearby Caltech – or that the encounter theory of the origin of the solar system, which made it somewhat plausible that the Earth is about as old as the universe, had been generally rejected in favour of a more gradual formation (Brush 1996b, c).

By 1958 the age of the universe was thought to be about 13 000 Ma, with an uncertainty of a factor of 2 (Sandage 1958), so even the lowest reasonable value was comfortably greater than the age of the Earth. Unfortunately Hubble died in 1953 so we don't know whether he would have retracted his opposition to the expanding universe in the light of this new evidence. His last published statement on the subject, edited by Sandage from the manuscript of his George Darwin Lecture on 8 May 1953, acknowledges the doubling of the distance scale and concluded that the age of the universe is 'likely to be between 3000 and 4000 million years, and thus comparable with the age of rock in the crust of the Earth' (Hubble 1954, p. 666). According to Sandage, he still rejected the velocity interpretation of redshifts and thus did not accept expansion at the time of that lecture (Sandage 1989, p. 357).

Finding the age of the Earth and the universe: by physics or by faith?

The 4500 Ma age of the Earth and some meteorites, found in the 1950s by Patterson, Holmes, Houtermans, Wasserburg & Hayden, Russell & Allan, Thompson & Mayne, Folinsbee, Lipson & Reynolds, Schumacher, Webster, Morgan & Smales, Masuda and others, has remained one of the most reliable fixed points in astronomy, as well as in geology, in the last half-century. In particular, it survived the so-called 'revolution in the Earth sciences' of the 1960s, which overturned many widely accepted assumptions about our planet, and several changes in the prevailing theories of the origin of the solar system (Brush 1996b, c). According to Albrecht Hofmann (2001) some of the assumptions about the composition of the Earth originally used to estimate its age (e.g. by Patterson) are no longer considered strictly valid, but the numerical value has, nevertheless, been confirmed by more recent research.

That makes it all the more surprising to find, at least in the USA, a strong 'young-Earth creationism' movement, started in the 1970s and still claiming the support of many Protestant fundamentalists and evangelicals (Numbers 1992; Morris (1995) gives a bibliography of more than 125 books advocating this doctrine). Although biological evolution is obviously their major target, the creationists define evolution science to include all geological theories (such as plate tectonics) based on a timescale longer than a few thousand years, and all astronomical theories (such as the Big Bang) based on such a timescale – in other words, they reject the entire foundations of modern geology and astronomy. This is quite explicit in the 1999 action of the Kansas School Board, which removed both biological evolution and the Big Bang from the required state public school curriculum.

But that's not all. In order to discredit geochronology based on radiometric dating, the creationists assert that radioactive decay rates have been much faster in the past so the age of the Earth has been overestimated by several orders of magnitude. Thus they also reject the findings of modern nuclear physics (Brush 1982). Perhaps that does not seem very radical to a creationist who has already revised the Second Law of Thermodynamics in order to ban biological evolution.

Why go to such extraordinary lengths, thereby alienating many scientists who are sceptical of Darwinism for religious or other reasons, but are unwilling to trash the physical sciences? I do not pretend to understand the motivations of the creationists so I will just quote one of their leaders, Henry Morris, who explains why the scientists' Earth is *too old* for him. (Remember that the 4004 BC date for the creation, constructed by Bishop Ussher, was added in the seventeenth century and no longer appears in most modern Bibles.)

> The question of the duration of geologic time is undoubtedly the most vexing problem confronting the Biblical creationist. Most geologists insist that the Earth is about five billion years old, that life evolved probably three billion years ago, and even human life at least a million years ago. Yet the Bible seems clearly to teach that all things were created only about six thousand years ago.

From six thousand to five billion – this is how much the Earth has 'aged' in little more than a century! If the Bible is really wrong on this, it amounts to almost a million-fold mistake. And if it is mistaken this much in its very foundation – the chronologic framework of history – then how can we rely on it anywhere else? Writers who are unable to record sober facts of history correctly are not likely to inspire confidence when they forecast events of the eternal future (Morris 1972, pp. 80–81).

Now changing the radioactive decay rates in order to shorten the age of the Earth is not completely ridiculous – even the famous physicist Paul Dirac thought he could do that (Dirac 1938, p. 204; see also Wyse Jackson 2001, on John

Joly), although not by six orders of magnitude. Moreover, as noted above, the Bible also says (according to the creationists) that the universe is younger that the Earth, so it must also be only about 6000 years old. Yet astronomers tell us that most stars are more than 6000 light years away, so they must have existed more than 6000 years ago in order to send out the light that we receive now. But Morris can deal with that objection too, using an idea similar to one suggested by Philip Gosse (1857):

> If the stars were made on the fourth day, and if the days of creation were literal days, then the stars must be only several thousand years old. How, then, can many of the stars be millions or billions of light-years distant, since it would take correspondingly many millions or billions of years for their light to reach the Earth?
>
> This problem seems formidable at first, but is easily resolved when the implications of God's creative acts are understood. ... The sun, moon, and stars were formed specifically to 'be for signs, and for seasons, and for days, and years,' and 'to give light upon the Earth' (Genesis 1:14, 15). In order to accomplish these purposes. they would obviously have to be visible on Earth. But this requirement is a very little thing to a Creator! Why is it less difficult to create a star than to create the emanations from that star ... It is even possible that the 'light' bathing the Earth on the first three days was created in space as en route from the innumerable 'light bearers' which were yet to be constituted on the fourth day.

Actually, real creation necessarily involves creation of 'apparent age'. Whatever is truly created – that is, called instantly into existence out of nothing – must certainly look as though it had been there prior to its creation (Morris 1972, pp. 61–62).

This proposal, that God tried to deceive us into thinking the world is old when it is really young – by creating it to look old – clearly cannot be countered by any rational argument. Nor would the declaration by Pope Pius XII, that the age of the oldest mineral is 'at the most five thousand million years' (Pius XII 1951), make any impression on American creationists, who are mostly Protestants. The scientist can reply only by quoting Einstein's famous statement: 'Subtle is the Lord, but malicious He is not' (Pais 1982, p. vi).

Conclusions

According to David Raup, an American palaeontologist, one aspect of the episode I have described is that 'geology has a curious moral authority over astrophysics' that now makes astrophysics somewhat more receptive to geological arguments in cases such as the K/T extinction controversy (Raup 1986, p. 132). This is perhaps an exaggeration, but it is reflected in the wording of astronomer George Field's phrase: 'current models for the sun, having an age of 4.5 billion years *as dictated by the radioactive dating of meteorites*' (Field 1986, p. 74, emphasis added). We have seen that some astrophysicists, when choosing between alternative cosmological models, did take into account compatibility with the age of the Earth, along with other facts such as the ages of stars. During the 1930s, after the long timescale (10^{12} to 10^{13} years) was discarded, the leading candidate for the age of the universe was the time since the beginning of the expansion characterized by Hubble's linear relation between redshifts and distances of distant nebulae (external galaxies). The simplest relativistic model for this expansion, with no repulsion to counteract gravitational attraction, implied an age of about 1200 Ma, only a fraction of the 2000 to 3000 Ma estimated for the Earth's crust. Models with repulsion (cosmological constant) to lengthen the expansion time, and a long metastable period with a finite size, were therefore favoured. Such a model was still being suggested as late as 1951 to evade the conflict with the age of the Earth (Finlay-Freundlich 1951). But Eddington, who had earlier advocated an attitude of co-operation with geologists, ignored the age of the Earth when he found he could deduce the Hubble time from basic principles (or what his critics called numerology).

Conversely, Hubble, renowned as the founder of the expanding-universe theory, seemed to abandon that theory when it failed to account for several observed properties of the universe – of which the Earth's age was not the most important.

The most dramatic effect of geochronology on cosmology was the motivation it provided for the introduction of the steady-state cosmology, which managed to outlive for a decade the resolution of the timescale problem. An indirect effect was the delayed acceptance of the Big Bang theory, which did not seem plausible until after resolution of that problem, and the empirical disproof of steady-state cosmology.

This paper is based on research supported by a Fellowship from the John Simon Guggenheim Memorial Foundation. Donald Osterbrock and Virginia Trimble provided useful suggestions but are not responsible for the views presented here.

References

BAADE, W. 1952. Extragalactic nebulae. Report to IAU Commission 28. *Transactions of the International Union of Astronomy*, **8**, 397–399.

BERNSTEIN, J. & FEINBERG, G. (eds) 1986. *Cosmological Constants: Papers in Modern Cosmology*. Columbia University, New York.

BOK, B. J. 1946. Reports on the progress of astronomy: The time-scale of the universe. *Monthly Notices of the Royal Astronomical Society*, **106**, 61–75.

BONDI, H. 1948. Review of cosmology. *Monthly Notices of the Royal Astronomical Society*, **108**, 104–120.

BONDI, H. 1952. *Cosmology*. Cambridge University, Cambridge.

BONDI, H. 1973. Setting the scene. *In:* JOHN, L. (ed.) *Cosmology Now*. BBC, London, 11–22.

BONDI, H. 1990a. *Science, Churchill and Me. The Autobiography of Hermann Bondi, Master of Churchill College Cambridge*. Pergamon, Oxford.

BONDI, H. 1990b. The cosmological scene 1945–1952. *In:* BERTOTTI, B., BALBINOT, R. & BERGIA, S. (eds) *Modern Cosmology in Retrospect*. Cambridge University, Cambridge, 189–196.

BONDI, H. 1992. The philosopher for science. *Nature*, **358**, 363.

BONDI, H., & GOLD, T. 1948. The steady state theory of the expanding universe. *Monthly Notices of the Royal Astronomical Society*, **108**, 252–270.

BONDI, H., GOLD, T. & HOYLE, F. 1995. Origins of steady-state theory. *Nature*, **373**, 10.

BONNOR, W. 1964. *The Expanding Universe*. Macmillan, New York.

BOWLER, P. J. 1983. *The Eclipse of Darwinism: Anti-Darwinian Evolution Theories in the Decades around 1900*. Johns Hopkins University, Baltimore.

BOWLER, P. J. 1986. *Theories of Human Evolution: A Century of Debate, 1844–1944*. Johns Hopkins University, Baltimore.

BRUSH, S. G. 1967. Thermodynamics and history: science and culture in the 19th century. *The Graduate Journal* [University of Texas], **7**, 477–565.

BRUSH, S. G. 1979. Looking up: the rise of astronomy in America. *American Studies*, **20**(2), 41–67.

BRUSH, S. G. 1982. Finding the age of the Earth – By physics or by faith? *Journal of Geological Education*, **30**, 34–58.

BRUSH, S. G. 1996a. *Nebulous Earth: The Origin of the Solar System and the Core of the Earth from Laplace to Jeffreys*. A History of Modern Planetary Physics, Vol. 1. Cambridge University, New York.

BRUSH, S. G. 1996b. *Transmuted Past: The Age of the Earth and the Evolution of the Elements from Lyell to Patterson*. A History of Modern Planetary Physics, Vol. 2. Cambridge University, New York.

BRUSH, S. G. 1996c. *Fruitful Encounters: The Origin of the Solar System and of the Moon from Chamberlin to Apollo*. A History of Modern Planetary Physics, Vol. 3. Cambridge University, New York.

BRUSH, S. G. 2000. Thomas Kuhn as a Historian of Science. *Science & Education*, **9**, 39–58.

BURCHFIELD, J. D. 1990. *Lord Kelvin and the Age of the Earth* [reprint of the 1975 edition with a new afterword]. University of Chicago, Chicago.

CHAMBERLIN, T. C. 1920. Diastrophism and the formative process. XIII. The bearings of the size and rate of infall of planetesimals on the molten or solid state of the Earth. *Journal of Geology*, **28**, 665–701.

CHRISTIANSEN, G. 1995. *Edwin Hubble: Mariner of the Nebulae*. Farrar, Straus and Giroux, New York.

COUDERC, P. 1952. *The Expansion of the Universe*. Faber & Faber, London 1952 [translated by J. B. Sidgwick from *L'Expansion de l'Universe*, 1950].

DALRYMPLE, G. B. 2001. The age of the Earth in the twentieth century: a problem (mostly) solved. *In:* LEWIS, C. L. E. & KNELL, S. J. (eds) *The Age of the Earth: from 4004 BC to AD 2002*. Geological Society, London, Special Publications, **190**, 205–221.

DEVORKIN, D. 1982. *The History of Modern Astronomy and Astrophysics: A Selected, Annotated Bibliography*. Garland, New York.

DIRAC, P. A. M. 1937. The cosmological constant. *Nature*, **139**, 323.

DIRAC, P. A. M. 1938. A new basis for cosmology. *Proceedings of the Royal Society*, **A165**, 199–208.

EDDINGTON, A. S. 1923. The Borderland of Astronomy and Geology. *Nature*, **111**, 18–21 [a lecture delivered before the Geological Society of London on 21 November 1922].

EDDINGTON, A. S. 1926 (reprinted 1988). *The Internal Constitution of the Stars*. Cambridge University, Cambridge.

EDDINGTON, A. S. 1931a. On the value of the cosmical constant. *Proceedings of the Royal Society*, **A133**, 605–615.

EDDINGTON, A. S. 1931b. [Contribution to British Association discussion on the Evolution of the Universe] *Nature*, **128**, 709.

EDDINGTON, A. S. 1932a. The recession of the extra-galactic nebulae. *Monthly Notices of the Royal Astronomical Society*, **92**, 3–7.

EDDINGTON, A. S. 1932b. 'The expanding universe'. Lecture at Royal Institution, 22 January 1932. [Reproduced in *The Royal Institution Library of Science – Astronomy*, Vol. II. American Elsevier, New York 1970, 305–313.]

EDDINGTON, A. S. 1933. *The Expanding Universe* [based on a lecture at the International Astronomical Union meeting, Cambridge (Mass), September 1932 and radio broadcasts in the USA shortly afterwards]. Cambridge University, New York.

EINSTEIN, A. 1945. *The Meaning of Relativity* [reprint of the second edition, with a new appendix]. Princeton University, Princeton.

FEAST, M. 2000. Stellar populations and the distance scale: The Baade-Thackeray correspondence. *Journal of the History of Astronomy*, **31**, 29–36.

FIELD, G. 1986. Astronomy of the twentieth century. *American Scientist*, **74**, 173–181.

FINLAY-FREUNDLICH, E. 1951. Cosmology. *In: International Encyclopedia of Unified Science*, Vol. 1, no. 8. University of Chicago, Chicago.

GAMOW, G. 1952. *The Creation of the Universe*. Viking, New York.

GARBER, E. 1999. *The Language of Physics: The Calculus and the Development of Theoretical Physics in Europe, 1750–1914*. Birkhäuser, Boston, Basel, Berlin.

GINGERICH, O. 1990. Shapley, Hubble, and cosmology. In: KRON, R. G. (ed.) *Evolution of the Galaxies, Edwin Hubble Centennial Symposium*, Astronomical Society of the Pacific Conference Series, **10**, 19–21.

GLEN, W. 1982. *The Road to Jaramillo: Critical Years of the Revolution in Earth Science*. Stanford University, Stanford.

GOSSE, P. H. 1857. *Omphalos: An Attempt to Untie the Geological Knot*. London [reprinted 1998 by Ox Bow, Woodbridge (Conn.)].

GRAY, G. W. 1937. *The Advancing Front of Science*. Whittlesey House/McGraw-Hill, New York.

GRAY, G. W. 1953. A larger and older universe. *Scientific American*, **188**(6), 56–66.

GRIBBIN, J. 2000. *The Birth of Time: How Astronomers Measured the Age of the Universe*. Yale University, New Haven.

HAAR, D. TER. 1950. Cosmogonical problems and stellar energy. *Reviews of Modern Physics*, **22**, 119–152.

HALDANE, J. B. S. 1945. A new theory of the past. *American Scientist*, **33**(3), 129–145, 188.

HETHERINGTON, N. S. 1971a. The measurement of radial velocities of spiral nebulae. *Isis*, **62**, A309–313.

HETHERINGTON, N. S. 1971b. *Edwin Hubble and a relativistic expanding model of the universe*. Astronomical Society of the Pacific, Leaflet No. 509.

HETHERINGTON, N. S. 1982. Philosophical values and observation in Edwin Hubble's choice of a model of the universe. *Historical Studies in the Physical Sciences*, **13**, 41–67.

HETHERINGTON, N. S. 1989. Geological time versus astronomical time: Are scientific theories falsifiable? *Earth Sciences History*, **8**, 167–169.

HETHERINGTON, N. S. 1990. Hubble's cosmology. *American Scientist*, **78**, 142–151.

HOFMAN, A. W. 2001. Lead isotopes and the age of the Earth – a geochemical accident. In: LEWIS, C. L. E. & KNELL, S. J. (eds) *The Age of the Earth: 4004 BC to AD 2002*. Geological Society, London, Special Publications, **190**, 223–236.

HOLMES, A. 1913. *The Age of the Earth*. Harper, London.

HOLMES, A. 1926. Radium uncovers new clues to Earth's age. *New York Times*, 6 June, Sect. IX, pp. 4f [reprinted 1976 in Sullivan, W. (ed.) *Science in the Twentieth Century*, Arno, New York, 175–177].

HOLMES, A. 1927. The problem of geological time. *Scientia*, **42**, 263–272.

HOLMES, A. 1931. Radioactivity and geological time. In: KNOPF, A., SCHUCHERT, C., KOVARIK, A. F., HOLMES, A. & BROWN, E. W. *The Age of the Earth*, Bulletin of the National Research Council, **80**, WASHINGTON, DC, 124–459.

HOLMES, A. 1947a. The construction of a geological time-scale. *Transactions of the Geological Society of Glasgow*, **21**, 117–152.

HOLMES, A. 1947b. The age of the Earth. *Endeavour*, **6**, 99–108.

HOLMES, A. 1949. Lead isotopes and the age of the Earth. *Nature*, **163**, 453–456.

HÖNL, H. 1949. Zwei Bemerkungen zum kosmologischen Problem. *Annalen der Physik* [6], **6**, 169–176.

HOUTERMANS, F. G. & JORDAN, P. 1946. Über die Annahme der zeitliche Veränderlichkeit des β-Zerfalls und die Möglichkeiten ihrer experimentellen Prüfung. *Zeitschrift für Naturforschung*, **1**, 125–130.

HOYLE, F. 1948. A new model for the expanding universe. *Monthly Notices of the Royal Astronomical Society*, **108**, 372–382.

HOYLE, F. 1950. *The Nature of the Universe*. Harper, New York [reprinted 1955 by New American Library, New York].

HOYLE, F. 1994. *Home is Where the Wind Blows*. University Science, Mill Valley, CA.

HUBBLE, E. P. 1929. A relation between distance and radial velocity among extra-galactic nebulae. *Proceedings of the National Academy of Sciences, USA*, **15**, 168–173.

HUBBLE, E. P. 1934. The realm of the nebulae. *Scientific Monthly*, **39** 193–202.

HUBBLE, E. P. 1936a. *The Realm of the Nebulae*. Yale University, New Haven.

HUBBLE, E. P. 1936b. Effects of red shifts on the distribution of nebulae. *Astrophysical Journal*, **84**, 517–554.

HUBBLE, E. P. 1936c. Effects of red shifts on the distribution of nebulae. *Proceedings of the National Academy of Sciences, USA*, **22**, 621–627.

HUBBLE, E. P. 1937a. *The Observational Approach to Cosmology*. Oxford University, Oxford.

HUBBLE, E. P. 1937b. Red shifts and the distribution of nebulae. *Monthly Notices of the Royal Astronomical Society*, **97**, 506–513.

HUBBLE, E. P. 1937c. Our sample of the universe. *Scientific Monthly*, **45**, 481–493.

HUBBLE, E. P. 1940a. The Hector Maiben Lecture, on 'problems in nebular research'. *Publications of the Astronomical Society of the Pacific*, **52**, 288–289.

HUBBLE, E. P. 1940b. Problems of nebular research. *Scientific Monthly*, **51**, 391–408.

HUBBLE, E. P. 1942. The problem of the expanding universe. *American Scientist*, **30**, 99–115.

HUBBLE, E. P. 1951. Explorations in space: The cosmological program for the Palomar telescopes. *Proceedings of the American Philosophical Society*, **95**, 463–470.

HUBBLE, E. P. 1954. The law of red-shifts. *Monthly Notices of the Royal Astronomical Society*, **113**, 658–666 [George Darwin Lecture 1953 May 8, edited by A. R. Sandage].

HUBBLE, E. P. & TOLMAN, R. C. 1935. Two methods of investigating the nature of the nebular red-shift. *Astrophysical Journal*, **82**, 302–337.

HUMASON, M. L. 1931. Apparent velocity-shifts in the spectra of faint nebulae. *Astrophysical Journal*, **74**, 35–42.

JEANS, J. 1929. *Astronomy and Cosmogony* (second edition). Cambridge University, Cambridge.

JEANS, J. 1932. *The Mysterious Universe* (new revised edition). Cambridge University, Cambridge.
JEANS, J., EDDINGTON, A. S. & MILNE, E. A. 1935. The age of the universe. *Observatory*, **58**, 108–114.
JEFFREYS, H. 1921. Age of the Earth. *Nature*, **108**, 24.
JEFFREYS, H. 1924. *The Earth: Its Origin, History, and Physical Constitution*. Cambridge University, Cambridge.
JEFFREYS, H. 1929. *The Earth: Its Origin, History, and Physical Constitution* (second edition). Cambridge University, Cambridge.
JEFFREYS, H. 1931. Physics of the Earth. *Nature*, **128**, 984–986.
JEFFREYS, H. 1948. Lead isotopes and the age of the Earth. *Nature*, **162**, 822–823.
JEFFREYS, H. 1949. Lead isotopes and the age of the Earth. *Nature*, **164**, 1046–1047.
JUNGNICKEL, C. & MCCORMMACH, R. 1986. *Intellectual Mastery of Nature, Theoretical Physics from Ohm to Einstein*. Volume 2, *The Now Mighty Theoretical Physics 1870–1925*. University of Chicago, Chicago.
KNOPF, A. 1931. Summary of principal results. *In*: KNOPF, A., SCHUCHERT, C., KOVARIK, A. F., HOLMES, A. & BROWN, E. W. *The Age of the Earth, Bulletin of the National Research Council*, **80**, Washington, DC, 3–9.
KNOPF, A., SCHUCHERT, C., KOVARIK, A. F., HOLMES, A. & BROWN, E. W. 1931. *The Age of the Earth, Bulletin of the National Research Council*, **80**, Washington, DC.
KRAGH, H. 1996. *Cosmology and Controversy: The Historical Development of Two Theories of the Universe*. Princeton University, Princeton, NJ.
KRAGH, H. 1999. *Quantum Generations: A History of Physics in the Twentieth Century*. Princeton University, Princeton.
KUHN, T. S. 1962. *The Structure of Scientific Revolutions*. University of Chicago, Chicago.
KUHN, T. S. 1978. *Black-Body Theory and the Quantum Discontinuity: 1894–1912*. Oxford University, New York.
LANG, K. R. & GINGERICH, O. (eds) 1979. *A Source Book in Astronomy and Astrophysics 1900–1975*. Harvard University, Cambridge (Mass.).
LEWIS, C. L. E. 2001. Arthur Holmes' vision of a geological timescale. *In*: LEWIS, C. L. E. & KNELL, S. J. (eds) *The Age of the Earth: from 4004 BC to AD 2002*. Geological Society, London, Special Publications, **190**, 121–138.
MACMILLAN, W. D. 1932. Velocities of the spiral nebulae. *Nature*, **129**, 93.
MCCREA, W. H. 1953. Cosmology. *Reports on Progress in Physics*, **16**, 321–363.
MCCREA, W. H. 1968. Cosmology after half a century. *Science*, **160**, 1295–1299.
MCVITTIE, G. C. 1937. *Cosmological Theory*. Methuen, London.
MILNE, E. A. 1940. Cosmological theories. *Astrophysical Journal*, **91**, 129–158.
MILNE, E. A. 1952. *Modern Cosmology and the Christian Idea of God*. Clarendon, Oxford.
MORRIS, H. M. 1972. *The Remarkable Birth of Planet Earth*. Dimension, Minneapolis.
MORRIS, H. M. 1995. *A Young-Earth Creationist Bibliography*. Acts & Facts, Impact No. **269**, Institute for Creation Research, El Cajon, i–iv.
NIER, A. O. 1938. Variations in the relative abundances of the isotopes of common lead from various sources. *Journal of the American Chemical Society*, **60**, 1571–1576.
NIER, A. O. 1939. The isotopic constitution of radiogenic leads and the measurement of geologic time. II. *Physical Review* [2], **55**, 153–163.
NORTH, J. D. 1965. *The Measure of the Universe: A History of Modern Cosmology*. Clarendon, Oxford.
NUMBERS, R. L. 1992. *The Creationists*. Knopf, New York.
NYE, M. J. 1999. *Before Big Science: The Pursuit of Modern Chemistry and Physics, 1800–1940*. Harvard University, Cambridge, (Mass.).
OMER, G. C., JR. 1949. A nonhomogeneous cosmological model. *Astrophysical Journal*, **109**, 164–176.
OORT, J. H. 1931. Some problems concerning the distribution of luminosities and peculiar velocities of extragalactic nebulae. *Bulletin of the Astronomical Institutes of the Netherlands*, **6**, 155–160.
ÖPIK, E. J. 1933. Meteorites and the age of the universe. *Popular Astronomy*, **41**, 71–79.
ÖPIK, E. J. 1954a. The age of the universe. *British Journal for the Philosophy of Science*, **5**, 203–214.
ÖPIK, E. J. 1954b. The time-scale of our universe. *Irish Astronomical Journal*, **3**(4), 89–108.
OSTERBROCK, D. E. 1997. Walter Baade, observational astrophysicist, (3): Palomar and Göttingen 1948–1960 (Part A). *Journal of the History of Astronomy*, **28**, 283–316.
OSTERBROCK, D. E. 1998. Walter Baade, observational astrophysicist, (3): Palomar and Göttingen 1948–1960 (Part B). *Journal of the History of Astronomy*, **29**, 345–377.
OSTERBROCK, D. E., GWINN, J. A. & BRASHEAR, R. S. 1993. Edwin Hubble and the expanding universe. *Scientific American*, **269**(1), 84–89.
PAIS, A. 1982. *'Subtle is the Lord ...': The Science and the Life of Albert Einstein*. Oxford University, Oxford and New York.
PATTERSON, C. C. 1953. The isotopic composition of meteoric, basaltic and oceanic leads, and the age of the Earth. *In: Proceedings of the Conference on Nuclear Processes in Geologic Settings, Williams Bay, Wisconsin, September 21–23 1953*, 36–40.
PIUS XII, POPE. 1951. The proofs for the existence of God in the light of modern science. *In: Discorso di sua santita Pio XII alla Pontificia Accademia delle Scienze, 22 November 1951*. Tipografia Poliglotta Vaticana, Rome, 39–51.
POPPER, K. R. 1934. *Logik der Forschung* [translated as *The Logic of Scientific Discovery*, Hutchinson, London 1959].
RAUP, D. M. 1986. *The Nemesis Affair*. Norton, New York.
REICHENBACH, H. 1938. *Experience and Prediction*. University of Chicago, Chicago.
ROBERTSON, H. P. 1933. Relativistic cosmology. *Reviews of Modern Physics*, **5**, 62–90.

ROBERTSON, H. P. 1940. The expanding universe. *In*: BAITSELL, G. A. (ed.) *Science in Progress* (second series). Yale University, New Haven.

ROMER, A. (ed.) 1964. *The Discovery of Radioactivity and Transmutation*. Dover, New York.

RUSSELL, H. N. 1921. A superior limit to the age of the Earth's crust. *Proceedings of the Royal Society*, **A99**, 84–86.

RUSSELL, H. N. 1935. *The Solar System and its Origin*. Macmillan, New York.

RUTHERFORD, E. 1929. Origin of actinium and the age of the Earth. *Nature*, **123**, 313–314.

SANDAGE, A. 1958. Current problems in the extragalactic distance scale. *Astrophysical Journal*, **127**, 513–526.

SANDAGE, A. 1989. Edwin Hubble, 1889–1953. *Journal of the Royal Astronomical Society of Canada*, **83**, 351–362.

SCRIVEN, M. 1954. The age of the universe. *British Journal for the Philosophy of Science*, **5**, 181–90.

SHAPLEY, H. 1944. Trends in the metagalaxy. *American Scientist*, **32**, 65–77.

SHAPLEY, H. 1969. *Through Rugged Ways to the Stars*. Scribner, New York.

SHAROV, A. S. & NOVIKOV, I. D. 1993. *Edwin Hubble*. Cambridge University, New York.

SITTER, W. DE. 1931. [Contribution to British Association discussion on the evolution of the Universe] *Nature*, **128**, 706–709.

SITTER, W. DE. 1932a. *Kosmos: A Course of Six Lectures on the Development of our Insight into the Structure of the Universe, delivered for the Lowell Institute in Boston, in December 1931*. Harvard University, Cambridge, (Mass.)

SITTER, W. DE. 1932b. On the expanding universe. *Proceedings of the Section of Sciences, Koninklijke Akademie van Wetenschappen te Amsterdam*, **35**, 596–607.

SITTER, W. DE. 1932c. The size of the universe. *Publications of the Astronomical Society of the Pacific*, **44**, 89–104.

SITTER, W. DE. 1933. On the expanding universe and the time scale. *Monthly Notices of the Royal Astronomical Society*, **93**, 628–634.

SKILLING, W. T. & RICHARDSON, R. S. 1947. *Astronomy* (revised edition). Holt, New York.

SMITH, R. W. 1982. *The Expanding Universe: Astronomy's 'Great Debate' 1900–1931*. Cambridge University, Cambridge.

STRÖMGREN, B. 1933. On the interpretation of the Hertzsprung-Russell diagram. *Zeitschrift für Astrophysik*, **7**, 222–248.

STRUVE, O. & ZEBERGS, V. 1962. *Astronomy of the 20th Century*. Macmillan, New York and London.

TOLMAN, R. C. 1934. *Relativity, Thermodynamics, and Cosmology*. Clarendon, Oxford.

TOLMAN, R. C. 1949. The age of the universe. *Reviews of Modern Physics*, **21**, 374–378.

TREDER, H.-J. 1969. Geophysik und Kosmologie. [Gerlands] *Beiträge zur Geophysik*, **78**, 1–26.

WHITROW, G. J. 1949. *The Structure of the Universe: An Introduction to Cosmology*. Hutchinson, London.

WHITROW, G. J. 1954a. The orthodox theory of the expanding universe. *Occasional Notes of the Royal Astronomical Society*, **3**(17), 81–88.

WHITROW, G. J. 1954b. The age of the universe. *British Journal for the Philosophy of Science*, **5**, 215–225.

WHITROW, G. J. 1959. *The Structure and Evolution of the Universe: An Introduction to Cosmology*. Harper, New York.

WHITROW, G. J. 1972. Hubble, Edwin Powell. *Dictionary of Scientific Biography*, **6**, 528–533.

WOOD, R. M. 1985. *The Dark Side of the Earth*. George Allen & Unwin, London.

WYSE JACKSON, P. N. 2001. John Joly (1857–1933) and his determinations of the age of the Earth. *In*: LEWIS, C. L. E. & KNELL, S. J. (eds) *The Age of the Earth: from 4004 BC to AD 2002*. Geological Society, London, Special Publications, **190**, 107–119.

ZWICKY, F. 1929. On the red shift of spectral lines through interstellar space. *Proceedings of the National Academy of Sciences*, **15**, 773–779.

ZWICKY, F. 1935. Remarks on the redshift from nebulae. *Physical Review*, **48**, 802–806.

The oldest rocks on Earth: time constraints and geological controversies

BALZ S. KAMBER[1], STEPHEN MOORBATH[2] & MARTIN J. WHITEHOUSE[3]

[1] *Department of Earth Sciences, University of Queensland, Brisbane, Queensland 4072, Australia*
[2] *Department of Earth Sciences, University of Oxford, Oxford OX1 3PR, UK*
(*e-mail*: stephen.moorbath@earth.ox.ac.uk)
[3] *Laboratory of Isotope Geology, Swedish Museum of Natural History, Box 50007, SE – 10405 Stockholm, Sweden*

Abstract: Ages in the range 3.6–4.0 Ga (billion years) have been reported for the oldest, continental, granitoid orthogneisses, whose magmatic precursors were probably formed by partial melting or differentiation from a mafic, mantle-derived source. The geological interpretation of some of the oldest ages in this range is still strongly disputed.

The oldest known supracrustal (i.e. volcanic and sedimentary) rocks, with an age of 3.7–3.8 Ga, occur in West Greenland. They were deposited in water, and several of the sediments contain ^{13}C-depleted graphite microparticles, which have been claimed to be biogenic.

Ancient sediments (c. 3 Ga) in western Australia contain much older detrital zircons with dates ranging up to 4.4 Ga. The nature and origin of their source is highly debatable. Some ancient (magmatic) orthogneisses (c. 3.65–3.75 Ga) contain inherited zircons with dates up to c. 4.0 Ga. To clarify whether zircons in orthogneisses are inherited from an older source region or cogenetic with their host rock, it is desirable to combine imaging studies and U-Pb dating of single zircon grains with independent dating of the host rock by other methods, including Sm-Nd, Lu-Hf and Pb/Pb.

Initial Nd, Hf and Pb isotopic ratios of ancient orthogneisses are essential parameters for investigating the degree of heterogeneity of early Archaean mantle. The simplest interpretation of existing isotopic data is for a slightly depleted, close-to-chondritic, essentially homogeneous early Archaean mantle; this does not favour the existence of a sizeable, permanent continental crust in the early Archaean.

By analogy with the moon, massive bolide impacts probably terminated on Earth by c. 3.8–3.9 Ga, although no evidence for them has yet been found. By c. 3.65 Ga production of continental crust was well underway, and global tectonic and petrogenetic regimes increasingly resembled those of later epochs.

On the Earth's surface, crust must have been forming soon after accretion. There is apparently no direct record of these rocks, although detrital zircons of up to 4.4 Ga (billion years) have been found in much younger sedimentary rocks. The age of the oldest terrestrial rocks is therefore not an issue of rock formation but of rock preservation, and the oldest known rocks must have formed in a way that allowed their preservation. Study of cratonic lithospheric mantle shows that the present continents grew together with a buoyant lithospheric mantle keel that helped resist the destructive forces of mantle convection. The age of the oldest preserved terrestrial rocks is therefore closely connected with the magmatic and tectonic style that operated on Earth.

Estimates for the age of the oldest known in-situ terrestrial rocks approach the age of the so-called 'late heavy' meteorite bombardment experienced by the inner solar system (3.8–3.9 Ga). The earliest lunar surface (>4 Ga), although heavily cratered, has partly survived the late heavy meteorite bombardment so that, by analogy, the absence of in-situ terrestrial rocks exceeding 4.0 Ga cannot be entirely attributed to eradication by impact. However, the close approach of conservative estimates for the age of the oldest in-situ terrestrial rocks (c. 3.8 Ga) and the end of the heavy meteorite bombardment may not be fortuitous. During the impact regime, massive thermal release from the Earth may not have permitted (at least periodically) the existence of a liquid hydrosphere, which is a pre-requisite for plate tectonics and for the existence of a biosphere. In our view, the birth of plate tectonics and the creation of a biosphere marked the time from which terrestrial rocks stood a good chance of being preserved.

Since publication of a previous review on the oldest terrestrial rocks (Moorbath et al. 1986), the research focus on the early Archaean geological record has broadened. While the exact growth history of the continents is still debated, the principal aim of many geochemical studies has shifted to establish whether the oldest terrestrial rocks carry a 'memory' of differentiation processes that might have occurred hundreds of millions of years before their formation. To this question has been added the issue of emergence of terrestrial life and the associated establishment of a permanent hydrosphere. Ratios of radiogenic isotopes are the most widely used tool to probe the time window for which there is no known rock record. Because these ratios are sensitive to accurate decay correction, establishing the correct age of the oldest rocks, as precisely as possible, is a prerequisite for meaningful geochemical studies. Most of the present controversies about the early Earth concern interpretation of isotope data, not the quality of the data. The age of the oldest terrestrial rocks has therefore lost none of its relevance.

In this review, we concentrate on the beginning of the observed geological and geochronological record, involving rocks older than 3.5–3.6 Ga. Most emphasis will be placed on the early Archaean (pre-3.6 Ga) rocks of southern West Greenland, which constitute the largest, most varied, best exposed and most closely investigated terrain of the oldest known terrestrial rocks.

Ancient rocks and modern dating methods

The oldest granitoid gneisses of magmatic origin (orthogneisses), widely regarded as typical of continental crust, are claimed to be in the range 4.0–3.6 Ga on several continents, although evidence within the older part of this range is still inadequate. The oldest known supracrustal (volcanic and sedimentary) rocks, from West Greenland, are debated to be between 3.9 and 3.7 Ga (for references see later). Most of these subsequently highly metamorphosed supracrustal rocks were originally deposited in water. They are enveloped by younger granitoid gneisses, but the basement on which they were deposited is not yet identifiable. However, there is some evidence (see later) that this basement was not sialic (continental) in character.

The former presence of even older rocks than the above is proved by the important discovery of detrital zircons, with high-resolution ion-microprobe U-Pb dates in the range 4.4–3.9 Ga, in much younger quartzites and conglomerates at Jack Hills and Mount Narryer in western Australia (for references see later). Furthermore, some ancient orthogneisses also show indisputable evidence for the presence of much older, inherited zircons.

Introduction of rapid and precise ion-probe dating of single zircon grains has supplemented and, for some workers, entirely replaced conventional dating techniques. This has proved to be a mixed blessing. Dates obtained from orthogneiss (and other) whole-rock regressions have been regarded by some as too imprecise or unreliable for adequate age resolution, whilst constraints on mantle and crust evolution imposed by initial Sr, Nd and Pb isotope ratios obtained from isochron and errorchron regressions have been disregarded, often for no convincing reason. In ancient orthogneisses, with a complex geological history, it has become fashionable to regard the oldest date(s) from the commonly observed wide range of zircon U-Pb dates as yielding the age of their magmatic precursors, sometimes without adequate (or indeed any) independent age evidence. This raises the question of possible zircon inheritance, because it has been known for many years from conventional zircon U-Pb measurements that many magmatic rocks contain inherited zircons derived from much older crustal sources (e.g. Pidgeon & Compston 1992; Mezger & Krogstad 1997; Miller et al. 2000).

Very precise zircon U-Pb dates are commonly used to back-calculate initial isotopic ratios (most often Nd) from Sm-Nd (and other) analyses of the host rock, on the further assumption that the rock has remained a closed system to parent and daughter isotope migration since the quoted zircon U-Pb age. It is shown later from published examples that reliable initial isotopic ratios can only be obtained where the zircon U-Pb age is identical with the independently known age of the host rock, and where the rock has remained a closed system to migration of parent and daughter isotope since time of rock formation. When these conditions are not met, a wide range of apparent initial isotopic ratios can be expected for a given rock suite, which yields no valid information whatever on the isotopic nature of the magmatic protolith of the zircon host rock. Despite these uncertainties, repeated claims have been made (for references see later) for extreme isotopic heterogeneity (especially in Nd) of the early Archaean mantle source region of the oldest magmatic rocks.

Such heterogeneity is unexpected because the early mantle was almost certainly hotter, less viscous and more vigorously convective than in later geological times. In contrast, a more rigorous combination of ages and initial isotopic ratios (especially in Nd) for early Archaean mantle-derived rocks suggests a small, homogeneous depletion of the mantle (high Sm/Nd) relative to the chondritic uniform reservoir (CHUR). This could be a genuine corollary to mantle differentiation and removal from the mantle of a small amount of crustal protolith material in earliest Archaean times (e.g. Nägler & Kramers 1998).

Despite the great practical advantages offered by single-grain ion-probe zircon U-Pb techniques, the problem of zircon inheritance in orthogneisses can only be addressed when ion-probe analysis is guided by precise identification of different age zones (if any) within a single grain, representing different crystallization episodes. Even with the inadequate transmitted-light photomicrography of former years, it was noticed that the oldest ion-probe dates resided in cores of grains, surrounded by younger magmatic or metamorphic zones. Even so, the oldest core ages were often taken to represent the age of the host rock. With recent routine introduction of effective imaging techniques, such as cathodoluminescence (CL) and back-scatter electron (BSE), it is possible to map out a single grain into separate zones of crystallization as a basis for high-spatial resolution ion-probe dating of each zone (e.g. Fig. 1). Combined with independent age work on the orthogneiss, this facilitates distinction between inherited, comagmatic and metamorphic sectors within a single zircon grain. Similarly, ion-probe work on imaged zircons is necessary for elucidating the crystallization history of detrital grains from ancient sediments and metamorphosed sediments, in order to derive an age for the source region and to set an upper age limit of deposition.

Progress is being made by combining ion-probe U-Pb dating with Hf isotope analysis on separate age zones of single zircon grains (Kinny et al. 1991; Amelin et al. 1999, 2000). Since zircon is the principal carrier of Hf in rocks, this combined technique is potentially more informative than combined, conventional Lu-Hf work on whole rocks and whole zircons, especially when their age and genetic relationships are not independently established.

We now review some early Archaean case histories which have been described and debated in the literature.

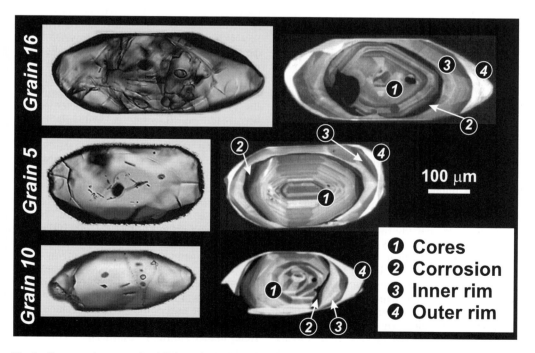

Fig. 1. Comparative transmitted light optical microscope images (left) and cathodoluminescence (CL) images (right) of three grains from sample GGU 110999, showing the much greater resolution of complex internal structure available from CL. These and similar images were used by Whitehouse et al. (1999) to guide the placement of their c. 30 μm diameter ion-probe analytical sites. In this particular sample, almost all zircon grains yield CL images showing the same internal subdivision into cores (>3.8 Ga), corrosion overgrowths (c. 3.74 Ga), inner rims (c. 3.65 Ga) and outer rims (c. 2.7 Ga). An age spectrum from this sample is given in Whitehouse et al. (1999).

Acasta gneisses, northwestern Canada: the oldest rocks on Earth?

We begin our case studies with the Acasta gneisses which, for the past ten years, have been widely quoted as the oldest rocks on Earth. The Acasta gneiss complex occupies an area of some 50 km^2 along the western margin of the Slave craton, an Archaean granite–greenstone terrain in the northwestern part of the Canadian Shield, with an area of 190 000 km^2. The Acasta gneisses provide a formidable test-case for the interpretation of ages and isotopes.

Bowring & Williams (1999) reported zircon U-Pb dates of 4.03–4.00 Ga for the Acasta orthogneisses, following earlier reports of a 3.96 Ga age (Bowring et al. 1989). They claim that these zircon dates reflect the time of magmatic crystallization of the tonalitic-to-granodioritic precursors. Such ancient samples of continental crust potentially carry much isotopic and trace element information to provide constraints on the earliest evolution of the Earth. Bowring & Housh (1995) and Bowring & Williams (1999) claim that whole-rock Nd isotope ratios of individual Acasta gneiss samples, when corrected back to the oldest zircon dates of 4.0–3.6 Ga, indicate extreme geochemical heterogeneity of early Archaean mantle, with initial ε_{Nd} (a measure of the initial ^{143}Nd/^{144}Nd ratio) values ranging from -4 (enriched) to $+4$ (depleted). Bowring and colleagues claimed, without discussion of physical implications, that separate, chemically distinct mantle domains existed for a few hundred million years before Acasta gneiss formation. Furthermore, they take this to imply that a chemically differentiated continental crust, comparable to its present mass, was in existence a few hundred million years after Earth accretion.

Moorbath et al. (1997) noted a striking co-linearity of published and new Sm-Nd whole-rock data on 34 samples of Acasta gneisses, yielding a regression age of 3371 ± 59 Ma (mean square weighted deviate (mswd) = 9.2) with an initial $\varepsilon_{Nd} = -5.6 \pm 0.7$. (The mswd is a statistical measure for the degree of scatter about a regression of data points; it does not provide a measure of the 'correctness' of any age or initial ratio derived from a regression line, as has recently been argued by Nutman et al. 2000.) Moorbath et al. (1997) argued that Nd isotope systematics had been set or reset at c. 3.37 Ga and could therefore not be used to draw any conclusions on the Nd isotope geochemistry of the Earth's mantle at c. 4.0 Ga, or on the amount of continental crust present at that time. Bowring & Williams (1999) countered by stating that they found no c. 3.4 Ga zircons in their analysed rocks and that the co-linearity of Sm-Nd data found by Moorbath et al. (1997) had no age significance. Bowring and colleagues did not publish their own Sm-Nd data plotted on isochron diagrams because they appear to regard the linear array as a mixing line, so that calculated ages and initial isotopic ratios have no geological age significance.

Ample U-Pb and Sm-Nd age evidence has recently emerged that major magmatic, metamorphic and migmatitic events occurred in and around the Acasta region at c. 3.38–3.35 Ga (Stern & Bleeker 1998; Yamashita et al. 2000), indistinguishable from the Sm-Nd regression age of c. 3.37 Ga of Moorbath et al. (1997). It is clear that the Sm-Nd isotopic system in the Acasta gneisses was set or reset at this time. We concur with Bowring & Williams (1999) that the problem of the true age of these complex, partially melted, highly tectonized, migmatitic gneisses is a semantic issue. Because we believe that, for the purpose of inferring mantle geochemistry, the term 'age' requires geochemical and isotopic integrity of the rocks (rather than only that of some of its components, i.e. zircon, as proposed by Bowring & Williams 1999) we maintain that 'the age of the Acasta gneisses is only c. 3.37 Ga, although both the zircon U-Pb ages (4.0 to 3.6 Ga) and the very negative initial ε_{Nd} of -5.6 obtained from the Sm-Nd regression provide incontrovertible evidence for the existence of a substantially older precursor' (Moorbath et al. 1997).

A further claim for the significance of a wide range of initial ε_{Nd} values in the Acasta gneisses for identifying c. 4.0 Ga mantle heterogeneity was made by Bowring & Williams (1999). However, Whitehouse et al. (2001) demonstrated that the interpretation of the Sm-Nd regression age of c. 3.37 Ga either as a Nd isotope homogenization event or as a mixing line precludes the application of Acasta gneiss Sm-Nd data to constrain the geochemical evolution of the Earth in its first 500 Ma. In this connection, Amelin et al. (1999, 2000) reported preliminary results of a combined Hf isotope and U-Pb age study on c. 3.6 Ga zircons from the Acasta gneiss. These zircons contained no evidence for a strongly depleted or enriched early Archaean mantle component.

Itsaq gneiss complex, southern West Greenland

This is the largest known and best exposed terrain of early Archaean rocks on Earth, on which an extensive, multidisciplinary literature has built up since the pioneering field work of

V. R. McGregor in the late 1960s (McGregor 1973). The earliest determinations provided an age of c. 3.75–3.65 Ga for the regional gneisses and supracrustal rocks (e.g. Black et al. 1971; Moorbath et al. 1972, 1973; Baadsgaard 1973). The first phase of the age work was reviewed by Moorbath et al. (1986). Only the salient features of the large amount of recent and ongoing work are reviewed here. Regional descriptions are omitted, but can be found in the references. Following the recommendation of Nutman et al. (1996), irrespective of its necessity, we here use the term Itsaq orthogneisses, or simply Itsaq gneisses, for the former Amîtsoq gneisses. These workers combined all the varied early Archaean rocks of this region into the so-called Itsaq gneiss complex.

Age of the Itsaq orthogneisses

The age of these lithologically complex, variably deformed, granitoid orthogneisses is currently the focus of much debate. We have regressed the published, conventional whole-rock isotopic data (Fig. 2) from the Itsaq gneisses, with the following results.

(i) Rb-Sr. Age = 3660 ± 67 Ma (million years), initial $^{87}Sr/^{86}Sr = 0.7006 \pm 9$ mean square weighted deviate (mswd) = 33, number of samples (n) = 78 (Moorbath et al. 1972, 1975, 1977; Baadsgaard et al. 1976, 1986a).
(ii) Sm-Nd. Age = 3640 ± 120 Ma, initial ε_{Nd} = 0.9 ± 1.4, mswd = 10, n = 26 (Baadsgaard et al. 1986b; Moorbath et al. 1986, 1997; Shimizu et al. 1988).
(iii) Pb/Pb (whole rock and selected feldspar). Age = 3654 ± 73 Ma, intersection of Pb/Pb regression with primary growth curves of Stacey & Kramers (1975) and Kramers & Tolstikhin (1997) are 3.65 Ga and 3.66 Ga respectively, mswd = 17.6, n = 83 (Moorbath et al. 1975; Gancarz & Wasserburg 1977; Griffin et al. 1980; Kamber & Moorbath 1998).

These regressions are by no means perfect isochrons, and the scatter of points about the regressions in excess of analytical error, as measured by the mswd, is due *either* to open-system behaviour of parent and/or daughter isotopes during regionally well-attested late Archaean and mid-Proterozoic metamorphism, *or* to a small degree of heterogeneity in initial Sr, Nd and Pb isotope ratios for different Itsaq gneiss components, *or* to a combination of these. However, we regard the weighted mean of 3655 ± 45 Ma from all three methods as a well-constrained emplacement age for the bulk of the magmatic precursors of the Itsaq orthogneisses. It is unlikely that agreement between the three methods is the result of pervasive regional resetting of ages during metamorphism, because of the very different geochemical behaviour of the three isotopic systems. Initial Sr, Nd and Pb isotopic constraints indicate derivation from a mantle-like source for the magmatic precursors at c. 3.65 Ga, and not from resolvably older reworked sialic crust.

Many Itsaq gneisses studied by the ion-probe have yielded at least some zircon U-Pb dates at c. 3.65 Ga, and this is seen as the age of a major crust-forming event by Nutman et al. (1996) and by McGregor (2000). However Nutman et al. have also reported many older zircon dates from which they conclude that the Itsaq gneisses had a complicated earlier history, having been added to and modified in several discrete events from c. 3.9 Ga to 3.6 Ga. No consideration was given to the possibility that dates >3.65 Ga might represent inherited zircons. Bennett et al. (1993) and Nutman et al. (1996) persistently claim:

(i)   that Itsaq gneiss samples in this age range cannot be resolved in age and initial ratio from amongst the scatter of analysed whole-rock regression points (see above);
(ii) that high-grade metamorphism and partial melting at c. 3.65 Ga has reset ages and initial isotopic ratios in older rocks, even to the extent of erasing previous isotopic memories in granitoid rocks;
(iii) that individual zircon dates in the range c. 3.9–3.6 Ga can be used to back-calculate reliable initial Nd isotope ratios for the host rock.

The arguments against these three points are as follows (noting that the claims in (ii) and (iii) above are, in any case, incompatible).

(i) It cannot be entirely discounted that the quoted whole-rock regressions (see above) contain individual points from some older rocks. However, careful and detailed whole-rock (plus selected mineral) regressions on individual rock units claimed to be ≥3.8 Ga would be quite easily resolvable from c. 3.65 Ga, especially in the Pb/Pb system. Regrettably, not many independent age data are yet available for rocks regarded by Bennett et al. (1993) and Nutman et al. (1996, 1999) as >3.65 Ga (but see below).

(ii) Regional-scale metamorphic resetting of Rb-Sr, Sm-Nd, Pb/Pb whole-rock systems in coarse-grained granitoid rocks is implausible and, in any case, cannot preserve mantle-like initial Sr, Nd and Pb isotope ratios in rocks

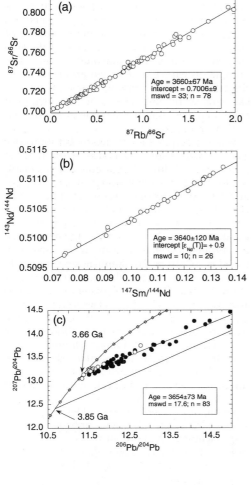

Fig. 2. Isochron diagrams for Itsaq orthogneisses (data sources given in Kamber & Moorbath 1998). (**a**) Rb-Sr regression line calculated for 78 Itsaq orthogneisses (samples with $^{87}Rb/^{86}Sr > 2$ omitted due to susceptibility to later isotope exchange) yields a 3.66 Ga age. Note the lack of scatter of the low Rb/Sr samples, arguing against derivation from variably fractionated reservoirs, including pre-existing continental crust. Rather, the initial Sr isotope ratio is compatible with derivation from a largely undepleted mantle. (**b**) Sm-Nd regression of 26 Itsaq orthogneisses yields a similar age of 3.64 Ga. The significance of this age and the initial isotope composition have been discussed by Moorbath et al. (1997). (**c**) Common Pb diagram for Itsaq orthogneisses in which whole-rock data points are shown as solid circles and leached feldspar data as open circles. The combined data set ($n=83$) defines a regression age of 3.65 Ga (solid line through data points). The important aspect of the regression is that plausible mantle evolution curves (here Kramers & Tolstikhin (1997) – open diamond symbols) are intersected at a time (3.66 Ga) indistinguishable from the regression age. Due to the relatively short half-life of ^{235}U (parent of ^{207}Pb), the $^{207}Pb/^{204}Pb$ and the $^{207}Pb/^{206}Pb$ ratios of the terrestrial reservoir evolved very quickly in early Archaean times. This is illustrated by the large difference in Pb isotope composition (see arrows) between 3.85 Ga mantle and 3.66 Ga mantle in the selected model. A hypothetical isochron of a 3.85 Ga mantle-derived sample suite (solid line starting at 3.85 Ga) is shown to highlight the magnitude of the difference between observed data and those expected if the Itsaq granitoid gneisses were indeed 3.85 Ga old.

as old as 0.2–0.3 Ga at the time of resetting, because of the very different Rb/Sr, Sm/Nd and U/Pb ratios of granitoid and mafic (mantle-like) rocks. In this respect, the Pb/Pb system is extremely sensitive, particularly for the Itsaq gneisses, because of their extremely low U/Pb ratios and correspondingly uniquely (i.e. for terrestrial rocks) unradiogenic present-day Pb isotope ratios (e.g. Black et al. 1971; Gancarz & Wasserburg 1977). Their Pb/Pb regression line (age = 3654 ± 73 Ma) intersects, by a short extrapolation, the Kramers & Tolstikhin (1997) terrestrial Pb isotope evolution curve at 3.66 Ga, providing no leeway for development in a significantly older crustal environment (Fig. 2c). As emphasized by Kamber & Moorbath (1998), the $^{207}Pb/^{206}Pb$ compositions of 3.85 Ga and 3.65 Ga mantle Pb on the primary growth curve differ by some 13%, far outside analytical uncertainties (typically c. 0.15%). There is as yet no hint from published Pb isotope ratios that any Itsaq gneiss began its existence in the crust as long ago as 3.85 Ga or, indeed, significantly in excess of c. 3.65 Ga.

(iii) Bennett et al. (1993) used a range of observed zircon U-Pb dates of c. 3.9–3.7 Ga for the Itsaq gneisses, in conjunction with Sm-Nd analyses, to calculate initial ε_{Nd} values in the range −4.5 to +4.5 for the mantle source region of their magmatic precursors. This implied a locally very heterogeneous mantle reservoir, with all its consequences for transient mantle–crust differentiation in earliest Earth history. For analogous reasons to those outlined for the Acasta gneisses earlier, these claims were criticized (Moorbath et al. 1997; Kamber & Moorbath 1998; Whitehouse et al. 1999) on the grounds that the oldest U-Pb dates might represent inherited zircons and that the true magmatic age of the protoliths of the Itsaq gneisses was close to c. 3.65 Ga, when they were differentiated from a relatively homogeneous mantle reservoir.

The continuing debate about inherited versus comagmatic interpretation of zircon U-Pb dates in the Itsaq gneisses came to a head on the small island of Akilia in the outer part of the Godthaabsfjord region (Fig. 3). Here a large enclave of the so-called Akilia supracrustals (McGregor & Mason 1977; see also next section) was claimed to be discordantly cut by a thin, gneissic granitoid sheet with ion-probe zircon U-Pb dates of up to 3.87 Ga (Mojzsis et al. 1996; Nutman et al. 1997a). Indeed, Nutman et al. (1996) had earlier published a range of zircon dates from 3.87 Ga to 3.62 Ga from a small region on and around Akilia island. The putative pre-3.87 Ga supracrustal enclave was of special interest, because apatite grains in a metamorphosed iron-rich sediment yielded isotopically light $\delta^{13}C‰$ values of −20 to −50, interpreted by Mojzsis et al. (1996) as biological in origin. These authors stated that

'... the late heavy bombardment (>3800 Ma), documented in the lunar record has been speculated to place an upper limit on the age of a continuous terrestrial biosphere (Chyba 1993). The evidence for life ... overlaps this critical time period and shows that if the accretion models are realistic, such a bombardment did not lead either to the extinction of life or the perturbation of the finely laminated >3850 Ma BIF (banded iron-formation) preserved on Akilia island' (Mojzsis et al. 1996, p. 59).

In a detailed Pb isotope study on whole rocks and feldspars from the presumed discordant granitoid sheet on Akilia island, Kamber & Moorbath (1998) failed to find any evidence for isotopic compositions significantly older than 3.65 Ga and surmised that the oldest U-Pb dates of >3.8 Ga referred to inherited zircons. This was supported in a new ion-probe study by Whitehouse et al. (1999), in which CL imaging was used for the first time in the early Archaean rocks of Greenland. These authors studied zircons from the crucial granitoid sheet (and other rocks) and revealed a previously undocumented zircon growth history. Several samples contained zircon grains with >3.8 Ga cores of magmatic origin, further magmatic overgrowths at 3.65 Ga, and some metamorphic overgrowth at c. 2.7 Ga. (Fig. 1). The >3.8 Ga cores were regarded by Whitehouse et al. (1999) as inherited, with apparent dates ranging down to c. 3.65 Ga reflecting differential Pb loss. The 3.65 Ga magmatic zones in zircon grains were interpreted as the true age of the host rock. This contrasts with the published views of Mojzsis et al. (1996) and Nutman et al. (1997a), interpreting >3.8 Ga dates as the magmatic age of the host rock, with all younger ages reflecting Pb loss and/or metamorphic recrystallization. Whitehouse et al. (1999) concluded that the true minimum age of the Akilia supracrustal rocks of possible biological significance is 3.65 Ga, and not 3.85 Ga,

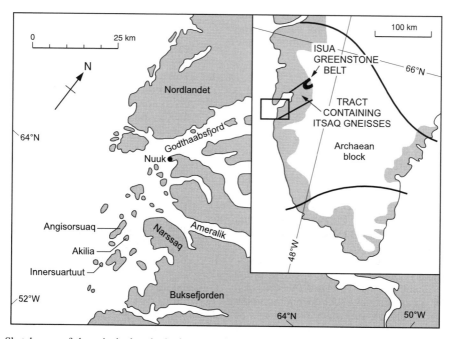

Fig. 3. Sketch-map of the principal early Archaean rock units and localities of southern West Greenland, discussed in the text.

thus making significant overlap with the >3.8 Ga period of lunar and presumed terrestrial bolide impact highly questionable.

Some heat may (or may not) be taken out of this particular debate, because detailed mapping in 1999 of this specific locality on Akilia island by Myers & Crowley (2000) has failed to find any convincing evidence for a discordant, intrusive relationship between the gneiss sheet and the Akilia supracrustal rocks. Myers & Crowley (2000, p. 110) state that 'the rocks on Akilia indicate a complex history of intense ductile deformation in which most geological features were attenuated, rotated and transposed into a new tectonic fabric'. This means that no conclusions can be drawn about the relative age of the gneiss sheet and the Akilia supracrustal rocks. Clearly it is essential for direct age measurements to be made on the Akilia rocks themselves (see next section). Myers & Crowley (2000, p. 114) furthermore consider that 'the possibility of any trace of early Archaean life that may have existed in southwestern Akilia surviving these multiple episodes of intense ductile deformation and high grade metamorphism, and being recognizable as early Archaean, appears miraculous'.

Despite the continuing debate, the existence of genuine in-situ Itsaq orthogneisses with an age >3.65 Ga is, of course, a distinct possibility. Thus, Nutman et al. (1999) report CL-imaged zircons dominated by single-component oscillatory-zoned prismatic grains with dates close to 3.8 Ga from a terrain of relatively undeformed tonalitic and quartz–dioritic orthogneisses south of the Isua greenstone belt (see later section), some 150 km northeast of Nuuk. Nutman et al. (1999) regard this terrain as the best-preserved suite of c. 3.8 Ga felsic igneous rocks yet documented.

A detailed lutetium–hafnium (Lu-Hf) isotope study by Vervoort & Blichert-Toft (1999) on juvenile rocks through time and their bearing on mantle–crust differentiation demonstrates that no early Archaean samples show evidence for derivation from a crustal reservoir (i.e. none has negative ε_{Hf} values). They suggest that relative lack of Hf isotope heterogeneity and absence of negative ε_{Hf} in early Archaean rocks argues against existence of present-day volumes of continental crust in the early Earth. Nevertheless, their comparison of initial Hf and Nd isotope ratios for the Itsaq gneisses may be unrealistically complicated. The apparent initial ε_{Nd} and ε_{Hf} values range from −4.5 to +4.5 and from 0 to +4 respectively. Although Vervoort & Blichert-Toft (1999) recognize that the wide dispersion of initial ε_{Nd} values is due to later disturbances in the Sm-Nd system and does not represent initial isotopic variations in the early Earth, their range of calculated ε_{Hf} values may still be unrealistically wide. Three factors could have caused overestimation of the range of recorded initial Hf isotope compositions.

(i) As outlined by Patchett (1983), despite the low diffusivity of Hf, zircon is not immune to metamorphic disturbance of its original Hf isotope composition. Because most Lu in a granitoid is generally hosted by minerals other than zircon, relatively radiogenic Hf will accumulate in the matrix and may be incorporated into recrystallized or newly grown zircons (Whitehouse et al. 1999). If sufficient time has elapsed between original magmatic zircon growth and metamorphic recrystallization, significant amounts of unsupported (radiogenic) Hf could be incorporated into zircons and cause artificial spread in recalculated initial Hf isotope ratios.

(ii) There is a possibility that at least some of the zircons in these rocks are inherited (see above), particularly in view of the fairly wide range of zircon U-Pb dates (3.82–3.64 Ga) of Bennett et al. (1993). Inherited zircons whose age exceeds that of the entraining melt by several hundred million years are expected to have evolved to a less radiogenic Hf isotope composition (generally negative ε_{Hf} values) than the whole rock or contemporaneous mantle. However, it is plausible that during incorporation into a 3.65 Ga melt, inherited zircons could have sequestered relatively radiogenic Hf from their surrounding older rock volume.

(iii) There is disagreement between the eucrite meteorite-based ^{176}Lu decay constant and determinations based on counting experiments. Whilst a slower decay, which is indicated by the most recent experiments (Nir-El & Lavi 1998), would not significantly reduce the spread in initial Hf isotope ratios for early Archaean low Lu-Hf samples, it would, nevertheless, result in less positive ε_{Hf} values (Patchett & Vervoort 2000). This in turn would indicate derivation from a less depleted mantle and hence the existence of only minor amounts of continental crust.

Age of the Akilia association

This sequence of rocks was first named by McGregor & Mason (1977), who described it as metamorphosed basic, ultrabasic and sedimentary rocks enclosed in the c. 3700 Ma old

Amîtsoq (now Itsaq) gneisses in the Godthaab region, West Greenland. McGregor & Mason furthermore consider that the rocks of the Akilia association are fragments of a greenstone-belt type of sequence that was intruded and disrupted by the granitic parents of the Itsaq gneisses.

As outlined in the previous section, Akilia metasedimentary rocks on Akilia island are important on account of possibly biogenic components (Mojzsis et al. 1996; Nutman et al. 1997a). Following on from earlier arguments, we now summarize direct evidence to suggest that the Akilia association rocks on Akilia island and nearby Innersuartuut island are probably no older than c. 3.7 Ga.

(i) A biotite-schist with a precursor of volcano-sedimentary origin gave an ion-probe zircon U-Pb date of 3685 ± 8 Ma (Schiøtte & Compston 1990). They regarded this as the original age of the Akilia association, and viewed an older group of zircons at 3756 ± 22 Ma as xenocrysts, i.e. derived from older rocks.

(ii) Sm-Nd data of Bennett et al. (1993) on four gabbroic rocks from the Akilia association plot on a statistically valid isochron (mswd < 1) and yield an age of 3677 ± 37 Ma (Moorbath & Kamber 1998). Note that Bennett et al. (1993) did not regress their own Sm-Nd data, because their ideas are constrained by the 3.87–3.78 Ga zircon U-Pb dates which are regarded as the true age of supposedly discordant granitoid sheets. Moorbath & Kamber (1998) regarded the Sm-Nd age of 3677 ± 37 Ma as a close estimate for the age of not only the gabbroic components of the Akilia association on these islands, but also for the closely associated banded iron formation lithologies containing apatite with graphite inclusions of possible biological origin (Mojzsis et al. 1996).

(iii) Pb/Pb analyses by Kamber & Moorbath (1998) on feldspars and whole rocks from a suite of mafic Akilia association rocks from Akilia island and Innersuartuut island yield extremely unradiogenic Pb isotope compositions which plot so close to the Itsaq gneiss regression and its intersection on the primary growth curve (see previous section) that an age resolution between Itsaq gneisses and Akilia association rocks is not possible. There is no apparent contribution from any ≥ 3.8 Ga crustal material.

In addition to these direct lines of evidence, we note that the putative early-life-bearing Akilia association rock has yielded c. 2.7–2.6 Ga low-Th/U zircons interpreted by Nutman et al. (1997a) as the result of late Archaean metamorphism, as well as c. 3.64–3.42 Ga zircons reported recently by Mojzsis & Harrison (1999). The latter show a range of Th/U ratios from c. 0.4 to <0.01 (which, we suggest, indicates more than one origin), with higher Th/U zircons overlapping in value with those interpreted by Schiøtte & Compston (1990) as being volcanic in origin, and low U/Th zircons probably being metamorphic in origin. If the Akilia association enclaves are >3.85 Ga, as proposed by Nutman et al. (1997a), and the adjacent Itsaq gneisses truly record a long history (c. 200–300 Ma) of magmatic and thermal events as claimed by Nutman et al. (1996), then the Akilia enclaves will undoubtedly also have experienced these events. Given the tendency for metamorphic zircons to grow during the first high-grade metamorphic event to affect a rock (e.g. Söderlund et al. in press), the apparent absence of >3.65 Ga metamorphic zircon in the Akilia association rocks is surprising. Such an absence may, however, be easily explained if the true age of the Akilia rocks is less than c. 3.70 Ga.

The evidence summarized above suggests that the protoliths of the Akilia association rocks on Akilia island were deposited between c. 3.70 and 3.65 Ga, and are thus some 150 to 200 Ma younger than proposed in previous work (Mojzsis et al. 1996; Nutman et al. 1997a). There is no overlap with any period of massive bolide impacts on the moon, widely regarded as terminating at c. 3.8 Ga (e.g. Wilhelms 1987; Ryder 1990). Indeed, recent work by Cohen et al. (2000) provides strong evidence from ^{40}Ar-^{39}Ar studies of lunar impact melts supporting the occurrence of a short, intense period of bombardment in the Earth–Moon system peaking at c. 3.9 Ga.

The Isua greenstone belt (IGB)

The IGB (also known as the Isua supracrustal belt) lies some 150 km northeast of Nuuk, the capital of Greenland (Fig. 3), within the regional early Archaean gneisses (Allaart 1976). It comprises rocks of volcanic, volcanoclastic, clastic (Fedo 2000) and chemical–sedimentary origin with a wide range of chemical compositions, most (if not all) of which were deposited in water. These are the oldest (3.7–3.8 Ga) known rocks of their type on Earth (if, as we argued above, the true minimum age for the Akilia association is 3.65 to 3.70 Ga, and not >3.85 Ga as claimed). The main metavolcanic rock unit in the IGB is a mafic (basaltic) rock, mostly metamorphosed to fine-grained talc–chlorite schists.

Two of the principal sedimentary rock types are banded iron formation (BIF), and a massive garnet–biotite ± hornblende schist probably derived from a mafic volcanogenic sediment. IGB rocks underwent intense multiple deformation, peak metamorphic recrystallization at 550°C (Boak & Dymek 1982) and widespread metasomatic alteration (Rose et al. 1996; Rosing et al. 1996) which have partly obscured original lithology and chemistry. Localized low-strain domains containing conglomerates, pillow lavas, discordant contacts and relict sedimentary structures are present (e.g. Nutman et al. 1984; Appel et al. 1998; Fedo 2000). The IGB is currently being remapped in detail within the framework of the Isua Multidisciplinary Research Project (Appel & Moorbath 1999). It was reported long ago that Isua graphite yields fractionated carbon isotope values somewhere between biogenic-C and carbonate-C, interpreted by Schidlowski et al. (1979) and Schidlowski (1988) as indicating the existence of a significant biomass by the time of Isua sedimentation. More recently, Rosing (1999) identified ^{13}C-depleted carbon microparticles of possible biogenic origin in what are described as turbiditic and pelagic marine IGB metasedimentary rocks.

Of all currently known early Archaean terrains the IGB is unique in that it contains true, physically recognizable metasediments (both clastic and chemical) and an abundance of metamorphosed mafic volcanic rocks, some of which preserve primary volcanological features. It has long been suspected that the geochemistry and particularly the isotope record of these lithologies might contain a memory of pre-depositional geological history of a time window not otherwise accessible. We first discuss age constraints on the deposition of the sediments and volcanic rocks themselves and then evaluate the more speculative issue of interpreted isotope geochemical features.

Direct age constraints on the deposition of IGB lithologies

There is general agreement that the plutonic rocks (now gneisses) immediately surrounding the IGB are younger than the belt itself and that some contacts are tectonically discordant or intrusive. No granitoid, gneissic basement to the IGB has been recognized, and none of the pebble conglomerates in the belt contain recognizable gneissic clasts. This precludes determination of an absolute maximum age of deposition and attempts at dating the IGB therefore rely on direct age information (summarized in Table 1) as well as minimum age constraints by dating cross-cutting igneous rocks.

Since the early 1970s, conventional dating with Rb-Sr, U-Pb, Pb/Pb and Sm-Nd methods (summarized by Moorbath et al. 1986) has given an age range of c. 3.75–3.70 Ga for IGB lithologies. More recent attempts at directly dating IGB rock units have yielded similar results. Moorbath et al. (1997) determined a more precise 58-point Sm-Nd regression date of 3776 ± 56 Ma for combined Isua metasediments. The approach of combining diverse metasediments was later criticized by Bennett & Nutman (1998). In response, Kamber et al. (1998) pointed out that a regression of schist data alone (24 points) yielded a very similar result of 3742 ± 49 Ma. Frei et al. (1999) determined a Pb/Pb regression age of 3691 ± 22 Ma for a magnetite-enriched fraction from the major BIF deposit at Isua. This is in excellent agreement with an earlier Pb/Pb isochron age of 3698 ± 70 Ma determined on bulk-rock BIF by Moorbath et al. (1973). While Frei et al. (1999) interpreted their Pb/Pb regression date as the age of earliest metamorphism, we prefer to interpret 3.7 Ga as the most likely age of deposition. The strongest evidence for a 3.71 Ga deposition age of the Isua BIF (as well as associated lithologies) was presented by Nutman et al. (1997b) who dated zircons (by ion-probe U-Pb) from BIF, schists and an associated felsic volcanic band. More recently, Blichert-Toft et al. (1999) obtained a Sm-Nd regression with an age of 3712 ± 26 Ma for a combined IGB dataset (amphibolites, schists and clastic sediments). The same rocks yielded a Lu-Hf regression date of 3593 ± 15 Ma which Blichert-Toft et al. (1999) interpreted to indicate Hf mobility during subsequent geological events. However, Villa et al. (2001) recalculated the Lu-Hf regression age with the more recently determined Lu decay constant by Nir-El & Lavi (1998) and obtained an age of 3729 ± 16 Ma, in perfect agreement with all other age constraints.

In an important study, Gruau et al. (1996) found that a suite of metabasalts from the IGB demonstrated such a high degree of open-system behaviour that even single outcrops did not yield uniform initial ε_{Nd} values. Blichert-Toft et al. (1999) and Albarède et al. (2000) had to screen their combined Sm-Nd and Lu-Hf datasets using geochemical criteria (i.e. omit data points) because of problems related to element mobility and partial isotope resetting, particularly in metabasalt samples. There is thus no direct age constraint for the sometimes pillowed metabasalts. Nutman et al. (1997b) interpreted weighted mean ^{207}Pb-^{206}Pb dates for two discordant tonalitic sheets which yielded ages of 3791 ± 14 and

Table 1. Recent and selected older radiometric age data for Isua greenstone belt

Lithology	Method	Age (Ma)	±	Source	Remarks
Volcano-sediment Unit B1	Ion-probe U-Pb zircon	3708	3	Nutman et al. (1997b)	Interpreted by original authors as dating the entire metabasic-dominated part of the belt
Clast-rich sample from graded meta-volcanic rock	Ion-probe U-Pb zircon	3718	4	Nutman et al. (1997b)	Sample is from base of graded unit immediately overlying Unit B1
'Dirty' meta-banded iron formation	Ion-probe U-Pb zircon	3707	6	Nutman et al. (1997b)	Average of three oldest dates, interpreted by original authors as dating zircons of volcanic origin
Meta-banded iron formation	Pb-Pb step-leaching regression	3691	22	Frei et al. (1999)	Interpreted by original authors as dating a metamorphic event immediately after deposition
Meta-banded iron formation	Pb-Pb regression	3698	70	Moorbath et al. (1973)	Interpreted by original authors to date deposition of sediment
Kyanite schist	Ion-probe U-Pb zircon	3711	6	Nutman et al. (1997b)	Sample also yielded morphologically distinct metamorphic zircons of late Archaean age
Mica schists	Sm-Nd whole rock	3742	49	Kamber et al. (1998)	Regression of compiled data interpreted to date age of deposition
Mica schists, amphibolites and clastic meta-sediments	Sm-Nd whole-rock regression Lu-Hf whole-rock regression	3712 3729	26 16	Blichert-Toft et al. (1999)	Interpreted by original authors to either have no age significance or to date later resetting. Note that Lu-Hf age is recalculated using Nir-El & Lavi (1998) Lu decay constant, following Villa et al. (2001)
Felsic (meta-volcanic?) rock, Unit A6	Ion-probe U-Pb zircon Conventional U-Pb zircon	3806 3813	2 9	Compston et al. (1986) Baadsgaard et al. (1984)	If this unit is of volcanic origin, it would constitute an earlier phase of magmatism (and sedimentation) than that recorded by the rest of the belt

3798 ± 4 Ma, respectively, as minimum ages for pillow basalt extrusion. There is no *a priori* reason to believe that the IGB could not contain lithological packages that are unrelated in time and space. Indeed, ductile shear zones transect the IGB into mappable tracts and domains (Fedo 2000) that may not share a common origin. The possibility that the IGB contains >3.79 Ga lithologies clearly remains (Nutman *et al.* 1997b) but is not directly relevant for the discussion of isotopic signatures which were obtained on samples that are 3.71 Ga in age.

Small sulphide mineralizations within the IGB contain galena (lead sulphide) which has the least uranogenic Pb isotope ratios (i.e. lowest $^{206}Pb/^{204}Pb$ and $^{207}Pb/^{204}Pb$) known on Earth (Appel *et al.* 1978; Richards & Appel 1987). The ratios are as low as $^{206}Pb/^{204}Pb = 11.15$ and $^{207}Pb/^{204}Pb = 13.04$, which demonstrate that this uranium-free galena Pb was separated from a source region with a finite U/Pb ratio some 3.75–3.80 Ga ago

Initial isotope constraints from IGB metasediments

While no direct evidence exists for the nature of the basement onto which the IGB was deposited, it appears reasonable to expect that clastic sediments might hold clues to its age and nature. For example, a 4.1 Ga granitoid basement would have evolved by 3.71 Ga to initial Nd, Sr, Pb and Hf isotope ratios resolvably different from juvenile crust or mantle. Because many of the aforementioned whole-rock isotope studies yield ages that are concordant with each other and with the 3.71 Ga U-Pb zircon age constraints, their regression intercepts should be valid approximations for the isotope compositions of the source rocks.

Moorbath *et al.* (1986) concluded, based on a survey of then-published isotope work, that initial Nd and Sr isotope ratios recorded by regression lines of IGB lithologies are similar to those of coeval mantle. In other words, these initial isotope ratios do not permit a lengthy (>100 Ma) crustal pre-history. Since then, further whole-rock regression lines for Sm-Nd and Lu-Hf (Moorbath *et al.* 1997; Kamber *et al.* 1998; Blichert-Toft *et al.* 1999) have confirmed this view and there is general agreement that a substantially older, evolved basement is not visible in the initial isotope ratios of clastic sediments. Nutman *et al.* (1997b) have tentatively identified rare siliceous rocks of possible clastic sedimentary origin. These contain zircons which range widely in age, but two samples yield age clusters between 3.8 and 3.9 Ga. Irrespective of whether these rocks are metasomatized metagranitoids or clastic sediments, their zircon age distribution indicates that there was some kind of a pre-3.71 Ga IGB substratum but that it was not sufficiently old to leave a discernible record in the initial isotope ratios of clastic rocks.

It is, of course, clear that a relatively wide range is obtained when initial isotope ratios are calculated for individual samples (using an assumed appropriate age of deposition). This range largely reflects uncertainty in assumed age, assumption of a constant daughter/parent ratio, and potential influence of later isotope resetting. Jacobsen & Dymek (1988) proposed that the Sm-Nd whole-rock system in the Isua 'sediments' remained closed during metamorphism, thus interpreting the spread in individually calculated initial ε_{Nd} as a primary sedimentary signature due to variable mixing of pre-3.8 Ga continental crust with 3.8 Ga juvenile crust with very different initial ε_{Nd} values. However, as pointed out by Gruau *et al.* (1996), the wide variation of apparent initial ε_{Nd} (−1.0 to +5.0) can instead be attributed to disturbance during late Archaean metamorphism, as recorded by mineral ages.

A survey of conventional dating methods on IGB rocks thus suggests that some rock units yield concordant multi-method regression ages that are within error identical with U-Pb zircon ages and that yield mantle-like initial isotope signatures. The view that IGB represents a relatively juvenile terrain is in apparent conflict only with Harper & Jacobsen's (1992) claim for the existence of a measurable ^{142}Nd anomaly in an Isua 'metabasalt'. Because of the short half-life (102 Ma) of the now-extinct ^{146}Sm, a true ^{142}Nd anomaly would most certainly indicate a source with an age in excess of 4.2 Ga. The analytical controversy about the reality of the effect has not been entirely resolved (Jacobsen & Harper 1996; Sharma *et al.* 1996). Here we point out that the relevant sample originates from a thick horizon of quartz–mica–garnet schist, a strongly metamorphosed sediment. The postulated ^{142}Nd anomaly, if real, of this rock would therefore imply erosion from a source that had remained isolated for at least 0.5 Ga. In the absence of other terrestrial rocks with ^{142}Nd anomalies and given that none of the other isotope systems support the effect at Isua, Harper & Jacobsen's (1992) observation cannot be used to question the deposition of the IGB onto a relatively juvenile basement.

In contrast to the clastic sediments of the IGB, which appear to be relatively locally derived,

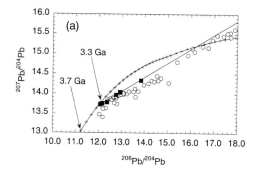
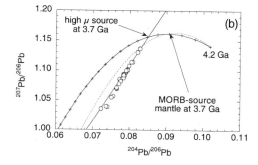
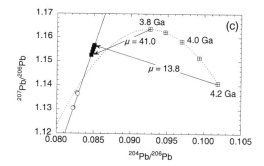

Fig. 4. Pb isotope diagrams. (a) IGB metasediments in $^{207}Pb/^{204}Pb$ versus $^{206}Pb/^{204}Pb$ isotope space. Undifferentiated data points (open circles) are garnet-mica schists (unpublished) and leached carbonates ($n = 30$; Table 2). The data define trends subparallel to the BIF data of Moorbath et al. (1973) – solid squares connected by regression line – which has a slope corresponding to c. 3.7 Ga. The important point is that regressions calculated for the chemical sediments (i.e. BIF and carbonate) intersect plausible mantle evolution curves (here Kramers & Tolstikhin (1997) – solid line connecting plus marks) at loci corresponding to much younger dates (i.e. c. 3.3 Ga). This indicates that the Pb which co-precipitated with these sediments had evolved for some time in an environment with a substantially higher μ than the mantle. (b) The same point is illustrated in a $^{207}Pb/^{206}Pb$ versus $^{204}Pb/^{206}Pb$ diagram. The Kramers & Tolstikhin (1997) mantle evolution curve (broken line shown from 4.2 Ga) evolves from right to left in this diagram. Also shown is a model reservoir (solid line connecting + marks) that separated from the mantle at 4.2 Ga and evolved with a μ of 13.8. After 500 Ma (i.e. at the time of deposition of BIF and carbonate, open circles), this reservoir would have evolved to a Pb isotope composition representing the initial for the present-day BIF isochron (solid line connecting open circles). (c) Blow-up of (b) highlighting the relationship between μ and separation time of the modelled high μ reservoir. Shown again is the Kramers & Tolstikhin (1997) mantle curve (broken line starting at 4.2 Ga) and the least radiogenic BIF data (open symbols) as well as the BIF isochron (solid line). Five different high-μ model reservoirs were calculated by using mantle separation ages of 4.2, 4.1, 4.0, 3.9 and 3.8 Ga (crossed squares). The 4.2 Ga separation model is identical to that shown in (b). With decreasing mantle separation age, the required μ increases in the steps 13.8, 15.3, 17.9, 23.3, 41.0. Note that all modelled initial Pb isotope compositions (solid squares) plot clearly above the mantle curve (i.e. to the left in the conventional diagram).

chemical sediments can potentially yield chemical and isotope signatures that reflect integration over a much larger area if they formed in a basin that was in contact with open ocean waters (provided, of course, that oceans existed). Moorbath et al. (1973), in their study of the Pb isotope systematics of BIF from the IGB, noted that the BIF recorded a distinctly higher first-stage $^{238}U/^{204}Pb$ (μ_1) of 9.9 compared to the Itsaq gneisses (8.6). They concluded that the significantly different two-stage time-integrated μ_1 values of 9.9 and 8.6 must indicate different immediate precursors. The full significance of their finding can only be retrospectively appreciated because of recent advances in the understanding of terrestrial Pb isotope evolution (Zartmann & Haines 1988; Kramers & Tolstikhin 1997; Kramers et al. 1998; Collerson & Kamber

1999; Elliott et al. 1999). Here we discuss Moorbath et al.'s (1973) BIF Pb isotope data and additional new Pb isotope data (Table 2) in the framework of modern terrestrial Pb isotope models.

In Figure 4a, a $^{207}Pb/^{204}Pb$ versus $^{206}Pb/^{204}Pb$ diagram, we plot the Pb isotope compositions of IGB sediments. Shown separately are the most significant published data points of the BIF unit for which the actual regression line is also drawn. The important observation, implicit in Moorbath et al.'s (1973) finding of a high first-stage μ_1, is that the BIF regression intersects possible mantle evolution curves (here Kramers & Tolstikhin 1997) at a much younger date (c. 3.3 Ga) than the deposition age (3.71 Ga). The most likely explanation for this observation is that the Pb that was incorporated into the

Table 2. Pb isotope data for Isua carbonate rocks

Sample	Method	$^{206}Pb/^{204}Pb$	±	$^{207}Pb/^{204}Pb$	±	$^{208}Pb/^{204}Pb$	±	Corr-Coeff (7/6)	Corr-Coeff (8/6)
248486/A	wr	12.814	0.039	13.931	0.044	32.610	0.103	0.949	0.998
248486/A	1M HCl	12.570	0.008	13.850	0.010	32.511	0.023	0.953	0.964
248486/A	6M HCl	12.606	0.020	13.866	0.022	32.626	0.052	0.975	0.992
248486/A	1M HNO$_3$	13.215	0.008	13.957	0.010	33.005	0.023	0.947	0.964
248486/A	HF	12.585	0.006	13.851	0.007	32.215	0.019	0.965	0.951
248486/B	wr	13.317	0.040	14.049	0.045	33.345	0.101	0.949	0.998
248486/B	1M HCl	12.759	0.008	13.898	0.010	33.186	0.024	0.956	0.966
248486/B	6M HCl	12.877	0.021	13.926	0.023	33.597	0.056	0.979	0.993
248486/B	1M HNO$_3$	15.019	0.020	14.236	0.020	35.425	0.050	0.978	0.990
248486/B	HF	13.048	0.008	13.940	0.009	32.571	0.022	0.961	0.962
248486/D	wr	12.719	0.038	13.885	0.044	32.395	0.098	0.949	0.998
248486/D	1M HCl	12.406	0.016	13.813	0.018	32.226	0.043	0.984	0.989
248486/D	6M HCl	12.504	0.023	13.844	0.026	32.518	0.060	0.980	0.994
248486/D	1M HNO$_3$	12.740	0.005	13.845	0.007	32.424	0.018	0.951	0.947
248486/D	HF	12.634	0.008	13.848	0.010	32.129	0.023	0.960	0.965
248486/G	wr	12.968	0.039	13.976	0.044	32.833	0.099	0.949	0.998
248486/G	1M HCl	12.559	0.164	13.813	0.181	32.507	0.424	0.994	1.000
248486/G	6M HCl	12.625	0.023	13.867	0.026	32.683	0.060	0.973	0.994
248486/G	1M HNO$_3$	13.493	0.007	14.013	0.008	33.204	0.020	0.940	0.953
248486/G	HF	12.727	0.008	13.910	0.010	32.373	0.023	0.949	0.965
248487/A	wr	13.090	0.039	14.007	0.044	33.098	0.100	0.949	0.998
248487/A	1M HCl	12.630	0.011	13.884	0.013	33.415	0.032	0.971	0.979
248487/A	6M HCl	12.733	0.031	13.900	0.035	33.819	0.083	0.979	0.997
248487/A	1M HNO$_3$	13.558	0.019	14.089	0.020	33.896	0.049	0.982	0.990
248487/A	HF	13.052	0.009	13.970	0.011	32.484	0.026	0.967	0.970

Pb isotopes were measured at the Universities of Oxford (whole rocks) and Queensland. Queensland data were obtained on a multi-collector TIMS in static mode using pyrometer temperature control (1300–1325°C). During the relevant period, NBS SRM 981 yielded the following reproducibilities: $^{206}Pb/^{204}Pb = 16.911 \pm 28$; $^{207}Pb/^{204}Pb = 15.458 \pm 36$; $^{208}Pb/^{204}Pb = 36.600 \pm 114$.
All reported ratios are fractionation corrected ($-0.057 \pm 17\%$/amu, $n = 12$; calculated from all ratios). Internal errors quoted at 2 sigma level are similar to or better than external reproducibilities.
Of each rock, a whole-rock (wr) split was measured and a separate aliquot was subjected to a step-wise dissolution in Teflon beakers with (i) 1M HCl (10 min at room temperature); (ii) 6M HCl (20 min at room temperature); (iii) 1M HNO$_3$ (24 h at 110°C); (iv) 12M HF (24 h at 130–150°C).

chemically precipitated BIF had a very different isotope composition from coeval mantle. Other possible chemical metasediments, such as carbonates, plot co-linear with the BIF data (Fig. 4a) and also indicate evolution of Pb in a high μ environment prior to incorporation in the carbonate. Regression lines through data from clastic sediments (e.g. garnet–mica schists), while also plotting subparallel, intersect the mantle evolution curve at $c.3.6$ Ga (much closer to the actual deposition age). In our view, the difference in apparent mantle evolution intersection age of the BIF data is significant and deserves further treatment.

Of critical importance is the question of how high a plausible μ of Earth's early Archaean surface could have been. In Figure 4b, a ^{207}Pb/^{206}Pb versus ^{204}Pb/^{206}Pb diagram, the evolution of one possible Pb isotope model is shown (i.e. defined to intersect the BIF regression line at 3.71 Ga). The two parameters that define this model are an early mantle separation age, here chosen as 4.2 Ga, and an elevated μ of 13.8 compared to that of model MORB-source mantle (after Kramers & Tolstikhin 1997). Obviously the younger the mantle separation age, the higher the μ in the reservoir from which the BIF Pb derived. Figure 4c shows that for a separation age of 3.8 Ga, the required μ is 41. Separation ages between 4.2 Ga and 3.8 Ga will yield μ values ranging from 13.8 to 41 (Fig. 4c). High model μ values (13.8–41) would normally be associated with continental crust, and marine high μ Pb could, in this case erroneously, be interpreted to argue that the early Archaean surface Pb was dominated by continental run-off. This is, however, not a possibility because the first terrestrial Pb paradox (e.g. Kramers & Tolstikhin 1997), Nb-Th-U systematics (Collerson & Kamber 1999) and further evidence (as outlined before) strongly argue against a voluminous early continental crust in the modern sense. An interesting solution is found by inspection of Pb isotope systematics of lunar rocks. It is, unfortunately, not possible to directly compare lunar rocks with terrestrial Pb isotope evolution because the Moon experienced a different mode of siderophile element fractionation (e.g. Kramers 1998) and because the bulk silicate Moon's initial μ is unknown (e.g. Tatsumoto et al. 1987). The important observation, however, is that the μ of different lunar lithologies is highly variable and that some of the highest μ values are found in mare basalts. Tatsumoto et al. (1987) argued that, while poorly defined, the Moon probably evolved with a μ in the range of 19 to 55, much lower than the μ of mare basalts (around 300). Nyquist & Shih (1992) highlighted the roles of ilmenite and trapped liquids as potential agents for strong U/Pb fractionation. Whether broadly comparable volcanism existed on Earth remains an elusive question but, if so, it would imply that substantial areas of the planet (such as covered by mare on the Moon) could have been covered by basaltic rock with a μ of at least six times the bulk silicate Moon. Scaled to the Earth's bulk silicate μ ($c.8$), this would translate to a potential surface μ of almost 50. The high ^{207}Pb/^{206}Pb ratio of the IGB chemical sediments could therefore be explained by weathering of basaltic crust (possibly produced by impact melting) with an age of $c.200$ Ma (approximately the culmination age of heavy meteorite bombardment, cf. Cohen et al. 2000) at the time of erosion.

In summary, geological, isotopic and geochemical records of the Earth's oldest identifiable sediments indicate the absence of a substantially older local basement. In contrast, chemical sediments offer a tantalizing glimpse of a Moon-like pre-3.8 Ga terrestrial surface.

Initial isotope constraints from IGB metavolcanic rocks

Basaltic volcanics are the major probes for reconstruction of mantle depletion history by most radiogenic isotope systems (Sm-Nd, Lu-Hf and Re-Os) and some geochemical proxies (e.g. Nb-Th). It is therefore not surprising that attempts have been made at using the IGB metabasalts for constraining the degree of depletion (or enrichment) of the early Archaean mantle. Unfortunately, such attempts are facing a multitude of problems that relate to unsystematic sampling strategies, deformational and metamorphic overprints, metasomatism, and a general lack of reliable trace element characterization of the rocks that are somewhat carelessly grouped under the term 'metabasalts' or 'supracrustals'. There is no doubt since publication of the meticulous reports by Gruau et al. (1996) and Rosing et al. (1996), respectively, that isotopic and geochemical resetting and metasomatic overprint, particularly in the depleted metabasalts, require utmost care in interpretation of back-corrected initial isotope values of these rocks. Two recent papers by Blichert-Toft et al. (1999) and Albarède et al. (2000) have attempted to use Isua 'supracrustals' to search for a memory of early differentiation processes. These authors claim, based on the same data set, that the 3.8 Ga mantle was geochemically structured and that this structure was partly inherited from the initial differentiation of the Earth. The claims of the first publication by Blichert-Toft et al. (1999) have since been shown to be invalid (Villa et al., 2001), due to: (i) use of an inadequate Lu decay constant; (ii) unsupported back-correction of parent isotope decay 100 Ma past the isochron age; (iii) inherent assumption of the existence of an isotopic mantle array; (iv) complete disregard of other geochemical and geochronological constraints derived from Isua and elsewhere.

Much of this criticism applies especially to the arguments of Albarède et al. (2000). These authors, too, back-calculate ^{147}Sm and ^{176}Lu decay to 3.8 Ga despite the fact that their samples define 3.7 Ga regression dates. Totally irrespective of the significance of the 3.7 Ga regression dates (i.e. rock formation ages versus isotope resetting ages), parent decay cannot under any circumstances be back-corrected past that time. Any conclusion drawn from such a violation of this simple isotopic principle is therefore by definition flawed and needs no further discussion here. Although geochemical study of basalts is naturally preferred over study of more evolved igneous rocks for much of the geological record, this intuitive choice may not be the best for the early Archaean. We suggest that because of the general lack of older continental basement and sediment, and their much higher incompatible trace element contents compared to basalts, evolved igneous rocks offer a better chance to reconstruct the degree of early mantle depletion.

Northern Labrador, Canada

The early Archaean rocks of the Saglek–Hebron area of northern Labrador have many similarities with the penecontemporaneous evolution of rocks in the Godthaabsfjord region of southern West Greenland, with which they were once contiguous (Bridgwater & Collerson 1976; Bridgwater et al. 1978; Collerson & Bridgwater 1979), but far less age and isotope work has been published than on the West Greenland rocks. Conventional dating by Rb–Sr, U–Pb, Pb/Pb and Sm–Nd methods, summarized by Schiøtte et al. (1986), clearly indicated an early Archaean age >3.5 Ga, but also showed that the so-called Uivak gneisses were much more disturbed by late Archaean events than the approximately equivalent Itsaq (then called Amîtsoq) orthogneisses of West Greenland. Ion-probe zircon U–Pb dates based on maximum ^{207}Pb/^{206}Pb ratios from three separate Uivak gneiss samples agree within error at 3732 ± 6 Ma, interpreted as age of emplacement of the magmatic precursors (Schiøtte et al. 1989). A few of these gneisses contain rounded zircon inclusions (observed by photomicrography) with dates up to 3863 ± 12 Ma, regarded as inherited from an older source region.

The early Archaean Nulliak (supracrustal) assemblage was broken up by intrusion of the protoliths of the Uivak gneisses and then deformed, metamorphosed and metasomatized (Nutman et al. 1989). The Nulliak assemblage is regarded as broadly equivalent in age and origin to the IGB and Akilia association rocks of West Greenland (see earlier). The three oldest ion-probe zircon U–Pb dates with a mean of 3776 ± 8 Ma were obtained on an acid metavolcanic rock from the Nulliak assemblage. This was regarded by Schiøtte et al. (1989) as the age of deposition of the volcanic precursor.

We comment here briefly on a Sm–Nd study of mafic/ultramafic gneisses by Collerson et al. (1991). These authors obtained geochemical and Sm–Nd isotope data on two suites of rocks. The first is a group of ultramafic gneisses that occur interleaved with 3.7 Ga Uivak I gneisses. Based on their major element composition, these ultramafic gneisses were interpreted as tectonically emplaced fragments of lithospheric mantle. They yielded a Sm–Nd isochron date of 3.82 ± 0.12 Ga with an intercept corresponding to a strongly positive ε_{Nd} of 3.0 ± 0.6. According to Collerson et al. (1991), the radiogenic initial Nd isotope composition would support an origin of the protoliths in the subcontinental lithospheric mantle, which is widely regarded as a depleted residual reservoir. Some later studies, however, have misquoted the large degree of depletion shown by these gneisses as indicating strong and early depletion of the depleted (MORB-source) mantle (Bowring & Housh 1995; McCulloch & Bennett 1994). Collerson et al. (1991) noted that the radiogenic initial Nd of these rocks is in conflict with their light rare-earth-element enrichment (relative to chondrite). This was explained in terms of re-enrichment shortly before tectonic emplacement at c. 3.7 Ga. In a more recent study, Wendt & Collerson (1999) analysed the same samples for Pb isotope compositions. Pb isotopes failed to shed further light on the early Archaean history as the U–Pb system in many of these samples appears to have been disturbed during subsequent geological events.

The second group of rocks studied by Collerson et al. (1991) comprised five subsamples of metakomatiite taken from a single outcrop. These depleted rocks, which differ in their major- and trace-element chemistry from the lithospheric mantle suite, yield an age of $3.83 \pm$

0.17 Ga with an initial ε_{Nd} of 0.23 ± 1.58. This implies that these komatiites could have been derived from a largely undepleted mantle source. Such an interpretation is supported by the 3.85 ± 0.16 Ga Pb-Pb regression age and model μ_1 of 7.9 obtained on these samples by Wendt & Collerson (1999). The Pb isotopes thus also indicate that the peridotite source of these melts was largely undepleted. Wendt & Collerson (1999) included in their regression a sample of metabasaltic komatiite from a different outcrop. In contrast to the U-Pb system, the Sm-Nd regression parameters change markedly when the metabasaltic komatiite sample is included. Collerson et al. (1991) obtained a combined regression date of 4.02 ± 0.19 Ga with an initial ε_{Nd} of -2.11 ± 2.33. While they noted that, within (the large) errors, the fitting parameters agreed with those of the metakomatiite proper regression, they appear to attach more significance to the combined result. The negative ε_{Nd} of the combined regression was thus interpreted to indicate contamination by continental crust (with low time-integrated Sm/Nd) of a melt from an ultradepleted peridotite source. We question the use of a single datum point, such as the metabasaltic komatiite, to calculate speculative early mantle parameters because: (i) Pb isotopes fail to show such mantle complexity, but agree with a younger age of the metakomatiites and an origin from an undepleted mantle; and (ii) the metabasaltic komatiite sample has an unreasonably negative ε_{Nd} of -4.12 at 3.8 Ga.

Other early Archaean regions

Only brief summaries are given here of regions where presence of in-situ early Archaean rocks is claimed or suspected. The list is not exhaustive, and we do not review rocks younger than 3.5 Ga. Published evidence may be insufficient or inconclusive, particularly when based solely on unimaged zircon grains from orthogneisses. The most frequent supporting evidence for claims of early Archaean ages from zircon dates consists of isolated maximum Sm-Nd model ages (T_{DM}, based on a conventional depleted-mantle model). Sometimes they are broadly concordant with zircon U-Pb dates, but any discordance may be due to uncertainty in interpretation of the zircon date, or to Sm-Nd open-system behaviour. In most of the cases summarized below, much more work is required in order to clarify the true geochronological and geological significance of the reported early Archaean dates.

Australia

The oldest terrestrial mineral ages yet reported until very recently are in the range 3.91 to 4.27 Ga for detrital zircons from much younger (c. 3.1 Ga) quartzites in the Mount Narryer complex and from the Jack Hills metasedimentary belt, both in the Yilgarn craton of western Australia (Froude et al. 1983; Compston & Pidgeon 1986; Maas & McCulloch 1991; Maas et al. 1992). The ion-probe data from Jack Hills were supported by precise U-Pb isotope dilution of detrital zircons (Amelin 1998), with an age range of 3.82–4.11 Ga.

Very recently, detrital zircon grains with U-Pb ages as old as 4.3 to 4.4 Ga have been reported from western Australia (Wilde et al. 2001; Mojzsis et al. 2001), with the oldest grain analysed yielding an age of 4404 ± 8 Ma. This whole topic is discussed in more detail in a later section.

The oldest known in-situ gneisses in western Australia do not approach the detrital zircon dates. The Manfred complex in the Yilgarn craton yields a zircon U-Pb date of 3.73 Ga (Kinny et al. 1988), together with approximate Sm-Nd and Pb/Pb regression dates of c. 3.68–3.69 Ga (Fletcher et al. 1988). The enclosing Meeberrie orthogneiss complex contains multiple generations of magmatic zircons at c. 3.6–3.7 Ga (Kinny & Nutman 1996).

In the ancient Pilbara craton, Nelson et al. (1999) quote a zircon U-Pb date of 3655 ± 6 Ma from a banded gneiss as the oldest age. From an angular unconformity beneath rocks of the c. 3.46 Ga Warrawoona Group, Buick et al. (1995) demonstrate a record of emergent, buoyant continental crust at c. 3.5 Ga.

South Africa and Zimbabwe

Numerous conventional and single zircon U-Pb measurements show that the Ancient Gneiss Complex of Swaziland, which forms the basement to the c. 3.5–3.2 Ga Barberton Greenstone Belt, contains tonalitic orthogneisses with an age of c. 3.65 Ga (Compston & Kröner 1988; Kröner & Todt 1988; Kröner et al. 1991, 1996; Kröner & Tegtmeyer 1994). Sporadic Sm-Nd (T_{DM}) model ages go back to 3.7–3.8 Ga.

Stratigraphical and structural aspects of the formation of the Kaapvaal craton of southern Africa have been summarized by De Wit et al. (1992), whilst chronological correlation between the Kaapvaal craton and the Pilbara craton of western Australia is described by Nelson et al. (1999). In both cases, crustal development began at c. 3.65 Ga.

The oldest gneisses of Zimbabwe yield conventional Rb-Sr and Pb/Pb ages of c. 3.5 Ga, with some Sm-Nd (T_{DM}) ages of up to c. 3.6 Ga (reviewed by Taylor et al. 1991). Horstwood et al. (1999) report zircon U-Pb evidence suggesting that c. 3.5 Ga basement (the 'Sebakwian protocraton') underlies much of the Zimbabwe craton. Ancient, detrital zircons of c. 3.7–3.8 Ga have been reported in much younger greenstone-belt sediments (Dodson et al. 1988; Nägler et al. 1997).

Of particular significance is a Re-Os study of chromites from age-constrained (2.7–3.5 Ga) ultramafic inclusions in the Zimbabwe craton (Nägler et al. 1997). The data show that the Zimbabwean subcontinental lithospheric mantle (SCLM) began to separate from asthenospheric mantle just before 3.8 Ga and grew quasi-continuously throughout the Archaean. Similar studies elsewhere have shown that in Archaean cratons the time of SCLM Re depletion broadly matches the formation ages of the overlying crust (Pearson et al. 1995a, b; Shirey & Walker 1998). No Os isotope evidence has yet been reported for pre-4.0 Ga separation of SCLM from the mantle.

West Africa

A 3.05 Ga orthogneiss from northern Nigeria contains zircons with inherited cores at 3.56 Ga (Bruguier et al. 1994). Zircon U-Pb dates of c. 3.5 Ga are reported from orthogneiss in the West African craton of Mauretania, whilst Sm-Nd (T_{DM}) model ages of 3.9–3.5 Ga are reported for adjacent metasediments (Potrel et al. 1996).

Antarctica

Black et al. (1986) reported ion-probe U-Pb data for zircons from a granulite orthogneiss from Mount Sones, Enderby Land. They regarded the oldest zircon data of 3927 ± 10 Ma as the magmatic age of the tonalite precursor of the orthogneiss. A Sm-Nd (T_{DM}) model age of c. 3.85 Ga is quoted by Black & McCulloch (1987). In our view, the true significance (i.e. whether comagmatic or inherited) of this frequently quoted zircon date is uncertain, and much more work is required.

United States

Xenocrystic zircon dates are reported from middle to late-Archaean gneisses in Wyoming by Aleinikoff et al. (1989). More recently, Mueller et al. (1996) report an ion-probe zircon U-Pb date of c. 3.5 Ga from a Wyoming trondhjemitic gneiss. Detrital zircon grains with an age of 3.96 Ga have been reported from a 3.3 Ga quartzite in Montana (Mueller et al. 1992).

China

Zircon U-Pb dates in the range 3.6–3.85 Ga have been reported from metasediments in the Qianxi complex of the Sino-Korean craton (Liu et al. 1990, 1992). Biao et al. (1996) present a model of crustal evolution from 3.8 Ga to 2.5 Ga in the Liaoning province of northeastern China, starting off with two ion-probe zircon U-Pb dates of 3804 ± 5 Ma and 3812 ± 4 Ma, which are interpreted as the age of intrusion of the trondhjemitic protolith of the Baijafen granite. Much more work is required!

Discussion

Age claims for the oldest terrestrial rocks are crucially dependent on the dating methods used. To achieve a convincing synthesis, clearly documented field relationships must be combined with appropriate geochemical and multi-method isotopic approaches (e.g. U-Pb, Pb/Pb, Sm-Nd, Lu-Hf, Rb-Sr). Zircon U-Pb dating in isolation may not be definitive in deciding whether zircon in an orthogneiss is comagmatic with its host rock, or inherited from an older source rock. This, in turn can lead to ambiguity in interpreting initial Nd and Hf isotope ratios in the host rock, not least when combined with subsequent open-system behaviour in the Sm-Nd system, particularly for low-Nd rocks. The Lu-Hf system is less prone to open-system behaviour but, since zircon is the main carrier of Hf in a rock (i.e. very low Lu-Hf), any inherited zircon will itself contain inherited Hf with a different isotopic composition from that in comagmatic zircon. Furthermore, since almost all Hf resides in zircon and almost all Lu resides in the matrix (surrounding the zircon), a later thermal event may result in diffusive transfer of radiogenic Hf from matrix to zircon, where it is unsupported by Lu decay.

We again emphasize the importance of Pb isotope modelling on feldspars and whole rocks for determining the timing of mantle–crust differentiation of the magmatic protoliths of orthogneiss complexes. The extremely high degree of age resolution in the Pb/Pb system is especially useful in those early Archaean orthogneisses (and other rocks) with such low U/Pb ratios that even their present-day Pb isotope composition lies close to the primary Pb/Pb growth curve (e.g. Kramers & Tolstikhin 1997; Kamber & Moorbath 1998).

Bearing the above factors in mind, we regard the following age and isotope scenarios for early Archaean rocks as plausible on the basis of the reviewed evidence. Detailed reference to published work is not repeated below; please refer to previous sections.

The oldest in-situ orthogneisses

By $c.3.65$ Ga, production of magmatic precursors of typical trondhjemite–tonalite–granodiorite (TTG) gneisses was in progress on a global scale, as confirmed by all dating methods. They are similar in lithology and chemistry to the voluminous TTG gneisses of later epochs and were probably formed by similar petrogenetic mechanisms, involving partial melting of the mantle or of mantle-derived mafic rocks, or differentiation of mantle-derived melts (e.g. Arth & Barker 1976; Martin 1986; Stern & Hanson 1991). This is largely borne out by initial Sr, Pb and Nd isotope ratios, frequently obtained from Rb-Sr, Pb/Pb and Sm-Nd whole-rock and feldspar (for Pb/Pb) regressions, which do not permit a long-term pre-magmatic crustal residence time. Despite frequent significant geological scatter, we place greater confidence in regression ages than some recent workers, especially where there is consistency between different methods and where mixing relationships are unlikely. In this connection, an argument often levelled at the whole-rock regression method is that it generates only average values for the emplacement parameters of a range of non-cogenetic orthogneisses (e.g. Nutman et al. 1996), although such arguments about cogenicity are themselves based on the assumption that interpretations of single-grain zircon U-Pb dates are correct.

Potentially promising claims (Nutman et al. 1999) for the existence of $c.3.8$ Ga in-situ orthogneisses south of the Isua region in West Greenland have been made solely from zircon U-Pb data. The claim of a $c.4.0$ Ga age for the Acasta gneiss of northern Canada (see earlier) needs to be assessed in the light of the intense reworking that these rocks underwent much later, although the existence of a $c.4.0$ Ga precursor is not in serious doubt from zircon U-Pb and Sm-Nd data. The evidence reported for other 3.8–3.9 Ga claims for in-situ orthogneisses (e.g. Antarctica, China: see earlier) is simply inadequate for a firm conclusion. Much more multi-isotopic work, in combination with detailed zircon imaging, is required to substantiate these claims.

The oldest in-situ supracrustal rocks

Certainly by 3.75–3.70 Ga, and possibly by 3.80–3.75 Ga, a wide range of volcanic rocks, volcanogenic sediments, clastic sediments and chemical sediments were being deposited under marine conditions, as exemplified by the IGB of West Greenland (see earlier). There is as yet no geological evidence for sialic basement to the IGB, nor any compelling age or isotopic evidence in IGB rocks for significant crustal residence time. However, possibly detrital zircons as old as 3.80–3.85 Ga have been reported (Nutman et al. 1997b), and Pb isotope ratios on chemical sediments suggest a contribution from a much older source (see pp. 189–191).

The picture that emerges from IGB rocks is of predominantly mafic volcanic centres, associated with deposition of shallow-water, coarse- to fine-grained, clastic volcanogenic sediments, as well as deeper-water chemical sediments, such as chert and magnetite-BIF. Carbon isotope evidence suggests that life might have been present (Rosing 1999). No evidence for major bolide impact has yet been found.

The IGB rocks and analogous Akilia and Nulliak associations (see earlier) may be typical of conditions on the Earth's surface at $c.3.70$–3.75 Ga. The preservation of banded iron formations and pillowed basalts documents the presence of a liquid hydrosphere. Magmatism was typically bimodal (dominantly mafic, subordinately felsic). By comparison with younger Archaean terrains, it appears that magmatic, sedimentary and tectonic styles that dominated until 2.5 Ga were already in operation by 3.7 Ga. Due to the strong metamorphic and metasomatic overprints on these early Archaean volcanic rocks, reliable trace element interpretations could be an elusive goal. Magmatic processes in this time range may be more readily delineated using the more pristine, highly evolved, regional granitoid plutonic rocks which surround the IGB.

The oldest inherited zircons

There is plenty of evidence for detrital zircons in the range $c.4.4$–3.8 Ga in younger sediments and metasediments from several regions, with the oldest and most spectacular ages occurring in western Australia. In this region, Amelin (1998) thinks that 'the old zircons might have been incorporated through assimilation of sedimentary rocks containing pre-3.9 Ga components by c. 3.4 Ga granitoid magmatism'.

The existence of such ancient zircons demonstrates the existence of evolved, intermediate- to high-silica rocks as far back as c. 4.4 Ga. The origin and true nature of this crust are unknown, but earlier workers (e.g. Maas & McCulloch 1991; Maas et al. 1992) favoured the view from geochemical work, based on rare-earth-element patterns and trace-element ratios, that the detrital zircons were derived from a differentiated continental source of substantial thickness rather than from felsic differentiates within dominantly mafic, oceanic-type crust. More recently, detrital zircons from western Australia in the range of 4.3–4.4 Ga are regarded (Wilde et al. 2001; Mojzsis et al. 2001) as evidence for the existence of full-scale continental crust. Moreover, an enriched ^{18}O isotope signature is regarded by these authors as the result of low-temperature interaction between the source rock (possibly basaltic rock?) for the original zircon-bearing granitoids and liquid water and, therefore, as evidence for oceans at 4.3–4.4 Ga. These preliminary conclusions remain to be confirmed and must be discussed in the context of the following observations.

(i) Zircon is the only datable accessory mineral that can survive a complex geological history. The survival of accessory minerals is therefore inherently biased towards silicic source rocks, and accessory minerals from very ancient mafic precursors either cannot be dated or did not survive weathering and erosion processes. From the discovery of ancient zircons alone, it cannot be excluded that their provenance could have been from felsic differentiates within a dominantly mafic crust, analogous, for example, to the felsic supracrustal rocks from the Isua greenstone belt at c. 3.71–3.81 Ga.

(ii) Amelin et al. (1999) reported Hf isotope data for 37 individual zircon grains with U-Pb ages up to 4.14 Ga from the Mount Narryer gneiss complex. They found that none of the grains had a depleted mantle signature, but that many were derived from a source with a Hf isotope composition similar to that of chondritic meteorites. This argues strongly against very early, large-scale differentiation of the mantle to produce major amounts of sialic crust.

(iii) Zircons may have crystallized from chemically evolved impact melts between c. 4.4 and 3.8 Ga. If any genuine pre-impact zircons survive from a particular event they might even show shock structures, as suggested by Amelin (1998).

(iv) As a possible analogy to the existence of pre-4.0 Ga terrestrial felsic rocks, it should be noted that zircon-bearing granites and quartz–monzodiorites have been recorded amongst the recovered lunar fragments (Nyquist & Shih 1992). Zircon U-Pb dates from these evolved rocks are in the range 4.32–3.90 Ga, but mostly fall in the upper parts of this range (Meyer et al. 1996). The quoted authors conclude that the formation of lunar granite appears not to have been restricted to a single episode in lunar history as, for example, during final solidification of the magma ocean. Rather, it appears to be the result of localized differentiation of individual plutons intruded into the lunar crust.

(v) The ^{18}O-rich isotopic composition in the most ancient (4.3–4.4 Ga) terrestrial zircon grains does not necessarily have to reflect a low-temperature hydrothermal history of the magma source. Halliday (2001) expressed concerns that subsequent diffusional exchange might have affected the O isotope composition of these grains. Furthermore, in view of the fact that the O isotope composition of even the post-3.8 Ga Archaean oceans remains to be determined, we regard the preliminary interpretation of hydrothermal exchange as speculative.

The above discussion has focused on ancient detrital zircons. As will be evident from previous sections, some of the oldest in-situ orthogneisses of magmatic origin contain inherited zircons dating back to 4.0–3.8 Ga, and it is very plausible that older ones may be discovered.

Mantle evolution and differentiation

Critical assessment of Nd and Hf isotope data for early Archaean mantle-derived rocks provides no compelling evidence for large-scale heterogeneities in the depleted mantle. This suggests that apparent enriched-to-depleted heterogeneities may be due to uncertainties in U-Pb, Sm-Nd and Lu-Hf interpretations or may relate to local mantle characteristics that are irrelevant for global differentiation. Based on the absence of evidence for very early depletion of the mantle (and, implicitly, very early growth of continental crust) we conclude, with posthumous apologies to Richard Armstrong, that the volume of early Archaean continental crust was small, and that the scarcity of known outcrops of early Archaean continental crust in part reflects that fact.

Armstrong's (1991) 'no-growth' model for continental crust volume versus time evolution apparently received renewed support from studies of Nb-U systematics of late Archaean basalts (Sylvester et al. 1997; Kerrich et al. 1999). These studies concluded, based on depletion of U (with respect to Nb) in these rocks, that the volume of continental crust 2.7 Ga ago was similar to the present. However, this interpretation is not supported by Nb-Th systematics, which are expected to show an equivalent degree of depletion (Collerson & Kamber 1999). The answer to this apparent paradox is found in the solution to the second terrestrial Pb paradox in which the present-day Th/U ratio of the mantle is much lower than the time-integrated ratio recorded by Pb isotope systematics of MORB. Kramers & Tolstikhin (1997) and Elliott et al. (1999), on the basis of terrestrial Pb isotope systematics, solved the second terrestrial Pb paradox by postulating a strong decrease in the depleted mantle's Th/U ratio after 2.0 Ga. This decrease reflects increased recycling of crustal U once a pandemic oxidizing atmosphere was established (at c. 2 Ga). Collerson & Kamber (1999) and Kamber & Collerson (2000), in a study of terrestrial Nb-Th-U systematics, elaborated on the implications of a dynamic mantle Th/U for previous claims of early growth of continental crust based on U depletion alone. They found that in the depleted mantle the Nb/Th ratio is the reliable proxy for the continental crust volume while the Nb/U ratio is additionally influenced by oxygenation of the atmosphere and cannot, therefore, be used to claim support for Armstrong's (1991) 'no-growth' model. On the contrary, Nb/Th, like critically assessed Nd and Hf isotope data, also points to a slow start for continental growth.

A third class of constraints for terrestrial differentiation involves forward transport models that aim at reproducing present-day isotope and trace-element characteristics of the silicate reservoirs (e.g. Zartman & Haines 1988). These models depend on a large number of input parameters, and do not provide unique solutions to differentiation. Nevertheless, some constraints are valid for a wide range of plausible input parameters. Notably, the combination of an average continental Nd isotope mantle extraction age of 2 to 2.5 Ga with the existence of the first terrestrial Pb isotope paradox (i.e. the fact that modern basalts and continental crust plot to the right of the meteorite isochron) is inconsistent with a 'no-growth' continental crust. This is because the large degree of recycling required to achieve the mean continental mantle extraction age would effectively eradicate the first Pb isotope paradox (Kramers & Tolstikhin 1997). Other constraints from forward modelling are more strongly dependent on input parameters and the solutions need to be evaluated not only with how well they fit present data but also with constraints on past evolution. The emerging picture from modelling Pb and Nd isotope systematics (Nägler & Kramers 1998) is that the continental volume increased with time as a roughly S-shaped function. This is fully supported by Nb-Th-U systematics of the mantle, which confirm the importance of changing erosion rates with time for the crust volume versus time curve.

A significant contribution towards understanding differentiation is the determination of the Re-depletion age of subcontinental lithospheric mantle (Shirey & Walker 1998). Such studies of Archaean cratons show that, on average, continents are of similar age to their underlying lithospheric mantle keel. This intuitively confirms geodynamical predictions (e.g. Bickle 1986) for the preservation potential of continental plates. It also appears that any pre-4.0 Ga crust, short of tectonic incorporation into younger continental lithosphere, will have stood little chance of preservation. The implication is that any possibly enriched pre-4.0 Ga crust (as on the Moon) will have been recycled back into convecting mantle without any lasting effect on further differentiation. It cannot be excluded that transition-zone (670 km and 2900 km) mantle heterogeneities could have formed and persisted if large-scale accumulation of majorite and/or perovskite accompanied pre-4.0 Ga crust extraction. Such mantle heterogeneities are alluded to in some isotope studies that recalculate initial isotope ratios for individual samples to the age of the 'most probable' deposition age determined by U-Pb zircon geochronology (e.g. Bennett et al. 1993; Bowring & Housh 1995; Blichert-Toft et al. 1999; Albarède et al. 2000). For reasons outlined throughout this review, we remain sceptical about the validity of these claims.

We still await convincing claims for earliest mantle heterogeneity. Because on the Moon (with which the early Earth may have had some similarities) the U-Pb system experienced a high degree of fractionation, we hope that such studies will include a treatment of Pb isotope systematics.

Summary

We now summarize briefly the salient features arising from the above review of the oldest terrestrial rocks.

The oldest, reliably dated, in-situ granitoid rocks of magmatic origin (orthogneisses), regarded as broadly representative of the type of continental crust formed throughout the rest of Earth history, mostly give ages in the range 3.65–3.75 Ga, and probably up to 3.81 Ga. Claims for older ages back to c. 4.0 Ga are still inadequately documented because they are based solely on U-Pb dates of the refractory, recyclable mineral zircon, which could have been inherited from older rocks of unknown type and origin. Such rocks may no longer exist in-situ, or they may not yet have been discovered. Published claims for the existence of true continental crust and oceans back to $c.4.4$ Ga based on very few detrital zircon grains in much younger metasedimentary rocks from western Australia require far more persuasive and detailed documentation and debate. At this stage of knowledge, alternative explanations for the origin of these zircons are equally, if not more, compelling.

The oldest reliable dates, from in-situ chemical and detrital sedimentary rocks and volcanic rocks (i.e. supracrustal rocks), give ages in the range 3.71–3.81 Ga. They show indubitable evidence for deposition in water (even the volcanic rocks frequently occur as pillow lavas). ^{13}C-depleted graphite microparticles in chemical sediments (subsequently strongly deformed and metamorphosed) from two groups of localities in West Greenland are claimed to be biogenic in origin. Here we summarize evidence to show that the probable age of deposition is in the range of 3.68–3.75 Ga, and not >3.87 Ga as claimed by some workers. Thus these rocks are $c.100$–200 Ma younger than termination of the putative terrestrial equivalent of the main lunar impact episode at $c.3.85$–3.90 Ga. Widely publicized claims for temporal overlap of the earliest putative biogenic components (the ^{13}C-depleted graphite microparticles) with massive global impacts are not required from the available age data.

Initial radiogenic isotope ratios (e.g. Nd, Pb, Hf) of reliably dated ancient (<3.8 Ga) orthogneisses from several regions are most simply interpreted as the result of magmatic differentiation from a slightly depleted, close-to-chondritic, essentially homogeneous early Archaean mantle. These results argue strongly against the existence of a sizeable, permanent continental crust prior to $c.3.8$ Ga.

B.S.K. is supported by Swiss National Science Foundation grant 8220-050352 and M.J.W. acknowledges support from the Swedish Natural Science Research Council (NFR). R. Goodwin provided expert technical assistance with the Greenland Pb-isotope analyses. We thank C. L. E. Lewis and N. Rogers for their helpful reviews.

References

ALBARÈDE, F., BLICHERT-TOFT, J., VERVOORT, J. D., GLEASON, J. D. & ROSING, M. 2000. Hf-Nd evidence for a transient dynamic regime in the early terrestrial mantle. *Nature*, **404**, 488–490.

ALEINIKOFF, J. N., WILLIAMS, I. S., COMPSTON, W., STUCKLESS, J. S. & WORL, R. G. 1989. Evidence for an early Archean component in the middle to late Archean gneisses of the Wind River Range, West-Central Wyoming – conventional and ion microprobe U-Pb data. *Contributions to Mineralogy and Petrology*, **101**, 198–206.

ALLAART, J. H. 1976. The pre-3760 old supracrustal rocks of the Isua area, central West Greenland, and the associated occurrence of quartz-banded ironstone. *In:* WINDLEY, B. F. (ed.) *The Early History of the Earth.* Wiley, London, 177–189.

AMELIN, Y. V. 1998. Geochronology of the Jack Hills detrital zircons by precise U-Pb isotope dilution analysis of crystal fragments. *Chemical Geology*, **146**, 25–38.

AMELIN, Y., LEE, D. C., HALLIDAY, A. N. & PIDGEON, R. T. 1999. Nature of the Earth's earliest crust from hafnium isotopes in single detrital zircons. *Nature*, **399**, 252–255.

AMELIN, Y., LEE, D. C. & HALLIDAY, A. N. 2000. Early-middle Archaean crustal evolution deduced from Lu-Hf and U-Pb isotopic studies of single zircon grains. *Geochimica et Cosmochimica Acta*, **64**, 4205–4225.

APPEL, P. W. U. & MOORBATH, S. 1999. Exploring earth's oldest geological record in Greenland. *EOS Transactions of the American Geophysical Union*, **80**, 257–264.

APPEL, P. W. U., MOORBATH, S. & TAYLOR, P. N. 1978. Least radiogenic terrestrial lead from Isua, West Greenland. *Nature*, **272**, 524–526.

APPEL, P. W. U., FEDO, C. M., MOORBATH, S. & MYERS, J. S. 1998. Recognizable primary volcanic and sedimentary features in a low-strain domain of the highly deformed, oldest known (3.7–3.8 Gyr) greenstone belt, Isua, West Greenland. *Terra Nova*, **10**, 57–62.

ARMSTRONG, R. L. 1991. The persistent myth of crustal growth. *Australian Journal of Earth Sciences*, **38**, 613–630.

ARTH, J. G. & BARKER, F. 1976. Rare-earth partitioning between hornblende and dacitic liquid and implications for genesis of trondhjemitic-tonalitic magmas. *Geology*, **4**, 534–546.

BAADSGAARD, H. 1973. U-Th-Pb dates on zircons from the early Precambrian Amîtsoq gneisses, Godthaab District, West Greenland. *Earth and Planetary Science Letters*, **19**, 22–28.

BAADSGAARD, H., LAMBERT, R. S. & KRUPICKA, J. 1976. Mineral isotopic age relationships in the polymetamorphic Amîtsoq gneisses, Godthaab District, West Greenland. *Geochimica et Cosmochimica Acta*, **40**, 513–527.

BAADSGAARD, H., NUTMAN, A. P., BRIDGWATER, D., ROSING, M., MCGREGOR, V. R. & ALLAART, J. H. 1984. The zircon geochronology of the Akilia association and Isua supracrustal belt. *Earth and Planetary Science Letters*, **68**, 221–228.

BAADSGAARD, H., NUTMAN, A. P. & BRIDGWATER, D. 1986a. Geochronology and isotopic variation of the early Archaean Amîtsoq gneisses of the Isukasia area, southern West Greenland. *Geochimica et Cosmochimica Acta*, **50**, 2173–2183.

BAADSGAARD, H., NUTMAN, A. P., ROSING, M., BRIDGWATER, D. & LONGSTAFFE, F. J. 1986b. Alteration and metamorphism of Amîtsoq gneisses from the Isukasia area, West Greenland: Recommendations for isotope studies of the early crust. *Geochimica et Cosmochimica Acta*, **50**, 2165–2172.

BENNETT, V. C. & NUTMAN, A. P. 1998. Extreme Nd isotope heterogeneity in the early Archean – fact or fiction? Case histories from northern Canada and West Greenland – Comment. *Chemical Geology*, **148**, 213–217.

BENNETT, V. C., NUTMAN, A. P. & MCCULLOCH, M. T. 1993. Nd isotopic evidence for transient, highly depleted mantle reservoirs in the early history of the Earth. *Earth and Planetary Science Letters*, **119**, 299–317.

BIAO, S., NUTMAN, A. P., DUNYI, L. & JIASHAN, W. 1996. 3800–2500 Ma evolution in the Anshan area of Liaoning Province, northeastern China. *Precambrian Research*, **78**, 79–94.

BICKLE, M. J. 1986. Implications of melting for stabilisation of the lithosphere and heat loss in the Archaean. *Earth and Planetary Science Letters*, **80**, 314–324.

BLACK, L. P. & MCCULLOCH, M. T. 1987. Evidence for isotopic equilibration of Sm-Nd whole-rock systems in early Archean crust of Enderby Land, Antarctica. *Earth and Planetary Science Letters*, **82**, 15–24.

BLACK, L. P., GALE, N. H., MOORBATH, S., PANKHURST, R. J. & MCGREGOR, V. R. 1971. Isotopic dating of the very early Precambrian amphibolite gneisses from the Godthaab District, West Greenland. *Earth and Planetary Science Letters*, **12**, 245–249.

BLACK, L. P., WILLIAMS, I. S. & COMPSTON, W. 1986. Four zircon ages from one rock – the history of a 3930 Ma-old granulite from Mount Sones, Enderby Land, Antarctica. *Contributions to Mineralogy and Petrology*, **94**, 427–437.

BLICHERT-TOFT, J., ALBARÈDE, F., ROSING, M., FREI, R. & BRIDGWATER, D. 1999. The Nd and Hf isotopic evolution of the mantle through the Archean. Results from the Isua supracrustals, West Greenland, and from the Birimian terranes of West Africa. *Geochimica et Cosmochimica Acta*, **63**, 3901–3914.

BOAK, J. L. & DYMEK, R. F. 1982. Metamorphism of the ca. 3800 Ma supracrustal rocks at Isua, West Greenland: implications for early Archaean crustal evolution. *Earth and Planetary Science Letters*, **59**, 155–176.

BOWRING, S. A. & HOUSH, T. 1995. The earth's early evolution. *Science*, **269**, 1535–1540.

BOWRING, S. A. & WILLIAMS, I. S. 1999. Priscoan (4.00–4.03 Ga) orthogneisses from northwestern Canada. *Contributions to Mineralogy and Petrology*, **134**, 3–16.

BOWRING, S. A., WILLIAMS, I. S. & COMPSTON, W. 1989. 3.96 Ga gneisses from the Slave province, Northwest Territories, Canada. *Geology*, **17**, 971–975.

BRIDGWATER, D. & COLLERSON, K. D. 1976. The major petrological and geochemical characters of the 3600 Ma Uivak gneisses from Labrador. *Contributions to Mineralogy and Petrology*, **54**, 43–59.

BRIDGWATER, D., COLLERSON, K. D. & MYERS, J. S. 1978. The development of the Archaean gneiss complex of the North Atlantic region. *In*: TARLING, D. H. (ed.) *Evolution of the Earth's Crust*. Academic Press, London, 19–69.

BRUGUIER, O., DADA, S. & LANCELOT, J. R. 1994. Early Archean component (>3.5 Ga) within a 3.05-Ga orthogneiss from northern Nigeria: U-Pb zircon evidence. *Earth and Planetary Science Letters*, **125**, 89–103.

BUICK, R., THORNETT, J. R., MCNAUGHTON, N. J., SMITH, J. B., BARLEY, M. E. & SAVAGE, M. 1995. Record of emergent continental crust c. 3.5 billion years ago in the Pilbara Craton of Australia. *Nature*, **375**, 574–577.

CHYBA, C. F. 1993. The violent environment of the origin of life: progress and uncertainties. *Geochimica et Cosmochimica Acta*, **57**, 3351–3358.

COHEN, B. A., SWINDLE, T. D. & KING, D. A. 2000. Support for the lunar cataclysm hypothesis from lunar meteorite impact melt ages. *Science*, **290**, 1754–1756.

COLLERSON, K. D. & BRIDGWATER, D. 1979. Metamorphic development of early Archaean tonalitic and trondhjemitic gneisses, Saglek area, Labrador. *In*: BARKER, F. (ed.) *Trondhjemites, Dacites and Related rocks*. Elsevier, Amsterdam, 205–273.

COLLERSON, K. D. & KAMBER, B. S. 1999. Evolution of the continents and the atmosphere inferred from Th-U-Nb systematics of the depleted mantle. *Science*, **283**, 1519–1522.

COLLERSON, K. D., CAMPBELL, L. M., WEAVER, B. L. & ZENON, A. P. 1991. Evidence for extreme mantle fractionation in early Archaean ultramafic rocks from northern Labrador. *Nature*, **349**, 209–214.

COMPSTON, W. & KRÖNER, A. 1988. Multiple zircon growth within early Archaean tonalitic gneiss from the Ancient Gneiss Complex, Swaziland. *Earth and Planetary Science Letters*, **87**, 13–28.

COMPSTON, W. & PIDGEON, R. T. 1986. Jack Hills, evidence of more very old detrital zircons in Western Australia. *Nature*, **321**, 766–769.

COMPSTON, W., KINNY, P. D., WILLIAMS, I. S. & FOSTER, J. J. 1986. The age and Pb loss behaviour of zircons from the Isua supracrustal belt as determined by ion microprobe. *Earth and Planetary Science Letters*, **80**, 71–81.

DE WIT, M., ROERING, C., HART, J. R. ET AL. 1992. Formation of an Archaean continent. *Nature*, **357**, 553–562.

DODSON, M. H., COMPSTON, W., WILLIAMS, I. S. & WILSON, J. F. 1988. A search for ancient detrital zircons in Zimbabwean sediments. *Journal of the Geological Society, London*, **145**, 977–983.

ELLIOTT, T., ZINDLER, A. & BOURDON, B. 1999. Exploring the kappa conundrum: the role of recycling in the lead isotope evolution of the mantle. *Earth and Planetary Science Letters*, **169**, 129–145.

FEDO, C. M. 2000. Setting and origin for problematic rocks from the >3.7 Ga. Isua Greenstone Belt, southern West Greenland: Earth's oldest coarse clastic sediments. *Precambrian Research*, **101**, 65–78.

FLETCHER, I. R., ROSMAN, K. J. R. & LIBBY, W. G. 1988. Sm–Nd, Pb–Pb and Rb–Sr geochronology of the Manfred Complex, Mount Narryer, Western-Australia. *Precambrian Research*, **38**, 343–354.

FREI, R., BRIDGWATER, D., ROSING, M. & STECHER, O. 1999. Controversial Pb-Pb and Sm-Nd isotope results in the early Archean Isua (West Greenland) oxide iron formation: preservation of primary signatures versus secondary disturbances. *Geochimica et Cosmochimica Acta*, **63**, 473–488.

FROUDE, C. F., IRELAND, T. R., KINNY, P. D., WILLIAMS, I. S., COMPSTON, W., WILLIAMS, I. R. & MYERS, J. S. 1983. Ion-microprobe identification of 4100–4200 Myr old terrestrial zircons. *Nature*, **304**, 616–618.

GANCARZ, A. J. & WASSERBURG, G. J. 1977. Initial Pb of the Amîtsoq gneiss, West Greenland, and implications for the age of the Earth. *Geochimica et Cosmochimica Acta*, **41**, 1283–1301.

GRIFFIN, W. L., MCGREGOR, V. R., NUTMAN, A., TAYLOR, P. N. & BRIDGWATER, D. 1980. Early Archaean metamorphism south of Ameralik, West Greenland. *Earth and Planetary Science Letters*, **50**, 59–74.

GRUAU, G., ROSING, M. D. B. & GILL, R. C. O. 1996. Metamorphism and Sm-Nd remobilization in the >3.7 Ga Isua supracrustal belt of West Greenland: implications for isotope models of early earth differentiation. *Chemical Geology*, **133**, 225–240.

HALLIDAY, A. N. 2001. In the beginning. *Nature*, **409**, 144–145.

HARPER, C. L. & JACOBSEN, S. B. 1992. Evidence from coupled ^{147}Sm-^{143}Nd and ^{146}Sm-^{142}Nd systematics for very early (4.5 Ga) differentiation of the Earth's mantle. *Nature*, **360**, 728–732.

HORSTWOOD, M. S. A., NESBITT, R. W., NOBLE, S. R. & WILSON, J. F. 1999. U-Pb zircon evidence for an extensive early Archean craton in Zimbabwe: A reassessment of the timing of craton formation, stabilization, and growth. *Geology*, **27**, 707–710.

JACOBSEN, S. B. & DYMEK, R. F. 1988. Nd and Sr isotope systematics of clastic metasediments from Isua, West Greenland: identification of pre-3.8 Ga differentiated crustal components. *Journal of Geophysical Research*, **93**, 338–354.

JACOBSEN, S. B. & HARPER, C. L. 1996. Comment on 'The issue of the terrestrial record of ^{146}Sm'. *Geochimica et Cosmochimica Acta*, **60**, 3747–3749.

KAMBER, B. S. & COLLERSON, K. D. 2000. Variability of Nb/U and Th/La in 3.0 to 2.7 Ga Superior Province ocean plateau basalts: implications for the timing of continental growth and lithosphere recycling – Comment. *Earth and Planetary Science Letters*, **177**, 337–339.

KAMBER, B. S. & MOORBATH, S. 1998. Initial Pb of the Amîtsoq gneiss revisited: implication for the timing of early Archaean crustal evolution in West Greenland. *Chemical Geology*, **150**, 19–41.

KAMBER, B. S., MOORBATH, S. & WHITEHOUSE, M. J. 1998. Extreme Nd-isotope heterogeneity in the early Archaean – fact or fiction? Case histories from northern Canada and West Greenland – Reply. *Chemical Geology*, **148**, 219–224.

KERRICH, R., WYMAN, D., HOLLINGS, P. & POLAT, A. 1999. Variability of Nb/U and Th/La in 3.0 to 2.7 Ga Superior Province ocean plateau basalts: implications for the timing of continental growth and lithosphere recycling. *Earth and Planetary Science Letters*, **168**, 101–115.

KINNY, P. D. & NUTMAN, A. P. 1996. Zirconology of the Meeberrie Gneiss, Yilgarn Craton, Western Australia: An early Archaean migmatite. *Precambrian Research*, **78**, 165–178.

KINNY, P. D., WILLIAMS, I. S., FROUDE, D. O., IRELAND, T. R. & COMPSTON, W. 1988. Early Archean zircon ages from orthogneisses and anorthosites at Mount Narryer, Western-Australia. *Precambrian Research*, **38**, 325–341.

KINNY, P. D., COMPSTON, W. & WILLIAMS, I. S. 1991. A reconnaissance ion-probe study of hafnium isotopes in zircons. *Geochimica et Cosmochimica Acta*, **55**, 849–859.

KRAMERS, J. D. 1998. Reconciling siderophile element data in the Earth and Moon; W isotopes and the upper lunar age limit in a simple model of homogeneous accretion. *Chemical Geology*, **145**, 461–478.

KRAMERS, J. D. & TOLSTIKHIN, I. N. 1997. Two terrestrial lead isotope paradoxes, forward transport modelling, core formation and the history of the continental crust. *Chemical Geology*, **139**, 75–110.

KRAMERS, J. D., NÄGLER, T. F. & TOLSTIKHIN, I. N. 1998. Perspectives from global modelling of terrestrial Pb and Nd isotopes on the history of the continental crust. *Schweizerische Mineralogisch-Petrographische Mitteilungen*, **78**, 169–174.

KRÖNER, A. & TEGTMEYER, A. 1994. Gneiss-greenstone relationships in the Ancient Gneiss Complex of southwestern Swaziland, southern Africa, and implications for early crustal evolution. *Precambrian Research*, **67**, 109–139.

KRÖNER, A. & TODT, W. 1988. Single zircon dating constraining the maximum age of the Barberton greenstone belt, southern Africa. *Journal of Geophysical Research*, **93**, 15 329–15 337.

KRÖNER, A., BYERLY, G. R. & LOWE, D. R. 1991. Chronology of early Archaean granite-greenstone evolution in the Barberton Mountain Land, South Africa, based on precise dating by zircon evaporation. *Earth and Planetary Science Letters*, **103**, 41–54.

KRÖNER, A., HEGNER, E., WENDT, J. I. & BYERLY, G. R. 1996. The oldest part of the Barberton granitoid-greenstone terrain, South Africa: Evidence for crust formation between 3.5 and 3.7 Ga. *Precambrian Research*, **78**, 105–124.

LIU, D. Y., SHEN, Q. H., ZHANG, Z. Q., JAHN, B. M. & AUVRAY, B. 1990. Archaean crustal evolution in China: U-Pb geochronology of the Qianxi Complex. *Precambrian Research*, **48**, 223–244.

LIU, D. Y., NUTMAN, A. P., COMPSTON, W., WU, J. S. & SHEN, Q. H. 1992. Remnants of ≥3800 Ma crust in the Chinese part of the Sino-Korean craton. *Geology*, **20**, 339–342.

McCulloch, M. T. & Bennett, V. C. 1994. Progressive growth of the Earth's continental crust and depleted mantle: Geochemical constraints. *Geochimica et Cosmochimica Acta*, **58**, 4717–4738.

McGregor, V. R. 1973. The early Precambrian gneisses of the Godthaab district, West Greenland. *Philosophical Transactions of the Royal Society of London*, **A273**, 343–358.

McGregor, V. R. 2000. Initial Pb of the Amîtsoq gneiss revisited: implications for the timing of early Archaean crustal evolution in West Greenland – Comment. *Chemical Geology*, **166**, 301–308.

McGregor, V. R. & Mason, B. 1977. Petrogenesis and geochemistry of metabasaltic and metasedimentary enclaves in the Amîtsoq gneisses, West Greenland. *American Mineralogist*, **62**, 887–904.

Maas, R. & McCulloch, M. T. 1991. The provenance of Archean clastic metasediments in the Narryer Gneiss Complex, Western Australia – trace-element geochemistry, Nd isotopes, and U-Pb ages for detrital zircons. *Geochimica et Cosmochimica Acta*, **55**, 1915–1932.

Maas, R., Kinny, P. D., Williams, I. S., Froude, D. O. & Compston, W. 1992. The Earth's oldest known crust: A geochronological and geochemical study of 3900–4200 Ma old detrital zircons from Mt. Narryer and Jack Hills, Western Australia. *Geochimica et Cosmochimica Acta*, **56**, 1281–1300.

Martin, H. 1986. Effect of steeper Archaean geothermal gradient on geochemistry of subduction zone magmas. *Geology*, **14**, 753–756.

Meyer, C., Williams, I. S. & Compston, W. 1996. Uranium-lead ages for lunar zircons: Evidence for a prolonged period of granophyre formation from 4.32 to 3.88 Ga. *Meteoritics and Planetary Sciences*, **31**, 370–387.

Mezger, K. & Krogstad, E. J. 1997. Interpretation of discordant U-Pb zircon ages: An evaluation. *Journal of Metamorphic Geology*, **15**, 127–140.

Miller, C. F., Hatcher, R. D., Ayers, J. C., Coath, C. D. & Harrison, T. M. 2000. Age and zircon inheritance of the Eastern Blue Ridge plutons, southwestern North Carolina and northeastern Georgia, with implications for magma history and the evolution of the Southern Appalachian orogen. *American Journal of Science*, **300**, 142–172.

Mojzsis, S. J. & Harrison, T. M. 1999. Geochronological studies of the oldest known marine sediments. *In: Ninth Annual, V. M. Goldschmidt Conference*. Lunar and Planetary Institute, Houston, Contribution No. 971, 201–202.

Mojzsis, S. J., Arrhenius, G., McKeegan, K. D., Harrison, T. M., Nutman, A. P. & Friend, C. R. L. 1996. Evidence for life on Earth before 3800 million years ago. *Nature*, **384**, 55–59.

Mojzsis, S. J., Harrison, T. M., & Pidgeon, R. T. 2001. Oxygen-isotope evidence from ancient zircons for liquid water at the Earth's surface 4,300 Myr ago. *Nature*, **409**, 178–181.

Moorbath, S. & Kamber, B. S. 1998. A reassessment of the timing of early Archaean crustal evolution in West Greenland. *Geology of Greenland Survey Bulletin*, **180**, 88–93.

Moorbath, S., O'Nions, R. K., Pankhurst, R. J., Gale, N. H. & McGregor, V. R. 1972. Further rubidium-strontium age determinations on the very early Precambrian rocks of the Godthaab district, West Greenland. *Nature*, **240**, 78–82.

Moorbath, S., O'Nions, R. K. & Pankhurst, R. J. 1973. Early Archaean age for the Isua Iron Formation, West Greenland. *Nature*, **245**, 138–139.

Moorbath, S., O'Nions, R. K. & Pankhurst, R. J. 1975. The evolution of early Precambrian crustal rocks at Isua, West Greenland – geochemical and isotopic evidence. *Earth and Planetary Science Letters*, **27**, 229–239.

Moorbath, S., Allaart, J. H., Bridgwater, D. & McGregor, V. R. 1977. Rb-Sr ages of early Archaean supracrustal rocks and Amîtsoq gneisses at Isua. *Nature*, **270**, 43–45.

Moorbath, S., Taylor, P. N. & Jones, N. W. 1986. Dating the oldest terrestrial rocks – fact and fiction. *Chemical Geology*, **57**, 63–86.

Moorbath, S., Whitehouse, M. J. & Kamber, B. S. 1997. Extreme Nd-isotope heterogeneity in the early Archaean – fact or fiction? Case histories from northern Canada and West Greenland. *Chemical Geology*, **135**, 213–231.

Mueller, P. A., Wooden, J. L. & Nutman, A. P. 1992. 3.96 Ga zircons from an Archean quartzite, Beartooth Mountains, Montana. *Geology*, **20**, 327–330.

Mueller, P. A., Wooden, J. L., Mogk, D. W., Nutman, A. P. & Williams, I. S. 1996. Extended history of a 3.5 Ga trondhjemitic gneiss, Wyoming province, USA: Evidence from U-Pb systematics in zircon. *Precambrian Research*, **78**, 41–52.

Myers, J. S. & Crowley, J. L. 2000. Vestiges of life in the oldest Greenland rocks? A review of early Archaean geology in the Godthaabsfjord region, and reappraisal of field evidence for >3850 Ma life on Akilia. *Precambrian Research*, **103**, 101–124.

Nägler, T. F. & Kramers, J. D. 1998. Nd isotopic evolution of the upper mantle during the Precambrian: Models, data and the uncertainty of both. *Precambrian Research*, **91**, 233–252.

Nägler, T. F., Kramers, J. D., Kamber, B. S., Frei, R. & Prendergast, M. D. A. 1997. Growth of subcontinental lithospheric mantle beneath Zimbabwe started at or before 3.8 Ga: A Re-Os study on chromites. *Geology*, **25**, 983–986.

Nelson, D. R., Trendall, A. F. & Altermann, W. 1999. Chronological correlations between the Pilbara and Kaapvaal cratons. *Precambrian Research*, **97**, 165–189.

Nir-El, Y. & Lavi, N. 1998. Measurement of the half-life of ^{176}Lu. *Applied Radiation and Isotopes*, **49**, 1653–1655.

Nutman, A. P., Allaart, J. H., Bridgwater, D., Dimroth, E. & Rosing, M. 1984. Stratigraphic and geochemical evidence for the depositional environment of the early Archaean Isua supracrustal belt, Southern West Greenland. *Precambrian Research*, **25**, 365–396.

Nutman, A. P., Fryer, B. J. & Bridgwater, D. 1989. The early Archean Nulliak (supracrustal) assemblage, Northern Labrador. *Canadian Journal of Earth Sciences*, **26**, 2159–2168.

NUTMAN, A. P., MCGREGOR, V. R., FRIEND, C. R. L., BENNETT, V. C. & KINNY, P. D. 1996. The Itsaq Gneiss Complex of southern West Greenland; the world's most extensive record of early crustal evolution (3,900–3,600 Ma). *Precambrian Research*, **78**, 1–39.

NUTMAN, A. P., MOJZSIS, S. J. & FRIEND, C. R. L. 1997a. Recognition of ≥3850 Ma water-lain sediments in West Greenland and their significance for the early Archaean Earth. *Geochimica et Cosmochimica Acta*, **61**, 2475–2484.

NUTMAN, A. P., BENNETT, V. C., FRIEND, C. R. L. & ROSING, M. T. 1997b. ~3710 and ≥3790 Ma volcanic sequences in the Isua (Greenland) supracrustal belt; structural and Nd isotope implications. *Chemical Geology*, **141**, 271–287.

NUTMAN, A. P., BENNETT, V. C., FRIEND, C. R. L. & NORMAN, M. D. 1999. Meta-igneous (non-gneissic) tonalites and quartz-diorites from an extensive ca. 3800 Ma terrain south of the Isua supracrustal belt, southern West Greenland: constraints on early crust formation. *Contributions to Mineralogy and Petrology*, **137**, 364–388.

NUTMAN, A. P., BENNETT, V. C., FRIEND, C. R. L. & MCGREGOR, V. R. 2000. The early Archaean Itsaq Gneiss Complex of southern West Greenland: the importance of field observations in interpreting age and isotopic constraints for early terrestrial evolution. *Geochimica et Cosmochimica Acta*, **64**, 3035–3060.

NYQUIST, L. E. & SHIH, C. Y. 1992. The isotopic record of lunar volcanism. *Geochimica et Cosmochimica Acta*, **56**, 2213–2234.

PATCHETT, P. J. 1983. Importance of the Lu-Hf isotopic system in studies of planetary chronology and chemical evolution. *Geochimica et Cosmochimica Acta*, **47**, 81–91.

PATCHETT, P. J. & VERVOORT, J. D. 2000. The poorly determined decay constant of ^{176}Lu. AGU Spring Meeting, Baltimore. American Geophysical Union, S445.

PEARSON, D. G., CARLSON, R. W., SHIREY, S. B., BOYD, F. R. & NIXON, P. H. 1995a. Stabilisation of Archaean lithospheric mantle: A Re-Os isotope study of peridotite xenoliths from the Kaapvaal craton. *Earth and Planetary Science Letters*, **134**, 341–357.

PEARSON, D. G., SNYDER, G. A., SHIREY, S. B., TAYLOR, L. A., CARLSON, R. W. & SOBOLEV, N. V. 1995b. Archean Re-Os age for Siberian eclogites and constraints on Archean tectonics. *Nature*, **374**, 711–713.

PIDGEON, R. T. & COMPSTON, W. 1992. A SHRIMP ion microprobe study of inherited and magmatic zircons from four Scottish Caledonian granites. *Transactions of the Royal Society of Edinburgh*, **83**, 473–483.

POTREL, A., PEUCAT, J. J., FANNING, C. M., AUVRAY, B., BURG, J. P. & CARUBA, C. 1996. 3.5 Ga old terranes in the West African Craton, Mauritania. *Journal of the Geological Society, London*, **153**, 507–510.

RICHARDS, J. R. & APPEL, P. W. U. 1987. Age of the least radiogenic galenas at Isua, West Greenland. *Chemical Geology*, **66**, 181–191.

ROSE, N. M., ROSING, M. T. & BRIDGWATER, D. 1996. The origin of metacarbonate rocks in the Archaean Isua supracrustal belt, West Greenland. *American Journal of Science*, **296**, 1004–1044.

ROSING, M. T. 1999. ^{13}C-depleted carbon microparticles in >3700-Ma sea-floor sedimentary rocks from west Greenland. *Science*, **283**, 674–676.

ROSING, M., ROSE, N. M., BRIDGWATER, D. & THOMSEN, H. S. 1996. Earliest part of Earth's stratigraphic record: a reappraisal of the >3.7 Ga Isua (Greenland) supracrustal sequence. *Geology*, **24**, 43–46.

RYDER, G. 1990. Lunar samples, lunar accretion and the early bombardment of the Moon. *EOS (Transactions of the American Geophysical Union)*, **71**, 322–323.

SCHIDLOWSKI, M. 1988. A 3800-Million-Year isotopic record of life from carbon in sedimentary rocks. *Nature*, **333**, 313–318.

SCHIDLOWSKI, M., APPEL, P. W. U., EICHMANN, R. & JUNGE, C. E. 1979. Carbon isotope geochemistry of the 3700 myr-old Isua sediments, West Greenland: Implications for the Archaean carbon and oxygen cycle. *Geochimica et Cosmochimica Acta*, **43**, 189–199.

SCHIØTTE, L. & COMPSTON, W. 1990. U-Pb age pattern for single zircon from the early Archaean Akilia association south of Ameralik fjord, southern West Greenland. *Chemical Geology*, **80**, 147–157.

SCHIØTTE, L., BRIDGWATER, D., COLLERSON, K. D., NUTMAN, A. P. & RYAN, A. B. 1986. Chemical and isotopic effects of late Archaean high-grade metamorphism and granite injection on early Archaean gneisses, Saglek-Hebron, northern Labrador. *In*: DAWSON, J. B., CARSWELL, D. A., HALL, J. & WEDEPOHL, H. H. (eds) *The Nature of the Lower Continental Crust*. Geological Society, London, Special Publications, **24**, 261–273.

SCHIØTTE, L., COMPSTON, W. & BRIDGWATER, D. 1989. Ion probe U-Th-Pb zircon dating of polymetamorphic orthogneisses from northern Labrador. *Canadian Journal of Earth Sciences*, **26**, 1533–1556.

SHARMA, M., PAPANASTASSIOU, D. A., WASSERBURG, G. J. & DYMEK, R. F. 1996. The issue of the terrestrial record of ^{146}Sm. *Geochimica et Cosmochimica Acta*, **60**, 2037–2047.

SHIMIZU, H., NAKAI, S., TASAKI, S., MASUDA, A., BRIDGWATER, D., NUTMAN, A. P. & BAADSGAARD, H. 1988. Geochemistry of Ce and Nd isotopes and REE abundances in the Amîtsoq gneisses, West Greenland. *Earth and Planetary Science Letters*, **91**, 159–169.

SHIREY, S. B. & WALKER, R. J. 1998. The Re-Os isotope system in cosmochemistry and high-temperature geochemistry. *Annual Reviews in Earth and Planetary Sciences*, **26**, 423–500.

SÖDERLUND, U., MÖLLER, C., ANDERSSON, J., JOHANSSON, L. & WHITEHOUSE, M. (in press). Zircon geochronology in polymetamorphic gneisses in the Sveconorwegian orogen, SW Sweden: ion microprobe evidence for 1.46–1.42 and 0.98–0.96 Ga reworking. *Precambrian Research*.

STACEY, J. S. & KRAMERS, J. D. 1975. Approximation of terrestrial lead isotope evolution by a two-stage model. *Earth and Planetary Science Letters*, **26**, 207–221.

STERN, R. A. & BLEEKER, W. 1998. Age of the world's oldest rocks refined using Canada's SHRIMP: The Acasta gneiss complex, Northwest Territories, Canada. *Geosciences Canada*, **25**, 27–31.

STERN, R. A. & HANSON, G. N. 1991. Archean high-Mg granodiorite – a derivative of light rare-earth-element-enriched monzodiorite of mantle origin. *Journal of Petrology*, **32**, 201–238.

SYLVESTER, P. J., CAMPBELL, I. H. & BOWYER, D. A. 1997. Niobium/uranium evidence for early formation of the continental crust. *Science*, **275**, 521–523.

TATSUMOTO, M., PREMO, W. R. & UNRUH, D. M. 1987. Origin of lead from green glass of Apollo-15426 – a search for primitive lunar lead. *Journal of Geophysical Research*, **92**, E361–E371.

TAYLOR, P. N., KRAMERS, J. D., MOORBATH, S., WILSON, J. F., ORPEN, J. L. & MARTIN, A. 1991. Pb/Pb, Sm-Nd and Rb-Sr geochronology in the Archean Craton of Zimbabwe. *Chemical Geology*, **87**, 175–196.

VERVOORT, J. D. & BLICHERT-TOFT, J. 1999. Evolution of the depleted mantle: Hf isotope evidence from juvenile rocks through time. *Geochimica et Cosmochimica Acta*, **63**, 533–556.

VILLA, I. M., KAMBER, B. S. & NÄGLER, T. N. 2001. The Nd and Hf isotopic evolution of the mantle through the Archean. Results from the Isua supracrustals, West Greenland, and from the Birimian terranes of West Africa – comment. *Geochimica et Cosmochimica Acta*, **65**, 2017–2021.

WENDT, J. I. & COLLERSON, K. D. 1999. Early Archaean U/Pb fractionation and timing of late Archaean high-grade metamorphism in the Saglek-Hebron segment of the North Atlantic Craton. *Precambrian Research*, **93**, 281–297.

WHITEHOUSE, M., KAMBER, B. S. & MOORBATH, S. 1999. Age significance of ion-microprobe U-Th-Pb zircon data from early Archaean rocks of West Greenland – a reassessment based on new combined ion-microprobe and imaging studies. *Chemical Geology*, **160**, 204–221.

WHITEHOUSE, M. J., NÄGLER, T. F., MOORBATH, S., KRAMERS, J. D., KAMBER, B. S. & FREI, R. 2001. Discussion: Priscoan (4.00–4.03 Ga) orthogneisses from northwestern Canada. *Contributions to Mineralogy and Petrology*, **141**, 248–250.

WILDE, S. A., VALLEY, J. W., PECK, W. H. & GRAHAM, C. M. 2001. Evidence from detrital zircons for the existence of continental crust and oceans on the Earth 4.4 Gyr ago. *Nature*, **409**, 175–178.

WILHELMS, D. E. 1987. *The Geologic History of the Moon*. United States Geological Survey Professional Papers, **1348**.

YAMASHITA, K., CREASER, R. A., JENSEN, J. E. & HEAMAN, L. M. 2000. Origin and evolution of mid- to late-Archean crust in the Hanikahimajuk Lake area, Slave Province, Canada; evidence from U-Pb geochronological, geochemical and Nd-Pb isotopic data. *Precambrian Research*, **99**, 197–224.

ZARTMAN, R. E. & HAINES, S. 1988. The plumbotectonic model for Pb isotopic systematics among major terrestrial reservoirs – A case for bi-directional transport. *Geochimica et Cosmochimica Acta*, **52**, 1327–1339.

The age of the Earth in the twentieth century: a problem (mostly) solved

G. BRENT DALRYMPLE

College of Oceanic and Atmospheric Sciences, 104 Ocean Admin Building, Oregon State University, Corvallis, OR 97330–5503, USA
(*email*: bdalrymple@attglobal.net)

Abstract: In the early twentieth century the Earth's age was unknown and scientific estimates, none of which were based on valid premises, varied typically from a few millions to billions of years. This important question was answered only after more than half a century of innovation in both theory and instrumentation. Critical developments along this path included not only a better understanding of the fundamental properties of matter, but also: (a) the suggestion and first demonstration by Rutherford in 1904 that radioactivity might be used as a geological timekeeper; (b) the development of the first mass analyser and the discovery of isotopes by J. J. Thomson in 1914; (c) the idea by Russell in 1921 that the age of a planetary reservoir like the Earth's crust might be measured from the relative abundances of a radioactive parent element (uranium) and its daughter product (lead); (d) the development of the idea by Gerling in 1942 that the age of the Earth could be calculated from the isotopic composition of a lead ore of known age; (e) the ideas of Houtermans and Brown in 1947 that the isotopic composition of primordial lead might be found in iron meteorites; and (f) the first calculation by Patterson in 1953 of a valid age for the Earth of 4.55 Ga, using the primordial meteoritic lead composition and samples representing the composition of modern Earth lead. The value for the age of the Earth in wide use today was determined by Tera in 1980, who found a value of 4.54 Ga from a clever analysis of the lead isotopic compositions of four ancient conformable lead deposits. Whether this age represents the age of the Earth's accretion, of core formation, or of the material from which the Earth formed is not yet known, but recent evidence suggests it may approximate the latter.

In 1904 George F. Becker, head of the division of chemical and physical research of the US Geological Survey, addressed the International Congress of Arts and Sciences (ICAS) on the unresolved and important problems in geophysics (Becker 1904). At the time there was probably no greater controversy in geology than the age of planet Earth and Becker gave the problem its due. After briefly discussing the results of Lord Kelvin (William Thomson) and Clarence King, both of whom had calculated ages of the Earth of less than 100 million years based on cooling from a set of presumed initial conditions, Becker commented:

> These researches, together with Helmholtz's [1856] investigation on the age of the solar system, which is incomplete for lack of knowledge of the distribution of density in the sun, have had a restraining influence on the estimates drawn from sedimentation by geologists. Many and perhaps most geologists now regard something less than 100 million years as sufficient for the development of geological phenomena. Yet the subject can not be regarded as settled until our knowledge of conductivities is more complete. An iron nucleus, for example, would imply greater conductivity of the interior and a higher age for the earth than that computed by King, though probably well within the range [of 20–400 Ma] explicitly allowed by Lord Kelvin in view of the uncertainty of this datum.

Later in his paper Becker, after commenting on the importance of knowledge of the thermal conductivity and specific heat in understanding deformation, remarked that: 'The data for constitution and thermal diffusivity will readily be applicable to the problem of the earth's age and will yield a corrected value of the probable lapse of time since the initiation of the consistentior status of the Protogaea'.

It seems relatively clear from these two statements that Becker thought that the method (cooling of the Earth) for solving the problem was in hand and that the solution only awaited

---

Parts of this paper are excerpts from Dalrymple, G. B., *The Age of the Earth*, Stanford University Press, © 1991 by the Board of Trustees of the Leland Stanford Junior University, and are included with permission.

the measurement of a few relatively simple geophysical parameters. Becker was wrong on both counts. The answer to the question of the age of the Earth would eventually come from methods based not on cooling, sedimentation or any of the other methods that were popular around the turn of the century, but on the newly discovered phenomenon of radioactivity. And to top it off, the answer would not be forthcoming for another half century.

The reason that it took five decades for scientists to solve the riddle of the age of the Earth is simple – it was that long before the concepts and instrumentation that were necessary to address the problem adequately were available. Between 1904 and the mid-1950s, when the answer was finally revealed, there were a number of developments that were critical to finding the age of the Earth and a substantial number of scientists contributed to the quest. The purpose of this paper is to review what, in the author's opinion, were the most significant advances in thought and instrumentation that finally allowed scientists to show convincingly that the age of the Earth is, to within an error of only 1% or so, 4.5 Ga.

The age of the Earth in the nineteenth century

In the 1800s and the early 1900s there were four principal methods used to calculate the age of the Earth and solar system. These included thermal calculations, orbital physics, change in ocean chemistry, and erosion and sedimentation. Thermal methods included calculating the time required for the Earth to cool from an initial (usually) molten state or for the Sun to exhaust its fuel by ordinary combustion. Orbital physics involved finding orbital lifetimes of planetary bodies, primarily the Moon, from known tidal effects. Methods based on changes in ocean chemistry were usually based on the increase over time of the concentration of an element, commonly sodium, in the oceans. And erosion and sedimentation methods usually involved estimating the time required for a carefully measured stratigraphic section to accumulate and then extrapolating the resulting value to all of geologic time.

These different methods, based as they were on uncertain assumptions, inadequate data, or both, produced a wide variety of values (Table 1). The results of thermal calculations, for example, ranged from the minimum 1.2 Ga for the Earth found by Haughton in 1865 to the 5 Ma (5×10^6 years) or so for the Sun calculated by Ritter in 1899. Calculations based on erosion and sedimentation yielded values from a low of 3 Ma published by Winchell in 1883 to a high of 15 Ga found by McGee in 1892.

Among the wide variety of results, however, none was more influential than the thermal calculations of Lord Kelvin and Clarence King. Kelvin was probably the most prolific and most honoured, arguably the most creative, and certainly the most influential scientist of his time (Burchfield 1975; Albritton 1980). In addition to devising the absolute temperature scale and fathering thermodynamics, for which he is well known, he invented the mirror galvanometer and siphon recorder used to receive telegraph signals, the stranded electrical conductor, the tide gauge, and the first mariner's compass that could be compensated for the magnetism of a steel ship. In addition, he supervised the laying of the first transatlantic telegraph cable in 1866. At the time of his retirement in 1899 he had authored more than 600 scientific papers and books and held some 70 patents. It is against this background of fame and influence that Kelvin's conclusions concerning the age of the Earth and Sun took on a decided air of authority (Thomson 1862, 1864, 1871, 1897).

Clarence King was also a scientist of great stature (Wilkins 1988). Leader of the Geological Survey Along the Fortieth Parallel and first director of the US Geological Survey, King was an energetic, ambitious and talented geologist who was highly regarded by the scientific community and by the public. King's (1893) contribution to the debate about the age of the Earth was to refine Lord Kelvin's calculations using improved data on the thermodynamic properties of diabase, which was then considered a reasonable analogue for rocks of the upper mantle, to estimate the present-day temperature distribution within the Earth, and to evaluate the assumed initial temperature gradient. King's result was 24 Ma, a value with which Kelvin took no issue (Thomson 1897).

Because of their pre-eminence in physics and geology, the results of Kelvin and King did, indeed, have a 'restraining influence' on many geologists who would have preferred a longer time to account for the thick accumulations of sedimentary rocks and for the numerous evolutionary changes evident in the fossil record. Kelvin's and King's results, however, did not go unchallenged. Among the challengers were T. H. Huxley (1869), John Perry (1895a, b; Shipley 2001), and T. C. Chamberlin (1899). Chamberlin, in particular, carefully dissected Kelvin's reasoning and mathematics and showed that Kelvin's assumptions, on which King's calculations were

Table 1. *Examples of early (pre-1950) estimates of the age of Earth*

Basis	Author	Year	Age of Earth (Ma)
Temperature			
Cooling of Earth	Comte de Buffon	1774	0.075
Cooling of Earth	Lord Kelvin	1862	20–400
Cooling of Earth	S. Haughton	1865	>1200
Cooling of Earth	C. King	1893	24
Cooling of Earth	G. F. Becker	1910	55–70
Cooling of Earth	Lord Kelvin	1897	20–40
Cooling of Sun	H. L. F. von Helmholtz	1856	22
Cooling of Sun	Lord Kelvin	1862	10–500
Cooling of Sun	S. Newcomb	1892	18
Cooling of Sun	A. Ritter	1899	4.4–5.8
Orbital physics			
Earth–Moon tidal retardation	G. Darwin	1898	>56
Earth tidal effects	P. G. Tait	1876	<10
Earth tidal effects	Lord Kelvin	1897	<1000
Change in eccentricity of Mercury	H. Jeffreys	1918	3000
Ocean chemistry			
Sulphate accumulation	T. M. Reade	1876	25
Sodium accumulation	J. Joly	1899	89
Sodium accumulation	J. Joly	1900	90–100
Sodium accumulation	J. Joly	1909	<150
Sodium accumulation	W. J. Sollas	1909	80–150
Sodium accumulation	G. F. Becker	1910	50–70
Sodium accumulation	A. Knopf	1931	>100
Erosion and sedimentation			
Limestone accumulation	T. M. Reade	1879	600
Limestone accumulation	A. Holmes	1913	320
Sediment accumulation	T. H. Huxley	1869	100
Sediment accumulation	S. Haughton	1871	1526
Sediment accumulation	A. Winchell	1883	3
Sediment accumulation	W. J. McGee	1892	15 000
Sediment accumulation	C. D. Walcott	1893	35–80
Sediment accumulation	J. Joly	1908	80
Sediment accumulation	J. Barrell	1917	1250–1700
Radioactivity			
Decay of U to Pb in crust	A. Holmes	1913	>1600
Decay of U to Pb in crust	H. N. Russell	1921	2000–8000
Decay of U to Pb in crust	A. Holmes	1927	1600–3000
Decay of U to Pb in crust	E. Rutherford	1929	3400
Decay of U to Pb in minerals	A. Knopf	1931	>2000
Pb isotopes in Earth	E. K. Gerling	1942	3940
Pb isotopes in Earth	A. Holmes	1946	3000
Pb isotopes in Earth	H. Jeffreys	1948	1340
Decay of Rb to Sr	A. K. Brewer	1938	<15 000

Not all of the values are ages for Earth. Some are for very early events in Earth's history, such as the age of the ocean, while others are for the age of the solar system or the age of matter. None of the methods gives the correct age of Earth. After Dalrymple (1991), who gives many other examples.

also based, were seriously flawed. Thus, despite Becker's seemingly casual statement, both the methods and the results of cooling calculations were highly suspect to those who were not intimidated by Kelvin's and King's statures.

By 1904 only the most preliminary groundwork for finding the age of the Earth had been laid and it was just being recognized for what it would eventually become. In 1896 the French physicist Henri Becquerel discovered that uranium salts spontaneously emitted invisible rays similar to X-rays, and two years later Marie Curie and her husband Pierre discovered that thorium also emitted a similar radiation. The

Curies determined that the new radiation was an atomic property and named the new phenomenon 'radioactivity.' In 1902 Ernest Rutherford and Frederick Soddy published the results of experiments on thorium compounds that led them to propose that the activity of a substance is directly proportional to the number of atoms present, to formulate a general theory predicting the rates of radioactive decay, and to suggest that helium might be the product of the decay of radioactive elements.

The study of radioactivity, the phenomenon that would eventually be used to find the age of the Earth, was in its infancy. It was not known how many elements were radioactive nor what their decay products might be, isotopes had not been discovered, the mass spectrometer had not been invented, and only Rutherford had suggested that radioactivity might be used as a natural clock to date rocks and minerals. For at that same ICAS meeting in 1904 addressed by Becker, Rutherford proposed for the first time that the age of minerals might be determined by radioactivity:

> If the rate of the production of helium by radium (or other radioactive substance) is known, the age of the mineral can at once be estimated from the observed volume of helium stored in the mineral and the amount of radium present (Rutherford 1905, p. 33).

It is impossible to know if Becker was in the audience during Rutherford's address, so he can hardly be blamed for not seeing how the question of the age of the Earth would eventually be resolved, much less that the answer would be more than two orders of magnitude greater than the 24 Ma calculated by Clarence King and endorsed by Lord Kelvin. Becker's natural proclivity was towards physics and chemistry, and he eventually participated in the age of the Earth debate using physical and chemical methods. In 1908 and again in 1910 he published ages for the Earth of 50–70 Ma based on the cooling of the Earth and on the accumulation of sodium in the oceans (Becker 1908, 1910), but he seems not to have heard the prescient message given that day by Rutherford.

In the half-century following 1904 there were a number of significant advances in both thought and instrumentation that led to methods to measure the age of the Earth using radioactivity. But they were slow in coming and it was not until the mid-1950s that all of the tools were in place, radiometric dating was a reality, and a reasonable basis for calculating the age of the Earth was available. The first of these advances was development of Rutherford's idea that radioactivity might be used to measure the ages of rocks and minerals.

A geological timekeeper

In 1905 Bertram Boltwood examined the composition of naturally occurring uranium minerals. Invariably, he noted, they contained lead and helium. Moreover, there was more lead and helium in the geologically older minerals than in the younger, from which he concluded that lead might be a decay product of uranium in addition to helium (Boltwood 1905). Then in March 1905, Rutherford, who was by now the Macdonald professor of physics at McGill University in Montreal, delivered the Silliman Lectures at Yale. In them he again offered the possibility of using radioactivity as a geological timekeeper (Rutherford 1906), but this time he presented two examples of the proposed radioactive method of calculating ages, using an estimate of the production rate of helium from uranium. The first was a sample of the mineral fergusonite, which yielded a U-He age of 497 Ma and the second a uraninite from Glastonbury, Connecticut, which yielded an age of about 500 Ma. Rutherford cautioned, however, that the values were minimum ages for some of the helium had probably escaped. Subsequent work by R. J. Strutt (1908), who compared the He/U ratios of 13 samples of phosphate nodules and phosphatized bone as a function of stratigraphic age, showed that helium was imperfectly retained and that U-He ages were minimum ages, as Rutherford had suggested.

Rutherford (1906) also suggested that age calculations based on lead might be superior to those based on helium:

> If the production of lead from radium is well established, the percentage of lead in radioactive minerals should be a far more accurate method of deducing the age of the mineral than the calculation based on the volume of helium for the lead formed in a compact mineral has no possibility of escape.

Over the next six years, a number of workers tested Rutherford's ideas and published ages for a variety of uranium minerals. These workers included B. B. Boltwood (1907), R. J. Strutt (1908) and Arthur Holmes (1911, 1913). The ages calculated by these pioneers in radiometric dating, although chemical uranium–lead ages rather than isotopic ages, were roughly in agreement with their relative stratigraphical ages and indicated that the calculations based on the old geological methods might be incorrect. Holmes, in particular, felt that the assignment of numerical ages to the Earth and to the subdivisions of the geological timescale would come only through the application of radioactivity methods to these problems (Lewis 2001), and he could not have been more correct.

Not all geologists greeted the new radioactivity method and its results with enthusiasm and Becker (1910) was among them, despite his proclivity for physics and chemistry. His own calculations, based on cooling and sodium accumulation, indicated that the Earth's age was 70 million years or less, from which he concluded: 'This being granted, it follows that radioactive minerals cannot have the great ages which have been attributed to them'. Becker was not alone, nor did the scepticism end quickly. Nearly 15 years after Becker's remark and even as the evidence for an old Earth from radioactivity data continued to accumulate, F. W. Clarke (1924), also of the US Geological Survey, commented:

> From chemical denudation, from palaeontological evidence, and from astronomical data the age [of the Earth] has been fixed with a noteworthy degree of concordance at something between 50 and 150 millions of years. The high values found by radioactive measurements are therefore to be suspected until the discrepancies shall have been explained.

For all their imperfections, the early and highly experimental mineral ages based on the decay of uranium to lead and helium were at least as firmly grounded in both theory and empirical evidence as those methods that relied on the cooling of the Earth, orbital physics, the accumulation of sodium in the ocean or the accumulation of sediment. Although they did not directly date the time of the Earth's origin, their importance to scientific thought about the age of the Earth cannot be overestimated. They were the first quantitative indication, based on physical principles rather than scientific intuition, that the Earth might be billions, rather than a few tens or hundreds of millions, of years old. In addition, these early results marked the birth, albeit in primitive form, of modern radiometric dating and thereafter the science of geology would never be the same.

Russell's age for the crust

Another key idea was that the age of the Earth, or at least a large planetary reservoir like the crust, might be measured from the relative abundances of a radioactive parent element and its daughter, specifically uranium and lead. The first calculation of this type appeared in 1921 in a paper authored by Henry N. Russell, a professor of astronomy at Princeton University (Russell 1921). Using published estimates of the amount of radium and lead in the crust, Russell estimated the amount of uranium and thorium in the crust and calculated the length of time required to form the lead from the decay of uranium and thorium. His value was 8 Ga. This, however, was an upper limit because lead may have been present initially in the crust and there was then speculation that uranium itself might be produced in the crust by the decay of some other element. The lower limit for the crust, Russell noted, must be considerably greater than 1.1 Ga, which was the approximate age of the oldest Precambrian minerals that had been dated by U-Pb methods. Russell (1921) concluded:

> Taking the mean of this and the upper limit found above from the ratio of uranium to lead, we obtain 4×10^9 years as a rough approximation to the age of the Earth's crust. The radio-active data alone indicate that this estimate is very unlikely to be in error in either direction by a factor as great as three. Indeed, it might be safe to say that the age of the crust is probably between two and eight thousand millions of years.

This estimate, Russell observed, was consistent with H. Jeffreys' (1918) age of 3 Ga for the age of the solar system, which was based on entirely different data relating to the eccentricities of the present orbit of Mercury (Table 1).

Like Boltwood's, Rutherford's and Holmes' early mineral ages, Russell's age for the crust was a chemical, rather than an isotopic, age. In addition, it was based on estimates of the crustal abundances of uranium and lead rather than on hard data. Nonetheless, Russell had provided the concept that a planetary body, or at least a significant part of it, might be treated as a single reservoir and dated by radioactive decay. Holmes (1927) revised Russell's calculation in a popular booklet using current estimates of crustal composition and concluded that the age of the Earth was between 1.6 Ga and 3.0 Ga, probably nearer the former than the latter.

At the time Holmes published his 1927 booklet, the second edition of a book originally published in 1913, there were only a handful of data on mineral ages in existence, and Holmes was able to summarize them all in a brief table that contained entries for only 23 localities. Yet these ages, ranging from 35 to 1260 Ma, were so consistent with the geologically determined ages of the localities that they were difficult to doubt.

In addition, if rocks in the Earth's crust were as old as one billion years, then Holmes' age of 1.6 to 3.0 Ga for the Earth was credible. In his final chapter Holmes tabulated the other physical evidence for the age of the Earth and concluded that it was consistent with the age based on radioactivity. Results for sodium accumulation and sediment thickness were relegated to insignificance; cooling calculations were not even listed. The methods so important to the pioneers in the search for Earth's age had been rendered obsolete by the new evidence from radioactivity. Any lingering doubts were put to rest in 1931 with the publication of a treatise on the age of the Earth by a committee (of which Holmes was a member) of the National Research Council of the National Academy of Sciences (Knopf et al. 1931). In the face of the data from radioactivity, the old methods and the results derived therefrom were shown to be untenable.

The discovery of new radioactive elements and the measurement of their rates of decay in the period between 1930 and 1950 provided the means for additional estimates of the maximum age of the Earth or, more exactly, of the Earth's matter (Table 1). The methods all followed more or less the general approach established by Russell, which involved estimating the crustal or earthly abundances of a radioactive isotope and its ultimate decay product and then calculating the time required for all of the product isotope to be generated by decay of the parent. It was, however, a method based on Russell's original idea – the decay of uranium to lead within a large reservoir – that finally provided a precise value for the age of the Earth and solar system.

The mass spectrometer

Scientific concepts are important, but the tools to make the necessary measurements are also important, and often those new tools lead to discoveries that modify existing concepts in major ways. The explosive growth of physics during the early part of the twentieth century resulted in the development of many new instruments to explore the nature of matter and its constituents. One of the most important of these was the mass spectrograph, a forerunner of the modern mass spectrometer, which led to the discovery of isotopes and eventually to modern radiometric dating and the solution to the puzzle of the age of the Earth.

'Positive rays' were discovered by Goldstein (1896) and two years later were shown by Wien (1898, 1902) to be streams of positively charged particles. These particles attracted the attention of J. J. Thomson, of the Cavendish Laboratories at Cambridge, who constructed an instrument he called the 'parabola mass analyser,' which was based on an earlier design used by Kaufman (1901) to investigate cathode rays (electrons), to investigate these particles (Thomson 1914). Using this apparatus Thomson was able to show that neon had two isotopes of masses 20 and 22, thus confirming a suggestion made by William Crookes in 1886 that atoms of an element might have several different whole-number weights (Faure 1977, p. 5). Thomson's experiment was not sensitive enough to measure the relative proportions of the two isotopes, but the large difference in their masses made them easily detectable.

Within a few years, F. W. Aston, working in Thomson's laboratory, redesigned the apparatus, built the first quantitative 'positive ray spectrograph', verified Thomson's result, discovered a third isotope of neon of mass 21 (Aston 1919, 1920), and set about determining the isotopes of a variety of elements. At about this same time, A. J. Dempster of the University of Chicago designed and built a somewhat different type of quantitative mass spectrograph and made accurate measurements of the abundances of the isotopes of magnesium (Dempster 1918, 1920). Thus, the search for isotopes began, and only ten years after the first isotope was discovered, 70 isotopes of 29 elements had been identified and their abundances measured. By the mid-1930s, and largely due to the indefatigable efforts of Aston, most of the isotopes of the known elements had been determined.

In 1927 Aston turned his attention to lead and made the first successful measurements of the isotopic composition of common lead, i.e. lead in minerals whose uranium content is negligible and so represent 'frozen' lead compositions. Aston (1927) showed that lead had three isotopes of masses 206, 207 and 208, in approximately the right proportions to account for the then-accepted atomic weight of lead. Two years later Aston (1929) measured the lead isotopic composition of a sample of uranium ore and found it greatly enriched in ^{206}Pb relative to ^{207}Pb. In that same year and from Aston's data, C. N. Fenner and C. S. Piggot (1929), of the geophysical laboratory of the Carnegie Institution of Washington, calculated the first isotopic age based on the decay of ^{238}U to ^{206}Pb, and isotopic dating was born. Geology would never be the same.

Rutherford (1929) used the new isotopic data to estimate the age of the Earth in a unique way. He was able to show that ^{207}Pb was probably the product of another isotope of uranium of mass

235; he estimated the half-life of ^{235}U, and estimated the age of the Earth to be 3.4 Ga presuming that ^{238}U and ^{235}U were equal in abundance when the Earth formed out of matter from the sun. All this when the actual discovery of ^{235}U was still six years in the future (Dempster 1935)!

Photographic plates were the standard means of detecting ions, measuring mass differences and determining relative abundances in the early mass spectrographs, but during the mid-1930s several workers began experimenting with electrical detectors. Among the foremost of these was A. O. Nier of the University of Minnesota, who, within a period of a few years, developed a precision instrument that incorporated electrical ion detection, a versatile monoenergetic ion source, and the latest advances in vacuum technology (Nier 1940). The modern mass spectrometers now used to measure precisely the isotopic abundances for radiometric dating and for petrologic studies are still based in large part on Nier's designs.

Gerling and the primordial lead connection

Russell had advanced the idea that the Earth, or more precisely the Earth's crust, might be dated from the accumulation of lead owing to the radioactive decay of uranium in a reservoir that essentially behaved like a closed system, and had also recognized the problem posed by the likely presence of initial lead. It was E. K. Gerling of the Radium Institute of the Academy of Sciences of the USSR, however, who formulated a working isotopic model that would eventually be used to accomplish Russell's goal of dating the Earth.

In the years 1938 to 1941, Nier and his colleagues carefully measured the isotopic composition of a score of elements with a precision previously unequalled. Among the more significant of these important contributions were their studies of lead, in which a non-radiogenic isotope (^{204}Pb) was discovered (Nier 1938) and systematic variations in the proportions of ^{206}Pb and ^{207}Pb relative to ^{204}Pb in uranium and lead ores were carefully documented (Nier 1938, 1939; Nier et al. 1941). They proposed that these variations were due to the admixture of 'primordial' and radiogenic lead. The former contained ^{204}Pb and was the lead present in the Earth when the Earth and solar system formed. The latter was due to the radioactive decay of uranium and thorium since the Earth formed, was a function of geologic time, and contained no ^{204}Pb. This concept of a two-component system for lead was essentially identical to that which Holmes had proposed for calcium isotopes in 1932 (Lewis 2001). With this one simple idea Nier and his colleagues had provided the basis for estimating the age of the Earth based on new principles. Gerling quickly seized the opportunity.

Gerling (1942) realized that an age for the Earth could be calculated from the isotopic composition of a lead ore of known age, provided that the composition of 'primordial' lead was known and assuming that the lead ore represented a 'fossil' sample of the lead composition of a single-stage reservoir within the Earth. Nier and his co-workers had found one lead ore sample, a galena from Ivigtut, Greenland, whose ^{206}Pb/^{204}Pb and ^{207}Pb/^{204}Pb ratios were extremely low relative to other measured ore leads and they speculated that the amount of radiogenic lead in this sample was small or negligible (Nier et al. 1941). Gerling used the lead isotope ratios in this Greenland sample to represent the composition of primordial lead.

Gerling's calculations were quite simple. He found the length of time required for the isotopic composition of lead to change from a value represented by the Greenland sample to a value represented by the average composition of seven more radiogenic lead ores. This time was 3.1 Ga, to which he added the average age of the radiogenic ores, about 130 Ma, to obtain an age for the Earth of 3.23 Ga. Gerling presented no graphics but his calculation is shown graphically in Figure 1. He had, in effect, determined a two-point isochron as it would have appeared at 130 Ma using the Greenland analysis for primordial lead and the average of the seven lead

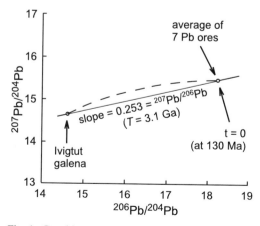

Fig. 1. Graphic representation of Gerling's (1942) calculation of a minimum age for the Earth using the Ivigtut, Greenland, galena to represent 'primordial' lead. His result was $3.1 \pm 0.13 = 3.23$ Ga. After Dalrymple (1991).

ores as the second datum. The slope of this isochron was equivalent to an age of 3.1 Ga, which was the age of the ore at 130 Ma.

Gerling's result was a minimum age for the Earth because, as Gerling recognized, the most primitive ore measured by Nier was not necessarily representative of true primordial lead. Gerling went through the calculation again, comparing a primitive galena from Great Bear Lake, whose age was 1.25 Ga, to the Greenland ore. This calculation yielded a minimum age for the Earth of 3.95 Ga. Gerling concluded: 'From these computations the age of the earth is not under $3 \times 10^9 - 4 \times 10^9$ years. This is certainly not too much, since the age of certain minerals, calculated with reference to AcD/RaG [^{207}Pb/^{206}Pb], was put at 2.2×10^9 years'.

Several assumptions are implicit in Gerling's calculations. Foremost of these are that: (1) the seven lead ores originated from the same homogeneous source (reservoir) whose initial lead isotopic composition was identical to the Greenland galena; and (2) all of the leads are single-stage leads, i.e. the leads all evolved in a reservoir that was effectively a closed system to uranium and lead. Gerling's results are minimum values for the age of the Earth and, while of the correct order of magnitude, they are too low primarily because the lead isotopes in the Greenland galena are not of primordial composition. Nonetheless, he had devised a brilliant and fruitful approach that is still the basis for finding the age of the Earth from the decay of uranium to lead.

The generalized model for the evolution of lead isotopes in the Earth is usually credited to Arthur Holmes of the University of Edinburgh and F. G. Houtermans of the University of Göttingen, who both pursued the technique with considerable vigour (Holmes 1946, 1947a, b, 1948; Houtermans 1946, 1947). Holmes and Houtermans developed the method independently and were unaware of Gerling's work, probably because Gerling's paper was in Russian and was not available in translation until many years later. Nonetheless, Gerling clearly has priority in the literature and it is unfortunate that the model is commonly known as the Holmes–Houtermans model: Gerling–Holmes–Houtermans would be far more appropriate (but see Lewis 2001).

The meteorite connection

Houtermans is well known for his part in developing the Gerling–Holmes–Houtermans model, especially for advancing the concept of lines of equal time, which he named 'isochrons'. In addition, he had pointed the way for future work by suggesting that a better value for primordial lead might be found by analysing iron meteorites (Houtermans 1947), a suggestion also made by Brown (1947). In 1953 Houtermans got his wish.

In that year C. C. Patterson, of the California Institute of Technology, and his colleagues determined the lead isotopic composition and the uranium and lead concentrations in both the iron–nickel phase and in troilite (FeS) from the iron meteorite Canyon Diablo, which excavated Meteor Crater some 50 000 or so years ago (Patterson et al. 1953a). The troilite was found to contain the lowest lead isotope ratios ever measured and was also exceedingly low in uranium relative to lead. The low ratio of uranium to lead meant that the lead isotopic composition could not have changed significantly since the meteorite, which was even then known to be an ancient object, was formed. Thus, suggested Patterson and his colleagues, the lead ratios in Canyon Diablo might record the composition of primordial lead.

Houtermans and Patterson were both quick to take advantage of the new lead data for Canyon Diablo. In December of 1953 Houtermans published a paper in which he calculated an age for the Earth that is very close to the presently accepted value (Houtermans 1953). Houtermans made two principal assumptions: (1) that the isotopic composition of lead at the time of formation of the Earth's lithosphere was represented by the values found in the troilite of Canyon Diablo; and (2) that certain of the Tertiary lead ores whose lead isotopic compositions had been measured formed by single-stage growth from a common time of origin up to the time of formation (i.e. extraction from the reservoir) of the lead ores. He chose lead ores of Tertiary age because for these young leads the calculated age of the Earth is relatively insensitive to errors in the geological age of the samples. At that time the literature contained lead isotopic compositions for 22 ores, from which Houtermans selected ten. For each of the ten ores he calculated a two-point isochron through the lead composition of the ore and the lead composition of Canyon Diablo troilite. He then averaged the ten results and used the result to find an age for the Earth of 4.5 ± 0.3 Ga.

Graphically, Houtermans' solution is nearly identical to Gerling's (Fig. 1) except that he used the Canyon Diablo values instead of the Greenland galena for primordial lead. In selecting the ten Tertiary leads, Houtermans rejected 12 others because their compositions were anomalous and their model ^{207}Pb/^{206}Pb ages were negative. He noted, however, that even if such data were included the results were not changed appreciably. Therefore, he concluded, as long as a large number of carefully selected samples was used any multi-staged leads inadvertently included in the calculations resulted in only small errors in the final result.

Patterson (1953) presented the results of calculations that were virtually identical to those of Houtermans at a conference held three months before Houtermans' 1953 paper was published. He used the meteoritic lead composition as the composition of primordial lead and two different types of materials to represent the composition of present-day lead. One calculation used the average lead composition of Recent oceanic sediment and a manganese nodule. The other used the composition of lead in Columbia River Basalt (Miocene). His results were 4.51 and 4.56 Ga, respectively. The publication date of the conference proceedings is commonly given in bibliographic references (as it is here) as 1953, the year of the conference, but the actual date of publication is not given in the proceedings volume and is unclear. Most probably, Patterson's paper was published early in 1954, a few months after Houtermans' paper (L. T. Aldrich pers. comm. 1990; C. C. Patterson pers. comm. 1990), but it is clear that Patterson made the first presentation of these important results to the scientific community and he is generally acknowledged as the first scientist to calculate the true age of the Earth.

Houtermans' and Patterson's 1953 results are notable not only because they are near the present-day value but also because they were the first calculation to link the age of the Earth to the age of meteorites, thereby implying a genetic connection. Neither Houtermans nor Patterson made any attempt to justify this assumption of co-genesis, but as we shall see, a reasonable case can be made for its validity.

Patterson and the meteoritic lead isochron

Houtermans and Patterson had assumed that there was a genetic connection between meteoritic lead and young terrestrial lead but had not provided any arguments for the validity of that assumption. Moreover, their calculations were each based essentially on only two points: troilite lead in the Canyon Diablo meteorite and young leads in either Tertiary ores, Recent ocean sediment or Miocene basalt. Patterson corrected both deficiencies three years later in a now-classic paper (Patterson 1956).

Patterson used the lead isotope analyses from three stone meteorites and the troilite phase of two iron meteorites and showed that these data fell precisely on an isochron (Fig. 2) whose slope indicated an age of 4.55 ± 0.07 Ga. Such co-linearity from a set of data with a wide range in isotopic composition, Patterson argued, strongly indicated that these five meteorites fulfilled the

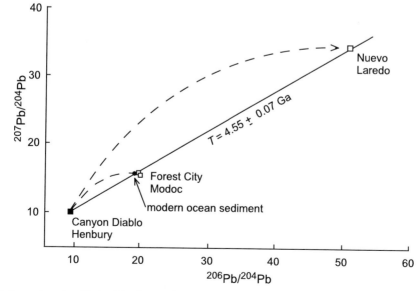

Fig. 2. Patterson's meteoritic lead isochron using three stone and two iron meteorites (squares). The lead composition of modern ocean sediment (filled circle) falls on the isochron, suggesting that meteorites and the Earth are genetically related and of the same age. The dashed lines are growth curves. After Patterson (1956).

assumptions of the method. Any meteorite that had a differentiation history that fractionated uranium relative to lead after its initial formation would not fall on the isochron. Therefore, Patterson concluded, the isochron age represents the time of initial formation and differentiation of meteorites. He also noted that the age was in agreement with the existing K-Ar and Rb-Sr ages for meteorites, which were then few in number and poor in quality.

Patterson's next step was to make the genetic connection between meteorites and the Earth. He used the newly found age for the Earth and the primordial lead ratios from Canyon Diablo troilite to predict the lead isotope ratios for any modern lead (i.e. lead of zero age) that belonged to the meteoritic system. This is simply another way of saying that if modern Earth lead falls on the meteoritic isochron, then it must have evolved, over the past 4.55 Ga, in a closed system from a primordial composition the same as that measured in Canyon Diablo troilite. Patterson realized that there were other ways that an Earth lead could have developed a composition on the isochron, but these required complicated and improbable mechanisms.

The problem of choosing a representative sample of modern Earth lead is not simple because Earth's crust is being continually created and destroyed, and so has a complicated history. Patterson proposed that modern sediment from the deep ocean might provide a reasonable sample of modern Earth lead because such sediment samples a wide volume of material from the present continents and thus represents average crustal lead. The lead isotopic composition of Pacific deep-sea sediment had been measured previously (Patterson et al. 1953b) and its lead composition satisfied Patterson's prediction very well (Fig. 2). And so Patterson had not only determined a precise age for meteorites but had also shown it probable that the Earth was part of the meteoritic lead system and, therefore, of the same age.

Six years later Patterson teamed with V. R. Murthy of the University of California at San Diego to refine Patterson's age of meteorites and to strengthen the hypothesis that the Earth was part of the meteoritic lead system (Murthy & Patterson 1962). Murthy and Patterson selected lead analyses of five stone meteorites thought, on the basis of other isotopic age data, most likely to have been closed systems since formation. To this array of data they added the composition of primordial lead, which they took to be the average composition of lead in troilite from five iron meteorites as measured by three different laboratories. The six meteorite lead compositions formed a linear array (Fig. 3). To determine the slope of the isochron, Murthy and Patterson calculated the mean of the five slopes defined by each of the stone meteorites and the primordial composition. This average slope represented an

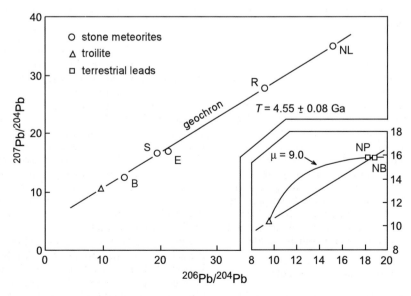

Fig. 3. Murthy and Patterson's meteoritic geochron, which is bracketed by terrestrial leads from North Pacific sediments (NP) and lead ores from Bathurst, New Brunswick (NB). The terrestrial leads lie on a growth curve of $\mu = (^{235}U/^{204}Pb) = 9.0$ that passes through meteoritic troilite. B = Beardsley, E = Elenovka, NL = Nuevo Laredo, R = Richardton, S = Saratov. After Murthy & Patterson (1962).

age for meteorites of 4.55 Ga and the isochron they named the meteoritic geochron.

To show the relationship between the meteoritic geochron and the terrestrial geochron, Murthy and Patterson used two samples of terrestrial lead (Fig. 3 inset). One, the average composition of lead in more than 100 samples of recent North Pacific sediments, should, they reasoned, lie to the right of the terrestrial geochron because the marine sediments are eroded from rocks of the upper layers of the crust. At the time of their formation, these source rocks contained lead compositions representative of the entire crust but were enriched in ^{238}U relative to ^{204}Pb by the crustal formation process. Thus, the marine leads should be displaced to the ^{206}Pb-enriched side of the average crustal lead composition. In other words, the average marine sediment lead composition should lie slightly to the right of the terrestrial (= crustal) geochron. On the left, the terrestrial geochron should be bracketed by single-stage ore leads that define the crustal growth curve. Murthy and Patterson chose the mean composition of ore leads from Bathurst, New Brunswick, which have a geological age of about 350 Ma and which were, at the time (but no longer), thought to be single-stage leads. These two points, they reasoned, should limit the position of the terrestrial geochron, and since they also bracketed the meteoritic geochron, the two geochrons must be very nearly the same, if not identical. Moreover, both the North Pacific and Bathurst leads lie on a primary (single-stage) growth curve that passes through the primordial (troilite) composition and satisfies what was then known about the average U/Pb ratio of the crust. Murthy and Patterson concluded, therefore, that meteorites and the Earth's crust are parts of the same Pb isotopic system and the age of meteorites and the age of the Earth are the same.

In addition to refining the Pb isotopic age of meteorites, Murthy and Patterson provided a sound basis for connecting Pb growth in the Earth, a body whose time of origin cannot be determined directly, with Pb growth in meteorites, whose ages can be precisely measured. Houtermans (1953) and Patterson (1953) had assumed that meteorites and the Earth were cogenetic. In one bold and clever stroke, Murthy and Patterson had shown that such an assumption was not only reasonable but probable.

The primary terrestrial growth curve

No discussion of the age of the Earth would be complete without some mention of the exhaustive attempt to reconstruct the primary terrestrial lead isotope growth curve – an ingenious idea with which nature was not entirely co-operative.

The idea of a primary terrestrial growth curve is based on the assumption that the Earth formed at the same time as the meteorites with its own ratio of U/Pb, has remained a closed system to uranium and lead since formation, and thus can be treated like a meteorite. It then follows that the lead isotopic composition of the Earth, too, must have evolved along a similar single-stage growth curve that originated, as do the growth curves for meteorites, at the primordial composition of lead. Such a growth curve need not necessarily involve the entire Earth but could apply to any uranium–lead reservoir within the Earth, such as the mantle or crust, so long as that reservoir formed at the same time or very shortly after the Earth and has remained a closed system to uranium and lead.

The hypothesis of a single growth curve for the Earth, or some substantial portion of it, was implicit in the early work of Russell, Gerling, Holmes and Houtermans and was more fully developed in the 1950s by R. D. Russell of the University of British Columbia and his colleagues (e.g. Collins *et al.* 1953; Russell 1956; Russell & Farquhar 1960). The hypothesis appears to be a reasonable approximation because modern terrestrial sediments and young lead ores plot very close to the meteoritic geochron. But a more convincing case could be made if the terrestrial growth curve actually could be 'traced' backward in time, i.e. reconstructed, and if it could be shown that the reconstructed curve passes through the composition of meteoritic troilite.

As was recognized by Gerling, lead ores represent the isotopic composition of lead in their parent rocks at the time the ores formed and thus represent the fossil lead isotopic record of some large uranium–lead reservoir within the Earth. The amount of lead withdrawn from the reservoir by the formation of the ore is so small that the withdrawal does not, for all practical purposes, violate the assumption of a closed system. Thus isotopic analyses of single-stage lead ores from this presumed reservoir should, theoretically, permit the evolution of lead isotopes in the Earth to be traced through time and the primary terrestrial growth curve to be thereby reconstructed.

One difficulty with this procedure involves the selection of appropriate samples. Since it is known that most lead ores are the products of multi-stage processes, how can single-stage lead ores be identified? A potential solution to this problem was suggested by Stanton & Russell (1959), who proposed that conformable, or stratiform, lead ore deposits were probably composed of single-stage leads. These stratiform ores are thought to form by deposition from seawater of lead produced by volcanic eruptions and so should be the same age as the sedimentary beds in which they occur. Stanton and Russell observed that the isotopic composition of leads from conformable deposits were ordinary and quite uniform, whereas those from other types of deposits commonly were anomalous and highly variable, and they found that the leads from nine conformable deposits fit a single-stage growth curve to within a few tenths of one per cent.

During the 1960s and early 1970s, the definition and refinement of the primary terrestrial growth curve was a major goal of lead isotope studies. Conformable ores are not numerous, but results from the dozen or so then-analysed deposits appeared to fit rather precisely a single-stage growth curve that passed through the composition of Canyon Diablo troilite and substantiated an age of 4.55 Ga for the Earth. Figure 4 shows an example of one such single-stage growth curve based on 14 conformable lead ores ranging in age from 0.1 to 2.2 Ga. Numerous authors of the period presented similar primary terrestrial growth curves, but all used the same basic data set, differing in only a few details, and came to the same basic conclusion that, to a first approximation, the source of ordinary leads has behaved as a single-stage system since formation of the Earth at approximately 4.55 Ga (e.g. Russell & Farquhar 1960; Ostic et al. 1963, 1967; Russell & Reynolds, 1965; Kanasewich 1968; Cooper et al. 1969; Stacey et al. 1969; Doe 1970; Russell 1972).

At the end of the 1960s the data from lead ores seemed to be in excellent accord with the independently measured radiometric ages of meteorites. But the beautiful concordance was partly fortuitous and was destined to degenerate with the more accurate measurement of the uranium decay constants and of the isotopic composition of lead in Canyon Diablo troilite.

Oversby (1974) was the first to publish a detailed analysis of the problems with the single-stage hypothesis for the evolution of conformable lead ores. One problem arose when the new and highly precise values for the uranium decay constants (Jaffey et al. 1971) and for the isotopic composition of lead in Canyon Diablo troilite (Tatsumoto et al. 1973) became available. With these better values in hand, it was no longer possible to construct a single-stage growth curve that passed through the conformable ores and through Canyon Diablo troilite with the age of the Earth at 4.55 Ga. The deviation of the data from a single-stage growth curve was especially pronounced for ores younger than about 2.5 Ga.

The second problem noted by Oversby was that the model Pb-Pb ages for the conformable ores, i.e. the age calculated from an 'isochron'

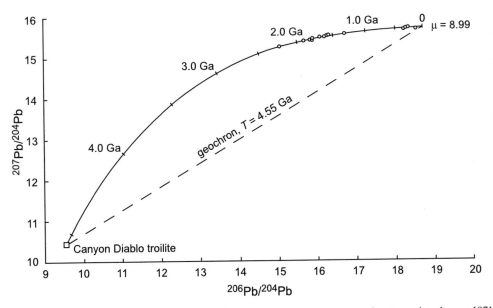

Fig. 4. The isotopic compositions of 14 conformable leads fit a single-stage growth curve using the pre-1971 uranium decay and abundance constants, the pre-1973 value for Canyon Diablo, and an age of 4.55 Ga for Earth. Data from a compilation by Kanasewich (1968).

drawn through Canyon Diablo troilite and through an individual ore datum, were 300–450 Ma younger than the known geological ages of the ore leads as measured by other means, both radiometric and stratigraphic.

The only solutions to this problem were either to accept an age of 4.43 Ga for the Earth, which was at odds with the ages of meteorites, or to admit that the reservoir for conformable leads had not behaved as a single-stage system since the Earth formed (Doe & Stacey 1974).

The abandonment of the concept of single-stage leads resulted in the loss of some of the uniqueness of growth curve solutions for the age of the Earth. Conformable ore lead data can be made to fit multi-stage models with the age of the Earth at 4.55 Ga quite precisely because such models have considerably more flexibility than a single-stage model (e.g. Sinha & Tilton 1973; Stacey & Kramers 1975; Cumming & Richards 1975). But even these models may oversimplify the evolution of lead isotopes in the Earth (e.g. Hofmann 2001). Thus, the primary terrestrial growth curve can be made consistent with the concept that the Earth and meteorites are part of the same lead isotopic system by assuming a multi-stage history, but it does not yield a unique numerical solution for the age of the Earth.

The age of the Earth in 2001

The presently accepted value for the age of the Earth is based on data from only a few very old conformable leads. The calculations utilize old leads because they presumably spent less time evolving in the lead reservoir than young leads and did so early in Earth's history. As a result, they are more likely to be, or to closely approximate, single-stage leads. In addition, any deviation from the single-stage assumption has less of an effect on the calculations involving a very old lead than it does on those that utilize younger leads.

There are two slightly different types of age-of-the-Earth calculations. The first type of calculation is virtually identical to the one developed by Gerling. It involves finding the length of time required for the composition of an ancient lead ore, whose age is known independently from radiometric dating, to evolve from the composition of Canyon Diablo troilite to its composition at the time it was separated from the Earth reservoir (which is the same as its composition now). This calculated time is then added to the independently known age of the lead ore to find the age of the reservoir (Earth). There are only three results of this type from stratiform ores (Table 2). They all give very nearly the same value for the age of the Earth and have a mean of 4.54 Ga.

In 1980, F. Tera, of the Carnegie Institution of Washington, developed a method of determining the age of the Earth from ancient conformable lead deposits that does not require that the ages of the lead deposits (t) to be known (Tera 1980, 1981). Instead, Tera's method is based on the assumption that age–composition profiles of galenas of different age, but from a common source, must have a single point of congruency that defines unique values for both the $^{238}U/^{204}Pb$ ratio (μ) in the source and for the age of reservoir (T).

Tera used lead isotope data for the four oldest conformable galenas known (Table 2). For each galena he assumed various values of T and calculated the corresponding values of t using the measured $^{207}Pb/^{206}Pb$ ratio of the galena and Canyon Diablo troilite for primordial lead. For each pair of T and t values he then calculated the corresponding value of μ_s, which is the present-day value of μ for the source. The calculated values of T and μ_s provided the data to construct a 'source profile' of T versus μ_s for the galena (Fig. 5). This profile is a unique function of the lead composition of the analysed galena, but only one point on the curve can be correct and from the data of a single galena it is not possible to determine this point. Tera assumed, however, that the four ancient galenas originated from the same source and that the individual profiles from these galenas would intersect at a point that represents the true age and composition of the source. For the four analysed galenas, this congruency point is 4.53 Ga.

Table 2. *Ages of the Earth based on lead isotope data from conformable Archean lead ores*

Locality	Age of ore (Ga)	Age of Earth (Ga)	Source
Timmons, Ontario	2.64	4.56	Bugnon *et al.* (1979)
Manitouwadge, Ontario	2.68	4.55	Tilton & Steiger (1965, 1969)
Barberton, South Africa	3.23	nd	Stacey & Kramers (1975)
Big Stubby, W. Australia	3.45	4.52	Pidgeon (1978)

See also Figure 5. After Dalrymple (1991).

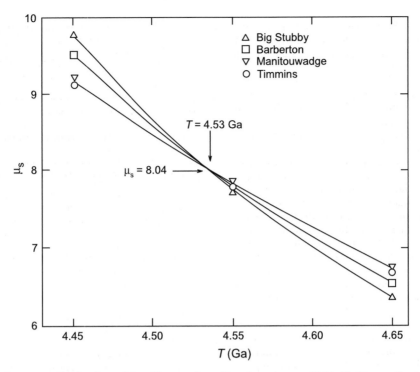

Fig. 5. Source profiles for four of the oldest conformable galenas known (Table 2). The profiles intersect at a congruency point corresponding to a parental source that is 4.53 Ga with $\mu_s = {}^{235}U/{}^{204}Pb = 8.04$. Because the galenas of the Abitibi Belt, to which both Timmons and Manitouwadge belong, seem to be homogeneous with regard to lead isotope composition, a single profile was constructed for these two localities. After Tera (1980).

The type of source profile shown in Figure 5 is only one of several that can be constructed from the same basic data. For example, Tera (1981) has also constructed source profiles for the same four ancient galenas by plotting the radiogenic $^{207}Pb/^{206}Pb$ versus the source composition expressed as $^{204}Pb/^{206}Pb$. In this analysis he obtained a congruency point that results in $T = 4.54$ Ga. Tera observed that it was probably significant that ancient galenas from three continents seemed to define a common source with a common age and lead composition and also that the age obtained is similar to the age determined for meteorites.

The precise nature of the 4.54 Ga event indicated by treating the lead isotopic data from old Earth leads and from troilite in iron meteorites as one lead isotopic system is not clear. Tera (1981) speculated that the age represents the time of uranium–lead fractionation in the primary materials from which the Earth formed. If this fractionation occurred at or very near the time Earth accumulated, then the age is the age of the Earth. If not, then it represents the age of the debris from which the Earth formed. Alternatively, the fractionation might be the result of separation of the Earth's materials into core and mantle. A distinct possibility is that Tera's result, as well as the results from other terrestrial lead models, does not precisely represent any particular event in the Earth's formation. Hofmann (2001) has argued rather persuasively that any sample of lead on Earth is unlikely to have come from a truly primitive source and, therefore, that terrestrial lead isotope data are incapable of providing a refined value for the age of the Earth within the broad limits of about 4.4–4.56 Ga. But all is not lost.

Recent and past studies of extinct radioactive isotope systems, such as $^{129}I/^{129}Xe$ ($t_{1/2} = 15.7$ Ma), $^{53}Mn/^{53}Cr$ ($t_{1/2} = 3.7$ Ma) and $^{182}Hf/^{182}W$ ($t_{1/2} = 9$ Ma) of meteorites as well as lunar and terrestrial samples, combined with increasingly precise Pb isotope studies of meteorites, have begun to clarify the timing of events in the early solar system (e.g. Allègre et al. 1995; Halliday & Lee 1999; Tera & Carlson 1999). It now appears as if the sequence of events, beginning with the condensation of solid matter from the solar nebula at 4.566 Ga, the age of calcium–aluminium inclusions in primitive meteorites (Allègre et al. 1995), and ending with the final accretion of the Earth, segregation of the Earth's core and formation of the Moon, occurred within an interval of c. 50 ± 10 Ma, with the latter events occurring about 4.51 Ga (Halliday & Lee 1999). This suggests that Tera's 1981 age of the Earth (4.54 Ga) may, as he first speculated, approximate the age of the material from which the Earth formed rather than the age of formation of the planet itself. Despite the present uncertainties, which are numerically small, there is little doubting that the age of the Earth (or at least its material) and the solar system exceed, by some small fraction, 4.5 Ga. That much has not changed significantly since Patterson's 1953 result. If knowledge of the early events in the solar system continues to increase at the current rate, then there is little doubt that a detailed chronology of those events will be known within this decade.

The approximate age of the Earth determined from isotopic evidence is substantiated by a large number of radiometric ages of different types on meteorites and lunar samples, the oldest of which are 4.5–4.6 Ga. In addition, an age of 4.51–4.55 Ga is consistent with ages determined for the oldest rocks and minerals on the Earth (4.0–4.4 Ga), the globular cluster stars in the Milky Way galaxy (14–18 Ga), the r-process elements (9–16 Ga), and the universe (7–20 Ga) (see Dalrymple (1991) for a detailed discussion of the evidence for the age of the Earth, meteorites, galaxy and universe).

There are still many interesting things to learn about the formation of the solar system and the age and early history of the Earth, but the age of the Earth is no longer the mystery it was when George Becker addressed the ICAS in 1904. We now know, to within 1% or better and from a variety of evidence, that the age of the Earth–Moon–meteorite system is about 4.51–4.55 Ga.

How would George Becker and his contemporaries view the progress made by the end of the twentieth century in solving this important problem that occupied so much of their time? That question is, of course, impossible to answer, but my guess is that it would be with both astonishment and pleasure.

I thank T. Stern, B. Glen, S. Moorbath and P. Wyse Jackson for their helpful reviews of the manuscript, as well as C. L. E. Lewis and A. W. Hofmann for valuable discussions about the subject.

References

ALBRITTON, C. C. JR. 1980. *The Abyss of Time*. Freeman, Cooper, San Francisco.

ALLÈGRE, C. J., MANHÈS, G. & GÖPEL, C. 1995. The age of the Earth. *Geochimica et Cosmochimica Acta*, **59**, 1445–1456.

ASTON, F. W. 1919. A positive ray spectrograph. *Philosophical Magazine*, **38**(6), 707–714.

ASTON, F. W. 1920. The constitution of atmospheric neon. *Philosophical Magazine*, **39**(6), 449–455.

ASTON, F. W. 1927. The constitution of ordinary lead. *Nature*, **120**, 224.

ASTON, F. W. 1929. The mass-spectrum of uranium lead and the atomic weight of protactinium. *Nature*, **123**, 313.

BECKER, G. F. 1904. Present problems of geophysics. *Science*, **20**, 545–556.

BECKER, G. F. 1908. Relations of radioactivity to cosmogony and geology. *Bulletin of the Geological Society of America*, **19**, 113–146.

BECKER, G. F. 1910. The age of the Earth. *Smithsonian Miscellaneous Collections*, **56**(6), 1–28.

BOLTWOOD, B. B. 1905. On the ultimate disintegration products of the radio-active elements. *American Journal of Science*, **20**, 253–267.

BOLTWOOD, B. B. 1907. On the ultimate disintegration products of the radio-active elements. Part II. The disintegration products of uranium. *American Journal of Science*, **23**(4), 77–88.

BROWN, H. 1947. An experimental method for the estimation of the age of the elements. *Physical Review*, **72**, 348.

BUGNON, M.-F., TERA, F. & BROWN, L. 1979. Are ancient lead deposits chronometers of the early history of the Earth? *Carnegie Institution of Washington Yearbook*, **78**, 346–352.

BURCHFIELD, J. D. 1975. *Lord Kelvin and the Age of the Earth*. Science History, New York.

CHAMBERLIN, T. C. 1899. On Lord Kelvin's address on the age of the Earth as an abode fitted for life. *Annual Report of the Smithsonian Institution*, **1899**, 223–246.

CLARKE, F. W. 1924. *Data of geochemistry*. US Geological Survey Bulletin, **770**.

COLLINS, C. B., RUSSELL, R. D. & FARQUHAR, R. M. 1953. The maximum age of the elements and the age of the earth's crust. *Canadian Journal of Physics*, **31**, 420–428.

COOPER, J. A., REYNOLDS, P. H. & RICHARDS, J. R. 1969. Double-spike calibration of the Broken Hill standard lead. *Earth and Planetary Science Letters*, **6**, 467–478.

CUMMING, G. L. & RICHARDS, J. R. 1975. Ore lead isotope ratios in a continuously changing Earth. *Earth and Planetary Science Letters*, **28**, 155–171.

DALRYMPLE, G. B. 1991. *The Age of the Earth*. Stanford University, Stanford.

DEMPSTER, A. J. 1918. A new method of positive ray analysis. *Physical Review*, **11**, 316–325.

DEMPSTER, A. J. 1920. Positive ray analysis of magnesium. *Science*, **52**, 559.

DEMPSTER, A. J. 1935. Isotopic composition of uranium. *Nature*, **136**, 180.

DOE, B. R. 1970. *Lead Isotopes*. Springer-Verlag, Berlin.

DOE, B. R. & STACEY, J. S. 1974. The application of lead isotopes to the problems of ore genesis and ore prospect evaluation: A review. *Economic Geology*, **69**, 757–776.

FAURE, G. 1977. *Principles of Isotope Geology*. Wiley, New York.

FENNER, C. N. & PIGGOT, C. S. 1929. The mass-spectrum of lead from Broggerite. *Nature*, **123**, 793–794.

GERLING, E. K. 1942. Age of the Earth according to radioactivity data. *Comptes Rendus (Doklady) de l'Academie des Sciences de l'URSS*, **34**, 259–261 [translation in: Harper, C. T. (ed.), 1973. *Geochronology: Radiometric Dating of Rocks and Minerals*, Dowden, Hutchinson & Ross, 121–123].

GOLDSTEIN, E. 1896. Über eine noch nicht unter suchte Strahlungsform an der Kathode inducirter Entlaclungen. *Sitzungsberichte der Königlich Preussischen Akademie der Wissenschaften zu Berlin*, 691–699.

HALLIDAY, A. N. & LEE, D.-C. 1999. Tungsten isotopes and the early development of the Earth and Moon. *Geochimica et Cosmochimica Acta*, **63**, 4157–4179.

HOFMANN, A. W. 2001. Lead isotopes and the age of the Earth – a geochemical accident. *In*: LEWIS, C. L. E. and KNELL, S. J. (eds) *The Age of the Earth: from 4004 BC to AD 2002*. Geological Society, London, Special Publications, **190**, 223–236.

HOLMES, A. 1911. The association of lead with uranium in rock-minerals, and its application to the measurement of geological time. *Proceedings of the Royal Society of London*, **85A**, 248–256.

HOLMES, A. 1913. *The Age of the Earth*. Harper, London.

HOLMES, A. 1927. *The Age of the Earth: An Introduction to Geological Ideas*. Ernest Benn (Benn's Sixpenny Library, No. 102).

HOLMES, A. 1946. An estimate of the age of the Earth. *Nature*, **157**, 680–684.

HOLMES, A. 1947a. A revised estimate of the age of the Earth. *Nature*, **159**, 127–128.

HOLMES, A. 1947b. The age of the Earth. *Endeavor*, **6**, 99–108.

HOLMES, A. 1948. The age of the Earth. *Annual Report of the Board of Regents of the Smithsonian Institution*, 227–239.

HOUTERMANS, F. G. 1946. Die Isotopenhaufigkeiten im Naturlichen Bloi und das Alter des Urans (The isotopic abundances in natural lead and the age of uranium). *Naturwissenschaften*, **33**, 185–186, 219 [translation in: Harper, C. T. (ed.), 1973. *Geochronology: Radiometric Dating of Rocks and Minerals*. Dowden, Hutchinson & Ross, 129–131].

HOUTERMANS, F. G. 1947. Das Alter des Urans. *Zeitschrift fur Naturforschung*, **2a**, 322–328.

HOUTERMANS, F. G. 1953. Determination of the age of the Earth from the isotopic composition of meteoritic lead. *Nuovo Cimento*, **10**(12), 1623–1633.

HUXLEY, T. H. 1869. The anniversary address of the President. *Quarterly Journal of the Geological Society of London*, **25**, xxviii–liii.

JAFFEY, A. H., FLYNN, K. F., GLENDENIN, L. E., BENTLEY, W. C. & ESSLING, A. M. 1971. Precision measurements of half-lives and specific activities of ^{235}U and ^{238}U. *Physical Review C*, **4**, 1889–1906.

JEFFREYS, H. 1918. On the early history of the Solar System. *Monthly Notices of the Royal Astronomical Society*, **78**, 424–441.

KANASEWICH, E. R. 1968. The interpretation of lead isotopes and their geological significance. *In*: HAMILTON, E. I. & FARQUHAR, R. M. (eds) *Radiometric Dating For Geologists*. Interscience, New York, 147–223.

KAUFMAN, W. 1901. Methode zur exakten Bestimmung von Ladung und Geschwindigkeit der Becquerelstrahlen. *Physikalische Zeitschrift*, **2**, 602–603.

KING, C. 1893. The age of the Earth. *American Journal of Science*, **45**(3), 1–20.

KNOPF, A., SCHUCHERT, C., KOVARIK, A. F., HOLMES, A. & BROWN, E. W. 1931. Physics of the Earth – IV: The Age of the Earth. *Bulletin of the National Research Council, Washington*, **80**.

LEWIS, C. L. E. 2001. Arthur Holmes' vision of a geological timescale. *In*: LEWIS, C. L. E. and KNELL, S. J. (eds) *The Age of the Earth: from 4004 BC to AD 2002*. Geological Society, London, Special Publications, **190**, 121–138.

MURTHY, V. R. & PATTERSON, C. C. 1962. Primary isochron of zero age for meteorites and the Earth. *Journal of Geophysical Research*, **67**, 1161–1167.

NIER, A. O. 1938. Variations in the relative abundances of the isotopes of common lead from various sources. *Journal of the American Chemical Society*, **60**, 1571–1576.

NIER, A. O. 1939. The isotopic constitution of radiogenic leads and the measurement of geological time. II. *Physical Review*, **55**, 153–163.

NIER, A. O. 1940. A mass spectrometer for routine isotope abundance measurements. *Review of Scientific Instruments*, **11**, 212–216.

NIER, A. O., THOMPSON, R. W. & MURPHEY, B. F. 1941. The isotopic constitution of lead and the measurement of geological time. III. *Physical Review*, **60**, 112–116.

OSTIC, R. G., RUSSELL, R. D. & REYNOLDS, P. H. 1963. A new calculation for the age of the Earth from abundances of lead isotopes. *Nature*, **199**, 1150–1152.

OSTIC, R. G., RUSSELL, R. D. & STANTON, R. L. 1967. Additional measurements of the isotopic composition of lead from stratiform deposits. *Canadian Journal of Earth Sciences*, **4**, 245–269.

OVERSBY, V. M. 1974. A new look at the lead isotope growth curve. *Nature*, **248**, 132–133.

PATTERSON, C. C. 1953. The isotopic composition of meteoritic, basaltic and oceanic leads, and the age of the Earth. *Proceedings of the Conference on Nuclear Processes in Geologic Settings, Williams Bay, Wisconsin, Sept. 21–23, 1953*, 36–40.

PATTERSON, C. C. 1956. Age of meteorites and the Earth. *Geochimica et Cosmochimica Acta*, **10**, 230–237.

PATTERSON, C. C., BROWN, H., TILTON, G. R. & INGHAM, M. G. 1953a. Concentration of uranium and lead and the isotopic composition of lead in meteoritic material. *Physical Review*, **92**, 1234–1235.

PATTERSON, C. C., GOLDBERG, E. D. & INGHAM, M. G. 1953b. Isotopic compositions of Quaternary leads from the Pacific Ocean. *Geological Society of America Bulletin*, **64**, 1387–1388.

PERRY, J. 1895a. On the age of the Earth. *Nature*, **51**, 224–227.

PERRY, J. 1895b. The age of the Earth. *Nature*, **51**, 582–585.

PIDGEON, R. T. 1978. Big Stubby and the early history of the Earth. *In*: ZARTMAN, R. E. (ed.), *Short Papers of the Fourth International Conference, Geochronology, Cosmochronology, Isotope Geology*. US Geological Survey Open-file Report **78–701**, 334–335.

RUSSELL, H. N. 1921. A superior limit to the age of the Earth's crust. *Proceedings of the Royal Society of London*, series A, **99**, 84–86.

RUSSELL, R. D. 1956. Interpretation of lead isotope abundances. *Nuclear Processes in Geological Settings*, National Academy of Science – National Research Council, **400**, 68–78.

RUSSELL, R. D. 1972. Evolutionary model for lead isotopes in conformable ores and in ocean volcanics. *Reviews of Geophysics and Space Physics*, **10**, 529–549.

RUSSELL, R. D. & FARQUHAR, R. M. 1960. *Lead Isotopes in Geology*. Interscience, New York.

RUSSELL, R. D. & REYNOLDS, P. H. 1965. The age of the Earth. *In*: KHITAROV, N. I. (ed.) *Problems of Geochemistry (Vinogradov Volume)*. Academy of Sciences of the USSR [English translation 1969], 35–48.

RUTHERFORD, E. 1905. Present problems in radioactivity. *Popular Science Monthly*, **67**, 5–34.

RUTHERFORD, E. 1906. *Radioactive Transformations*. Scribner's, New York.

RUTHERFORD, E. 1929. Origin of actinium and age of the Earth. *Nature*, **123**, 313–314.

SHIPLEY, B. C. 2001. 'Had Lord Kelvin a right?': John Perry, natural selection and the age of the Earth, 1894–1895. *In*: LEWIS, C. L. E. and KNELL, S. J. (eds) *The Age of the Earth: from 4004 BC to AD 2002*. Geological Society, London, Special Publications, **190**, 91–105.

SINHA, A. K. & TILTON, G. R. 1973. Isotopic evolution of common lead. *Geochimica et Cosmochimica Acta*, **37**, 1823–1849.

STACEY, J. S. & KRAMERS, J. D. 1975. Approximation of terrestrial lead isotope evolution by a two-stage model. *Earth and Planetary Science Letters*, **26**, 207–221.

STACEY, J. S., DELEVAUX, M. E. & ULRYCH, T. J. 1969. Some triple-filament lead isotope ratio measurements and an absolute growth curve for single-stage leads. *Earth and Planetary Science Letters*, **6**, 15–25.

STANTON, R. L. & RUSSELL, R. D. 1959. Anomalous leads and the emplacement of lead sulfide ores. *Economic Geology*, **54**, 588–607.

STRUTT, R. J. 1908. On the accumulation of helium in geological time. *Proceedings of the Royal Society of London*, **81A**, 272–277.

TATSUMOTO, M., KNIGHT, R. J. & ALLÈGRE, C. J. 1973. Time differences in the formation of meteorites as determined from the ratio of lead-207 to lead-206. *Science*, **180**, 1279–1283.

TERA, F. 1980. Reassessment of the 'Age of the Earth'. *Carnegie Institution of Washington Year Book*, **79**, 524–531.

TERA, F. 1981. Aspects of isochronism in Pb isotope systematics – application to planetary evolution. *Geochimica et Cosmochimica Acta*, **45**, 1439–1448.

TERA, F. & CARLSON, R. W. 1999. Assessment of the Pb-Pb and U-Pb chronometry of the early Solar System. *Geochimica et Cosmochimica Acta*, **63**, 1877–1889.

THOMSON, J. J. 1914. Rays of positive electricity. *Proceedings of the Royal Society of London*, **89A**, 1–20.

THOMSON, W. (LORD KELVIN). 1862. On the age of the sun's heat. *Macmillan's Magazine*, **March**, 388–393 [reprinted in: Thomson & Tait, 1890. *Treatise on Natural Philosophy, Part II*. Cambridge University Press, 485–494].

THOMSON, W. (LORD KELVIN). 1864. On the secular cooling of the Earth. *Royal Society of Edinburgh Transactions*, **23**, 157–170 [reprinted in: Thomson and Tait, 1890. *Treatise on Natural Philosophy. Part II*. Cambridge University Press, 468–485].

THOMSON, W. (LORD KELVIN). 1871. On geological time. *Transactions of the Geological Society of Glasgow*, **3** (1), 1–28.

THOMSON, W. (LORD KELVIN). 1897. The age of the Earth as an abode fitted for life. *Annual Report of the Smithsonian Institution*, 337–357.

TILTON, G. R. & STEIGER, R. H. 1965. Lead isotopes and the age of the Earth. *Science*, **150**, 1805–1808.

TILTON, G. R. & STEIGER, R. H. 1969. Mineral ages and isotopic composition of primary lead at Manitouwadge, Ontario. *Journal of Geophysical Research*, **74**, 2118–2132.

WIEN, W. 1898. Die electrostatische und magnetische Ablenkung der Canalstrahlen. *Verhandlungen der Physikalischen Gesellschaft in Berlin*, 10–12.

WIEN, W. 1902. Untersuchengen über die elektrische Entladung in verdünnten Gasen. *Annalen der Physik*, ser. 4, **8**, 244–266.

WILKINS, T. 1988. *Clarence King: A Biography*. University of New Mexico, Albuquerque.

Lead isotopes and the age of the Earth – a geochemical accident

ALBRECHT W. HOFMANN

Max-Planck-Institut für Chemie, Postfach 3060, 55020 Mainz, Germany
(*email*: hofmann@mpch-mainz.mpg.de)

Abstract: The assumptions underlying the models used in the literature for obtaining the age of the Earth from terrestrial lead isotopes are severely violated by the complex evolution of the Earth, particularly the extreme chemical fractionation occurring during crust–mantle differentiation. Young conformable lead deposits are isotopically very similar to young sediments, the erosion products derived from the Earth's most highly fractionated large-scale reservoir, the upper continental crust. Therefore, ancient conformable lead deposits are also likely to track continental compositions rather than the composition of any truly primitive reservoir.

Although the specific enrichment mechanisms during crust formation are both extreme and quite different for U and Pb, the net enrichments in the crust, as well as the corresponding depletions in the residual mantle, are on average very similar for the two elements. It is because of this geochemical coincidence that the time-integrated U/Pb ratios of conformable lead deposits, which integrate and average large volumes of crustal lead, are very close to average mantle values. For the same reason, the isotopic evolution of these conformable lead deposits follows an apparently (nearly) closed system evolution path of a 4.4 to 4.5 Ga-old U-Pb system rather closely, even though that system was actually very far from remaining chemically closed during its history. From these considerations I conclude that terrestrial Pb isotopes do not furnish a suitable tool for determining a refined estimate of the age of the Earth within the broad bounds of 4.4 and 4.56 Ga limits, which are given by other types of evidence, such as the ages of meteorites, the Moon, and the formation intervals of the Earth's core and atmosphere derived from the decay products of short-lived, now-extinct nuclides.

> All the king's horses and all the king's men
> Couldn't put Humpty together again.
> (English nursery rhyme)

The scientific breakthroughs achieved by Holmes, Gerling, Houtermans and Patterson in using lead isotopes to determine the ages of meteorites and the Earth, which culminated with Patterson's classic paper published in 1956 'Age of meteorites and the Earth' (Patterson 1956), have led to the persistent and widely held belief, even today, that lead isotopes provide the best means for determining the age of the Earth. The scientific story of this approach has been traced in the superb, scholarly treatise of Dalrymple (1991; see also Dalrymple 2001).

The original idea of the Holmes–Houtermans approach was that the Earth was, at the time of its formation, permanently differentiated into different reservoirs characterized by different U/Pb ratios. This reflected a general view of the Earth that was essentially static, with no plate tectonics, no mantle convection, and no sea-floor spreading. A sampling of these different reservoirs at the present time would yield lead isotope systematics that could be described by straight-line isochrons on a diagram (referred to as the 'Holmes–Houtermans diagram') showing ^{207}Pb/^{204}Pb versus ^{206}Pb/^{204}Pb, which is still seen in present-day textbooks on isotope geology (Faure 1986). Indeed, the combined data for meteorites and terrestrial oceanic sediments conformed to such an isochron (see figs 2 and 3 in Dalrymple 2001). However, subsequent sampling of terrestrial lead sulphide ores of a given age showed that they did not yield any purely terrestrial isochrons. Instead, most of them appeared to follow a single isotopic growth curve (see fig. 4 in Dalrymple 2001), which could be explained if the ores were derived from a single reservoir with a single U/Pb ratio. This reservoir was thought to be the Earth's mantle, then believed to be chemically and isotopically homogeneous. Crustal rocks were thought to be extracted more or less directly from this reservoir at various times, without significantly affecting the composition of the (residual) mantle. In this case, the growth curve of lead ores could be used to date the ages not only of the ore deposits, but also that of their source reservoir, the Earth's mantle.

Given this general view of the Earth, much of the older literature concerning lead isotopes and their relationship to the age of the Earth was centred on the concepts of 'ordinary' and 'anomalous' lead and 'conformable' lead ores (Stanton & Russell 1959; Russell & Farquhar 1960). A review of these concepts is beyond the scope of this paper, and the reader is referred to Dalrymple (2001). For the present purpose, it is sufficient to say that the central idea developed by Stanton, Russell and others was that the isotopic compositions of 'ordinary' leads follow a single, unique growth curve which defines the age of the Earth. 'Anomalous' leads, on the other hand, deviated significantly from this growth curve. These authors were aware of the dangers of circularity in defining what is ordinary and what is anomalous. They argued, however, that 'ordinary' leads could be identified on geological grounds. Thus, geologically 'conformable' sulphide ores occurring in volcanic–sedimentary associations, and usually forming geologically conformable lenses or layers in the sediments, were found to have remarkably homogeneous ('ordinary') lead isotopic compositions, whereas vein-type, discordant deposits are isotopically much more heterogeneous and thus 'anomalous'. The observations of local isotopic uniformity and the remarkable uniqueness of the isotopic evolution curve with geological time led these workers to infer that conformable leads are derived from a chemically and isotopically primitive uniform reservoir beneath the crust, namely the mantle. Although the concept of 'conformable leads' is no longer particularly fashionable, the notion that common lead isotopes in lead ores and in other rocks derived from the crust or the mantle can define the age of the Earth is by no means dead. For example, a standard set of lead isotope data from conformable ores has been used by Stacey & Kramers (1975) to propose a two-stage model for the evolution of the Earth, which is still widely used today. A nearly identical set of data was used by Albarède & Juteau (1984) and by Galer & Goldstein (1996) to throw light on the earliest history (namely the first 100 million years) of the Earth. Tera (1980, 1981) used lead isotope data from Archaean crustal rocks to put closer constraints on the age of the Earth, and Allègre *et al.* (1995) attempted to use the isotopic composition of Recent mid-ocean ridge basalts to establish an upper limit of the age of the Earth of <4.45 Ga.

In this paper, I will argue that lead isotopes, by defining the ages of meteorites, do indeed provide a firm limit for the age of the Earth of about 4.56 Ga. However, the Earth is very likely significantly younger than 4.56 Ga, and lead isotopes cannot provide a definitive 'best estimate' of the actual age of the Earth and only a very approximate lower limit for this age. The reason for this is simple: the original assumptions for both the Gerling–Holmes–Houtermans method and the conformable-lead method are fundamentally inconsistent with our current knowledge of the Earth's evolution. The assumptions of both models were that one could sample at least one primitive reservoir within the Earth, whose U/Pb ratio has not changed since the formation of the Earth, and which has not exchanged lead or uranium with other, less primitive reservoirs. The developments in geochemistry since the advent of plate tectonics have led to a new understanding of crust–mantle evolution, which effectively rules out the possibility that a truly primitive reservoir is available for sampling anywhere on the Earth's surface. It is conceivable that nearly primitive reservoirs did exist during Archaean time and were sampled by lead deposits (Tera 1980, 1981), but the early Archaean crust–mantle evolution is far from being completely understood, and most workers argue, on the basis of Nd isotopes, that even the oldest continental crust was formed from a mantle reservoir that was already at least slightly fractionated with respect to Sm and Nd, for example (see, however, Nägler & Kramers 1998). Given this situation, it seems unlikely, and in any case it cannot be demonstrated by any kind of independent evidence, that any early Archaean lead deposit was formed from a reservoir that was truly primitive with respect to U and Pb.

Although the emphasis of this paper is somewhat different from that of Galer & Goldstein (1996) and of Kramers & Tolstikhin (1997), its conclusions are entirely consistent with the results of their thorough and elaborate analyses of Pb isotope data from several present-day and ancient Pb reservoirs and individual samples. The 'age' of the Earth, as measured by lead isotopes, is essentially the time when the present-day U/Pb ratio was established in the silicate portion of the Earth. This ratio was lowered relative to its value in the primitive solar nebula and stony meteorites, the Earth's basic building material, by factors of 10 to 100. This dramatic change was caused by a combination of loss of the relatively volatile lead to space and transfer of lead from the silicate mantle to the liquid metallic core. Thus, slow accretion of the Earth, lasting perhaps up to 100 Ma (Wetherill 1986), and relatively late core formation will yield lead isotope 'ages' for the Earth significantly younger than the ages of the oldest, most primitive meteorites.

Galer & Goldstein (1996) emphasized that *all* of the accessible present-day lead isotope 'reservoirs' (continental crust estimates, depleted mantle estimates, and bulk Earth estimates based on the combination of the former two reservoirs), as well as individual samples from Archaean rocks, yield accretion times lasting longer than about 40 Ma (and less than about 250 Ma). Kramers & Tolstikhin (1997) constructed a sophisticated (forward) Earth evolution model, in which they reconciled estimates of lead isotopic compositions of several mantle and crustal reservoirs of various ages with Earth accretion scenarios constrained by other information such as the age of the Moon and core segregation models (see also Halliday 2000). Thus, these investigations aimed to explain the composition of observed terrestrial lead isotope reservoirs in terms of independent information on the age and history of accretion, rather than determining an 'age of the Earth.' My aim in the present paper is consistent with their approach, but emphasizes the point that none of the lead samples or reservoirs used in connection with Gerling–Holmes–Houtermans models is likely to have remained a closed U-Th-Pb system since the end of accretion, so that none of these reservoirs is suitable for a detailed analysis of the age of the Earth. The observation that both the average present-day upper continental crust and the average oceanic crust (and upper mantle) have similar mean isotopic Pb compositions is a 'geochemical accident'. Far from actually being primitive, these reservoirs nevertheless yield apparent Pb isotope ages that are roughly consistent with what we do know about the age of the Earth.

Present-day lead

Patterson's initial, ingenious idea was that marine sediments might yield a well-mixed, representative terrestrial lead reservoir. By showing that lead isotopes of such sediments essentially fall on the 4.5 Ga isochron defined by meteorite samples, he had good reasons to think that this provides powerful confirmation for the hypothesis that the sediments are indeed representative of the whole Earth, and, equally important, that the age of the Earth is identical to that of the meteorites.

Sediments and conformable lead

Our present understanding is that marine sediments are derived from the upper continental crust. This crust has a mean age of about 2 Ga, far younger than the primary ages of meteorites. In addition, we know that the crust is the product of an extreme fractionation process of the mantle. Highly incompatible elements such as U, Th, Ba and Rb are enriched in the crust by factors of 50 to 100 relative to the initial ('primitive') mantle values (Rudnick & Fountain 1995). Consequently, the continental crust now contains (very roughly) 50% of the total terrestrial inventory of these elements, even though it constitutes only about 0.5% of the mass of the total silicate portion of the Earth. Moreover, this enrichment process has separated ('fractionated') these elements from one another. For example, the Th/U ratios of the continental crust and the oceanic crust differ by about a factor of two. To complicate matters further, the continental crust has become differentiated internally, and much of the uranium has been transported from the lower to the upper half of the crust, so that U/Pb ratios in these reservoirs differ considerably (Rudnick & Goldstein 1990). Finally, U is oxidized in near-surface environments to its rather soluble hexavalent oxidation state. As a result, this element becomes mobile and is partly removed from the bulk crust by dissolution during weathering and erosion. In summary, the continental crust represents a chemical reservoir that is as different from any primitive Earth reservoir as one can imagine. Given this knowledge, probably no geochemist would today choose the erosion product of this most highly differentiated of all terrestrial reservoirs, namely marine sediments, to represent the lead isotopes of the Earth system (or any primordial subreservoir of that system).

In spite of the overwhelming geological and geochemical evidence for their 'non-primitiveness', marine sediments do come amazingly close to falling on the meteorite isochron (although they are actually displaced to the right-hand side by a small amount). This is illustrated in Figure 1 which shows data for young conformable lead sulphide ores, oceanic and river sediments, and Pacific and Atlantic mid-ocean-ridge basalts (MORB). Also shown are the 4.566 Ga meteorite isochron, and two zero-age isochrons ('geochrons') for hypothetical primary Earth reservoirs of ages 4.50 and 4.43 Ga. (These younger isochrons are, strictly speaking, valid only for the case where the initial lead isotopic composition of the accreting material did not change between the time of meteorite formation, 4.56 Ga ago, and the time of instantaneous formation of the Earth, 4.50 or 4.43 Ga ago (see also Galer & Goldstein 1996).) Two primary growth curves are given for a 4.43 Ga

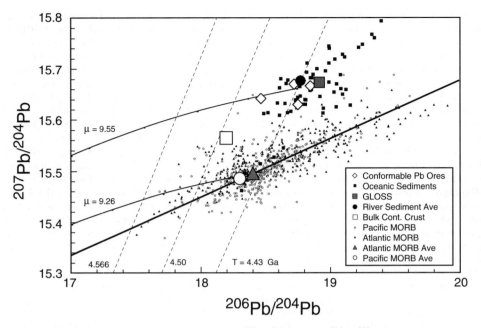

Fig. 1. Holmes–Houtermans-type lead isotope diagram (^{207}Pb/^{204}Pb versus ^{206}Pb/^{204}Pb) showing oceanic sediments (Ben Othman et al. 1989; Plank & Langmuir 1998), the global average of subducting sediments (GLOSS) estimated by Plank & Langmuir (1998), average river sediments (Asmerom & Jacobsen 1993), present-day values of conformable ore leads (large open diamonds) (Cumming & Richards 1975; Stacey & Kramers 1975; Albarède & Juteau 1984; Davies 1984), an estimated composition of the bulk continental crust (Rudnick & Goldstein 1990), mid-ocean ridge basalts (MORB) from the Pacific (small circles) and Atlantic (small triangles) oceans (for references see Hofmann 1997), as well as their respective average values (large open circle and large shaded triangle). The heavy, low-angle line is a linear regression through the Pacific MORB. The dashed lines are zero-age isochrons ('geochrons') for 4.566, 4.50 and 4.43 Ga old U-Pb systems, starting from the primordial Pb isotopic composition of the Canyon Diablo meteorite (see Dalrymple 2001). Also shown are two isotopic growth curves, starting 4.43 Ga ago with Canyon Diablo lead, and corresponding to U/Pb ratios described by μ ($\equiv (^{238}$U/^{204}Pb$)_{t=0}$) = 9.26 and 9.55.

old Earth, corresponding to two slightly different U/Pb ratios. Because U/Pb ratios necessarily change with time as a result of radioactive decay, they are conventionally expressed in terms of present-day values as a common reference frame. By this convention, a parameter expressing the U/Pb ratio is defined as $\mu \equiv (^{238}$U/^{204}Pb$)_{today}$. The two growth curves in Figure 1 are for μ = 9.55 for conformable lead and average recent sediments, and μ = 9.26 for average Pacific and Atlantic MORB.

Figure 1 shows that young conformable lead ores are essentially identical in Pb isotopes to average marine and river sediments, and subtly but significantly different from MORB. This figure also demonstrates that the inference of Stanton, Russell and others, namely that conformable leads are derived directly from the mantle, cannot be correct. The isotopic composition of the upper mantle is best represented by MORB, and it has systematically lower ^{207}Pb/^{204}Pb values than conformable lead ores

(and sediments). In the 1950s, MORB had not been sampled, and there was no independent way to know the isotopic composition of the upper mantle. Today, we know that oceanic sediments have lead concentrations roughly 50 times higher than those of MORB. It is therefore not so surprising that the lead found in the volcanic-hosted ores is primarily derived from the sediments, not the upper mantle. None of this could have been anticipated by Stanton and co-workers. Derivation from sediments also helps to explain the uniformity of lead isotopes in conformable ores. Oceanic sediments have lead isotopic compositions that are far more uniform than those of MORB (or 'unprocessed' crustal rocks). This must be the result of extensive mixing during erosion, transport and deposition of these sediments.

Figure 1 also shows that marine and river sediments actually cluster around a 4.43 Ga geochron of the Holmes–Houtermans type, not identical but remarkably close to the measured age (4.56 Ga) of primitive meteorites (Lugmair & Galer 1992; Göpel *et al.* 1994). This is true in spite of the extreme differentiation of incompatible elements into, and within, the continental crust. Even more remarkable is the fact that the mean values of lead isotopes of MORB also lie close to the 4.43 Ga isochron, in spite of the enormous range of individual sample compositions. (Fig. 1 shows lead isotopes for MORB from two of the three major ocean basins, the Atlantic and Pacific oceans; Indian Ocean MORB are even more heterogeneous and have been omitted for clarity.)

The classic interpretation of the relationships seen in Figure 1 in terms of the Holmes–Houtermans model would be that the two reservoirs, the upper mantle (which produces MORB) and the upper continental crust (which produces sediments), were separated about 4.43 Ga ago and have evolved separately since that time. The primary separation age would then be called the 'age of the Earth'. The slope of the correlation lines, shown here for Pacific MORB as an example, would simply reflect the mean age of internal differentiation of the Pacific MORB source (in this case about 1.9 Ga). From the record of crustal ages, we know that this simple interpretation cannot be correct, not only because the mean age of the crust is much younger than 4.43 Ga, but also because crust and mantle have not been closed systems but have been exchanging some material throughout traceable geological history. The distribution of formation ages of the continental crust, recently reviewed by Condie (1998) (see also Kamber *et al.* 2001), shows no continental masses older than 4 Ga, and even the advocates of constant-continental-volume models (e.g. Armstrong 1991) do not argue that the present-day continental crust is as old as the Earth, but is newly formed and destroyed through plate tectonic processes throughout Earth history. Thus, irrespective of whether the crust grew progressively in time or existed in a steady state of growth and destruction, the continental crust is not a reservoir that can be used to date the formation of the Earth.

Upper mantle and mid-ocean ridge basalts (MORB)

The mantle region that produces MORB is the depleted residue of the continental crust. It is internally quite heterogeneous in terms of lead isotopes (see Fig. 1) because of internal differentiation and exchange with the crust. Nevertheless, Allègre *et al.* (1995) attempted to use the mean isotopic composition of MORB from the three major ocean basins to calculate an *upper* limit of 4.45 Ga for the age of the Earth. Their argument was based on the assertion that the mean depleted mantle (as sampled by MORB) has a higher ^{207}Pb*/^{206}Pb* ratio[1] than the continental crust and the 'bulk silicate Earth' (meaning the total assemblage of mantle and crust). This ^{207}Pb*/^{206}Pb* ratio is defined as the slope of a single-stage isochron (or 'geochron'), such as those shown on Figure 1 formed by connecting the present-day isotopic composition of a given reservoir with that of primordial lead. The ^{207}Pb*/^{206}Pb* ratio can therefore be directly related to a geochron age. Figure 1 shows that present-day Pacific and Atlantic Ocean MORB lie close to a 4.43 Ga geochron. Inclusion of the Indian Ocean MORB would raise the overall average slightly to approximately 4.45 Ga. However, this is by no means a generally accepted maximum geochron age obtained for the major present-day terrestrial reservoirs. For example, the geochron age of the bulk continental crust as calculated by Allègre & Lewin (1989) is 4.43 Ga, but the corresponding age of the bulk crust estimated by Rudnick & Goldstein (1990) is 4.50 Ga. Even the bulk-silicate Earth age of Allègre *et al.* (1988) and Allègre & Lewin (1989) of 4.46 Ga is slightly higher than the upper limit claimed by Allègre *et al.* (1995). Clearly, this type of model calculation is no more applicable to the depleted mantle than it is to the crust, because there are good reasons to believe that the depleted mantle is strongly coupled, and largely complementary, to the continental crust (e.g. Hofmann 1988).

Bulk silicate Earth

If we knew the present-day lead isotopic composition of the bulk silicate Earth (BSE) (total mantle and crust) and its initial value (essentially from meteorites), we could determine the age of this silicate portion of the Earth directly according to the Holmes–Houtermans method. Galer & Goldstein (1996) have compiled eight published estimates of the lead isotopic composition of BSE. They are all rather similar, mostly because of the remarkable similarity between sediments, conformable lead ores and MORB, as discussed above. They have ^{207}Pb*/^{206}Pb*

---

[1] The asterisks signify an abbreviated notation for the 'radiogenic ^{208}Pb/^{206}Pb ratio'. It is defined as follows:

$$^{208}Pb*/^{206}Pb*$$
$$= ((^{208}Pb/^{204}Pb)\text{today} - (^{208}Pb/^{204}Pb)\text{primordial})/$$
$$((^{206}Pb/^{204}Pb)\text{today} - (^{206}Pb/^{204}Pb)\text{primordial}).$$

values ranging from 0.566 to 0.606, corresponding to single-stage ages of 4.42 to 4.52 Ga. These BSE estimates (rather than just conformable leads) constitute the main reason why many workers would agree today that lead isotope data require a 'young' age of the Earth. The apparent strength of this argument is that it is in rough agreement with the estimates derived from both the upper continental crust and the depleted mantle. In other words, it appears that all the accessible reservoirs point in the same direction.

One weakness of this argument is that it still leaves an uncertainty of about 100 Ma. Another, perhaps more serious weakness lies in the fact that the above BSE estimates are heavily dependent on the type of data shown in Figure 1 which represent a depleted mantle reservoir and the upper continental crust as sampled by sediments. Depending on the general geochemical Earth model one uses, the lower mantle may or may not be isotopically rather different from the MORB data shown in Figure 1 (see, for example Hofmann 1997).

A probably even larger element of uncertainty is introduced by our poor knowledge of the composition of the lower continental crust, which is generally not sampled by erosion and sediment production. Thus, while we do have outcrops of lower-crustal granulites as well as xenoliths from regions where there is no outcrop, crustal rocks are notoriously heterogeneous. In general, the lower crust is variably depleted in uranium relative to lead, and, depending on the age of the particular crustal province, this causes the lower crust to be less radiogenic than the upper crust. On diagrams of the type shown in Figure 1 the lower crust lies to the left of the upper crust, and this causes the estimate of the BSE composition to shift in the same direction, and therefore to older values. Rudnick & Goldstein (1990) have taken this into account in estimating the composition of the bulk continental crust, by using data from lower-crustal xenoliths in conjunction with a specific crustal growth model. Their effort is probably the best that is available in the literature, but it is still model-dependent and it is based on only nine xenolith localities worldwide, in which the observed Pb isotopic compositions scatter widely, ranging from $^{206}Pb/^{204}Pb = {<}16$ to >19. In comparison, the much more extensive data for oceanic sediments have a total range from 18.3 to 19.3 (see Fig. 1), less than one-third of the xenolith range. Additional uncertainties are added by the largely unknown composition of the subcontinental lithosphere, for which very few lead isotope data exist, and which is also part of the required bulk silicate Earth budget.

All in all, substantial uncertainties remain about the total inventory of the bulk silicate Earth. Present indications are that BSE does lie to the right of, and is therefore younger than, the 4.56 Ga meteorite isochron. Indications are further that BSE is older than about 4.4 Ga, but the extreme internal heterogeneity of the accessible terrestrial reservoirs and the partial to complete inaccessibility (to sampling) of some reservoirs make it impossible to use these isotope data to narrow this range significantly. No individual, demonstrably primitive, geochemical reservoir is available today to which we could apply the Holmes–Houtermans model, and no amount of averaging, conducted by nature or on a computer, is likely to reconstruct reliably the composition of the present-day bulk silicate Earth reservoir from rock samples. As the nursery rhyme from *Mother Goose's Tales* quoted at the beginning of the paper suggests, no matter how hard we try, we cannot put Humpty together again.

Ancient lead

If present-day lead reservoirs are too evolved and fractionated with respect to U and Pb, as I have argued above, it is still conceivable, at least in principle, that examination of ancient rocks might reveal the existence of primitive reservoirs that have undergone only single-stage U-Pb evolution, and from these it might be possible to infer an 'age of the Earth'. This is the idea behind using ancient conformable lead deposits to obtain a unique growth curve, which would also define its initial starting time. In particular, Tera (1981) has investigated the systematics of Archaean lead with this objective in mind (see also Dalrymple 2001). However, any such effort must necessarily contain a circular element, as long as the choice of samples is made on the basis of the lead isotopic relationships themselves. If, on the other hand, evidence for a geochemically primitive mantle source reservoir could be found in some suite of Archaean rocks, then we could use initial isotopic compositions of these rocks to infer the age of this primitive reservoir (for an explanation of initial isotope ratios, see Lewis 2001).

Figure 2 shows the evolution of lead isotopes in conformable lead deposits as a function of age. For this purpose, it is useful to follow the approach of Albarède & Juteau (1984), which replaces the somewhat complex three-isotope plots used in the Holmes–Houtermans models and its more recent derivatives (e.g. Tera 1980, 1981), by a diagram plotting the evolution of a single isotope ratio directly as a function of age.

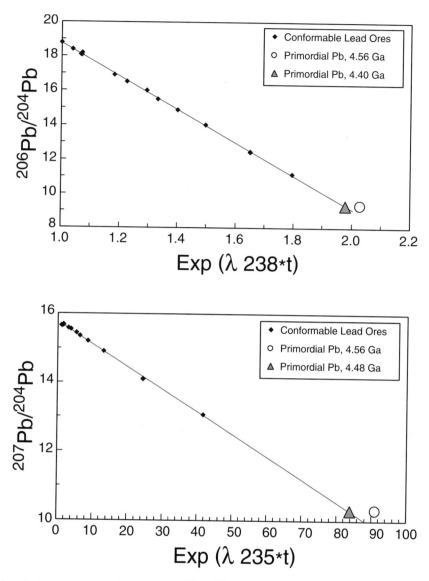

Fig. 2. Linearized evolution plots for radiogenic ^{206}Pb/^{204}Pb and ^{207}Pb/^{204}Pb in conformable lead ores as functions of $\exp(\lambda t)$, following the approach of Albarède & Juteau (1984). Extrapolation to the composition of primordial lead from the Canyon Diablo meteorite (Tatsumoto 1973) would yield 'ages' of 4.40 Ga for ^{206}Pb/^{204}Pb and 4.48 Ga for ^{207}Pb/^{204}Pb (Albarède & Juteau 1984).

The only complication here is that the half-lives of ^{238}U and ^{235}U (4.5 and 0.7 Ga respectively) are similar to and less than the age of the Earth, unlike those for ^{87}Rb and ^{147}Sm, which are ten and 20 times the age of the Earth. Because of these relatively short half-lives, the isotope ratios plotted against age form strongly curved evolution lines, following the exponential decay equation. However, by plotting the exponential term e^t, or $e^{(\lambda t)}$ following Albarède & Juteau (1984), where λ is the decay constant and t is the age, it is easy to linearize the closed-system growth curves of both ^{206}Pb/^{204}Pb and ^{207}Pb/^{204}Pb perfectly, as can be seen in Figure 2.

Lead isotopes from conformable lead sulphide deposits appear to follow single (closed-system) growth trajectories rather closely. However, if the starting composition of these growth lines is taken to be primordial lead as measured in the Canyon Diablo meteorite (Tatsumoto 1973), then these lines originate at two different ages, both of which are significantly younger than the meteorite age of 4.56 Ga. For the ^{206}Pb/^{204}Pb evolution line, this starting age is 4.40 Ga, whereas for ^{207}Pb/^{204}Pb it is 4.48 Ga. (This result of the analysis of Albarède & Juteau (1984) should formally be modified by adjusting the initial primordial isotopic composition for some amount of radiogenic growth during the time interval of 4.56 to 4.40 Ga ago (Galer & Goldstein 1996).)

There is a debate in the literature whether the combined ^{207}Pb and ^{206}Pb growth can be reconciled with a single, closed-system growth curve, which would define a sort of 'age of the Earth', or whether two-stage or multi-stage models are needed to explain the conformable lead data. What has not been sufficiently considered in this debate is the circumstance that the expectation of a single growth curve for lead isotopes from a truly primitive reservoir is unlikely from a geochemical point of view.

It was shown above that young conformable lead ores are predominantly derived from the continental crust, not the mantle. This is probably also true for ancient ores. However, ancient ores might be derived from crust that was rather 'juvenile' in that it had been separated from the mantle quite recently. In that case, the ancient lead ores might indeed track the isotopic composition of the ancient mantle. We can look for a comparison with isotope data for neodymium from ancient rocks. Unlike U/Pb, the relevant parent/daughter ratio, Sm/Nd, is universally believed to be identical in chondritic meteorites and in the Earth. We therefore assume that we know the evolution of the radiogenic daughter ratio, ^{143}Nd/^{144}Nd, in the primitive mantle quite precisely from measurements on chondritic meteorites. Deviations of ^{143}Nd/^{144}Nd from the meteoritic value are conventionally expressed in epsilon units, so that ε(Nd) = 1 represents a deviation of one part in 10 000 from the chondritic value for a given age. A review of the enormous literature on Nd isotopes is beyond the scope of this paper. It is sufficient to note here that the majority of ancient crustal rocks have initial ε(Nd) values ('initial' meaning at the time of their formation; see also Lewis 2001) that are greater than zero. This means that the mantle from which these ancient crustal rocks were derived was no longer truly primitive but had already been differentiated, and thereby changed its Sm/Nd ratio, by previous geological events. This conclusion has recently been challenged by Nägler & Kramers (1998) who suggested that the difference in ε(Nd) between ancient crustal rocks and meteoritic values might be caused by uncertainties in chondritic composition, analytical artefacts and interlaboratory bias. However, this issue is not resolved at present. The search for the elusive truly primitive mantle reservoir will no doubt go on, but at present, we cannot rely on having such a reservoir available for sampling.

The U-Pb coincidence

Why do the time-integrated U/Pb ratios of the average continental crust and the MORB-source reservoirs (indicated on Fig. 1 by the growth curves for $\mu = 9.55$ and 9.26) differ by only about 3%? The reason is neither obvious nor completely understood. If we assume, for the moment, that BSE is actually 4.43 Ga old, then marine sediments and their source rocks, in order for their lead isotopes to coincide approximately with this isochron, must have had U/Pb ratios nearly identical (i.e. within about 3%) to that of BSE, at least on average and in a 'time-integrated' sense. One way or the other, lead and uranium must be enriched in the continental crust (relative to the mantle) by large but very similar factors. This is not at all what one would expect from a conventional geochemical point of view. The geochemically relevant properties of lead, its ionic charge ($Z = +2$) and radius ($r = 1.26 \times 10^{-10}$ m) in silicate structures, are similar to those of strontium ($Z = +2$, $r = 1.21 \times 10^{-10}$ m), and these properties normally govern the degree of compatibility of an element with mantle silicates, the degree of enrichment in mantle-derived melts, and ultimately the enrichment in the continental crust (Taylor & McLennan 1985; Hofmann 1988). Because of its slightly greater ionic radius, one would expect Pb to be only slightly more enriched in the continental crust than Sr. Taylor & McLennan (1985) estimate the enrichment factors of Rb and Sr in the continental crust over BSE to be 53 and 13, respectively; Rudnick & Fountain (1995) estimate enrichments factors of 97 for Rb and 16 for Sr. This means that the average Rb/Sr ratio of the crust is four to six times greater than that of the primitive mantle. Uranium, in spite of its lower ionic radius (1.12×10^{-10} m), is also highly incompatible because of its greater ionic charge ($+4$). Its enrichment factor in the crust (70) (Rudnick & Fountain 1995) is only slightly lower than that of Rb. These considerations lead to the expectation that continental crust should have much higher U/Pb ratios and much more radiogenic lead than the mantle from which it is made.

To illustrate how far the actual lead isotope data are from what a geochemist might expect, we assume, for the moment, that both U and Pb behave geochemically like ordinary incompatible lithophile elements. Using ionic radius as a

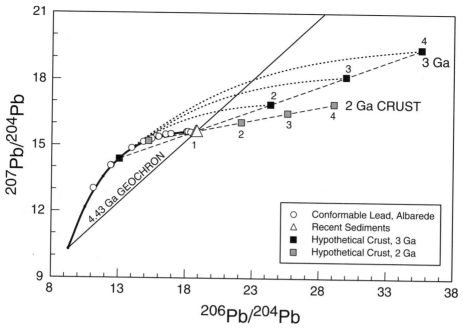

Fig. 3. Geochemically expected evolution of a hypothetical, average continental crust, assuming that U and Pb are partitioned into the crust in a manner similar to Rb and Sr, respectively. Dotted lines are evolution curves assuming U/Pb ratios that are two, three and four times greater than the primitive-mantle value. Black squares represent calculated, hypothetical, present-day compositions of 3 Ga old crust; shaded squares represent 2 Ga old crust. Also shown are conformable lead data (open circles), using the selection of Albarède & Juteau (1984) and recent sediments (open triangles), representing the actual upper continental crust, as in Figure 1.

guide, we further assume that the partitioning behaviour of U and Pb is similar to that of Rb and Sr, respectively, so that the crustal U/Pb ratio (or μ value) should be two to four times higher than the primitive BSE value. Figure 3 shows the hypothetical Pb isotopic compositions of average present-day crust, assuming that the crustal U/Pb ratio is greater than the primitive mantle ratio by factors of two, three and four, and for mean crustal ages of 2 and 3 Ga. This shows that the Pb isotopic composition of such a crust would be expected to be far more radiogenic (with $^{206}Pb/^{204}Pb$ values ranging from 22 to 36) than the observed values of recent sediments and conformable lead deposits ($^{206}Pb/^{204}Pb < 19$), and also quite far from any present-day geochron.

Therefore, the actual enrichment factor for lead must be three to five times higher than that of strontium. The reasons for this anomalously high enrichment of lead in the continental crust are not completely understood. However, it is quite likely that they are related to 'metasomatic' processes, specifically the hydrothermal transfer of lead from oceanic crust to sediments near ocean ridges (Peucker-Ehrenbrink *et al.* 1994) and hydrothermal transfer from subducted lithosphere into island-arc magmas and ultimately into the continental crust (Hofmann 1997). Uranium is also mobilized within the continental crust. It is depleted in the lower crust and enriched in the upper crust (Rudnick & Goldstein 1990). Ultimately, a significant portion of the total crustal U inventory is lost by oxidation and dissolution during weathering and erosion. The fortuitous, net result of these different processes is that upper continental crust, average depleted mantle, and probably the primitive mantle all have Pb isotopic compositions that lie close to the 'geochron' shown in Figure 1 rather than far to the right of it as shown on Figure 3.

One might be tempted to explain away the isotopic similarity of continental crust and depleted mantle by invoking mixing processes between the two reservoirs. However, while mixing can reduce the difference between the extreme end members, it will not fundamentally change the slope of a line drawn between these end members in $^{207}Pb/^{204}Pb$–$^{206}Pb/^{204}Pb$ space. In other words, extensive mixing might make the two reservoirs more similar to each other, but if the mean differentiation age between these end members is on the order of 2 Ga, then mixing alone cannot generate an apparent 4.4 or 4.5 Ga isochron. How can such an apparent 'isochron', giving an age clearly in excess of the actual mean differentiation age, be generated from an initially (isotopically) uniform system?

Figure 4 shows an example of a class of possible models for the origin of continental crust, all of which require at least three stages of Pb isotope evolution, including at least one second stage of high μ (U/Pb expressed as $\mu \equiv (^{238}U/^{204}Pb)_{today}$) and a third stage of relatively low μ. The earlier, high-μ stage accelerates the growth in $^{207}Pb/^{204}Pb$, and the later, low-μ stage retards the growth of $^{206}Pb/^{204}Pb$ while preserving the elevated $^{207}Pb/^{204}Pb$. Figure 4 shows a highly simplified, three-stage model for generating the Pb isotopic composition of bulk continental crust (Rudnick & Goldstein 1990), produced 2.5 Ga ago from a depleted mantle reservoir. For simplicity, this mantle is assumed to evolve with a single value of $\mu = 9.26$ to the present-day MORB composition, starting with primordial lead 4.43 Ga ago. Juvenile continental crust with a much higher value of $\mu_2 = 14.35$ was formed 2.5 Ga ago. It evolved for 500 Ma until 2.0 Ga ago, when it was differentiated internally and lost a sufficient amount of uranium (presumably through internal differentiation followed by oxidation, weathering, dissolution and transport of U into the ocean), so that its U/Pb ratio was reduced to $\mu_3 = 7.0$. (This simplified example is intended for illustration purposes only and should not be taken as model for any actual crust–mantle evolution.) Without the low-μ stage 3 this crust would have evolved to present-day Pb isotope ratios of $^{206}Pb/^{204}Pb = 21$ and $^{207}Pb/^{204}Pb = 5.9$ (end point of the dashed evolution curve in Fig. 4), quite different from any reasonable estimate of the actual composition. Only by reducing μ to a very low value ($\mu_3 = 7$) can the actual isotopic composition of bulk continental crust (as estimated by Rudnick and Goldstein) be reached. This example shows that the evolution of continental crust and MORB mantle, having generally very similar isotopic compositions but slightly different $^{207}Pb/^{204}Pb$ ratios, actually requires a very complex geological history. The final position of the continental crust (and of conformable lead ores) at elevated $^{207}Pb/^{204}Pb$, but close to the geochron of 4.5 Ga, thus requires a highly fortuitous interplay of growth parameters.

It is this *accidental* geochemical behaviour of the U/Pb ratio which is responsible not only for

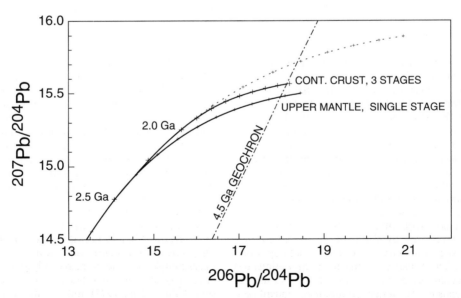

Fig. 4. Three-stage lead evolution model to explain the actual isotopic composition of the bulk continental crust (as estimated by Rudnick & Goldstein 1990). The lowermost curve, (labelled UPPER MANTLE) is a single-stage mantle evolution line which produces average MORB mantle in a 4.43 Ga old Earth with $\mu = 9.26$ (see Fig. 1). The uppermost, dashed curve represents a two-stage model forming a hypothetical crust 2.5 Ga ago with an elevated U/Pb ratio corresponding to $\mu_2 = 14.35$. The intermediate curve (labelled CONT. CRUST) requires three stages, where the third stage departs from the two-stage curve 2.0 Ga ago, when uranium was lost from the crust and U/Pb was reduced to $\mu_3 = 7.0$. This complex, three-stage model with a rather fortuitous sequence of μ values produces a bulk continental crust which lies approximately on a 4.5 Ga geochron (dash–dot line), similar to present-day MORB average which lies on a 4.43 Ga geochron.

Patterson's early perceived success of matching the present-day Earth with the meteorite isochron, but also for a voluminous geochemical literature on plumbology. This includes the concepts of the 'lead paradox' introduced by Allègre (1969), the long-term survival of the hypothesis of a single Pb isotope growth curve for lead deposits, particularly the isotopic evolution of 'conformable' lead deposits (Stanton & Russell 1959), the subsequent multi-stage evolution models (Cumming & Richards 1975; Stacey & Kramers 1975), the 'plumbotectonics' models of Zartman and co-workers (Zartman & Doe 1981), and the efforts of Tera (1980, 1981) to refine the estimate of the age of the Earth. All these concepts and developments are to a significant extent based on, or would not have been possible without, this 'geochemical accident', which produced the approximate equality of time-integrated U/Pb in crust and mantle. This accident probably does provide an approximately correct estimate of the age of the Earth within the broad bounds of uncertainty of about 150 Ma. This rough age estimate might still be in error, if the differentiation of the Earth did produce one or more additional, unsampled (or poorly sampled), isotopically very different reservoirs of lead in some region such as the lower continental crust, the subcontinental lithosphere, the lower mantle or the ocean island basalt source reservoir. In any case, if the reservoirs that we can sample well are not truly primitive, as I have argued above, then we cannot invert the Pb isotope data of terrestrial rocks to refine our estimates of the age of the Earth to closer limits than those given by other, independent information.

Other constraints on the age of the Earth

We have seen that lead isotopes provide only one powerful constraint on the age of the Earth, namely the upper bound of 4.56 Ga, the age of primitive meteorites, which can be measured rather precisely by U-Pb and by Pb-Pb methods. A lower bound is best determined by other methods, because chemical fractionation has obscured the precise record of Pb, Nd and Sr isotopes in the mantle, so that independent confirmation of any precise age derived from a particular selection of lead samples, 'conformable' or otherwise, fails. There are, however, other firm limits on the age of the Earth. A detailed discussion of the current best estimates of these limits is beyond the scope of this paper, and only a brief summary will be given here. For a more in-depth discussion of the early accretion history of the Earth, the reader is referred to the recent literature (Wetherill 1986; Allègre et al. 1995; Galer & Goldstein 1996; Harper & Jacobsen 1996; Halliday & Lee 1999; Ozima & Podosek 1999; Halliday 2000).

The oldest minerals found in the Earth's crust were, until recently, 4.27 Ga-old detrital zircons from western Australia (Compston & Pidgeon 1986). Most recently, further investigation of the same population of zircons yielded a single grain with a concordant U-Pb age of 4.40 Ga (Wilde et al. 2001). If confirmed, this will provide a firm lower limit of 4.40 Ga for the age of the Earth. Current theory about the accretion history of terrestrial planets suggests that the accretion process lasted at least several tens of millions of years (Wetherill 1986). Therefore it may be impossible, or in any case not meaningful, to give an 'age of the Earth' with a very much higher precision than, say, 10 to 50 Ma. Given a total timespan between the age of the oldest meteorites and the age of the oldest terrestrial minerals of only about 150 Ma, we may hope to refine the estimate of the 'true' age of the Earth a little further, but probably by less than an order of magnitude. As Claude Allègre once remarked in a lecture, most of us know on which day we were born, some of us know during which hour we were born, but none of us knows the exact second.

One way of determining a reasonable lower limit for the age of the Earth is to accept the widely held view that the Moon was created as a result of a very large impact on Earth at a stage when it was already close to its present size. In this case the oldest lunar rocks provide another lower limit for the age of the Earth. The compilation of ages of the oldest lunar rocks provided by Dalrymple (1991) shows several values ranging in age from 4.0 to 4.51 Ga. Dalrymple concluded from a careful review of the evidence that 'even the most conservative interpretation of the data ... leads to the conclusion that the Moon's age must exceed 4.4 Ga'. His overall conclusion of this review is, however, that 'the Moon is at least 4.5 Ga in age'. This conclusion is consistent (within the quoted uncertainty) with the more recently determined Sm-Nd age of 4.56 ± 0.07 Ga of a lunar anorthosite (Alibert et al. 1994) and a combined U-Pb and Pb-Pb age of 4.51 ± 0.01 Ga of another lunar anorthosite (Hanan & Tilton 1987).

The decay products of short-lived, now extinct radioactive nuclides also provide constraints on the formation history of the Earth. These nuclides were produced by nucleosynthesis in stars just prior to the condensation of the solar nebula. They were still 'alive' at the time of formation of meteorites 4.566 Ga ago, and they can be used to decipher a chronology of the earliest history of the solar system.

One of these short-lived nuclides is ^{182}Hf, which has lithophile affinities and decays to ^{182}W (with siderophile affinities) with a half-life of only 9 Ma. Halliday and co-workers have discovered that terrestrial rocks and carbonaceous chondrites have identical relative abundances of ^{182}W, whereas iron meteorites are deficient in ^{182}W (Halliday & Lee 1999). This deficiency means that tungsten was sequestered into the metal cores of ancient planetesimals before all of the ^{182}Hf had decayed. If the Earth's core had formed at the same time as the iron meteorites, namely within a few million years after the formation of chondritic meteorites 4.566 Ga ago, it would have the same deficiency in ^{182}W as the iron meteorites, and the silicate portion of the Earth would have a complementary relative excess. On the other hand, if the segregation of the core occurred several half-lives later (about 50 Ma or more), after nearly all of the ^{182}Hf had decayed, then the tungsten must have the same isotopic composition in the core and in the mantle. This would be identical to the tungsten isotopic composition of chondritic meteorites, and this is what Halliday and co-workers observed. In principle, this could mean that core formation was delayed relative to the formation of the Earth. However, assuming that current slow Earth accretion models (Wetherill 1986) are correct, it seems more plausible to infer that the protracted accretion process itself caused the delay in core formation (Halliday 2000).

Other evidence for either delayed or slow formation of the Earth comes from xenon isotopes. Ozima & Podosek (1999) have recently reviewed the evidence derived from the differences in the abundances of ^{129}Xe, the decay product of ^{129}I ($t_{1/2} = 16$ Ma), and ^{136}Xe, the decay product of spontaneous fission of ^{244}Pu ($t_{1/2} = 83$ Ma). The Earth's atmosphere is deficient in both ^{129}Xe and ^{136}Xe relative to meteorites, and this can be explained if iodine and plutonium were incorporated into the Earth some 50 to 100 Ma later than the age of meteorite formation and if the Earth's present-day atmosphere was produced by subsequent outgassing of the mantle.

In detail, all the considerations based on extinct radioactivity are somewhat model-dependent and cannot be expected to give highly precise estimates of the age interval between the formation of meteorites and that of the Earth. Nevertheless, all of the different lines of evidence appear to be consistent with the interpretation that the Earth reached a size close to its present value between 4.50 and 4.40 Ga ago. The fact that this conclusion is not very different from Patterson's original estimate and from the subsequent, more detailed evaluations of conformable lead ores is the result of a 'geochemical accident', namely the amazing similarity of time-integrated U/Pb ratios in averaged mantle and crustal reservoirs. This similarity is all the more remarkable as the compositions of both reservoirs are internally very heterogeneous. It is this stroke of luck which has given us such remarkably close estimates of the age of the Earth from lead isotopes, even though hindsight clearly reveals that the necessary assumption of the Holmes–Houtermans model and the conformable lead model, namely the survival of one or more closed systems of primordial age, has been grossly violated during the Earth's evolution.

I wish to thank G. Wasserburg for drawing my attention to G. B. Dalrymple's admirable book *The Age of the Earth*, Dalrymple for writing it, and C. Lewis for inviting me to help Celebrate the Age of the Earth, and for encouraging me to write this article. G. B. Dalrymple, S. Galer, A. Halliday, C. Hawkesworth, I. Kramers, S. Moorbath and I. Tolstikhin helped by discussing important issues and/or reviewing an earlier version of this manuscript. I thank them all for their efforts and their positive feedback, which, both in spite and because of the great diversity of opinions, helped enormously in preparing the final draft. None of these people should be held responsible for any blunders the reader may still find in this article.

References

ALBARÈDE, F. & JUTEAU, M. 1984. Unscrambling the Pb model ages. *Geochimica et Cosmochimica Acta*, **48**, 207–212.

ALIBERT, C., NORMAN, M. D. & MCCULLOCH, M. T. 1994. An ancient Sm-Nd age for a ferroan noritic anorthosite clast from lunar breccia 67016. *Geochimica et Cosmochimica Acta*, **58**, 2921–2926.

ALLÈGRE, C. J. 1969. Comportement des systèmes U-Th-Pb dans le manteau supérieur et modèle d'évolution de ce dernier au cours de temps géologiques. *Earth and Planetary Science Letters*, **5**, 261–269.

ALLÈGRE, C. J. & LEWIN, E. 1989. Chemical structure and history of the Earth: evidence from global non-linear inversion of isotopic data in a three-box model. *Earth and Planetary Science Letters*, **96**, 61–88.

ALLÈGRE, C. J., LEWIN, E. & DUPRÈ, B. 1988. A coherent crust-mantle model for the uranium-thorium-lead system. *Chemical Geology*, **70**, 211–234.

ALLÈGRE, C. J., MANHÈS, M. & GÖPEL, C. 1995. The age of the Earth. *Geochimica et Cosmochimica Acta*, **59**, 1445–1456.

ARMSTRONG, R. L. 1991. The persistent myth of crustal growth. *Australian Journal of Earth Science*, **38**, 613–630.

ASMEROM, Y. & JACOBSEN, S. B. 1993. The Pb isotopic evolution of the Earth: inferences from river water suspended loads. *Earth and Planetary Science Letters*, **115**, 245–256.

BEN OTHMAN, D., WHITE, W. M. & PATCHETT, J. 1989. The geochemistry of marine sediments, island arc magma genesis, and crust-mantle recycling. *Earth and Planetary Science Letters*, **94**, 1–21.

COMPSTON, W. & PIDGEON, R. T. 1986. Jack Hills, evidence of more very old detrital zircons in Western Australia. *Nature*, **321**, 766–769.

CONDIE, K. C. 1998. Episodic continental growth and supercontinents: a mantle avalanche connection? *Earth and Planetary Science Letters*, **163**, 97–108.

CUMMING, G. L. & RICHARDS, J. R. 1975. Ore lead isotope ratios in a continuously changing Earth. *Earth and Planetary Science Letters*, **28**, 155–171.

DALRYMPLE, G. B. 1991. *The Age of the Earth.* Stanford University, Stanford.

DALRYMPLE, G. B. 2001. The age of the Earth in the twentieth century: a problem (mostly) solved. *In*: LEWIS, C. L. E. & KNELL, S. J. (eds) *The Age of the Earth: from 4004 BC to AD 2002.* Geological Society, London, Special Publications, **190**, 205–221.

DAVIES, G. F. 1984. Geophysical and isotopic constraints on mantle convection: An interim analysis. *Journal of Geophysical Research*, **89**, 6017–6040.

FAURE, G. 1986. *Principles of Isotope Geology.* Wiley, New York.

GALER, S. J. G. & GOLDSTEIN, S. L. 1996. Influence of accretion on lead in the Earth. *In*: BASU, A. & HART, S. R. (eds) *Earth Processes: Reading the Isotopic Code.* American Geophysical Union, Washington, Geophysical Monograph **95**, 75–98.

GÖPEL, C., MANHÈS, G. & ALLÈGRE, C. J. 1994. U-Pb systematics of phosphates from equilibrated ordinary chondrites. *Earth and Planetary Science Letters*, **121**, 153–171.

HALLIDAY, A. N. 2000. Terrestrial accretion rates and the origin of the Moon. *Earth and Planetary Science Letters*, **176**, 17–30.

HALLIDAY, A. N. & LEE, D.-C. 1999. Tungsten isotopes and the early development of the Earth and Moon. *Geochimica et Cosmochimica Acta*, **63**, 4157–4179.

HANAN, B. B. & TILTON, G. R. 1987. 60025: Relict of primitive lunar crust? *Earth and Planetary Science Letters*, **84**, 15–21.

HARPER, C. L. JR & JACOBSEN, S. B. 1996. Noble gases and Earth's accretion. *Science*, **273**, 1814–1818.

HOFMANN, A. W. 1988. Chemical differentiation of the Earth: the relationship between mantle, continental crust, and oceanic crust. *Earth and Planetary Science Letters*, **90**, 297–314.

HOFMANN, A. W. 1997. Mantle geochemistry: the message from oceanic volcanism. *Nature*, **385**, 219–229.

KAMBER, B. S., MOORBATH, S. & WHITEHOUSE, M. J. 2001. The oldest rocks on Earth: time constraints and geological controversies. *In*: LEWIS, C. L. E. & KNELL, S. J. (eds) *The Age of the Earth: from 4004 BC to AD 2002.* Geological Society, London, Special Publications, **190**, 177–203.

KRAMERS, J. D. & TOLSTIKHIN, I. N. 1997. Two terrestrial lead isotope paradoxes, forward transport modelling, core formation and the history of the continental crust. *Chemical Geology*, **139**, 75–110.

LEWIS, C. L. E. 2001. Arther Holmes' vision of a geological timescale. *In*: LEWIS, C. L. E. & KNELL, S. J. (eds) *The Age of the Earth: from 4004 BC to AD 2002.* Geological Society, London, Special Publications, **190**, 121–138.

LUGMAIR, G. W. & GALER, S. J. G. 1992. Age and isotopic relationships among the angrites Lewis Cliff 86010 and Angra dos Reis. *Geochimica et Cosmochimica Acta*, **56**, 1673–1694.

NÄGLER, T. F. & KRAMERS, J. D. 1998. Nd isotopic evolution of the upper mantle during the Precambrian: Models, data and the uncertainty of both. *Precambrian Research*, **91**, 233–252.

OZIMA, M. & PODOSEK, F. A. 1999. Formation age of Earth from $^{129}I/^{127}I$ and $^{244}Pu/^{238}U$ systematics and the missing Xe. *Journal of Geophysical Research*, **104**, 25493–25499.

PATTERSON, C. C. 1956. Age of meteorites and the earth. *Geochimica et Cosmochimica Acta*, **10**, 230–237.

PEUCKER-EHRENBRINK, B., HOFMANN, A. W. & HART, S. R. 1994. Hydrothermal lead transfer from mantle to continental crust: the role of metalliferous sediments. *Earth and Planetary Science Letters*, **125**, 129–142.

PLANK, T. & LANGMUIR, C. H. 1998. The chemical composition of subducting sediment and its consequences for the crust and mantle. *Chemical Geology*, **145**, 325–394.

RUDNICK, R. L. & FOUNTAIN, D. M. 1995. Nature and composition of the continental crust: a lower crustal perspective. *Reviews of Geophysics*, **33**, 267–309.

RUDNICK, R. L. & GOLDSTEIN, S. L. 1990. The Pb isotopic compositions of lower crustal xenoliths and the evolution of lower crustal Pb. *Earth and Planetary Science Letters*, **98** 192–207.

RUSSELL, R. D. & FARQUHAR, R. M. 1960. *Lead Isotopes in Geology.* Interscience, New York.

STACEY, J. S. & KRAMERS, J. D. 1975. Approximation of terrestrial lead isotope evolution by a two-stage model. *Earth and Planetary Science Letters*, **26**, 207–221.

STANTON, R. L. & RUSSELL, R. D. 1959. Anomalous lead and the emplacement of lead sulphide ores. *Economic Geology*, **54**, 588–607.

TATSUMOTO, M. 1973. Time differences in the formation of meteorites as determined from the ratio of lead-207 to lead-206. *Science*, **180**, 1279–1283.

TAYLOR, S. R. & MCLENNAN, S. M. 1985. *The Continental Crust: its Composition and Evolution.* Blackwell, Oxford.

TERA, F. 1980. Reassessment of the 'Age of the Earth'. *Carnegie Institution of Washington Yearbook*, **79**, 524–531.

TERA, F. 1981. Aspects of isochronism in Pb isotope systematics – application to planetary evolution. *Geochimica et Cosmochimica Acta*, **45**, 1439–1448.

WETHERILL, G. W. 1986. Accumulation of the terrestrial planets and implications concerning lunar origin. *In*: HARTMANN, W. K., PHILLIPS, R. J. & TAYLOR, G. J. (eds) *Origin of the Moon*. Lunar Planetary Institute, Houston, 519–550.

WILDE, S. A., VALLEY, J. W., PECK, W. H. & GRAHAM, C. M. 2001. Evidence from detrital zircons for the existence of continental crust and oceans on the Earth 4.4 Gyr ago. *Nature*, **409**, 175–178.

ZARTMAN, R. E. & DOE, B. R. 1981. Plumbotectonics – the model. *Tectonophysics*, **75**, 135–162.

Fossils as geological clocks

JOHN H. CALLOMON

University College London, 20 Gordon Street, London WC1H 0AJ, UK

Abstract: To reconstruct the history of the Earth we need to know *what* happened and *when* – *events* and their *dates* – and we should like to know *how* it happened and *why* – *processes* and their *rates*. To date a historical event we need a timescale for reference – a *calendar* – and a means of placing events in this timescale – a *clock*. Direct access to the primary physical calendar, of time measured in years by means of elemental radiometry as clock, is possible in only a minority of geological problems. By far the richest historical source in the Phanerozoic Eon has been the stratigraphical analysis of sedimentary rocks by means of fossils, the approach pioneered by William Smith. The succession of fossil biotae found in the rocks is used to construct the calendar of *relative* time, the familiar geological calendar defining the standard chronostratigraphical timescale still in process of refinement today. Rocks are then dated through time correlations with this scale by means of their *guide* fossils (von Buch) as clocks. The power to measure the rates of geological processes then depends on the time *resolution* achievable by means of fossils, the time *intervals* between distinguishable events, the *finesse* of the calendar.

The present-day state of play is reviewed, both in the refinement of the geological calendar and the finesse that has been attained. Comparison of the geological calendar with our familiar human historical calendar reveals some illuminating parallels as well as some important differences. Illustrative examples are taken from the Jurassic Period (170 Ma BP) and its ammonites as clocks.

The last two centuries have seen the detailed exploration not just of the geological history of the Earth – of what happened, when and how – but also of the development of the timescales on which this history is based – the geological calendar – and of the methods of measuring geological ages – geological clocks. In the first of these two centuries, following the times of James Hutton and William Smith, the geological calendar, as it was described already by Buckman (1898), rapidly grew into the form in which we still use it today, filled with an ever more detailed chronology of events. It was a chronology largely based on the biostratigraphy of fossils as clocks. So the ages based on them were only relative; the rocks that could be dated were restricted to fossiliferous strata, which excluded the Precambrian; and the question of what were their 'true ages', in years, could lead to little more than speculation. The only positive outcome of the lively debate it engendered was to raise geological ages from the thousands into the realm of millions of years.

The discovery of radioactivity and its exploitation in radiometric age determination, forever coupled with the immortal name of Arthur Holmes, introduced a second class of geological clocks based on time-dependent physical properties and processes. These can lead to 'true ages' directly and things have not been the same in geology ever since. An increasing number of further physical methods, based on properties of rocks such as their remnant palaeomagnetism or the stable-isotope ratios of their elements, has given us today an impressive armoury of techniques for dating an ever-widening range of rocks and geological events. But, to what extent do newer methods replace older ones or merely complement them? And by what criteria may the relative merits of different methods be compared?

To do justice to this fascinating subject would take volumes and is far beyond the scope of these pages. I should like therefore to return here to a discussion of the original methods of dating rocks, those based on the stratigraphy of fossils in layered rocks pioneered by William Smith. The advent of physical methods has in fact in no way diminished the importance of biostratigraphy as a tool for geochronology. Intensive and sustained activity has greatly broadened the range of fossils drawn in as geochronometers, the types and ages of rocks that can be dated with them and the precision with which this can be done. To cover all these here in any detail is not possible. I shall therefore base the discussion largely on an example with which I am particularly familiar. The fossils are ammonites and the period in which they lived was the Jurassic. The arguments are, however, quite general and may be applied equally well to other groups and periods (with variable success). The emphasis will be on two leading criteria of merit: firstly, how precisely and geographically extensively can rocks be dated by means of fossils? And secondly, how closely in time can successive events be distinguished: what are the shortest time intervals that can be resolved? The biostratigraphy of fossils is, however, one step removed from the time factors derived from it – biostratigraphical ages are relative – and so it is worthwhile perhaps to begin with a brief review of the relationships between the conventional geological calendar and the primary physical timescale on which we like to express geological ages. Some of these relationships are subtle and perhaps not always appreciated.

Time, clocks and calendars

Time is one of the four physical dimensions in which we perceive the conscious world. To locate an *event* at a *point* in time, as the common expression has it, we must give the dimension a metric and scale – a *time-scale* – and to choose, or devise, a *clock* by means of which time can be *measured*. The clock that defines the scale is the *primary* clock and the scale that it defines is its *standard*, or *calendar*. A calendar is then the frame of reference for *dating* events and a record of events thus dated is their *history*. Dates are measures of relative time – relative in the calendar to one or more chosen *fixed points*. One of these fixed points is usually given the value zero, marking the *origin* of the scale. To determine the date of a particular event then requires a means of relating it to the calendar of the primary clock. Such means have to be observational devices – *secondary* clocks – whose kind can depend strongly on the nature of the event to be dated, whose *precision* is limited and whose relation to the primary clock is determined by *calibration*. A special case of time measurement commonly encountered is that of following the time *evolution* of a *system* relative to some arbitrary beginning, of its state at successive moments during a *period* of the 'passage of time' or, if the state remains constant during that period, of that period's *duration*. The time of a moment after the beginning of the evolution of the system is then its *age* at that moment. Finally, the precision with which a date can be determined by means of a secondary clock depends on its power of time *resolution*: the minimum interval between events that it is able to distinguish.

These points seem so obvious that to spell them out in such detail may appear pedantic. They seem obvious because we are so familiar with them from everyday experience and in the domains of our study of 'history' – history as preserved in human records or prehistory as deduced from archaeology. They become less obvious, however, when we stray outside the bounds of human experience into other historical domains, such as those of historical geology and its corollary, historical biology or palaeontology. The literature in these fields reveals the persistence of considerable confusion. It arises largely because of the nature of the clocks that have to be used in reconstructing their past. These clocks, both primary and secondary, differ fundamentally from those used in human history. Nevertheless, there are valuable insights to be gained from comparisons of the methods used in the different domains. They reveal some interesting analogies, as well as highlighting some important differences.

The historical calendar

Figure 1 shows a graphical representation of our traditional calendar and how a certain historical event is located in it. Setting aside the modifications introduced by modern physics and some relatively minor changes of convention, the calendar is based in fact on three primary clocks. They exploit three constants of solar planetary motion, characterized by highly regular and conveniently well-separated cyclical periods: the orbital motion of the Earth about the Sun – the year; the orbital motion of the Moon about the Earth – the lunar month; and the spin of the Earth – the day. The independence of the first and third is familiarly reflected in the need for leap-years, that of the second in the persistence of Easter as a moveable feast.

The periods of time of interest in human history vary over a very wide dynamical range of magnitudes. It becomes convenient, therefore, to subdivide historical time in a *hierarchy* of successively smaller units. A part of this hierarchy is incorporated in Figure 1 and its levels are numbered I–VII. The primary units lie at levels IV, V and VI. Larger units are then constructed as decadic multiples of the year, going up to the level appropriate for the chronicle of human history as a whole, the millennium (I). The decade (III) serves to enumerate human generations and the 'three score years and ten' of human lifetime. The year (IV) provides a convenient numerical metric in which the origin of the timescale is defined. Where placed is arbitrary and its

THE HISTORICAL CALENDAR

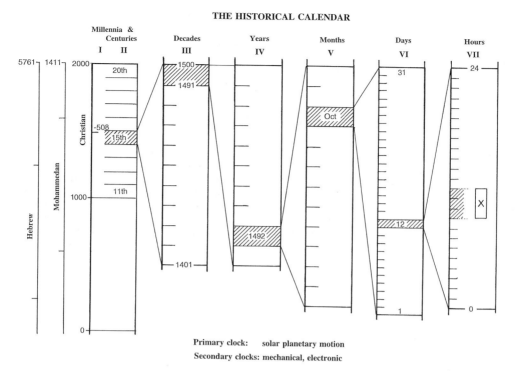

Fig. 1. The human historical calendar and its hierarchical subdivision. The two primary astronomical clocks still in practical use are based on the Earth's orbital motion, the year (level IV), and its spin, the day (level VI). The location in this calendar of a famous historical event (the late morning of 12 October 1492 (Julian calendar, local time), when Christopher Columbus landed in America) is picked out by diagonal shading.

positions at the BC/AD boundary in the Christian calendar relative to that of two other historical calendars widely used in the past are also indicated in Figure 1. These alternative calendars, together with others such as the Chinese, were also geographically separated, being used in different parts of the world. Yet they were all based on the same primary clock, that of the rotation of the Earth. Their time *correlation* therefore presents no special problems.

Smaller units in the hierarchy down to subdivisions of the day into hours are also shown, but further extensions into minutes (VIII) and seconds (IX) are of course familiar in everyday life. The second of time is appropriate for the description of the briefest events distinguishable by the human senses – the 'blink of an eye' or, more precisely, the persistence of vision, of about one-twentieth of a second, which governs the design of cinema and television projectors of moving pictures. (For physicists, the second is now also the primary SI unit of time, based on a natural internal oscillation of an atomic caesium clock having a period of 1.088×10^{-10} s. Modern technology enables us to measure times down to very small intervals, but the same principles as those outlined above apply. A radar speed-trap, to secure a conviction, has to measure a time difference between two periods of about one second with a precision of about one-thousandth of a microsecond).

Secondary clocks, essentials of our daily lives, are too diverse to enumerate but fall today into two classes: mechanical and electronic. The dynamical ranges of domestic timepieces vary from the simplest (and cheapest), a two-hands wristwatch telling the hours and minutes, to the more sophisticated, recording the years down to seconds, found in a personal computer. The accuracy of such a timekeeper is controlled by calibration against widely available broadcast time signals, but an important point to note is that it, and hence the accuracy of any event timed with it, depends also on the precision of its ability to measure time, which in turn depends on its temporal resolving power. A watch with a second-hand can tell the time to within only a second or so. This point becomes particularly important in geological time measurements, discussed below.

The geological calendar

Methods

As is well known, the ages of rocks can in principle be determined directly in terms of the primary clock of the historical calendar, the year, by radiometric methods. But it is equally well known that such 'direct methods' are only applicable under special circumstances to a small, albeit rapidly increasing, subset of geological problems (see e.g. Callomon 1984). Perusal of the geological literature immediately shows that the overwhelming proportion of geological ages continue to be cited by reference to the hallowed geological calendar that has been in use since the days of Smith, Sedgwick, Murchison, Lyell and Phillips, and it is this calendar that will be discussed here.

The most striking difference between the historical and geological calendars is that the latter has no primary clock outside itself. The geological calendar, as a frame of reference for dating geological events, has to be constructed by the same methods as those then used to date such events, to locate them in the calendar.

The methods that have been used and that continue to be the most powerful, diverse and widely applicable are those of stratigraphy, the study of successions of layered rocks. There are four steps in the argument. The first and basic *observation* is that of beds specified by relative heights in a succession – what lies above what. Description of their thicknesses and compositions is their *lithostratigraphy* and, if composition includes fossils, their *biostratigraphy*. The next step is to introduce Steno's (1669) 'Principle of Superposition', which states that in a normal succession of strata, the higher-lying are the younger. This transforms a static description of relative height into a dynamical one of relative time and is an *interpretation*. (Appearances can be deceiving, as in igneous sills intruded into sediments, or at imperceptible thrust-faults in fold-belts.) Specification of rocks in a stratal succession according to their relative ages is their *chronostratigraphy*. The third step involves the linking of local successions through time correlations These allow the ages of rocks at one place to be compared with those at another – the same, older or younger. The fourth step then becomes the synthesis of a standard time-ordered succession of rocks and its conjugate timescale, correlation with which allows any local rock to be dated in terms of the standard – the geological, chronostratigraphical calendar. This timescale is one of *relative* time. The age of any point on it is relative to those before and after. There is no *a priori* numerical measure either of time durations or of time intervals. The timescale does, however, have a fixed point as origin – the present.

We see therefore that time correlation plays a much more important role in historical geology than it does in human history because it is central not only to the dating of individual events but also to the construction of the timescale of reference itself. The same clocks are used for both. As in the recording of human history, the refinement of time measurement in geology has prompted the exploration of a widely diverse range of secondary clocks. Leading among these are fossils, and the manifestation of what may be the closest analogue of a primary physical clock in biostratigraphy: their biological evolution.

Time correlation by means of fossils depends on what may be called the 'Principle of biosynchroneity', usually ascribed to William Smith: beds with similar fossils are of the same age (see Knell 2000, Ch.1, p. 148). But fossil species have ranges. 'The same' therefore means 'more or less the same', depending on these ranges and on how closely species can be identified. Some fossil species are clearly better for time correlations than others, and those that are 'good' are selected as *guide* fossils ('Leit-Muscheln' of von Buch, 1839). 'More or less the same' implies that time correlations by means of fossils are approximations. 'Good' measures the minimum ranges of uncertainty in these approximations. Conversely, the degrees of approximation are set by the *temporal resolving powers* of the guide fossils, by the minimum time intervals between successive geological events that can be distinguished by means of such fossils. Then, to complete the circle, this leads back to the *finesse* of the standard geological calendar itself, constructed from the biostratigraphy of the guide fossils. This reciprocal interrelation of biostratigraphy and the construction of the standard calendar clearly form a basis for continual refinement. To claim that the two centuries since the days of William Smith have seen progress may sound trite. But how far have we come? How good is our calendar today? How closely can rocks be dated by means of fossils? And, finally, how well can our relative geological age determinations be correlated with their 'absolute' ages, in years, determined radiometrically? Should we today bother to retain our geological calendar at all?

The calendar

The standard geological calendar is shown diagrammatically in Figures 2 and 3. As in the historical calendar, periods of interest in the Earth's history also cover a very wide dynamical range of magnitudes. The total timespan from the present

THE GEOLOGICAL CALENDAR

I	II	III	
PHANEROZOIC (Chadwick 1930)	CAINOZOIC (Phillips 1841)	RECENT	(Lyell 1873)
		PLEISTOCENE	(Lyell 1839)
		PLIOCENE	(Lyell 1833)
		MIOCENE	(Lyell 1833)
		OLIGOCENE	(Beyrich 1854)
		EOCENE	(Lyell 1833)
		PALAEOCENE	(Schimper 1874)
	MESOZOIC (Phillips 1841)	CRETACEOUS	(Omalius d'Halloy 1822)
		JURASSIC	(Brongniart 1829)
		TRIASSIC	(Alberti 1834)
	PALAEOZOIC (Phillips 1840-41)	PERMIAN	(Murchison 1841)
		CARBONIFEROUS	(Conybeare 1822)
		DEVONIAN	(Sedgwick/Murchison 1839)
		SILURIAN	(Murchison 1833)
		ORDOVICIAN	(Lapworth 1879)
		CAMBRIAN	(Sedgwick 1835)
PRECAMBRIAN	PROTEROZOIC (Emmons 1888)	skeletal macrofossils appear — (Gunflint Formation) —	
	ARCHAEOZOIC/ ARCHAEAN (Dana 1872)		

Fig. 2. The geological calendar at the top three levels in its hierarchical subdivision, drawn on an equal-Period/System approximation.

back to the earliest days of our Earth of which we have any record has therefore also been subdivided in a hierarchy of successively finer units and, for ease of reference, the levels of this hierarchy are again numbered, I–VII. Figure 2 shows the first three levels, I–III, and, as an example, one of the units at level III, the Jurassic, is taken in Figure 3 to the limits of time resolution attainable by means of one group of fossils – ammonites. (Level III also represents the limit to which the geological calendar has penetrated the body of public general knowledge, to be tested in television programmes such as *Mastermind* or *University Challenge*, or used to set the scene in films such as *Jurassic Park*.)

In looking at these figures, some important points should be noted. Firstly, the columns of boxes can be read in two ways, reflecting the conjugate representations of rock–time duality (Table 1). In the first representation, the vertical co-ordinates are in each case those of time and the horizontal lines delimiting the boxes, which represent periods of time, mark instants – turning points – of time. Although the representation

THE GEOLOGICAL CALENDAR

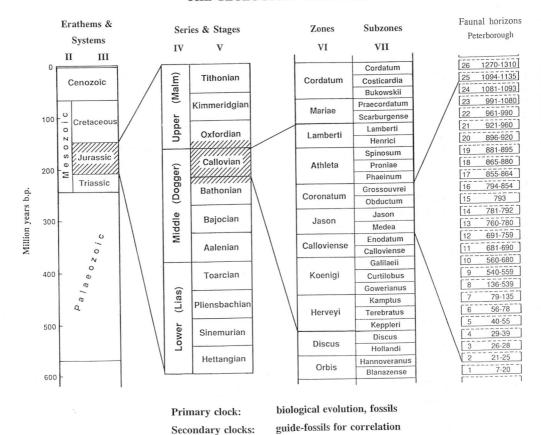

Fig. 3. The geological calendar and its finer subdivisions taken to the current limits of the standard hierarchy, at level VII Subzones/Subchrons, illustrated by an example in the Jurassic. The two principal columns are drawn at equal-Stage/Age and equal-Subzone/Subchron approximations. The column at the left is drawn on the geochronometric Cambridge timescale of Harland *et al.* (1990). The column at the right represents the distinguishable ammonite faunal horizons in the Oxford Clay of Peterborough recorded by Brinkmann (1929), drawn on an equal-time-interval approximation. They are labelled successively from 1 to 26 for convenience, together with their heights in centimetres above an arbitrary zero at the base of the section.

of time is continuous and complete, it has, with one exception, no scale: it is not linear, and relative thicknesses of time boxes are not proportional to real-time durations. The exception is the left-hand column in Figure 3, which incorporates *post hoc* information considered separately below. Column III in Figure 2 is drawn on an equal-Period representation, column V in Figure 3 on an equal-Age representation, and so on. On the evidence of chronostratigraphy alone, such representations are as valid as any other.

In the second of the dual representations, the columns represent what is commonly referred to as 'the standard geological column', a hypothetical stratigraphical column of rock whose formation was continuous and which therefore records the geological action of all time. There is of course no such complete succession at any one place, but discontinuous parts of it are what we observe at many places. The dividing lines between boxes do retain a clear meaning, however. They are ubiquitous *time planes* in the rocks. Whether we can identify one of them at any place other than at a type-locality, a section in which it has been typologically defined at some precise level in a stratified succession, is another matter: generally we cannot. But it must remain true that every piece of rock we collect in stratal succession must lie either below or above every time plane. If we can identify two time planes *between* which

Table 1. *Rock–time duality and the hierarchy of units of subdivision of standard chronostratigraphy and geochronology*

Level	Rock	Time
I	Eonothem	Eon
II	Erathem	Era
III	System	Period
IV	Series	Epoch
V	Stage	Age
VI	Zone	Chron
VII	Subzone	Subchron

a given piece of rock lies, we assign it to the chronostratigraphic rock-unit defined by those planes: we cite its age by the name of the unit. When we say *Tyrannosaurus rex* is Cretaceous, we say no more than that its remains are found in rocks that lie between the Jurassic–Cretaceous and Cretaceous–Tertiary boundaries. Similarly, standard chronostratigraphic units in general are identified in the field not by means of their defining boundary time planes but by means of what lies between them – such as guide fossils. Standard chronostratigraphic units are therefore all the rocks in the world lying between defining time planes, but their local representations are usually fragmentary and highly incomplete. And the '(in)completeness of the geological record' continues to preoccupy us in many geological and palaeontological problems, not least in mapping the phylogeny of biological evolution.

This distinction between the definition and recognition of standard chronostratigraphical units may present the apparent paradox that seems to puzzle so many members of commissions on stratigraphy (see e.g. Aubry *et al.* 1999, Remane 1996): that the ages of rocks can almost always be successfully determined by assignment to chronostratigraphical units of appropriate rank despite the fact that, of the vast number of these units in everyday use, the number so far formally defined in terms of typologically fixed bounding time planes is tiny. (You can usually tell, in finding your way to Burlington House, whether you are in Piccadilly or not without first having to find out where it formally starts and ends. Street-names helpfully displayed on placards at intervals can be useful guide signs.) And the corollary is that the closer together the time planes, the shorter the time intervals they define, the shorter the age spans of the rock-units between them, the more precise (less uncertain!) the dating of the rocks found in them.

Figure 2 shows the broader features of the geological calendar and brings out a number of points. Firstly, it indicates its early historical development. Previous classifications of rocks as Primary, Secondary and Tertiary, based largely on lithological characteristics only vaguely related to age, were largely abandoned as the results of more detailed mapping emerged. Emphasis shifted to litho- and biostratigraphy, and this is reflected in the names of the units. (Ghosts of earlier times linger in the frequent use of Secondary or Tertiary as exact synonyms of Mesozoic and Cainozoic (Cenozoic), with its subsequent extension to Quaternary.)

Secondly, it is remarkable how quickly the calendar stabilized into its present global form, following the relatively brief period of geological exploration spanning the first half of the nineteenth century: geological explorations, moreover, confined largely to a very small part of the Earth's terrestrial surface, that of classical Europe. Had the cradle of geology lain elsewhere, in North America, say, or in China, the calendar would have looked rather different. But the fact that later geological exploration of the rest of the world did not lead to the abandonment of the 'European' calendar at these higher levels, or even to any substantial modification of it, tells us that it conveys some geological truths about the Earth's history that are globally recognizable. And this universality rests today of course on the use of fossils as clocks.

Thirdly, at the highest level, the calendar makes use primarily of only one turning point in the Earth's history: that of the sudden appearance of macroscopic fossils at the base of the Cambrian. (A second, more recently recognized turning point was marked by the discovery of prokaryotic microfossils in the fine-grained cherty Gunflint Formation in the Precambrian Animikean rocks of southern Ontario, which for many years provided the oldest positive evidence of life on Earth.) The puzzle of what to do about the Precambrian persisted for a century and continues to challenge us today. Sparseness of interpretable evidence tended to reduce it almost to a footnote, a source of frustration and embarrassment. This is not helped by two relatively recent discoveries by radiometric methods that it accounts for seven-eighths of the age of the Earth, and by largely chemical methods that there was prokaryotic life already during most of it. And how sudden was 'sudden' also cannot be answered at the finesse of level III in Figure 2.

Refinements

Subsequent refinement of the geological column is illustrated in Figure 3 through the example of the level III System, the Jurassic, that has traditionally led the way in the development of the principles and techniques of bio- and chronostratigraphy that were needed. These were recently reviewed at some length (Callomon 1995) in yet another attempt to allay the confusion that persists (e.g. Whittaker *et al.* 1991). Levels IV and V were introduced by von Buch (1839) and d'Orbigny (1850) respectively and, most importantly, Zones (VI) by Oppel (1856–1858). The hierarchy of top-down subdivision has in the Jurassic gone one step further, to the level of Subzones (VII). As an illustration, the way the whole of the British Jurassic has now been chronostratigraphically classified down to this level may be seen in the Geological Society's correlation charts (Cope 1980*a, b*). A similarly detailed classification has been published for the Jurassic of France (Cariou & Hantzpergue 1997).

This degree of attainable time resolution owes its success to a special circumstance: the availability of exceptional guide fossils, the ammonites. The reason why they are so good depends on the fact that the morphology of their shells evolved over long periods of time more rapidly than that of almost any other marine organisms leaving fossil remains, for reasons that are still wholly unknown (Callomon 1985). (No environmental selection pressure to which this evolution could be plausibly ascribed as an adaptive response has been identified. Recourse to 'genetic drift' as an explanation of last resort would probably also be tautological, but if in fact a viable cause, it would account for the observation that the exceptionally rapid evolution of Ammonoidea has been so constant, even across divergent phylogenetic clades, for so long – from the Devonian to the latest Cretaceous – but limited in range and largely iterative: similar forms occur repeatedly.) The shells are in the majority of cases also strongly sculptured, so that very small evolutionary changes, over short periods of time, are readily discernible. How the reliability of ammonites as time-diagnostic guide fossils can be assured and exploited has also recently been discussed in some detail (Callomon 1995). Because the chronostratigraphy at zonal levels is based almost exclusively on biostratigraphy, the units are named after characteristic index species found in them following the convention introduced by Oppel. Those shown in Figure 3 are named after species of ammonites. Stages (V) are named after places, a convention introduced by d'Orbigny.

Limitations and alternatives

There is a price to be paid for high-resolution chronostratigraphy based on fossils, which can restrict their usefulness. There are commonly three possible causes. The first is bioprovincial endemism of the guide fossils even within an accessible and otherwise physically compatible environment, limiting their geographical distribution. The second is unsuitability of facies, e.g. non-marine sediments in the case of ammonites. The third is constraints imposed by available methods of sampling, e.g. in the exploration of subsurface formations by means of continuously chip-drilled boreholes, which rule out macrofossils.

The best-known example in the Jurassic of the first cause lies in its topmost Stage, the Tithonian. Its ammonites, equally prolific, became segregated in Europe into three faunal provinces so mutually exclusive that the zonation (levels VI–VII) constructed in any one of them is wholly unrecognizable in the other two. We have therefore three parallel standard calendars of Zones in three more or less contemporary Stages: the Tithonian, for central and southern Europe broadly south of the Alpine fold-belts; the Portlandian, for NW Europe and into the Arctic; and the Volgian, for the Russian and north Siberian platforms. Correlation of these three calendars continues to be problematic, but their temporal finesses are comparable. The selection of one as *primary standard* become a matter of convention. Generally speaking, recognition of most Stages by means of ammonites can be done globally. Ammonite Zones are recognizable over continental distances, more or less; and Subzones over formerly connected shelf-sea sedimentary basins, distances of from hundreds to a few thousand kilometres.

As illustration, two favourable cases are shown in Figures 4 and 5. Figure 4 samples the distribution of the characteristic species of the basal Keppleri Subzone of the Callovian (Fig. 3), *Kepplerites keppleri* (Oppel, 1857). Figure 5 shows some of the known occurrences of *Cardioceras martini* Reeside, 1919 (type from Alaska) and its more widely used synonym *Cardioceras bukowskii* Maire, 1938 (type from Poland), characterizing the Bukowskii Subzone of the Lower Oxfordian Cordatum Zone (Fig. 3). Both genera are bioprovincially restricted to the Boreal Realm, which occupied the higher temperate to polar latitudes in the northern hemisphere, and are wholly absent from the Jurassic equatorial belt and southern hemisphere, replaced there by other groups.

Problems of the second kind, of inimical facies, can be less tractable. Most acute is that of terrestrial sediments, such as the Carboniferous Coal Measures, the Red Beds of the Germanic Trias, or the Karoo of Mesozoic Gondwana. All one can say here is that one must use whatever there is, but that it is unlikely to be very good.

Kepplerites keppleri Oppel

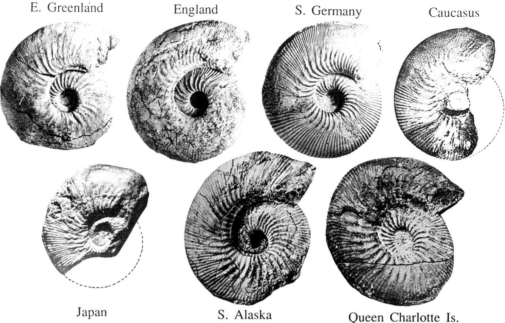

Fig. 4. Correlation by means of ammonites: the recorded distribution of the index of the Keppleri Subzone at the base of the Callovian Stage (Fig. 3). Recent additions include the Volga Basin on the Russian Platform.

Time resolution at level IV may be the closest attainable, or level V with luck.

Thirdly, the methods and problems of dating borehole logs are well known. They have spawned the immense multinational industry of oil-company micropalaeontology. Many chronostratigraphical calendars based on foraminifera, calcareous nannofossils, palynomorphs and pollen have been constructed. An interesting comparative compilation has been published by Cox (1990). (The problems of correlating these zonations are largely circumvented by the fact that most of them continue to be unpublished, if not confidential.) The time resolutions attained usually lie at Stage or Substage level, level V or a little better.

The limits of time resolution by means of fossils

Mesozoic ammonites

Does the time resolution implied by the lowest hierarchical level in the standard calendar represent the limit of what can be achieved by means of fossils? The answer is 'No'. Even within the timespan of a Subzone, further successive distinguishable evolutionary steps in an evolving lineage – the *transients* – can often be recognized. The strata providing these ultimately resolvable temporal snapshots of the state of the fossil clock, recording their closest resolvable *moments* of time, have come to be called simply characteristic faunal, floral or fossil *biohorizons* (Callomon 1964), characterized by a transient fossil taxon or assemblage. But the basic idea and its application go back to a landmark in the history of biostratigraphy set by S. S. Buckman (1893), who coined the term *hemera* for a moment of time distinguishable in the record of evolution of ammonites (or, more generally, in 'palaeobiology' as he termed it). His fundamental contributions and their subsequent development were also reviewed recently (Callomon 1995).

What continues to be one of the earliest and finest examples of the application of these ideas is a classical biostratigraphical analysis by Brinkmann (1929) of the evolution of the ammonite genus *Kosmoceras* through the Lower Oxford Clay exposed in the brick-pits around Peterborough. He collected some 3000 ammonites centimetre by centimetre through 13 m of finely stratified sediment and analysed their morphological characters biometrically. The main result

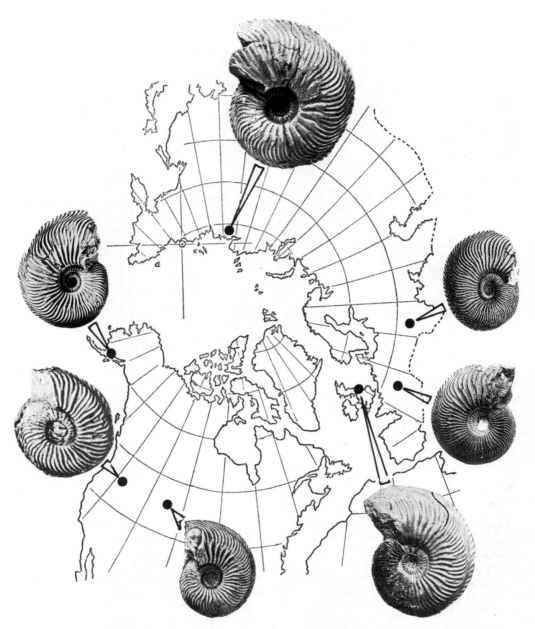

Fig. 5. Correlation by means of ammonites: the recognized distribution of the Bukowskii Subzone of the Cordatum Zone of the Oxfordian Stage (Fig. 3).

of interest from the present point of view was that the succession could be resolved into some 26 beds, sharply bounded by lithological discontinuities, each with an ammonite assemblage that can be distinguished visually and in part biometrically from its neighbours. These 26 faunal horizons are indicated on the right in Figure 3, labelled by their heights in centimetres in the section; and pictures of their characteristic ammonites at some of these levels are shown in Figure 6. How these horizons fit into only four subzones is also indicated. That they are not only of purely local significance is reflected in the fact that many of them, perhaps as many as half, can be traced across country over considerable distances, to Dorset in the southwest, and in some cases as far as Brora on the east coast of Sutherlandshire in the north. Conversely, additional

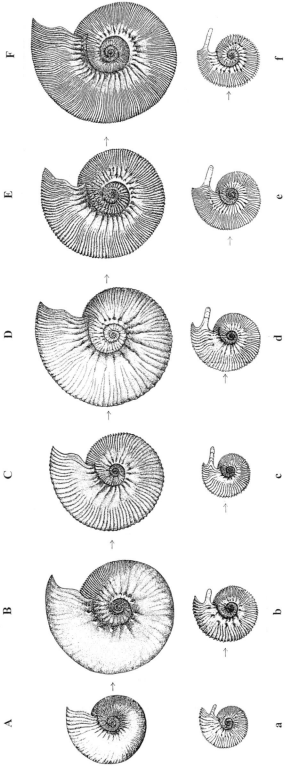

Fig. 6. Illustration of six transients in the evolving lineage of the ammonite genus *Kosmoceras* at Peterborough (see Fig. 3: after Brinkmann 1929, drawings by Mme P. Kyropoulos). In common with most other Jurassic ammonites, the shells of this genus were strongly sexually dimorphic. *Macroconchs*: (**A**) Jason Zone, Medea Subzone, horizon 4 (Fig. 3); (**B**) Jason Subzone, horizon 7; (**C, D**) Coronatum Zone, Obductum Subzone, horizons 8 and 10; (**E, F**) Grossouvrei Subzone, horizons 17 and 23. *Microconchs*: (**a**) horizon 1; (**b**) horizon 7; (**c, d**) horizons 8 and 9; (**e, f**) horizons 18 and 24. Note the marked change from **B**, b to **C**, c; this reflects what is probably a considerable non-sequence at level 135 cm at Peterborough.

horizons become inserted as the succession thickens southwestwards, to Oxford and beyond, showing that this rich succession is still incomplete even at Peterborough.

Faunal horizons are in general still a long way from being a continuous sedimentary or evolutionary record and are separated by time gaps of unknown durations. They are therefore shown in Figure 3 not as further top-down subdivisions of subzones as parts of a continuous column, but as rock equivalents of disjointed time slices of unknown but probably short durations. The gaps between them are still there, waiting to be filled by new discoveries. The process of inserting new faunal horizons as they are discovered, hence adding to the number of distinguishable events in the timespan of a subzone, is one of bottom-upward synthesis. One of Buckman's most significant discoveries was that in fact the more complete the fossil record becomes, the more incomplete the sedimentary record turns out to be. The ultimate paradox is that fossils can show not only the ages of the rocks that are present in a section but, perhaps even more importantly, the ages of the rocks that are *not* present – the gaps in the record. And the ability to do so depends strongly on the temporal resolving power of the guide fossils used. Some comparative examples were discussed in a review of Buckman's contributions (Callomon 1995). The temporal resolving power of ammonites is sufficient to have revealed the presence of major time gaps – major on the timescale of their temporal resolution – in almost every sedimentary succession in which they have been closely studied. (This prompts another interesting comparison between the historical and geological records. Sparseness of dated events in the historical calendar always implies failure to record, not the presence of time gaps in history. Sparseness of dated geological events can in certain circumstances be ascribed positively to gaps in the stratigraphical record, the all-too-common disconformal non-sequences.)

Radiometric calibration of the geological calendar

Radiometric dating of rocks within the domain of applicability of the method has become routine and revised 'absolute' timescales of the Phanerozoic appear almost annually. One of these, chosen arbitrarily, the 'Cambridge timescale' of Harland *et al.* (1990), is shown at the left of Figure 3, and the Systems/Periods of the Mesozoic at levels II and III of the hierarchy are drawn with vertical thicknesses now probably very closely proportional to their true time durations.

When we descend the hierarchy, however, the picture becomes less cheerful. Figure 7 shows an excerpt from a recent compilation (Palfy 1995) of timescales that have been proposed for the stages of the Jurassic by various authorities in the last 15 years. Clearly, we are still some way from finality. The rise and fall of the Callovian (Fig. 3), CLV in this chart, is typical. The reason for these uncertainties is that the number of securely dated points to which these timescales are anchored in the Mesozoic, and in the Triassic–Jurassic in particular, are so far still very few. The rest is done by interpolation in one way or another, not excluding simply counting the numbers of Zones or Subzones in a stage (Fig. 3). The appearance in print of a numerical scale, sometimes quoted even to one decimal place in Ma (units of a million years), as for example, in Harland *et al.* (1990) and the widely used instructional charts based on the 'Exxon' global scales of Haq *et al.* (1988 and revisions) can raise an illusory sense of accuracy in the minds of the unwary (see discussion by Miall 1997, Ch. 13, 14). But no harm is done as long as it continues to be associated with a representation of the standard chronostratigraphical calendar as primary frame of reference. A temptation to assume that numerical scales can now be used by themselves as life-simplifying substitutes for the bewildering plethora of arcane names that adorn the calendar can, however, be counterproductive. A valuable recent review (Howarth & McArthur 1997) of a powerful new method for dating marine sediments by means of the stable-isotope ratio $\rho(87:86)$ of their contained strontium shows fine curves of the change with time of this ratio in the interval from Recent to Jurassic. The temporal resolving power of this method can, in the Mesozoic, approach that of ammonites, level VI or even VII, and the samples used to construct these curves, largely made up of belemnites, were themselves collected from beds precisely dated at level VII. Yet the timescales plotted on the abscissae are all given linearly in millions of years (Ma): their chronostratigraphical ages are not shown at all. Only a casual sentence reveals that the numerical ages used were those of Gradstein *et al.* (1994), not reproduced. To use the curves, therefore, to date a new rock sample in familiar terms, one has first to obtain a copy of this timescale to convert back to the practical units in common use with which the curves were constructed in the first place.

Rates of geological processes

The precise values of radiometric ages are in fact not of much practical interest in stratigraphy.

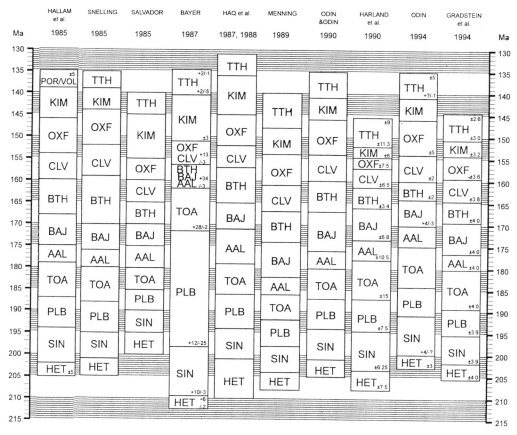

Fig. 7. A comparison of ten radiometric timescales for the stages of the Jurassic proposed in the last 15 years (from Pálfy 1995, with permission; an updated version may be found in Palfy et al. 2000).

It matters very little, for instance whether the horizon of *Kepplerites keppleri* defining the base of the Callovian (Fig. 3) is thought to lie at 157 Ma BP (Fig. 7: Haq et al. 1988) or at 165 Ma BP (Gradstein et al. 1994) – a difference of 5%. What are interesting are average values of time *intervals* – age differences. The ammonites of the Jurassic are again taken as an example, in Table 2. The reason is that these averages allow us to estimate the rates of many of the innumerable processes that make the geological structures we see. Most of these processes are cyclic or repetitive and so a hierarchy of orders of cyclicity was introduced by Vail et al. (1977), roughly in powers of ten of time duration in years. A useful classification of processes was compiled by Einsele et al. (1991) and is reproduced by Miall (1997, Ch. 3). Major continental rearrangements, such as the assembly and subsequent dispersal of Pangea, and their associated orogenic phases, such as those that dominated the study of geology a century ago – Caledonian, Hercynian, Alpine – are first-order processes. Long-ranging cycles of world-wide (eustatic) sea-level change that governed the characteristic sedimentary histories of whole systems, such as the Triassic–Jurassic–Cretaceous triplet of the Mesozoic, at level III in the chronostratigraphical hierarchy, are of second order. Processes at third to sixth order take us down the range from a million years (1 Ma) to a thousand years (1 ka), and in this range the shift of interest is more and more towards sedimentary processes determining increasingly local lithostratigraphy and its interpretation in genetic basin analysis, now called sequence stratigraphy. In this field, time controls are crucial and Table 2 indicates how good guide fossils can provide such time controls down to fourth-order processes. And, as already mentioned, one of the discoveries has been that already at this level of time resolution, the sedimentary record is often far less complete than the proponents of sequence stratigraphy would like us to believe. Finally, sedimentary records of tidal flow take us off-scale to Tuesday afternoon events, lasting 10^{-3} years.

Table 2. *Time resolution by means of fossils: an example based on ammonites in the Jurassic system (average age 175 Ma BP)*

Units	Number	Mean durations or intervals
III – System	1	60 Ma
V – Stages	11	5 Ma
VI – Zones:	76	860 ka
VII – Subzones	c. 160	375 ka
– faunal horizons	c. 450	130 ka

Time resolution: 1 part in 1300

Other eras, other clocks

The emphasis has here been on Mesozoic ammonites. How fares it with other groups of fossils and in other Eras? The extinction of the ammonites at the end of the Cretaceous (65 Ma BP) left the Tertiary without any guide fossils of comparable temporal resolving power and geographical extensions. The geological calendar is therefore rather fragmented into more regional, parallel schemes based on local circumstances. Many different guide fossils are used, most often microfossils of various kinds in short runs. Fortunately, intercorrelations are helped by physical methods, notably those of global magnetostratigraphy, stable-isotope-ratio stratigraphy and perhaps the litho- and biostratigraphical expressions of Milankovič cyclicity. Overall time resolutions and correlations can come down to 0.5–1 Ma in favourable circumstances, perhaps a little better locally. In the Palaeozoic (see a compilation in Callomon 1995), ammonites hold the lead as far down as the Devonian. The goniatites of the Carboniferous can in parts of it resolve 0.8 Ma; the ammonoids of the Middle–Upper Devonian, 0.4 Ma. But the most striking advances in the last half-century have been in the high-resolution biostratigraphy of three other groups: conodonts, graptolites and trilobites.

Conodonts have been known for 150 years: they are common, occur widely, range from mid-Cambrian to latest-Triassic, and look like small teeth (c. 1 mm). But whether they were teeth, and if so the teeth of what, remained a mystery until only some 20 years ago (Briggs *et al.* 1983). The animal to which they belonged was probably free-swimming, looked like a small chordate eel, and the conodonts were its teeth. The assemblages are rich in morphology and hence the ability to follow small collective changes with time also makes them powerful guide fossils. Their zonation in the Upper Devonian can give time resolutions of 0.5 Ma. Graptolites are fossils of conjunctive structures of small colonial animals, probably planktonic, found in rocks of Late Cambrian to Early Carboniferous ages, and also of somewhat uncertain systematic affinities. Again, small changes in complex, ornate structures make them highly time-diagnostic. Their acme as guide fossils was in the Silurian, giving there the closest available time controls at resolutions down to 1 Ma. Both conodonts and graptolites make the point that for purposes of time measurement in stratigraphy, it is possible to use guide fossils without knowing anything of the biology of the organisms of which they are the remains. (Like life today, one can use clocks as timekeepers without knowing anything about their mechanisms.) Trilobites ruled the Cambro-Ordovician and have given time resolutions of as little as 0.5–1 Ma. They were mostly benthic and hence they and graptolites largely complement each other in rocks of mutually exclusive facies.

Efforts formally to incorporate the advances in our knowledge of Palaeozoic geochronology into lower levels of the standard chronostratigraphical calendar are at present still largely concentrated on codification of Stages (Cowie *et al.* 1986), at level V. The evidence needed to go further, to the Zones of level VI, is in many cases already there, as the brief review in the previous paragraph indicates, and references to 'zones' occur frequently. But the units so labelled are in most cases best regarded as being still no more than in the category of biohorizons rather than standard chronozones.

Conclusion

The measurement of geological time has come a long way since the days of William Smith's *Strata Identified by Organized Fossils* (1816–1819). Many new methods of dating rocks based on their non-biostratigraphic, physical attributes have been discovered, and in unravelling the longest, Precambrian part of the Earth's history they provide the only means we have. In this brief review I have tried to indicate how at the same time the Phanerozoic geological calendar based on the use of fossils as clocks has also been refined. Although the new physical methods may increasingly claim to challenge traditional biostratigraphical methods even in the Phanerozoic, they do not yet surpass or displace them. Our venerable geological calendar seems safe for the foreseeable future.

References

AUBRY, M.-P., BERGGREN, W. A., VAN COUVERING, J. A. & STEININGER, F. 1999. Problems in chronostratigraphy: stages, series, unit and boundary stratotypes, global stratotype section and point and tarnished golden spikes. *Earth-Science Reviews*, **46**, 99–148.

BRIGGS, D. E. G., CLARKSON, E. N. K. & ALDRIDGE, R. J. 1983. The conodonts animal. Lethaia, **16**, 1–14.

BRINKMANN, R. 1929. Statistisch-biostratigraphische Untersuchungen an mittel-jurassischen Ammoniten über Artbegriff und Stammesentwicklung. *Abhandlungen der Gesellschaft der Wissenschaften zu Göttingen, mathematische-physikalische Klasse, Neue Folge*, **13**, 3.

BUCKMAN, S. S. 1893. The Bajocian of the Sherborne district: its relation to subjacent and superjacent strata. *Quarterly Journal of the Geological Society of London*, **49**, 479–522.

BUCKMAN, S. S. 1898. On the grouping of some divisions of so-called 'Jurassic' time. *Quarterly Journal of the Geological Society of London*, **54**, 442–462.

CALLOMON, J. H. 1964. Notes on the Callovian and Oxfordian Stages. *In*: MAUBEUGE, P. L. (ed.) *Colloque du Jurassique à Luxembourg 1962*. Publications de l'Institut grand-ducal, section des Sciences naturelles, physiques et mathématiques, Luxembourg, 269–291.

CALLOMON, J. H. 1984. The measurement of geological time. *Proceedings of the Royal Institution of London*, **56**, 65–99.

CALLOMON, J. H. 1985. The evolution of the Jurassic ammonite family Cardioceratidae. Palaeontological Association, London, *Special Papers in Palaeontology*, **33**, 49–90.

CALLOMON, J. H. 1995. Time from fossils:, S. S. Buckman and Jurassic high-resolution geochronology. *In*: LE BAS, M. J. (ed.) *Milestones in Geology*. Geological Society, London, Memoir **16**, 127–150.

CARIOU, E. & HANTZPERGUE, P. (eds) 1997. Biostratigraphie du Jurassique Ouest-Européen et Méditerranéen. Groupe Français d'Étude du Jurassique, *Bulletin du Centre de Recherche, Elf Éxploration et Production*, Bordeaux, Mémoir **17**.

COPE, J. C. W. (ed.). 1980a. *A Correlation of the Jurassic rocks in the British Isles. Part 1: Introduction and Lower Jurassic*. Geological Society, London, Special Reports, **14**.

COPE, J. C. W. (ed.). 1980b. *A Correlation of the Jurassic Rocks in the British Isles. Part 2: Middle and Upper Jurassic*. Geological Society, London, Special Reports, **15**.

COWIE, J. W., ZIEGLER, W., BOUCOT, A. J., BASSETT, M. G. & REMANE, J. 1986. *Guidelines and Statutes of the International Commission on Stratigraphy. Courier des Foschungsinstitut Senckenberg*, Frankfurt, **83**.

COX, B. M. 1990. A review of Jurassic chronostratigraphy and age indicators for the UK. *In*: HARDMAN, R. F. P. & BROOKS, J. (eds) *Tectonic Events Responsible for Britain's Oil and Gas Reserves*. Geological Society, London, Special Publication, **55**, 169–190.

EINSELE, G., RICKEN, W. & SEILACHER, A. 1991. Cycles and events in stratigraphy – basic concepts and terms. *In*: EINSELE, G., RICKEN, W. & SEILACHER, A. (eds) *Cycles and Events in Stratigraphy*. Springer, Berlin, 1–19.

GRADSTEIN, F. D. M., AGTERBERG, F. P., OGG, J. G., HARDENBOL, J., VAN VEEN, P., THIERRY, J. & HUANG, Z. 1994. A Mesozoic time scale. *Journal of Geophysical Research*, **99**, 24 051–24 074.

HAQ, B. U., HARDENBOL, J. & VAIL, P. R. 1988. Mesozoic and Cenozoic chronostratigraphy and cycles of sea level change. *Society of Economic Paleontologists and Mineralogists, Special Publication* **42**, 71–108.

HARLAND, W. B., ARMSTRONG, R. L., COX, A. V., CRAIG, L. E., SMITH, A. G. & SMITH, D. G. 1990. *A Geologic Time Scale*. Cambridge University, Cambridge.

HOWARTH, R. J. & MCARTHUR, J. M. 1997. Statistics for strontium isotope stratigraphy: a robust LOWESS fit to the marine Sr-isotope curve for 0 to 206 Ma, with look-up table for derivation of numeric age. *Journal of Geology*, Chicago, **105**, 441–456.

KNELL, S. J. 2000. *The Culture of English Geology, 1815–1851: A Science Revealed Through Its Collecting*. Ashgate, Aldershot.

MAIRE, V. 1838, Contribution à la connaissance des Cardiocératidés. *Mémoires de la Société géologique de France*, N.S., **34**.

MIALL, A. D. 1997. *The Geology of Stratigraphic Sequences*. Springer, Berlin.

OPPEL, A. 1856–1858. *Die Juraformation Englands, Frankreichs und des südwestlichen Deutschlands*. Ebner & Seubert, Stuttgart.

ORBIGNY, A. D' 1850. *Paléontologie Française. Terrains Jurassiques. I. Céphalopodes*. Résumé géologique, 600–623, in livraison 50, 521–632.

PÁLFY, J. 1995. Development of the Jurassic geochronologic scale. *Hantkeniana*, Budapest, **1**, 13–25.

PALFY, J., SMITH, P. L. & MORTENSEN, J. K. 2000. A U-Pb and $^{40}Ar/^{39}Ar$ time scale for the Jurassic. *Canadian Journal of Earth Sciences*, **37**, 923–944.

REMANE, J. 1996. The revised Guidelines of ICS and their bearing on Jurassic chronostratigraphy. *In*: RICCARDI, A. C. (ed.) *Advances in Jurassic Research*. GeoResearch Forum, Transtec, Zürich, **1–2**, 19–22.

REESIDE, J. B. 1919. *Some American Jurassic ammonites of the genera* Quenstedtoceras Cardioceras *and* Amoeboceras, *Family Cardioceratidae*. Professional Papers of the US Geological Survey, **118**.

SMITH, W. 1816–1819. *Strata Identified by Organized Fossils*. Parts 1–2, 1816; Part 3, 1817; Part 4, 1819. Aarding, London.

STENO, N. 1669. *De solido intra solidum naturaliter contento dissertationis prodromus*. Florence.

VAIL, P. R., MITCHUM, R. M., TODD, R. G. ET AL. 1977. Seismic stratigraphy and global changes of sea level. *American Association of Petroleum Geologists, Memoir* **26**, 49–212.

VON BUCH, L. 1839. Über den Jura in Deutschland. *Physikalische Abhandlungen der königlichen Akademie der Wissenschaften zu Berlin*, Jahrgang **1837**, 49–135.

WHITTAKER, A., COPE, J. C. W., COWIE, J. W. *ET AL.* 1991. A guide to stratigraphical procedure. *Journal of the Geological Society, London,* **148**, 813–824 [also published as Geological Society Special Report, **20**].

Time, life and the Earth

AUBREY MANNING

Institute of Cell, Animal and Population Biology, Ashworth Laboratories, West Mains Road, Edinburgh EH9 3JT, UK (e-mail: amanning@ed.ac.uk)

Abstract: Modern developments in the Earth sciences have revealed, for the first time, how our planet actually 'works.' For biologists, they have brought a fresh understanding of the interwoven histories of the physical and the living worlds. In turn, new biological discoveries show how early in Earth's history life arose and how it was able to flourish even under the apparently hostile conditions of the young solar system. In particular the high ocean temperatures following meteor impacts would not have precluded the successful progress of so-called hyperthermophilic prokaryotic organisms.

The early problems which evolutionary biologists faced, when presented with estimates of the age of the Earth as a few tens of millions of years, have now been replaced by arguments concerning mechanisms underlying the variable rates and patterns of evolution.

Life and the Earth have reciprocally interacted over billions of years and repeatedly planetary processes have led to mass extinctions whose dramatic results can be seen in the fossil record even though the details of such events remain problematical. So far mass extinctions have resulted in increased opportunities for the survivors; the current human-induced extinction is unlikely to do so.

It is a great honour for me to represent biologists in this volume, for the determination of the age and formation of the Earth is surely one of the great scientific achievements of the twentieth century and has had a profound influence on our study of life. Through the BBC and *Earth Story*, I have recently had the good fortune to meet a number of distinguished Earth scientists who have helped me begin to 'see' the planet as they do. This experience has certainly changed and enriched my own thinking about the history and pattern of life on Earth. It is some of these impressions which I want to share here.

The achievements of palaeontology have long revealed how a combination of biological and geological approaches and techniques can enlighten both sciences. As a student at University College, London, in the late 1940s I was taught zoology by D. M. S. Watson, a renowned palaeontologist of the period. Consequently we paid a lot of attention to the Palaeozoic and how, from his studies on the acanthodian fishes, he had been able to unravel the wonderful way in which vertebrate jaws had evolved from the gill arches of ostracoderms. So far as we were concerned then, Cambrian fossils were obscure and as rare as 'hen's teeth' and there was absolutely nothing Precambrian! Now we know better, but it is not simply that the fossil record has been greatly extended. With the extraordinary refinements of geophysical and geochemical techniques we can go far, far back in time and far beyond conventional palaeontology to explore the earliest appearance of life on Earth. It is possible to detect in very ancient Archaean rocks – perhaps back almost four billion years – carbon isotope ratios which provide a kind of fingerprint of life (but see Kamber *et al.* 2001) even if we cannot discern any structural signs of such early organisms, although now, in somewhat younger rocks, more and more of them are being identified.

The origin and early history of life

The origin of life itself is a field attracting a great deal of varied research. Our increasing knowledge of the environment on Earth soon after its formation is enabling a much more focused approach than in early work when various chemical mixtures were bombarded by electrical discharges. For example, we know that the hot water emerging through mid-ocean vents is often rich in iron sulphides which can form hollow spheres within which more complex molecules may be 'protected' from dilution and dispersal. It is always this 'containment' problem which must be addressed if we are to understand how self-perpetuating life emerged. Paul Davies (1998) in *The Fifth Miracle* provides an excellent account of modern ideas in this field.

Following on from such work, we can now understand how life could begin so amazingly early in Earth's history. It has been generally accepted for some time that hot springs rich in dissolved minerals were likely sites for life's origins. Those around volcanic areas on land and the upwellings at mid-ocean vents are plentifully inhabited by bacteria today. Much more recently has come the discovery of a far more extensive range of bacteria on the surface, in deep waters and in the Earth's crust itself which are living successfully at extremes of temperature, pressure, pH or salinity (Gross 1998). Thus various so-called thermophilic bacteria live and reproduce at a temperature of 110°C and cannot grow below 60°C! Bacteria are also found living at great depths in the interstices of rocks: in basalts near the Columbia river in Washington State, USA, at a depth of 1500 m, and several kilometres deep in granites in Canada and Sweden (Kaiser 1995). Abundant bacteria were found in oil deposits in Alaska and elsewhere, 1.5 km down at 70°C and 160 atmospheres' pressure (L'Haridon et al. 1995; Magot 1996). In all such cases, the evidence suggests that these organisms have been living in the crust for many millions of years, perhaps since the rocks first formed or the oil deposits were laid down.

Comparative analysis of the RNA/DNA structure of these and other prokaryotes suggests that many of them come from a very ancient stock. Figure 1 shows one simple version of the numerous 'family trees' which have been constructed from such DNA data. Nisbet & Fowler (1996) speculate that thermophilic bacteria may have been the very first forms of life, and the fact that various metallic ions such as copper and nickel are now involved in the make-up of key enzymes common to many forms of life may reflect the ionic environment in which these bacteria evolved on the young Earth. Certainly we can conclude that such forms of life could develop and – more crucially – survive as soon as the most intense period of bombardment from space was over and the moon was in place. Even if some meteor strikes were still causing the oceans to reach boiling point, provided they were not all vaporized, these thermophiles might hold on! The ubiquity and vast biomass of these Archaea, and the recognition that they are still existing as they have for perhaps four billion years, have taken many biologists by surprise. Amongst other speculations, it must surely shorten the odds on life having got started on Mars before it froze up! We may well find traces of Martian archaeans locked up in its permafrost. A stimulating review by Nisbet & Sleep (2001) brings together a whole range of such issues relating to the origins and earliest history of life.

However diverse life appears today, all family trees like those of Figure 1 suggest a single root. The universality of the RNA/DNA genetic code forces this conclusion. There are still some who

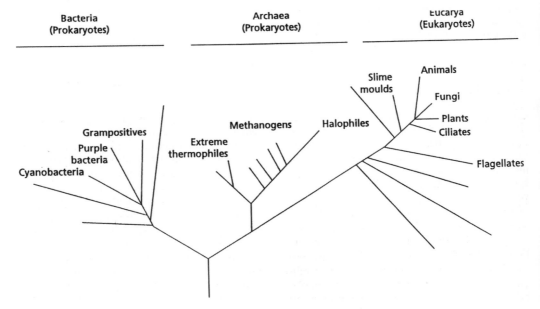

Fig. 1. A 'tree of life' based on RNA/DNA affinities. Many such trees have been constructed but they agree that all living forms have a common root and must be divided into three domains as shown. The Archaea include prokaryotes which live under extreme conditions of heat, as the labelling suggests. Note how 'close', relatively speaking, are all the multicellular lifeforms – fungi, plants and animals – with which we are most familiar.

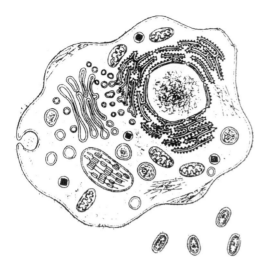

Fig. 2. Typical prokaryotic and eukaryotic cells showing the contrast in size and complexity. Linear dimensions range between 1 and 10 micrometres in prokaryotes, 10 and 100 micrometres in eukaryotes. Many of the organelles included within the eukaryote are derived from once free-living prokaryotes and are now crucial elements for eukaryotic metabolism (from Gross 1998).

suggest the 'seeding' of this life style from extra-terrestrial sources, but whether home-grown or not there can be no doubting our descent from a common ancestor. Once our form of life had got started there would never be any hope for a second – it would instantly be 'eaten' – metabolized by the archaean hordes!

The early life forms were all of the primitive prokaryotic type in which there is no nuclear membrane and the genetic material is not organized into multiple chromosomes. Here the word 'primitive' is used in the original sense, to mean that which came first. The enormous and continuing success of prokaryotes requires no more emphasis here, but we must note that for at least a billion years life existed in no other form. The next great step came with the evolution of eukaryotic cells and Figure 2 illustrates the nature of this change. Such cells are usually far larger, perhaps 1000 times the volume of prokaryotes. They have a nuclear membrane which encloses the chromosomes, and their cytoplasm includes a number of complex organelles with specialized functions, such as mitochondria which handle respiration and – in plant cells – chloroplasts which handle photosynthesis. In a brilliant intuitive leap, Margulis (1981) suggested that eukaryotic cells arose by a symbiotic union of various cells and that the organelles are descendants of once free-living prokaryotes; that the chloroplasts were once photosynthesizing cyanobacteria; the mitochondria purple bacteria, and so on. Although some cell biologists will not go so far as Margulis in the later developments of her symbiosis theory, few doubt that the basics are absolutely right. After all, intracellular symbiosis between eukaryotes and prokaryotes is familiar enough today. For example, the Pacific deep-sea vent worm *Riftia* has no gut, but a digestive gland whose cells are packed with the local sulphur-metabolizing bacteria. More familiarly, many coral polyps benefit from the photosynthetic activity of the protistan dinoflagellates which they shelter inside their cells.

It took a long time for life to achieve this step. The figure of one billion years was mentioned above. This comes from recent findings in Australia (Brocks *et al.* 1999) where traces of molecules indicative of eukaryotic cell membranes are found in rocks dated at 2.7 billion years. Until recently nothing certainly eukaryotic was known before 2.1 billion years when there are actual fossils too large to be prokaryotic. The 'difficulty' of this transition may have been related to cells having to abandon the protective cell wall of prokaryotes. Despite the advantages for exchange of gases and nutrients across the eukaryotic cell membrane, it must have been hazardous at first! Smith & Szathmary (1999) discuss in detail this and other evolutionary 'hurdles' which life has surmounted.

Now followed another long period during which living organisms, though certainly highly diverse, remained at the single-cell level of organization. Then, about a billion years ago, multicellular life emerged – fragments of what are probably seaweeds are found – and, considerably later again, a range of animal types – the Ediacara – whose origins and relationships within the animal kingdom are still fairly obscure. Becoming multicellular enabled organisms to become much larger and far more complex. Before, all life processes could only be carried out within the confines of a single tiny cell but now there could be 'division of labour' and the development of specialized tissues and organs. Within the Cambrian we now can study the marvellous fossil assemblages of the Burgess Shale, Chengjiang and Sirius Passet. Here, around 515 million years ago, we can recognize for the first time plants and animals with clear links to today's flora and fauna which lived together as an ecosystem in which food chains – plants, herbivores, carnivores, some microphagous, some macrophagous, some scavenging etc. – can be readily identified. Complex life, whose wonderful story has been revealed by palaeontology, was on its way. Conway Morris

Fig. 3. A reconstruction of burgeoning animal life in the mid-Cambrian shallow seas, as revealed by fossils from the Burgess Shale of British Columbia; a reproduction of a painting by Yukio Sato from Conway Morris (1998), by permission of Oxford University Press. As he describes it: 'In the foreground *Anomalocaris* has captured a hapless trilobite, seized in its anterior giant appendages which are manoeuvring the prey towards the armoured mouth'. Other strange animals are portrayed, most of which have distant affinities with sponges, annelids and arthropods alive today.

(1998) describes these early ecosystems most vividly and Figure 3 is taken from his book. It is beautifully reconstructed, but for the omission of plants which must have been there in some profusion. As Isiah pointed out, 'all flesh is grass'!

Evolution and questions of time

Determining the age of the Earth was indeed the first major scientific problem presented to biologists as the theory of evolution developed during the second half of the nineteenth century. During this period estimates of the age of the Earth mostly ranged from ten to a hundred million years. Darwin, Wallace, Huxley and their colleagues realized that this limited timescale required rates of evolutionary change which stretched credibility, although they mostly acknowledged that it was the geological estimates which must be the yardstick and biology would have to accept their dating. Darwin, in particular, was sorely troubled by the limitation of time. Observing the complexity of Silurian fossils he argued:

> ... if my theory be true, it is indisputable that before the lowest Silurian stratum was deposited long periods elapsed, as long as, or probably far longer than, the whole interval from the Silurian age to the present day; and that during these vast yet quite unknown periods of time the world swarmed with living creatures (Darwin 1859, pp. 306–307).

The trouble was, of course, that most estimates were nowhere near vast enough for the biologists, and indeed some physicists such as William Thomson (later Lord Kelvin) (Dalrymple 2001; Shipley 2001) almost seemed to take pleasure in pushing estimates as low as 30 million years under the noses, as it were, of both geologists and biologists! These decades, before the arrival of Rutherford and radioactive methods, were a period of great ferment both for the Earth sciences and biology, with much theorizing on the age of the Earth, estimated from rates of sedimentation or a cooling planet. The fascination of this period in the history of science is beautifully brought to life in Lewis' book on Arthur Holmes (Lewis 2000), one of the great pioneers of Earth dating (see also Lewis 2001), and by a most splendid contemporary review – Edward B. Poulton's presidential address to the Zoology Section of the British Association for the Advancement of Science in 1896 (Poulton 1896). Poulton sets out with a charming modesty, the difficulties presented to biologists by a 'young' Earth, accepting that Earth scientists must have the last word, but also offering some biological contributions to the discussion. For example, one debate centred around estimates of sedimentation rates – could geologist's assume that these had been roughly constant across time? Perhaps water flow had fluctuated, or the turbulence of the seas varied? Poulton pointed out the biological evidence from fossils of marine organisms which live attached to the substrate, such as seaweeds and crinoid echinoderms. They have stems whose dimensions closely approximate to those of living forms, elegantly suggesting that turbulence, and hence rates of sedimentation, have changed little throughout geological time.

Now, after a century or more of work, the quantitative side of the debate is largely over – we know that although the evolution of life has been constrained by many things, for the most part time is not one of them! We must assume that the RNA/DNA code became established during the earliest history of the prokaryotes, before they began to diverge into different niches. Smith & Szathmary (1999) point out that RNA, as well as encoding information, can act as an enzyme and may have been crucial in the initial assembly of self-replicating molecules. There are still a few who may argue that the speed with which the genetic code would have had to evolve, stretches probability too far. Thus we should not ignore the possibility of bacterial spores or some such seeding of the young Earth from a comet or other extraterrestrial source. Even Lord Kelvin himself suggested this at one time (Poulton 1896, p. 818) – long before the genetic code argument was invoked – perhaps because he seems to have had some feelings of guilt about the problems which his age estimates presented to the evolutionists! For the most part, such arguments hold little force today. Within the huge timespan of life's history on Earth, the evolutionary issues which remain are concerned with the causes of extinction and survival, and the mechanisms behind the great flourishing of biodiversity which we observe. Some of these issues are hotly contested; not without cause is Brown's (1999) book entitled *The Darwin Wars*.

It is perhaps inevitable that creationists have seized upon disputes over mechanisms and present them to their gullible audiences as hard evidence that scientists are arguing amongst themselves whether 'the theory of evolution' is true! This is a problem which biologists have had to face continuously ever since publication of *The Origin of Species*. In the 1860s the theory of evolution was perceived as an affront to conventional religious beliefs, but we delude ourselves if we think this issue is now dead. Attacks from those I may loosely group as 'creationists' have never ceased and, even in our modern culture which so much relies on science and technology, they persist. Most of us are insulated within our scientific communities, but 'out there' creationist ideas flourish. Anybody who doubts this should read *New Scientist*'s special issue of 22 April 2000 (No. 2235), and take note of Short (1994) and Downie & Barron (2000), who surveyed the beliefs of students about to set out on medical or biological courses. Short, teaching in Australia, and Downie & Barron in Glasgow, both found that about 10% of students starting a university biology course, rejected ideas of biological evolution.

In *The Blind Watchmaker*, a most stimulating account of neo-Darwinism, Richard Dawkins (1988) expresses some puzzlement that evolutionary theory is under constant attack, whereas quantum mechanics or general relativity, which might seem equally dangerous to creationist views, are largely ignored. In part it must be a question of understanding: physical theories require some sophistication in mathematics, while evolution seems deceptively simple, but that cannot be all. Whilst it is possible to distance oneself from difficult ideas about the physical universe, Darwin's theory forces us very obviously to accept that we humans are one end point amongst many of the processes of natural selection. In every generation there will be people who find this idea completely unacceptable and hence our problem continues. It seems that the creationists may always be with us.

Earth science theories too seem largely immune from attack, which is also puzzling, since one would have expected that Hutton's musing on 'no vestige of a beginning, no prospect of an end' would cause some outrage today, as it did many decades before *The Origin of Species*, although Brush (2001) does discuss areas where geologists too have been confronted by the creationists. Now geological studies have revealed the real 'workings' of our planet and the enormous timescales that that has required, geologists cannot remain on the side-lines of this issue. Hence the purpose of this digression is to summon the aid of Earth scientists in the battle to put across in a non-threatening way, the truth and the beauty of scientific ideas relating to the history of the Earth and its living inhabitants. It is essential to combat ignorance and irrationality.

Evolution's course: continuous or punctuated equilibria?

Many arguments among biologists are certainly relevant to any consideration of how life and the Earth have interacted. One contested point can be presented, rather simplistically, as two alternatives. Has evolution been a gradual process with natural selection operating each generation to shape organisms, leading to what Darwin called 'descent with modification'? Alternatively, has evolution been something of a stop/go process with long periods of stasis interrupted by periods of rapid change? Many will recognize the second alternative as one particularly espoused by Stephen Jay Gould who uses the term 'punctuated equilibrium' to describe it. Few would doubt that particular evolutionary events to match both alternatives can be found and neither side would deny this. Nevertheless there are distinctly different flavours to the two approaches, and historians of the Earth sciences will recognize analogies with the debates between uniformitarianism and catastrophism in nineteenth century geology. Gould's (1988) book *Time's Arrow, Time's Cycle* brings out this connection clearly.

It may well have been difficult for many geologists to accept that, quite apart from the sudden effects of volcanoes and earthquakes, enormous changes to landscapes can sometimes occur with great rapidity. Gould (1980, Ch. 19) recounts vividly how long it took the American field geologist J. Harlen Bretz to convince others that the extraordinary landscapes of the scablands of Washington State around Dry Falls were not the customary work of slow erosive forces. They were formed in a week or two by the impact of colossal flood waters let loose when glacial dams in Montana broke, releasing the water from Lake Missoula!

As one approach to sudden evolutionary change, biologists have toyed on and off with the possibility of large mutations – the emergence of 'hopeful monsters' – but they have rarely been convinced. Only a minority of mutations, even small ones, will be beneficial. Sometimes small populations of animals or plants will become isolated, a higher than normal degree of inbreeding will follow and by chance some genes may become fixed in the population, while others are lost, so it may change quite rapidly by this so-called 'genetic drift'. Of course the population will only survive if the genes fixed happen to be favourable or effectively 'neutral' and without natural selection operating, this will not often be the case. Consequently gradual change by small steps is much more in conformation with what we know of the subtle ways in which genes and environment interact during an organism's development.

However, even under continuous natural selection, evolution cannot proceed at anything like a constant rate. If environments are relatively constant for a period then selection will tend to be 'normalizing' – selecting against extremes – and evolution, especially as it may be judged from the fossil record, will slow or stop. By contrast, a period of rapid cooling, a lake drying up, a large basalt flow, will require a rapid response from the local flora and fauna. Those that can respond will leave survivors and something resembling a step will be marked in the evolutionary record.

Simpson's (1944) classic text *Tempo and Mode in Evolution* illustrated abundantly the varying rates of change at different geological periods amongst different groups. He points out that we can observe gradual change even in rapidly evolving groups where we have a good and more or less uninterrupted fossil record, such as the horses. He dealt there directly with the issue of whether reasonable assumptions about population size, generation interval, numbers of gene loci involved in the development of a character, and mutation rate could produce the change we observe. The horses grew much bigger as they evolved and their teeth grew even more in proportion as they switched from browsing to grass as a diet. Taking tooth crown height as an example, Simpson calculates that the accumulation of small genetic changes could easily have resulted in the growth we observe. The 60 million years between *Hyracotherium* of the Eocene and modern *Equus* could have seen tooth growth accomplished in about 300 tiny 'steps' of 0.1 mm each!

More recently, Nilsson & Pelger (1994) have critically examined one of Darwin's most exacting tests for 'gradualism' which he boldly tackled in Chapter 6 of *The Origin of Species*, entitled 'Difficulties of the Theory'. How long would be required for the gradual evolution of a well-developed eye, such as that of an octopus or a vertebrate, with its focusing lens, iris diaphragm, etc.? Certainly this example more than any other is held up for creationist ridicule and disbelief. Nilsson & Pelger start with the easily accepted hypothesis that there has been a constant advantage to forming better and better images of the outside world. Given this they calculate that even with the most pessimistic assumptions at every stage, the shift from a flat patch of light-sensitive cells to an advanced eye could be achieved in a few hundred thousand years.

Thus whilst accepting that marked fluctuations in rates of evolution occur, most neo-Darwinians are convinced that these are responses to environmental circumstances and there is no need to invoke forces other than natural selection acting on small, random mutations to bring them about. There is time enough, and apparent leaps in the fossil record may yet encompass thousands of generations for many organisms.

Evolution's course: contingency or selection?

There is another important dispute regarding the course of evolution and the mechanisms underlying it which also relates to the history of the Earth itself. How far has contingency played a part in the survival of particular groups and hence determined the pattern of descent? How far has survival been the result of natural selection operating to favour those best fitted? Once again, we can set up such a dichotomy but it is clear that examples of both are easy to find, and that the two mechanisms are not necessarily mutually exclusive. Gould has certainly tended to emphasize contingency in the outcomes of evolution. In a memorable passage at the end of his account of the Burgess Shale, *Wonderful Life*, he muses on the fact that from an assemblage with many casualties, one of the survivors was *Pikaia*, a chordate:

> Wind back the tape of life to Burgess times, and let it play again. If Pikaia does not survive in the replay, we are wiped out of future history – all of us, from shark to robin to orangutan. And I don't think that any handicapper, given Burgess evidence as known today, would have granted very favorable odds for the persistence of Pikaia (Gould 1991, p. 323).

Probably few would disagree with this conclusion, just as we all recognize that without the cataclysm which led to the demise of the dinosaurs, the mammals would not have achieved their present dominance amongst the vertebrates. However, many biologists feel that, if taken too far, this line of argument inevitably leads to an underestimate of what natural selection can and has achieved. Probably one reason why Gould puts emphasis on the role of chance, is to counteract the idea that *progress* is inherent in evolution and culminates in the emergence of human beings! This is indeed a view very commonly held and I share with him a distaste for the anthropocentrism which it implies, a failing shared with so-called 'creation science' which denies evolution – it is a distortion of the history of life. Nevertheless, it is impossible to ignore certain aspects of natural selection which suggest that, in fact, growth in *complexity* is well nigh inevitable in many cases and from some points of view this represents progress.

Commonly, complexity increases when species compete with one another. Such interactions are almost universal because nothing exists in a monoculture of its own type. In every ecosystem there will be competition between species as well as between individuals; parasites interacting with their hosts for example, or 'food chains' which we can see more simply in predator/prey interactions. Insects eat the leaves of plants, which respond by developing chemicals, making the leaves distasteful or hard to digest. The insects develop enzymes which break down these chemicals, and so on. Such competition results in changes to the structure and metabolism of both sides which do add complexity but are undoubtedly costly – an excellent human analogy is the 'arms race.' This term is commonly used by biologists now and we can often see it reflected in the fossil record. Prey animals develop armour, their predators develop teeth, claws and grasping limbs – an arms race which we can follow to amazing extremes both in early fish and in the dinosaurs. Prey can escape by running or becoming camouflaged and their predators must out-run them or out-detect them. Usually there is no ultimate victor and the arms race continues until costs exceed benefits and a kind of stalemate is attained with each side winning some and losing some. Cheetahs and Thomson's gazelles have been fine-tuned by selection through their contests over many thousands of generations. Certainly both are more complex in structure and physiology and behaviour – they have progressed one might say – in comparison with their rather clumsy ancestors of the early Tertiary.

Thus in complex ecosystems, natural selection may continue to drive organisms along evolutionary paths towards more complex adaptations even if the physical aspects of the environment remain very constant. This leads to another phenomenon which shows us that contingency cannot be the whole story of who survives. Natural selection seems to be remarkably good at

Fig. 4. Sabre-toothed animals from the late Tertiary of South America. These two are only the most distant of mammalian relatives. *Thylacosmilus* (above) is a marsupial, *Smilodon* (below) a eutherian cat (from Conway Morris 1998, reproduced from Nitecki 1981, courtesy of Academic Press Inc.).

carving out the best solution to an adaptational problem and can manage to do so from very diverse starting points. The phenomenon of parallel or convergent evolution is widespread. Figure 4 offers a dramatic example. Several orders of large herbivorous mammals evolved in South America which sought to protect themselves by growing very large and developing thick skins. They posed problems for carnivores, and indeed some modern equivalents such as elephants and rhinos are largely immune from predators, at least as adults. One of the carnivores' responses in this arms race was to become large also – though of course carnivores can never become as big as the largest prey animals – and to develop enormously long canine teeth, becoming 'sabre-toothed tigers'. Several times the true cats evolved sabre-toothed types during the late Tertiary: *Smilodon*, one of the most famous, is shown in Figure 4 (bottom). The other sabre-toothed animal has obviously evolved a very similar set of adaptations to meet a very similar life style, preying on large pachyderms. However, not only is it not a cat, it is not even a eutherian mammal. It is the marsupial *Thylacosmilus* whose last common ancestors with *Smilodon* would have been small, opossum-like animals living in the Upper Cretaceous!

Such examples, and they are abundant, demonstrate that contingency alone does not determine who flourishes. *Smilodon*'s ancestors could have been eliminated by some chance mishap, but it seems highly probable that another sabre-tooth type would have emerged to replace it, filling that particular niche. So long as the lumbering South American pachyderms remained as potential prey, there was a way of life there going vacant and natural selection would have been likely to draw some other carnivorous stock along that path. Gould's quotation above suggests, I am sure correctly, that the existence of *Homo sapiens* per se, is due to chance events in Earth's history. But if *Pikaia* had not made it, there would have been many vacant niches which other stocks would have evolved to fill and, because of the inevitable arms races, the 'human type' of adaptation – large brain size, high intelligence and very flexible behaviour – would almost certainly have emerged because of the big selective advantage it confers. Today, octopuses are the brightest of the invertebrates: the alternative world might have been inhabited by intelligent, land-living cephalopod molluscs!

Dynamic Earth and the history of life

Whatever the extraordinary force of natural selection, no biologist is going to deny that chance events have, from time to time, overwhelmed it. Dynamic changes during the development of the planet have had a major influence on the course of evolution. As the continental plates have split apart or merged, the area of shallow oceans, so crucial for early life, has waxed and waned. Plate tectonics has split apart populations and thus enabled them to evolve in isolation; it has forced populations, hitherto isolated, into contact and sometimes brought about wholesale extinction of a less competitive group as a result. The presence of life has, in turn, affected the history of the Earth: the original injection of oxygen into the oceans, eventually the atmosphere, the sequestration of carbon in plants, the action of bacteria, fungi and plant roots on the break-up of rocks, the movement of water through the atmosphere to fall as rain, the erosion of soils and the exposure of rocks to weathering – the list is pervasive.

James Lovelock (1995) has developed, in a most compelling way, studies of the reciprocal interactions between Earth and its biosphere in forming his Gaia hypothesis. They reveal how many such interactions are homeostatic in their effects and have, for example, maintained liquid water on the surface and an equable atmosphere over billions of years. In this way the Earth has provided an amenable stage on which life's story could be played out. But the story has not always been gentle. There have been violent events in our planet's history which have had equally dramatic results on the course of evolution.

So far as multicellular organisms are concerned, the processes of specialization and adaptation have led to an overall increase in biodiversity, which we may very roughly define as number of different 'types', since Cambrian times. However, and most crucially, this increase has not been uninterrupted. Figure 5 shows one of many versions of this story. It concerns only marine animals, which are well represented in the fossil record, so that their presence is unlikely to be missed. It takes 'number of families' as the measure of biodiversity; families are fairly large-scale and arbitrary units which depend on human judgement, but they can be applied with reasonable consistency. Figure 5 shows the overall increase with time but also five sharp falls which represent the so-called 'mass extinctions'. They turn up with the same timings no matter what measure of biodiversity is taken and there can be no question that they represent periods when conditions on Earth were hostile to multicellular life, sometimes almost lethal. This plot dates from the 1980s and more modern estimates suggest that the extinctions were even more severe than indicated here, particularly during the greatest extinction of all, that at the Permian–Triassic boundary. Then over 90% of marine species, including the last of the trilobites, disappeared and there were also huge reductions in terrestrial forms.

For all these events there remain considerable problems of interpretation, because the biological evidence is largely negative. All palaeontologists can do is observe that previously abundant types disappear as they come to a particular horizon in the rocks, and that above that horizon, it is fewer and different types which take their place. Can we discover what was happening that had such disastrous effects on life? Here of course biologists must look to the Earth sciences for clues, because the one certainty is that the primary cause of each mass extinction was 'geological' in the broadest sense.

For example, the most famous clue to the most famous extinction – that which resulted in the demise of the dinosaurs – is a thin, but consistent

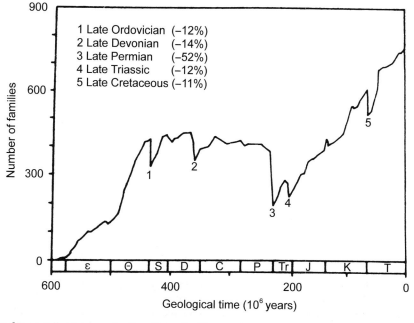

Fig. 5. The five so-called 'mass extinctions' over the history of multicellular life. Here the number of families well-represented in the marine fossil record is taken as the measure of diversity. Time runs back to the first great rise in diversity at the beginning of the Cambrian. The later extinctions will have involved terrestrial life also, but the timing of these major events was shared (reprinted with permission from Raup & Sepkoski 1982, © 1982 American Association for the Advancement of Science). Figures in brackets indicate the percentage loss of marine families which occurred during each 'mass extinction'.

iridium layer in rocks at the Cretaceous–Tertiary (K/T) boundary. This has been linked to a large meteorite strike in the Gulf of Mexico, that left traces of a huge crater in the Chicxulub region. We can be certain that such a colossal impact would have had pan-global effects – fires, dust clouds, tsunamis, all on a gigantic scale. Light and climate must have been immediately and drastically affected. Discussing the various types of evidence, Frankel (1999) cites estimates of pitch-darkness over the entire world for two months, followed by prolonged twilight, a 10°C drop in temperature, persistent sulphuric acid precipitation and so on. For months photosynthesis must have diminished almost to zero both on land and from the phytoplankton of the oceans. We can be sure that death came rapidly to much life on land, though perhaps more slowly to marine life far from the impact. But the time course of such an extinction still remains mysterious because the geological horizons from which we are extracting evidence may represent thousands of years or even longer. McGhee (1996), analysing the late Devonian mass extinction, recognizes that it may have taken place over *several million* years. So while we know that no land animals weighing more than about 25 kg survived the K/T event, just how long did the dinosaurs hold on? Could it have been decades or even centuries before the last of the giants succumbed? Biologists can deduce the hugely stressful conditions for life then by observing that burrowing animals and detritus or carrion feeders do better than carnivores or plant eaters. The plants must have held on through dormant seeds or root systems until some glimmers of sunlight returned. Even so, many mysteries remain.

For each of the five great extinctions identified in the geological record, and many less dramatic, there are geological clues as to what happened. Frankel (1999) believes that meteor impacts may have been involved more often than formerly supposed, and it is interesting to note that some new evidence of a gigantic impact associated with the Permo-Triassic extinction has just emerged (Kerr 2001). But impacts cannot be the whole story. There were also huge basalt flows near the K/T boundary (the Deccan traps) and at the Permo-Triassic boundary (the Siberian traps) which must have decimated photosynthesis for long periods. Earlier, the Earth almost froze out life during the Ordovician extinction when the continental plates were amassed together near the South Pole.

The remarkable fact is that life not only managed to survive these colossal physical blows, each time over millions of years it recovered to reach continuing new heights of biodiversity. To a great extent, each mass extinction left a world where competition was reduced and survivors had enhanced possibilities. The most often quoted example is the great radiation of the mammals which followed the decimation of large reptiles. Recent molecular evidence suggests that

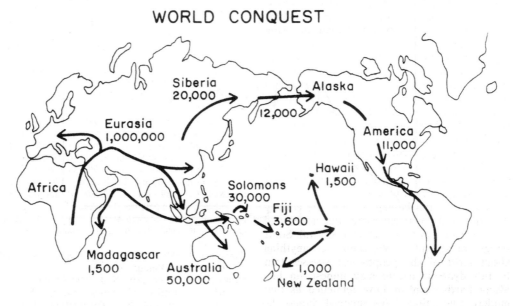

Fig. 6. The approximate timing (years before present) of our ancestors' emigration from Africa across the planet (from *The Rise and Fall of the Chimpanzee* by Jared Diamond 1992, published by Hutchinson. Used by permission of The Random House Group Ltd.).

they and the birds may have begun to diversify even before the K/T event (Cooper & Penny 1997), but even so, the absence of large reptiles must have had major liberating effects!

There is a crucial modern relevance to the history of these past extinctions, for the sixth mass extinction is now in progress! This is most certainly *not* the result of geological processes. It follows the huge increase of the human population and the demands which this is putting on space and natural resources of all types. Rates of change are now rising rapidly but our species has been a most powerful agent of extinction for many millennia. Figure 6 from Jared Diamond's (1992) *The Rise and Fall of the Third Chimpanzee* – Man – indicates the history of our spread across the globe from our African origins. The evidence is clear that with the entry of humans came the rapid exit of many large animals and well before history we had lost mammoths, woolly rhinoceros, cave bears and giant sloths amongst a host of other less conspicuous creatures. Now the insidious changes to the atmosphere and the oceans caused by human activities are having global effects; these and the destruction of habitats as we clear for food production are resulting in the decimation of many plant and animal species. May, Lawton & Stork (1995), authorities not given to exaggeration in these matters, estimate that impending rates of extinction are some *ten thousand times* the background rates observed in the fossil record. The huge biodiversity which has evolved over billions of years is in grave danger. Nor will the survivors be offered fresh opportunities because such huge proportions of the Earth's resources are being monopolized by our single species; we are becoming a voracious monoculture. Already there are signs that the recuperative powers of the biosphere are coming under strain with the pollution of oceans, deforestation, soil loss and desertification. Since the equitable conditions which support life – including human life – depend on the health of ecosystems everywhere on the planet, in the long term we reduce biodiversity at our own peril. This is to leave aside the practical, aesthetic and moral aspects of driving any unique species to extinction.

Slowly, modern human societies are coming to terms with such environmental issues and the threat they pose to our own survival. Some things are beyond us. We cannot do anything about most of the geophysical stresses that Earth's dynamic nature may impose on life. Many Earth scientists have reminded us that modern civilizations have emerged during the remarkably tranquil last ten thousand years. Huge basalt flows, huge earthquakes and new glaciations will inevitably come again: we have to accept this. The timing of such events is uncertain: they may be imminent, they may be hugely distant compared with a human lifespan. In the end humanity, like all life, will have to take its chance with our planetary home. But we *can* protect the current biosphere and we have to do so, trying to put our house in order for the immediately succeeding generations. We must come to live with the Earth more sustainably or our species will not survive in any acceptable way. Once again it is a matter of time. I judge the next 50 to 100 years will be crucial – a geological picosecond! The collaboration of Earth scientists and biologists has achieved a new holistic understanding of how the Earth and its life forms have evolved together over time. It is one of humanity's greatest intellectual achievements. It ought to help us to persuade human societies to look beyond short-term issues alone, to where the real keys to the future may lie.

I am most grateful to M. Dawkins, C. Lewis and S. Conway Morris for helpful comments on an earlier draft of this chapter.

References

BROCKS, J. J., LOGAN, G. A., BUICK, R. & SUMMONS, R. E. 1999. Archaean molecular fossils and the early rise of eukaryotes. *Science*, **285**, 1033.

BROWN, A. 1999. *The Darwin Wars*. Simon & Schuster, London.

BRUSH, S. G. 2001. Is the Earth too old? The impact of geochronology on cosmology, 1929–1952. *In*: LEWIS, C. L. E. & KNELL, S. J. (eds) *The Age of the Earth: from 4004 BC to AD 2002*. Geological Society, London, Special Publications, **190**, 157–175.

CONWAY MORRIS, S. 1998. *The Crucible of Creation*. The University Press, Oxford.

COOPER, A. & PENNY, D. 1997. Mass survival of birds across the Cretaceous-Tertiary boundary: molecular evidence. *Science*, **275**, 1109–1113.

DALRYMPLE, G. B. 2001. The age of the Earth in the twentieth century: a problem (mostly) solved. *In*: LEWIS, C. L. E. & KNELL, S. J. (eds) *The Age of the Earth: from 4004 BC to AD 2002*. Geological Society, London, Special Publications, **190**, 205–221.

DARWIN, C. 1859. *On the Origin of Species* (first edition). John Murray, London.

DAVIES, P. 1998. *The Fifth Miracle*. Penguin, London.

DAWKINS, R. 1988. *The Blind Watchmaker*. Penguin, London.

DIAMOND, J. 1992. *The Rise and Fall of the Third Chimpanzee*. Vintage, London.

DOWNIE, J. R. & BARRON, N. J. 2000. Evolution and religion: attitudes of Scottish first year biology and medical students to the teaching of evolutionary biology. *Journal of Biological Education*, **34**, 139–146.

FRANKEL, C. 1999. *The End of the Dinosaurs: Chicxulub Crater and Mass Extinctions.* Cambridge University, Cambridge.

GOULD, S. J. 1980. *The Panda's Thumb.* Norton, New York.

GOULD, S. J. 1988. *Time's Arrow, Time's Cycle.* Penguin, London.

GOULD, S. J. 1991. *Wonderful Life.* Penguin, London.

GROSS, M. 1998. *Life on the Edge.* Plenum, New York.

KAISER, J. 1995. Can deep bacteria live on nothing but rocks and water? *Science*, **270**, 377.

KAMBER, B. S., MOORBATH, S. & WHITEHOUSE, M. J. 2001. The oldest rocks on Earth: time constraints and geological controversies. *In*: LEWIS, C. L. E. & KNELL, S. J. (eds) *The Age of the Earth: from 4004 BC to AD 2002.* Geological Society, London, Special Publications, **190**, 177–203.

KERR, R. A. 2001. Whiff of gas points to impact mass extinction. *Science*, **291**, 1469–1470.

LEWIS, C. 2000. *The Dating Game.* Cambridge University, Cambridge.

LEWIS, C. L. E. 2001. Arthur Holmes' vision of a geological timescale. *In*: LEWIS, C. L. E. & KNELL, S. J. (eds) *The Age of the Earth: from 4004 BC to AD 2002.* Geological Society, London, Special Publications, **190**, 121–138.

L'HARIDON, S., REYSENBACH, A.-L., GLENAT, P., PRIEUR, D. & JEANTHON, C. 1995. Hot subterranean biosphere in a continental oil reservoir. *Nature*, **377**, 223–224.

LOVELOCK, J. 1995. *The Ages of Gaia* (second edition). Oxford University, Oxford.

MCGHEE, G. R. 1996. *The Late Devonian Mass Extinction: the Frasnian/Famennian Crisis.* Columbia University, Columbia, Ohio.

MAGOT, M. 1996. Similar bacteria in remote oil fields. *Nature*, **379**, 681.

MARGULIS, L. 1981. *Symbiosis in Cell Evolution.* Freeman, San Francisco.

MAY, R. M., LAWTON, J. H. & STORK, N. E. 1995. Assessing extinction rates. *In*: LAWTON, J. H. & MAY, R. M. (eds) *Extinction Rates.* Oxford University, Oxford, 1–24.

NILSSON, D.-E. & PELGER, S. 1994. A pessimistic estimate of the time required for an eye to evolve. *Proceedings of the Royal Society of London*, **256B**, 53–58.

NISBET, E. G. & FOWLER, C. M. R. 1996. Some liked it hot. *Nature*, **382**, 404–405.

NISBET, E. G. & SLEEP, N. H. 2001. The habitat and early nature of life. *Nature*, **409**, 1083–1091.

NITECKI, M. H. (ed.) 1981. *Biotic Crisis in Ecological Evolutionary Time.* Academic Press Inc., Florida.

POULTON, E. B. 1896. Presidential Address, Section D – Zoology: A naturalist's contribution to the discussion upon the age of the Earth. *Report of the British Association for the Advancement of Science, Liverpool, 1896*, 808–828.

RAUP, D. A. & SEPKOSKI, J. J. 1982. Mass extinction in the marine fossil record. *Science*, **215**, 150–153.

SHIPLEY, B. C. 2001. 'Had Lord Kelvin a right?': John Perry, natural selection and the age of the Earth, 1894–1895. *In*: LEWIS, C. L. E. & KNELL, S. J. (eds) *The Age of the Earth: from 4004 BC to AD 2002.* Geological Society, London, Special Publications, **190**, 91–105.

SHORT, R. V. 1994. Darwin, have I failed you? *Trends in Ecology & Evolution*, **9**, 275.

SIMPSON, G. G. 1944. *Tempo and Mode in Evolution.* Columbia University, New York.

SMITH, J. M. & SZATHMARY, E. 1999. *The Origins of Life.* Oxford University, Oxford.

Dating the origin of modern humans

CHRISTOPHER B. STRINGER

Department of Palaeontology, The Natural History Museum, London SW7 5BD, UK
(*email*: c.stringer@nhm.ac.uk)

Abstract: One of the most serious past difficulties facing realistic tests of evolutionary models for modern human origins was the lack of widely applicable dating procedures that could reach beyond the practical limits of radiocarbon dating. Moreover, the amount of fossil material that had to be sacrificed to obtain a conventional radiocarbon date meant that human fossils could generally only be dated indirectly through supposedly associated materials. Because of these limitations, the transition from Neanderthals to modern humans in Europe was believed to have occurred about 35 000 radiocarbon years ago, but detailed reconstruction of the processes involved (for example, evolution or population replacement) was not practicable. In the Levant, the transition period from Neanderthals to modern humans was believed to lie only slightly beyond this 35 000-year watershed. Elsewhere, in Africa, east Asia and Australasia, the chronology for modern human origins was even more difficult to establish. However, over the last fifteen years, radiocarbon and non-radiocarbon physical dating techniques such as luminescence and electron spin resonance have been increasingly refined, leading to a revolution in our understanding of the timescale for human evolution, particularly for the last 200 000 years. Although each dating method has its own strengths and weaknesses, the picture that now emerges is one of a gradual evolution of Neanderthal morphology in Europe, in parallel with a similar evolution of modern humans in Africa. Modern humans also appear surprisingly early in the Levant (*c.* 100 ka ago) and Australia (*c.* 60 ka ago). However, many uncertainties still surround the process of establishment of our species globally.

The fossil record of human evolution has grown appreciably in the last few years, ranging from African material representing some of the earliest stages of our evolution, through evidence of the first known human colonizers of Eurasia to a sample of over 2000 primitive Neanderthal fossils from the cave of Atapuerca in Spain (e.g. Klein 1999). New study techniques have also been introduced to investigate this growing record, such as computerized tomography (CT) scanning (e.g. Hublin *et al.* 1996) and genetic studies that have included the extraction of DNA from Neanderthal fossils (e.g. Ovchinnikov *et al.* 2000).

But arguably the greatest recent advances in our abilities to reconstruct the course of human evolution have come through the development and application of new chronometric dating techniques. When Kenneth Oakley, my distinguished predecessor at The Natural History Museum, published the final edition of his book *Frameworks for Dating Fossil Man*, in 1969, many of these dating methods were not yet in use or were still in their infancy. As Oakley explained, so-called *relative dating* then provided the main means of estimating the age of most of the human and pre-human fossil record, and in fact it still does. A fossil or artefact could be dated as earlier, later or contemporaneous in relation to another fossil, artefact, stratum or geological event. But estimating the actual *age* of important fossils often owed more to guesswork than to science. Few reliable methods of *absolute* or *chronometric dating* were then available and by far the most widely applied was radiocarbon dating. However, potassium–argon dating was also beginning to make an impact, and had doubled the estimated timescale of early human evolution at a stroke in 1960 through its application to volcanic deposits underlying an australopithecine skull and early stone tools at Olduvai Gorge, Tanzania, assigning them an age of about 1.75 Ma (e.g. Klein 1999).

These two techniques are quite different in character and application. Potassium–argon dating (and the related argon–argon technique) is most useful for ancient volcanic rocks, and therefore has great potential for dating the earliest, African, stages of human evolution, but it is of much less use in dating the later stages of human evolution in regions such as Europe. In contrast, radiocarbon dating depends on measuring the

Table 1. *A comparison of dates BP for fossil human material from Oakley (1969) with present data*

Site: fossil species	Age estimate (Oakley 1969)	Present estimate	Dating method
Zhoukoudian Lower Cave Locality I (China): *H. erectus*	200–400 ka	300–550 ka	ESR, U-S
Tighenif (Ternifine) Algeria: *H. erectus*	Lower Amirian (= 'early Mindel')	*c.* 800 ka	Correlation
Mauer Germany: *H. heidelbergensis*	Gunz-Mindel *c.* 500 ka	Late Cromerian *c.* 500 ka	Correlation
Swanscombe England: *H. heidelbergensis* or *neanderthalensis*	Hoxnian *c.* 250 ka	OIS 11 (?Hoxnian) *c.* 400 ka	Correlation
La Chapelle-aux-Saints France: *H. neanderthalensis*	Würm II	47–56 ka	ESR
Le Moustier France: *H. neanderthalensis*	Würm II	40–56 ka	TL, ESR
Pech de l'Aze France: *H. neanderthalensis*	Würm II	54–59 ka	ESR
Neanderthal Germany: *H. neanderthalensis*	Würm 35–70 ka	39–40 ka*	RC
Ehringsdorf Germany: *H. neanderthalensis*	Eemian 60–120 ka	150–250 ka	U-S
Monte Circeo (Guattari) Italy: *H. neanderthalensis*	Würm 35–70 ka	45–62 ka	ESR, U-S
Banolas Spain: *H. neanderthalensis*	Pleistocene	45 ka	U-S
Petralona Greece: *H. heidelbergensis*	Würm	>150 ka	U-S, ESR Correlation
Starolselye, Crimea: *H. sapiens*	Late Würm I *c.* 35 ka	Holocene	(Intrusive?)
Krapina Croatia: *H. neanderthalensis*	Mid-Würm 30–45 ka	*c.* 130 ka	ESR, U-S
Jebel Qafzeh Israel: *H. sapiens*	Early Würm *c.* 60 ka	92–120 ka	TL, ESR, U-S
Amud Israel: *H. neanderthalensis*	Würm	50–70 ka	TL
Skhul Israel: *H. sapiens*	Mid Würm	81–119 ka	ESR, TL
Tabun Israel: *H. neanderthalensis*	Mid Würm *c.* 40 ka	30–130 ka	U-S, ESR
Zuttiyeh Israel: ?*H. heidelbergensis*	Late Eeemian *c.* 70 ka	*c.* 148 ka	U-S
Ngandong Indonesia: *H. erectus*	Late Pleistocene	30–200 ka	ESR, U-S
Jebel Irhoud Morocco: early *H. sapiens*	Soltanian *c.* 40 ka	87–190 ka	ESR
Broken Hill Zambia: *H. heidelbergensis*	Upper Pleistocene *c.* 40 ka	>200 ka	Correlation
Florisbad S. Africa: early *H. sapiens*	Upper Pleistocene *c.* 35 ka	259 ka	ESR, OSL
Engis Belgium: *H. sapiens*	Würm 'Aurignacian'	4.6 ka*	RC

Table 1. (*continued*)

Site: fossil species	Age estimate (Oakley 1969)	Present estimate	Dating method
Gough's England: *H. sapiens*	Late Würm	12 ka* (Creswellian) 9 ka* (GC 1)	RC
Kent's Cavern England: *H. sapiens*	Late Würm	KC1 8 ka* KC4 31 ka*	RC
Langwith England: *H. sapiens*	Würm *c.* 10 ka	2.5 ka*	RC
Paviland Wales: *H. sapiens*	Würm *c.* 18 ka	26 ka*	RC
Whaley England: *H. sapiens*	Würm	3.5 ka*	RC
Cro-Magnon France: *H. sapiens*	Würm III *c.* 20 ka	32 ka*	Correlation
Sungir Russia: *H. sapiens*	Late Würm *c.* 15 ka	24 ka*	RC
Singa Sudan: early *H. sapiens*	Upper Pleistocene *c.* 23 ka	=/>133 ka	ESR, U-S
Mumbwa Zambia: *H. sapiens*	Upper Pleistocene Middle Stone Age *c.* 23 ka	=/>75 ka	Correlation
Laguna Beach USA: *H. sapiens*	*c.* 17 ka	5.1 ka*	RC
Natchez USA: *H. sapiens*	Late Wisconsin	5.6 ka*	RC
Tepexpan Mexico: *H. sapiens*	Late Wisconsin *c.* 11 ka	0.9–2 ka*	RC

* Radiocarbon years.
Present data mainly taken from Stringer & Gamble (1993), Klein (1999), Taylor (2000) and the Oxford radiocarbon accelerator date lists published in *Archaeometry* since 1985.
ESR, electron spin resonances; OSL, optically stimulated luminescence; RC, radiocarbon; TL, thermoluminescence; U-S, uranium series.

remaining amount of a radioactive isotope, not its daughter products, in organic materials, and hence is most useful in dating the recent past, back to about 35 000 years ago. Yet another difference, and one which has become increasingly apparent, is that methods such as potassium–argon dating use accurate physical clocks that can provide age estimates in calendar years. However, radiocarbon dating is a secondary method that depends on the original proportion of radiocarbon in the environment at the time of death of an organism. It was long suspected, and has now been demonstrated, that the rate of production of radiocarbon in the upper atmosphere has fluctuated in the past, which means that radiocarbon dates are ages in radiocarbon years, not calendar years. For example, comparisons of radiocarbon dates with varves and ice cores dated independently to between about 12 000 and 14 000 calendar years BP suggest that radiocarbon is underestimating real ages by over two thousand years at this particular period at the end of the Pleistocene (e.g. Taylor & Aitken 1997).

Non-radiocarbon physical dating techniques have been increasingly refined and available over the last decade, leading to many more applications of these methods, and a consequent revolution in our ability to calibrate events in human evolution beyond 40 000 years ago. However, these additional techniques have both strengths and weaknesses compared to the relatively straightforward application of potassium–argon and radiocarbon dating, and it must be admitted that their use has not always clarified the chronology of sites to which they have been applied. Radiometric clock techniques such as uranium-series methods measure the presence of daughter products from the radioactive decay of uranium. In substances such as cave stalagmites, these daughter products accumulate and are conserved, approximating a simple closed system. However, when applied to bones or teeth (for example in gamma-ray dating), open-system behaviour requires consideration of the rate of uranium uptake, or even loss. Techniques such as luminescence (e.g. thermoluminescence (TL) and optically stimulated luminescence (OSL)) and electron spin resonance (ESR) are less direct methods that measure the accumulated *effect* of environmental radiation on crystalline substances such as quartz, flint or tooth enamel. Here, the environmental dose history has to be reconstructed, and in the case of materials such as tooth enamel, uranium uptake also has to be modelled. For further details of these techniques, see Taylor & Aitken (1997).

As Oakley well appreciated in 1969, one of the most serious difficulties facing realistic tests of evolutionary models concerning the origins of modern humans was the lack of any widely applicable dating procedures which could reach beyond the practical limits of radiocarbon dating. Because of these limitations, the European transition from Neanderthals and their associated Middle Palaeolithic (Middle Old Stone Age) tools to modern *Homo sapiens* (the Cro-Magnons) and the associated Upper Palaeolithic was believed to have occurred about 35 000 radiocarbon years ago, but the nature and extent of evolution or overlap could not be determined. In the Levant, a similar transition period from Neanderthals to *Homo sapiens* was believed to lie only slightly beyond this 35 000-year-old watershed. In Africa, the locally equivalent transition from the Middle Stone Age (= Middle Palaeolithic) to Later Stone Age (= Upper Palaeolithic) was generally believed to date from only about 12 000 years ago, and hence African cultural and physical evolution was believed to have lagged behind that of Europe and western Asia. This was reinforced by the belief that archaic humans such as those known from sites such as Broken Hill (Zambia) and Florisbad (South Africa) dated from only about 40 000 years ago. In the Far East, the pattern of human evolution was even more difficult to discern in 1969. Some Asian fossils and stone industries were dated through attempted correlation with the Alpine glacial–interglacial sequence as far back as 400 000 years ago. However, the arrival of humans in Australia was apparently a very late event, dated by Oakley at perhaps 10 000–15 000 years ago. Some important localities or fossils, and the dates assigned to them in Oakley (1969), are shown in Table 1.

It is necessary at this point to discuss briefly the classification of fossil humans, although this remains a very controversial area. Some workers argue that our species *Homo sapiens* had a recent evolutionary origin, and that there were many other earlier human species, such as *Homo ergaster*, *Homo erectus*, *Homo antecessor*, *Homo heidelbergensis*, *Homo neanderthalensis* and *Homo helmei*. A few others have argued in the opposite direction, that our species has deep roots, extending even into the Pliocene, in which case the term *Homo sapiens* would encompass all of the previously mentioned species. However, for the rest of this paper, I will restrict myself to four species names: *erectus*, *heidelbergensis*, and the inferred two daughter species of *heidelbergensis* – *neanderthalensis* and *sapiens*.

Changing chronologies: Europe

The European hominid sequence can now be interpreted as showing a gradual appearance of Neanderthal characteristics during the later Middle Pleistocene. This is perfectly exemplified by the large skeletal sample from the Sima de los Huesos at Atapuerca, dated at *c.* 300 ka BP. For the Spanish workers who have described the material (Arsuaga *et al.* 1997), they represent late members of the species *H. heidelbergensis*, whereas I prefer to classify them as an early form of the probable descendent species *H. neanderthalensis*. After this time, Neanderthal features continued to accrue so that by about 125 ka ago (Rink *et al.* 1995), specimens such as those from Krapina (Croatia) and Saccopastore (Italy) are quite comparable to examples from the last glaciation. However, during this latter period, as mentioned above, there is a major change in the European pattern when modern humans make their first appearance in Europe, in the form of the Cro-Magnons.

New dating using burnt flint (luminescence) and mammalian tooth enamel (electron spin resonance), in concert with accelerator radiocarbon dating (which requires much smaller samples of organic material than conventional methods) has generally confirmed the previous picture of the Middle/Upper Palaeolithic sequence in Europe, but with some additional complexity. Upper Palaeolithic industries such as the Aurignacian, associated with early modern humans, have been dated in parts of Europe (e.g. northern Spain and Hungary) by luminescence, uranium series or radiocarbon accelerator methods to about 40 ka BP, and Middle Palaeolithic (Mousterian) industries, actually or presumably associated with Neanderthals, start to disappear from some areas of Europe from about this time (Mellars 1998; Klein 1999). But the old favoured models of rapid *in situ* evolution of Neanderthals into Cro-Magnons or a rapid replacement of Neanderthals by them can now both be shown to be invalid. Late Neanderthal levels at French sites such as Le Moustier and Saint-Césaire have been dated in the range 35–40 ka BP (Mercier *et al.* 1991), while those at Arcy have been radiocarbon dated at about 34 ka BP (Mellars 1998). These dates may well be compatible, given that radiocarbon dates at this period may underestimate calendar ages by some 10% (Taylor & Aitken 1997). However, Neanderthal fossils have also now been dated to less than 30 000 radiocarbon years BP in areas such as southern Spain, Croatia and the Caucasus, and regions such as southern Iberia and the Crimea show a similar persistence of Middle Palaeolithic industries (e.g. Hublin *et al.* 1995; Smith *et al.* 1999; Ovchinnikov *et al.* 2000). If these dates and associations are accurate, it appears that Neanderthals survived quite late in some regions, and had a potential coexistence with the Cro-Magnons of at least ten millennia. The apparent clear picture of the Middle–Upper Palaeolithic archaeological interface in Europe has also become cloudier since the 1979 discovery of Neanderthal remains in Châtelperronian (early Upper Palaeolithic) levels at Saint-Césaire, and the identification of further Neanderthal skeletal material associated with this industry, together with bone artefacts (including some interpreted as pendants) at Arcy (France). Moreover, it has been suggested that other industries with supposed Upper Palaeolithic affinities in eastern Europe (Szeletian) and Italy (Uluzzian) were also the handiwork of late Neanderthals. Thus some workers now claim that Neanderthal technological and symbolic capabilities were actually, or potentially, the same as that of the Cro-Magnons.

It has been suggested that a pattern of regionalization reflects the final fragmentation of the formerly continent-wide range of the Neanderthals, while in contrast, the wide distribution of the Aurignacian reflects the penecontemporaneous dispersal of incoming early modern humans across much of Europe (Mellars 1998, 1999). However, present dating evidence no longer clearly supports a wave of advance of the Aurignacian, since its oldest manifestations appear to be as ancient in northern Spain as in the east of the continent. The assumed external source for the Aurignacian and its manufacturers is now unclear, and it remains possible that *H. sapiens* arrived with a pre-Aurignacian, even Middle Palaeolithic, technology. A possible precursor industry that might mark the appearance of early modern pioneers, although currently without diagnostic fossil material, is the Bohunician of eastern Europe, possibly dating beyond 40 000 radiocarbon years BP.

Workers such as Zilhao and Trinkaus have proposed still greater complexity in the European picture (e.g. Duarte *et al.* 1999). To them, Middle–Upper Palaeolithic transitions are indicative of complex and changing population dynamics as incoming Cro-Magnons mixed and merged with native Neanderthals over many millennia. In this scenario, the Neanderthals were arguably as culturally advanced as the Cro-Magnons, and were simply absorbed into a growing Cro-Magnon gene pool. It is even claimed that a hybrid child has been discovered at Lagar Velho in Portugal, dated to about 25 000 radiocarbon years BP (Duarte *et al.* 1999), but this claim remains unresolved until more detailed studies have been completed. Whatever the outcome of that particular proposal (and I remain to be convinced that this is not just an unusually stocky modern human child), the impact of new dates and discoveries in Europe show that many different population interactions between the last Neanderthals and the first Cro-Magnons could, and perhaps did, occur, ranging from violence to possible interbreeding. Nevertheless, the end result of these processes was the extinction of the Neanderthals after a long period of survival in the challenging and unstable climates of Pleistocene Europe.

The Levant

This region has probably seen the greatest upheavals in previous thinking about modern human origins brought about by the application of new dating techniques. Even as late as 1985 it was believed by most workers, including this author, that the pattern of population change in this area followed that of Europe, or rather preceded it, by a small amount of time. Thus Neanderthals at Israeli sites such as Tabun and Amud evolved into, or gave way to, early modern humans such as those known from Skhul and Qafzeh by about 40 ka BP (e.g. Trinkaus 1984). For some workers there had been interlinked technological and biological changes leading to the evolution of modern humans in the region, and it was postulated that these early moderns could then have migrated into Europe, giving rise to the Cro-Magnons. One of the first applications of the newer chronometric techniques (thermoluminescence applied to burnt flint) seemed to reinforce this pattern, dating a recently discovered Neanderthal burial at Kebara in the anticipated time range of about 60 ka BP (Valladas *et al.* 1987).

However, shortly afterwards, the first application of these techniques was made to the site of the Qafzeh early modern material, giving a then-astonishing age estimate of about 90 ka, more than twice the generally expected figure. Further applications of non-radiocarbon dating methods have amplified the pattern suggested by the age estimates for Qafzeh and Kebara (e.g. Grün & Stringer 1991). It seems likely that the early modern burials at Qafzeh and Skhul date from more than 90 ka BP, and some may be as old as 130 ka. The Neanderthal burials at Kebara and Amud date younger than this figure, in the range 50–60 ka BP. As the intervening period approximates the transition from the supposedly predominantly interglacial stage 5 to predominantly glacial stage 4 this has led to a proposed scenario where Neanderthals only appeared in the Levant after the onset of glaciation further north.

However, it has not so far been possible to establish the age of the Tabun Neanderthal burial with any certainty, for two different reasons. First, while age estimates for the stratigraphy at Tabun based on ESR and TL both considerably stretch the late Pleistocene timescale previously proposed for the site, into the Middle Pleistocene, the methods do not give compatible results. Luminescence estimates from burnt flint excavated from the rear of the cave are much older than ESR estimates from mammal teeth from correlated levels nearer the mouth of the cave (compare Grün *et al.* 1991 with Mercier *et al.* 1995). Second, more than 60 years after its excavation, the stratigraphic position of the Tabun burial cannot be established with certainty, giving further doubt about its actual age (Garrod & Bate 1937; Bar-Yosef & Callander 1999). Recent direct non-destructive gamma-ray (uranium-series) dating of the mandible and leg bones from this skeleton has suggested a surprisingly young age of less than 40 ka (Schwarcz *et al.* 1998). However, this estimate remains controversial (Millard & Pike 1999; Alperson *et al.* 2000) and direct ESR dating of a tooth enamel fragment from a molar on the mandible gives a much older age estimate of about 120 ka (Grün & Stringer, in press). Thus the age of this particular specimen remains in dispute. The extent of possible Neanderthal–early modern overlap in the Levant in the period 90–130 ka BP is therefore still an open question, but given that the region lies in the potential overlap zone of range expansions of either the evolving African *sapiens* lineage or that of the Eurasian Neanderthals, some co-existence of the two lineages

was certainly likely (Stringer 1998). Yet after this time, the Neanderthals appear to have predominated in the region until the development of technologies and adaptations that may finally have allowed early modern humans to successfully challenge the Neanderthals in their home territory.

Africa

The pattern of human evolution in Africa is still less well understood than that of Europe, but the developing picture suggests that there are parallels between the two continents. In 1969 Oakley followed the prevailing view, based mainly on radiocarbon dating, that although the Palaeolithic may have begun in Africa, subsequent human development lagged behind that of Europe. Thus the Lower Palaeolithic was believed to have continued in Africa until about 50 ka BP, while the subsequent Middle Stone Age may have only given way to the Later Stone Age at about 12 ka BP, some 25 000 years later than the equivalent Middle–Upper Palaeolithic transition in Europe. The hominid sequence was thought to be comparably retarded, with the very archaic Broken Hill (Zambia) cranium dated at only about 40 ka and the somewhat less archaic Florisbad (South Africa) specimen dated about 35 ka BP.

The situation now is dramatically different. Argon–argon dating has shown that stone-tool making began in Africa by at least 2.3 Ma BP, and the whole timescale of the African Palaeolithic has been stretched back in time (Klein 1999). The Middle Stone Age is now believed to have begun by at least 250 ka BP and the transition to the Later Stone Age began prior to 45 ka BP. Thus the African record can now be seen to be in concert with, or even in advance of, the record from Eurasia. The hominid record has been similarly reassessed. Biostratigraphic correlation suggests that the Broken Hill cranium probably dates from at least 350 ka BP (Klein 1999), while a combination of ESR dating on human tooth enamel and luminescence dating of sediments suggest that the Florisbad cranium actually dates from about 260 ka BP rather than the former estimate of about 35 ka BP (Grün *et al.* 1996). Fossils that are perhaps transitional to modern humans from Guomde (Kenya) and Singa (Sudan) are now dated by gamma rays, and a combination of ESR and uranium series, respectively, to more than 130 ka BP (Bräuer *et al.* 1997; McDermott *et al.* 1996). Of this age, or somewhat younger, are early modern fossils such as Omo Kibish 1

(Ethiopia), Border Cave 1 (South Africa) and those from the Middle Stone Age levels of the Klasies River Mouth Caves (South Africa), but much of this material is fragmentary and also difficult to date more precisely (Klein 1999). Overall, the picture of human evolution in Africa over the last 300 000 years can now be seen to parallel that of Europe. Both regions appear to show a gradual transition from *Homo heidelbergensis* to a more derived species, in Europe *H. neanderthalensis* and in Africa *H. sapiens*. While the temperate-cold climates of western Eurasia may well have influenced the evolution of the Neanderthals, it is still unclear what drove the evolution of *H. sapiens*. The large habitable area of Africa combined with dramatic changes in precipitation and vegetation might have forced evolutionary change through isolation and adaptation, and there is growing evidence of the early appearance of some aspects of modern human behaviour, such as symbolism, during the Middle Stone Age.

China

H. erectus was probably in China by about 1 Ma ago (Klein 1999), and the famous material of 'Peking Man' from the Zhoukoudian Lower Cave is now dated at *c.* 300–550 ka BP by uranium series and ESR (Grün *et al.* 1997). The southern Chinese *erectus* material from Hexian is of similar or somewhat younger age (Grün *et al.* 1998), and these populations were then apparently succeeded by more derived populations represented by fossils from sites such as Jinniushan and Dali. Their affinities are still unclear, with some workers seeing them as descended from local *erectus* antecedents while others regard them as eastern representatives of *H. heidelbergensis* (Klein 1999). But the arrival of *H. sapiens* in the region is still poorly dated, although it must precede the modern human fossils known from the Upper Cave at Zhoukoudian dated by radiocarbon on possibly associated fauna to *c.* 30 ka BP (tragically, almost all of the original human fossils from Zhoukoudian were lost during the Second World War). However, the evolutionary origins of the present-day peoples of eastern Asia and the Americas remain enigmatic. There are a few late Pleistocene fossils from China and Japan relevant to this question, but in the case of the crania from Upper Cave and Liujiang (China), and Minatogawa (Japan), none seem very closely related to recent populations in the region (Brown 1999; Stringer 1999*b*).

Australasia

In Indonesia, some *H. erectus* fossils have recently been indirectly dated to about 1.7 Ma BP using argon–argon dating on volcanic sediments (Swisher *et al.* 1994), but some workers doubt that the early human fossils have been correctly associated with the dated rocks (Culotta 1995). Other *erectus* fossils have been dated by combinations of argon–argon, palaeomagnetics and biostratigraphy to between 500 ka and 1 Ma BP (Klein 1999). The Ngandong and Sambungmacan fossils have been much more controversially dated to less than 50 ka BP by ESR and uranium series, which would imply a survival of *Homo erectus* in Indonesia as late as Neanderthals survived in Eurasia (Swisher *et al.* 1996). However, other workers have argued that these dates must be underestimates (Grün & Thorne 1997). The date of arrival of modern humans in the region is still uncertain, but given the evidence from Australia discussed below it must lie before 60 ka BP. Known fossils, such as Wajak from Java and Niah from Sarawak, remain poorly dated, but may date from the late Pleistocene.

Exactly when people first arrived in Australia has remained unclear. The site of Jinmium was claimed to have red-ochre use dating from about 75 ka BP, and stone-tool use at perhaps twice that age (Fullagar *et al.* 1996). However, but the ancient luminescence dates obtained on the Jinmium sands have not been confirmed by further analyses using luminescence and radiocarbon, which instead suggest that the site was only occupied within the last 20 ka (Roberts *et al.* 1998). Nevertheless, the rock shelters of Malakunanja II and Nawalabila do appear to contain artefacts dating from about 50–60 ka BP, based on luminescence dates (Roberts *et al.* 1990, 1994). However, none of these sites contains ancient human remains, thus leaving the nature of the first Australians uncertain.

These doubts may now be assuaged because a fossil from southeastern Australia, the Mungo 3 burial, has been redated to approximately double the age originally estimated from radiocarbon (Bowler & Thorne 1976). Using a combination of the techniques of gamma-ray uranium-series dating on skull fragments, ESR on a piece of tooth enamel, uranium series on attached sediment, and OSL applied to the sands containing the burial, an estimated age of about 62 ka was obtained (Thorne *et al.* 1999; Grün *et al.* 2000). The new dates for the Mungo burial have wider implications for the debate about early modern human dispersals from Africa (Stringer 1999*a*). If they are accurate, then the oldest known burial associated with possible symbolic use of red ochre was conducted far from Africa, over 10 000 years before the posited 'human revolution' of the Later Stone Age and Upper Palaeolithic had begun. Moreover, as Australia has never had a land connection to Asia even during the lowest sea levels of the Pleistocene, sea-going craft must have been in use to reach Australia long before there is definite archaeological evidence of their existence.

Genetic data

Genetic data have lately assumed increasing importance in reconstructions of recent human evolution. Autosomal (normal nuclear) DNA, Y-chromosome DNA (inherited through males) and mitochondrial DNA (inherited through females) sampled from present-day peoples have been used to estimate coalescent (last common ancestral) dates for various widely dispersed gene systems; many of these range between 100 and 300 ka, and are consistent with a recent African origin (see for example Tishkoff et al. 1996; Zietkiewicz et al. 1998; Jorde et al. 2000). Mitochondrial DNA has also been extracted from Neanderthal fossils, suggesting a separation time of their lineage from that leading to modern humans of about 600 ka (Krings et al. 1999; Ovchinnikov et al. 2000). However, these estimates provide *maximum* ages for evolutionary separation, since any population and species separations would inevitably postdate the first genetic (mitochondrial) divergence by an unknown amount of time. But they are consistent with fossil evidence of an effective separation date of the *H. neanderthalensis* and *H. sapiens* lineages at about 300 ka and also with subsequent genetic divergence among modern humans beginning less than 200 000 years ago (Stringer 1998).

Conclusions

The last thirty years have seen remarkable progress in our ability to calibrate the most recent stages of human evolution. As Table 1 demonstrates, some specimens have had their estimated ages broadly confirmed – for example, the *Homo erectus* fossils from Zhoukoudian Lower Cave (China), and the Neanderthal cranium from Guattari (Monte Circeo (Italy): Fig. 1). Others have been shown to be much older than previously suspected, such as the fossils from Skhul (Israel), Florisbad (Zambia) and Singa (Sudan). In yet other cases, direct radiocarbon accelerator determinations have shown that supposed Pleistocene fossils were wrongly dated or were intrusive, such as Langwith (England) and Engis1 (Belgium) in Europe, and Tepexpan in Mexico. Such negative results have been important in removing potentially misleading data from the fossil record. However, many fossils are still not reliably dated. In some cases this is because they are isolated finds, whose original context has been lost, such as the Neanderthal skull from Forbes' Quarry in Gibraltar, found in 1848. This also appeared to be the case for the Neander Valley skeleton (Germany) until 1998, but remarkable detective work by German archaeologists led them to relocate the original quarry spoil from the Feldhofer grotto (long since quarried away) and find artefacts, fossils mammal bones and even some additional fragments of the Neanderthal skeleton itself (Schmitz & Thissen 2000)! Some of these fragments have been accelerator-dated to about 40 ka BP, and will be subjected to analysis for ancient DNA to add to that already obtained (Krings et al. 1999). So even 150 years later, it is sometimes possible to deal with seemingly intractable chronological problems.

In yet other cases, such as the famous Neanderthal skeleton from Tabun cave in Israel, the application of physical dating methods has not been successful in resolving longstanding uncertainties (compare Schwarcz et al.

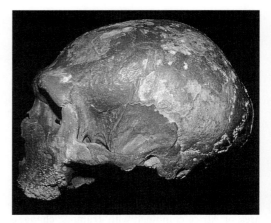

Fig. 1. The Guattari 1 Neanderthal cranium from Monte Circeo, Italy. Originally thought to be a relic of Neanderthal ritual behaviour, it is now believed to have been scavenged by hyenas. Oakley (1969) dated this fossil to the last ('Würm') glaciation. Now, through a combination of uranium series and ESR, it has been dated more precisely to between 45 and 62 ka ago (for discussion of the site and its dating see *Quaternaria Nova* 1 (whole volume) 1990–1991). Photograph taken by permission of Museo Preistorico ed Etnografico 'L. Pigorini'.

(1998) and Grün & Stringer (2000)). In that case, old excavation techniques and site complexities are partly to blame, but the problems of modelling uranium uptake (or even loss) in the essentially open-system behaviour of bones are also responsible. This points to the need for further experimental work on methods such as gamma-ray dating, where non-destructive procedures are valuable, but uranium mobilization may create difficulties (Barton & Stringer 1997; Millard & Pike 1999). Physical dating procedures are still evolving, and the new millennium will undoubtedly see further re-evaluation of existing chronologies, as well as the development and application of completely new dating techniques to the fossil human record. No doubt many more chronological surprises lie in store for the next generation of palaeoanthropologists!

I would like to thank the organizers of the conference 'Celebrating the Age of the Earth', particularly C. L. E. Lewis, for the opportunity to participate and to publish this contribution. I would also like to thank the two referees for their constructive comments on my manuscript.

References

ALPERSON, N., BARZILAI, O., DAG, D., HARTMAN, G. & MATSKEVICH, Z. 2000. The age and context of the Tabun I skeleton: a reply to Schwarcz et al. *Journal of Human Evolution*, **38**, 849–853.

ARSUAGA, J. L., BERMUDEZ DE CASTRO, J. M. & CARBONELL, E. (eds) 1997. The Sima de los Huesos Hominid site. *Journal of Human Evolution*, **33**, 105–421.

BARTON, J. & STRINGER, C. 1997. An attempt at dating the Swanscombe skull bones using non-destructive counting. *Archaeometry*, **39**, 205–216.

BAR-YOSEF, O. & CALLANDER, J. 1999. The woman from Tabun: Garrod's doubts in historical perspective. *Journal of Human Evolution*, **37**, 879–885.

BOWLER, J. & THORNE, A. 1976. Human remains from Lake Mungo: discovery and excavation of lake Mungo III. *In*: KIRK, R. & THORNE, A. (eds) *The Origin of the Australians*. Australian Institute of Aboriginal Studies, Canberra,127–138.

BRÄUER, G., YOKOYAMA, Y., FALGUERES, C. & MBUA, E. 1997. Modern human origins backdated. *Nature*, **386**, 337–338.

BROWN, P. 1999. The first modern East Asians?: another look at Upper Cave 101, Liujiang and Minatogawa 1. *In*: OMOTO, K. (ed.) *Interdisciplinary Perspectives on the Origins of the Japanese*. International Research Center for Japanese Studies, Kyoto, 105–131.

CULOTTA, E. 1995. Asian hominids grow older. *Science*, **270**, 1116–1117.

DUARTE, C., MAURÍCIO, J., PETTITT, P. B., SOUTO, P., TRINKAUS, E., VAN DER PLICHT, H. & ZILHÃO, J. 1999. The early Upper Paleolithic human skeleton from the Abrigo do Lagar Velho (Portugal) and modern human emergence in Iberia. *Proceedings of the National Academy of Science (USA)*, **96**, 7604–7609.

FULLAGAR, R., PRICE, D. & HEAD, L. 1996. Early human occupation of northern Australia: archaeology and thermoluminescence dating of Jinmium Rock Shelter, Northern Territory. *Antiquity*, **70**, 751–773.

GARROD, D. & BATE, D. 1937. *The Stone Age of Mount Carmel. Vol. 1*. Oxford University, Oxford.

GRÜN, R. & STRINGER, C. 1991. Electron spin resonance dating and the evolution of modern humans. *Archaeometry*, **33**, 153–199.

GRÜN, R. & STRINGER, C. 2000. Tabun revisited: revised ESR chronology and new ESR and U-series analyses of dental material from Tabun C1. *Journal of Human Evolution*, **39**, 601–612.

GRÜN, R. & THORNE, A. 1997. Dating the Ngandong humans. *Science*, **276**, 1575–1576.

GRÜN, R., STRINGER, C. & SCHWARCZ, H. 1991. ESR dating of teeth from Garrod's Tabun cave collection. *Journal of Human Evolution*, **20**, 231–248.

GRÜN, R., SPOONER, N. A., THORNE, A. ET AL. 2000. Age of the Lake Mungo 3 skeleton, reply to Bowler & Magee and to Gillespie & Roberts. *Journal of Human Evolution*, **38**, 733–741.

GRÜN, R., BRINK, J., SPOONER, N., TAYLOR, L., STRINGER, C., FRANCISCUS, R. & MURRAY, A. 1996. Direct dating of Florisbad hominid. *Nature*, **382**, 500–501.

GRÜN, R., HUANG, P.-H., WU, X., STRINGER, C., THORNE, A. & MCCULLOCH, M. 1997. ESR analysis of teeth from the palaeoanthropological site of Zhoukoudian, China. *Journal of Human Evolution*, **32**, 83–91

GRÜN, R., HUANG, P.-H., HUANG, W., MCDERMOTT, F., STRINGER, C., THORNE, A. & YAN, G. 1998. ESR and U-series analyses of teeth from the palaeoanthropological site of Hexian, Anhui Province, China. *Journal of Human Evolution*, **34**, 555–564.

HUBLIN, J.-J., BARROSO RUIZ, C., MEDINA LARA, P., FONTUGNE, M. & REYSS, J. 1995. The Mousterian site of Zafarraya (Andalucia, Spain): dating and implications on the Palaeolithic peopling processes of Western Europe. *Comptes Rendus Académie des Sciences, Paris*, IIa, **321**, 931–937.

HUBLIN, J.-J., SPOOR, F., BRAUN, M., ZONNEVELD, F. & CONDEMI, S. 1996. A late Neanderthal associated with Upper Palaeolithic artefacts. *Nature*, **381**, 224–226.

JORDE, L., WATKINS, W., BAMSHAD, M., DIXON, M., RICKER, C., SEIELSTAD, M. & BATZER, M. 2000. The distribution of human genetic diversity: a comparison of mitochondrial, autosomal, and Y-chromosome data. *American Journal of Human Genetics*, **66**, 979–988.

KLEIN, R. 1999. *The Human Career*. University of Chicago, Chicago.

KRINGS, M., GEISERT, H., SCHMITZ, R., KRAINITZKI, H. & PAABO, S. 1999. DNA sequence of the mitochondrial hypervariable region II from the Neandertal type specimen. *Proceedings National Academy Sciences, USA*, **96**, 5581–5585.

McDermott, F., Stringer, C., Grün, R., Williams, C. T., Din, V. & Hawkesworth, C. 1996. New Late-Pleistocene uranium-thorium and ESR dates for the Singa hominid (Sudan). *Journal of Human Evolution*, **31**, 507–516.

Mellars, P. 1998. The fate of the Neanderthals. *Nature*, **395**, 539–540.

Mellars, P. 1999. The Neanderthal problem continued. *Current Anthropology*, **40**, 341–350.

Mercier, N., Valladas, H., Joron, J. L., Reyss, J. L., Leveque, F. & Vandermeersch, B. 1991. Thermoluminescence dating of the late Neanderthal remains from Saint-Césaire. *Nature*, **351**, 737–739.

Mercier, N., Valladas, H., Valladas, G., Reyss, J.-L., Jelinek, A., Meignen, L. & Joron, J.-L. 1995. TL dates of burnt flints from Jelinek's excavations at Tabun and their implications. *Journal of Archaeological Science*, **22**, 495–509.

Millard, A. R. & Pike, A. W. G. 1999. Uranium-series dating of the Tabun Neanderthal: a cautionary note. *Journal of Human Evolution*, **36**, 581–585.

Oakley, K. 1969. *Frameworks for Dating Fossil Man*. Weidenfeld & Nicolson, London.

Ovchinnikov, I., Anders, G., Götherström, A., Romanova, G., Kharitonov, V., Lidén, K. & Goodwin, W. 2000. Molecular analysis of Neanderthal DNA from the northern Caucasus. *Nature*, **404**, 490–493.

Rink, W. J., Schwarcz, H. P., Smith, F. H., & Radovcic, J. 1995. Ages for Krapina Hominids. *Nature*, **378**, 24.

Roberts, R., Jones, R. & Smith, M. 1990. Thermoluminescence dating of a 50,000 year old human occupation site in northern Australia. *Nature*, **345**, 153–156.

Roberts, R., Jones, R., Spooner, N., Head, M., Murray, A. & Smith, M. 1994. The human colonization of Australia: optical dates of 53,000 and 60,000 years bracket human arrival at Deaf Adder Gorge, Northern Territory. *Quaternary Science Reviews (Quaternary Geochronology)*, **13**, 575–583.

Roberts, R., Bird, M., Olley, J. et al. 1998. Optical and radiocarbon dating at Jinmium Rock shelter in northern Australia. *Nature*, **393**, 358–362.

Schmitz, R. & Thissen, J. 2000. *Neandertal: Die Geschichte geht weiter*. Spektrum, Heidelberg.

Schwarcz, H. P., Simpson, J. J. & Stringer, C. B. 1998. Neanderthal skeleton from Tabun: U-series data by gamma-ray spectrometry. *Journal of Human Evolution*, **35**, 635–645.

Smith, F., Trinkaus, E., Pettitt, P., Karavanic, I. & Paunovic, M. 1999. Direct radiocarbon dates for Vindija G1 and Velika Pećina Late Pleistocene hominid remains. *Proceedings National Academy of Sciences, USA*, **96**, 12 281–12 286.

Stringer, C. 1998. Chronological and biogeographic perspectives on later human evolution. *In*: Akazawa, T., Aoki, K. & Bar-Yosef, O. (eds) *Neandertals and Modern Humans in Western Asia*. Plenum, New York, 29–37.

Stringer, C. 1999a. Has Australia backdated the Human Revolution? *Antiquity*, **282**, 876–879.

Stringer, C. 1999b. The origin of modern humans and their regional diversity. *Newsletter Interdisciplinary Institute for Study of the Origins of Japanese Peoples and Cultures*, **9**, 3–5.

Stringer, C. & Gamble, C. 1993. *In Search of the Neanderthals*. Thames and Hudson, London.

Swisher, C., Curtis, G., Jacob, T., Getty, A., Suprijo, A. & Widiasmoro. 1994. Age of the earliest known hominids in Java. *Science*, **263**, 1118–1121.

Swisher, C., Rink, W., Anton, S., Schwarcz, H., Curtis, G., Suprijo, A. & Widiasmoro. 1996. Latest *Homo erectus* of Java: potential contemporaneity with *Homo sapiens* in Southeast Asia. *Science*, **274**, 1870–1874.

Taylor, R. 2000. The contribution of radiocarbon dating to New World Archaeology. *Radiocarbon*, **42**, 1–21.

Taylor, R. & Aitken, M. (eds) 1997. *Chronometric Dating in Archaeology*. Plenum, New York.

Thorne, A., Grün, R., Mortimer, G. et al. 1999. Australia's oldest human remains: age of the Lake Mungo 3 skeleton. *Journal of Human Evolution*, **36**, 591–612.

Tishkoff, S., Dietzsch, E., Speed, W. et al. 1996. Global patterns of linkage disequilibrium at the CD4 locus and modern human origins. *Science*, **271**, 1380–1387.

Trinkaus, E. 1984. Western Asia. *In*: Smith, F. & Spencer, F. (eds) *The Origins of Modern Humans*. Liss, New York, 251–293.

Valladas, H., Joron, J., Valladas, G. et al. 1987. Thermoluminescence dates for the Neanderthal burial site at Kebara in Israel. *Nature*, **330**, 159–160.

Zietkiewicz, E., Yotova, V., Jarnik, M. et al. 1998. Genetic structure of the ancestral population of modern humans. *Journal of Molecular Evolution*, **47**, 146–155.

Understanding the beginning and the end

MARTIN J. REES

King's College, Cambridge CB2 1ST, UK (email: mjr@ast.cam.ac.uk)

Abstract: This paper attempts to set the Earth in a cosmic perspective. It discusses the Sun's life cycle in the context of stellar evolution, and the ideas of stellar nucleosynthesis, as an explanation of the origin of the atoms on the Earth. Current ideas on how planetary systems form are mentioned, along with data on recently discovered planets around other stars. The origin of matter itself can be traced back to a 'Big Bang': corroboration of this model comes from the microwave background and from observed helium and deuterium abundances. There is now a concordance between stellar ages and the cosmic evolutionary timescale inferred from cosmology, from which we derive an age for the universe. Recent progress brings into focus a new set of questions about the ultra-early universe. Until these can be answered, it will remain a mystery why the universe is expanding in the observed fashion, and why it contains the measured mix of atoms, radiation and dark matter.

Whilst this planet has been cycling on according to the fixed law of gravity, from so simple a beginning, forms most wonderful ... have been and are being evolved.

These are the famous closing words of Darwin's *On the Origin of Species*, but astronomers aim to go back before his 'simple beginning', to set our entire Earth and Solar system in a broader context, stretching back to the birth of our galaxy, perhaps even to the initial instants of a 'Big Bang' that set our entire universe expanding. Darwin guessed that it would have required hundreds of millions of years to have transformed primordial life (formed, he surmised, in a 'warm little pond') into the amazing varieties of creatures that crawl, swim or fly on Earth. And this concept did not overly concern him, because such timespans had already been invoked by geologists to account for the laying down of rocks and moulding of the Earth's surface features.

Other contributors to this volume (Dalrymple 2001; Lewis 2001; Shipley 2001) have described how Kelvin estimated the Earth's age. His inferences about the age of the Sun were actually rather more firmly based than those about the Earth: if the gravitational energy released by its continuing contraction was supplying the heat radiating away, it would deflate in ten million years. Kelvin's views carried great weight. But it was perhaps fortunate for his reputation that he included an escape clause: his conclusion regarding the age of the Sun only held good, he said, provided that there was no other power source 'prepared in the storehouse of creation' (Thomson 1862, p. 393).

As was realized in the 1930s, fusion of hydrogen is sufficient to sustain the Sun for ten billion years. Our knowledge of atomic and nuclear physics is now sufficient to give us a (broadly uncontroversial) quantitative picture of the Sun's life cycle. The proto-Sun condensed from a cloud of diffuse interstellar gas. Gravity pulled it together until its centre was squeezed hot enough to trigger nuclear fusion of hydrogen into helium at a sufficient rate to balance the heat shining from its surface. (Any deuterium in the original cloud would have been burnt at an earlier stage in the contraction.) Less than half the Sun's central hydrogen has so far been used up: it is already 4.5 billion years old, but will keep shining for a further 5 billion years. It will then swell up to become a red giant, large and bright enough to engulf the inner planets, and to vaporize all life on Earth. During this 'red giant' phase, lasting some five hundred million years, hydrogen will continue to burn in a shell around the helium core. Next, the Sun will undergo a more rapid convulsion, triggered by the onset of helium fusion in its core. This blows off some outer layers – about a quarter of the Sun's mass altogether. The residue will become a white dwarf – a dense 'stellar cinder' no larger than the Earth, which will shine with a bluish glow, no brighter than today's full Moon, on whatever remains of the solar system.

Our Sun has more time ahead than has so far elapsed; as explained later, our entire universe could have an infinite future ahead of it. So we may still be near Darwin's 'simple beginning': if life is not prematurely snuffed out, our remote progeny will surely – in the aeons that lie

From: LEWIS, C. L. E. & KNELL, S. J. (eds). *The Age of the Earth: from 4004 BC to AD 2002*. Geological Society, London, Special Publications, **190**, 275–283. 0305-8719/01/$15.00 © The Geological Society of London 2001.

ahead – spread far beyond this planet. Even if life is now unique to the Earth, there is time enough for it to spread through the entire galaxy, and even beyond.

The structure and life cycle can be computed for a star of any mass. The output of such calculations allows us to infer the ages of star clusters, which contain a population of coeval stars of different masses. The key idea is that heavier stars use up their core hydrogen fuel, and 'burn out', more quickly than lower-mass stars. The older a system is, the fainter (and lower-mass) will be the brightest stars that are still in the hydrogen-burning phase. Particularly interesting in this regard are the so-called 'globular clusters' – each a swarm of up to a million stars, of different sizes, held together by their mutual gravity – which are believed to be the oldest stellar systems of all. The uncertainties in the age estimates stem partly from the theoretical models themselves, but also from the difficulty of inferring stellar masses from observed brightness and colours. Estimated ages are up to 13 billion years, with, however, an uncertainty of at least 10%. Such estimates are of course crucial to cosmology, because it would be embarrassing if the inferred age of the entire universe (in other words the time since the Big Bang) were not comfortably higher than the age of the oldest stars. As I shall comment later, there now seems no such paradox.

Not everything happens slowly. Massive stars end their lives violently by exploding as supernovae. The closest supernova of the twentieth century flared up in February 1987. Its subsequent fading has been followed by all the techniques of modern astronomy. Theorists were given a chance to check the elaborate computer calculations they had developed over the previous decade. In about 1000 years its remnant will look like the Crab Nebula – the expanding debris from an explosion recorded by Chinese astronomers in 1054 AD; in a few thousand more years, it will have merged into the general interstellar medium. Supernovae fascinate astronomers, but why should Earth-bound scientists – geologists in particular – care about stellar explosions thousands of light years away? The answer, of course, is that supernovae made the atoms that the Earth is made of: without them we would not be here. On Earth, for every ten atoms of carbon, there are about 20 of oxygen and five each of nitrogen and iron. But gold is a million times rarer than oxygen; platinum and mercury are rarer still. These exploding stars suggest the reason why (Salpeter 1999).

Stars more than ten times heavier than the Sun use up their central hydrogen hundreds of times quicker than the Sun does – they shine much brighter in consequence. Gravity then squeezes them further, and the centres get still hotter, until helium atoms can themselves stick together to make the nuclei of heavier atoms. A kind of 'onion skin' structure develops: a layer of carbon surrounds one of oxygen, which in turn surrounds a layer of silicon. The hotter inner layers have been transmuted further up the periodic table and surround a core that is mainly iron. When their fuel has all been consumed, big stars face a crisis. A catastrophic infall compresses the stellar core to neutron densities, triggering a colossal explosion – a supernova.

The outer layers of a star, by the time a supernova explosion blows them off, contain the outcome of all the nuclear alchemy that kept it shining over its entire lifetime. Elements beyond the Fe peak can be built up during the explosion itself. Work over the last 40 years – taking account of different types of stars, different nuclear reactions – has shown that the calculated 'mix' of atoms is gratifyingly close to the proportions now observed in our solar system. This story is well authenticated by detailed modelling of the expected relative abundances of elements and isotopes, and also by spectroscopic evidence from the oldest stars (which contain less processed material). It is also found that the abundances are higher in gaseous environments like the galactic centre, where reprocessing would be fast, and lower in locations like the Magellanic Clouds where it is slower.

Our galaxy is like an ecosystem, recycling gas through successive generations of stars, gradually building up the entire periodic table. Before our Sun even formed, several generations of fast-burning heavy stars could have been through their entire life cycles, transmuting pristine hydrogen into the basic building blocks of life – carbon, oxygen, iron and the rest. We are literally the ashes of long-dead stars.

Planets around other stars?

One fascinating question, of course, is whether other worlds are orbiting other stars. For a long time astronomers have suspected planetary systems to be common, because protostars, as they contract from rotating clouds, spin off around them discs of dusty gas. These have now been seen in the Orion Nebula and elsewhere. In such discs, dust particles would stick together to make rock-sized lumps, which can in turn merge to make planets (the volatiles being lost from the inner planets, but not from the outer giant planets).

Evidence for actual planets orbiting other ordinary stars is harder to find, but it came in 1995. In that year, two Swiss astronomers, Michel Mayor and Didier Queloz, found that the Doppler shift of 51 Pegasi, a nearby star resembling our Sun, was varying sinusoidally by 50 m/s. They inferred that a planet weighing a thousandth as much was circling it at 50 km/s, causing the star to pivot around the combined centre of mass. Five years later, similar periodic wobbles in other stars have been detected; the number of inferred planets now exceeds fifty. Several research groups have contributed to that total, Marcy & Butler (2000) in California being the 'champions'. Some systems have two or even three planets. But the inferred planets are all big ones – like Jupiter or Saturn. These may be the largest members of other planetary systems like our own, but Earth-like planets would a hundred times harder to detect.

Planets on which life could evolve, as it did here on Earth, must be rather special. Their gravity must pull strongly enough to prevent an atmosphere from evaporating into space; they must be neither too hot nor too cold, and therefore the right distance from a long-lived and stable star. Most of Marcy & Butler's systems are not propitious – they contain Jupiters on eccentric orbits close to the star, which would preclude any stable orbit for a planet in the zone where water neither always freezes nor always boils. There is still theoretical dispute about how these (presumably hydrogen-rich) planets achieved orbits so close in. It is possible that the typical system formed with two or three 'Jupiters'. Interactions between them would expel one, leaving the others on eccentric orbits. Maybe our solar system is unusual in having only one Jupiter. But planetary systems are (we believe) so common in our galaxy that Earth-like planets should be numbered in millions

A search for Earth-like planets is now a main thrust of NASA's programme (NASA = National Aeronautics and Space Administration, USA). It is a long-term technical challenge – requiring large telescope arrays in space – but it is not as crazy as it sounds. Once a candidate had been identified, several things could be learnt about it. Suppose an astronomer forty light years away had detected our Earth – it would be, in Carl Sagan's phrase, a 'pale blue dot', seeming very close to a star (our Sun) that outshines it many million times. If Earth could be seen at all, its light could be analysed, which would reveal that it had been transformed (and oxygenated) by a biosphere. The shade of blue would be slightly different, depending on whether ocean or land mass was facing us. Distant astronomers could therefore, by repeated observation, infer the Earth was spinning, and learn the length of its day, and even infer something of its topography and climate.

The concept of a 'plurality of inhabited worlds' is still the province of speculative thinkers, as it has been through the ages. The year 2000 marks the fourth centenary of the death of Giordano Bruno, burnt at the stake, in Rome for his belief that:

> There are countless constellations, suns and planets; we see only the suns because they give light; the planets remain invisible, for they are small and dark. There are also numberless earths circling around their suns, no worse and no less than this globe of ours.

Only within the last five years has this conjecture been vindicated. But Bruno went on to say:

> For no reasonable mind can assume that heavenly bodies which may be far more magnificent than ours would not bear upon them creatures similar or even superior to those up-on our human Earth (Bruno 1584).

Here we are still in the speculative realm. But of course what motivates NASA and the American public, is whether there is life on any of the other Earths. Even in a propitious environment, what is the chance that 'simple' organisms emerge? Even when they do what is the chance they evolve into something that can be called intelligent? These questions are for biologists – they are too difficult for astronomers. There seems no consensus among the experts. Intelligent life could be 'natural'; or it could have involved a chain of accidents so surpassingly rare that nothing remotely like it has happened anywhere else in our galaxy.

The extragalactic universe: back towards the beginning

Let us now get back to the (relative) simplicity of the inanimate world. I have mentioned briefly how the atoms of the periodic table are made – that we are the 'nuclear waste' from the fuel that makes stars shine. But where did the original hydrogen come from? To answer this question, we must extend our horizons to the extragalactic realm. Our galaxy, with its hundred billion stars, is similar to millions of others visible with large telescopes. Andromeda, the nearest big galaxy to our own, lies about two million light years away. Its constituent stars are orbiting in a disc, seen obliquely. In others, like the Sombrero galaxy, ten billion stars are swarming around in more random directions, each feeling the gravitational pull of all the others.

The nearest few thousand galaxies – those out to about 300 million light years – have been mapped out in depth. They are irregularly distributed into clusters and superclusters. Are there, you may ask, clusters of clusters of clusters *ad infinitum*? It does not appear so. Deeper surveys show a smoother distribution: our universe is not a fractal. If it were, we would see equally conspicuous clumps, on ever-larger scales, however deep into space we probed. But even the biggest superclusters are still small in comparison with the horizon that powerful telescopes can reach. So we can define the average 'smoothed-out' properties of our observable universe. An analogy may be helpful here. If you are in the middle of an ocean, you may be surrounded by a complex pattern of waves, but even the longest-wavelength ocean swells are small compared to the horizon distance; you can therefore define 'average' properties of the waves. In contrast, if you are on land, in mountainous terrain, a single peak may dominate the entire view, and it makes less sense to define averages. Cosmology has proved a tractable subject because, in terms of this analogy, our universe resembles a seascape rather than a mountain landscape.

The overall motions in our universe are simple too. Distant galaxies recede from us with a speed proportional to their distance, as though they all started off packed together 10–15 billion years ago. But this does not imply we are in a special location. Suppose that all clusters of galaxies were joined by rods, which all lengthened at a rate proportional to how long they were. Then any two clusters would recede from each other with a speed proportional to their distance. That is what seems to be happening in our universe: there is no preferred centre, and an observer on any cluster would see an isotropic expansion around them.

What we actually see is modified by the fact that light takes a long time to reach us from distant places. As we probe deeper into space, towards our horizon, we see the universe as it was when it was younger and more closely packed. And we can now see very far back. Some amazing pictures taken with the Hubble Space Telescope each show a small patch of sky, less than a hundredth of the area covered by a full Moon. Viewed through a moderate-sized telescope, these patches would look completely blank. But these ultra-sensitive long exposures reveal many hundreds of faint smudges of light – a billion times fainter than any star that can be seen with the unaided eye. But each is an entire galaxy, thousands of light years across, which appears so small and faint because of its huge distance.

A huge span of time separates us from these remote galaxies. They are being viewed at the time when they were only recently formed, before they settled down into steadily spinning 'pinwheels' like Andromeda. Some consist mainly of glowing diffuse gas that has yet to condense into stars. When we look at Andromeda, we sometimes wonder if there may be other beings looking back at us. Maybe there are. But surely not on these remote galaxies. Their stars have not had time to manufacture the chemical elements. They would not yet harbour planets, and presumably no life. Astronomers can actually see the remote past. But what about still more remote epochs, before any galaxies had formed?

Georges Lemaître and George Gamow pioneered the idea that everything had 'exploded' from an initial dense state, which Lemaître called the 'primeval atom', and Gamow the 'ylem'. Neither of these names have stuck: we now talk about the 'Big Bang', a phrase introduced by Fred Hoyle as an insulting description of a theory he disliked. In 1948 Hoyle had developed, in collaboration with Thomas Gold and Hermann Bondi, the concept of a steady-state universe (Brush 2001): new atoms (and new galaxies) were postulated to be continuously created, so that, despite the expansion, the cosmos persisted with constant mean density in a statistically unchanging state.

There was boisterous and inconclusive debate in the 1950s (for more on this see Brush 2001), centred on whether the statistics of the 'radio galaxies' were compatible with a steady state, but clinching evidence for a Big Bang came in 1965. Intergalactic space is not completely cold. It is pervaded by weak microwaves, which have now been measured by the COBE satellite at many different wavelengths to a precision of one part in 10 000. This spectrum is just what you would expect if these microwaves are indeed an 'afterglow' of a pre-galactic era when the entire universe was hot, dense and opaque. The expansion has cooled and diluted the radiation, and stretched its wavelength. But this primordial heat is still around – it fills the Universe and has nowhere else to go!

And there is another 'fossil' in the universe – helium. When the entire universe was squeezed hotter than a star, there would have been nuclear reactions, but the temperature is only that high for the first three minutes – not enough time (fortunately) to convert everything into iron. (However, if that had happened, stars could still have formed, but they would exist only for Kelvin's timescale mentioned earlier, and there would be no hydrogen, carbon, oxygen and silicon to make the Earth.) During the expansion that immediately followed the Big Bang, reactions between protons and neutrons would have resulted in 23% of the material emerging as helium. This is gratifying because the theory of stellar nucleogenesis, which accounted so well for the build-up of most of the periodic table, was hard-pressed to explain why helium was much more uniform in its abundance than the heavier elements. The only other nucleus that is a relic of the Big Bang (apart, maybe, from lithium) is deuterium. This is an intermediate product in the primordial fusion of helium: the predicted deuterium abundance of a few parts in a hundred thousand is concordant with observations. Deuterium is destroyed in the course of stellar evolution, so its attribution to the Big Bang solves a long-standing problem.

The extrapolation back to the stage when the universe had been expanding for a few seconds (when the helium formed) deserves, I think, to be taken as seriously as anything that geologists or palaeontologists tell us about the early history of our Earth. Their inferences are just as indirect (and less quantitative). Moreover, there are several discoveries that might have been made, which would have invalidated the hypothesis, and which have not been made. For instance:

(i) astronomers might have discovered an object with helium abundance zero, or at least much less than 23%;
(ii) the microwave background was first observed, with poor spectral accuracy, in the 1960s. It might have turned out not to have a 'black body spectrum';
(iii) according to the Big Bang theory, photons outnumber baryons by a factor of about a billion. Moreover, in the first second of cosmic expansion, photons would come into equilibrium with neutrinos. There would therefore be roughly as many neutrinos as photons (the number differs by a modest numerical factor, because neutrinos and photons obey different quantum statistics, and the photon density is boosted by electron–positron annihilation). If physicists had found, experimentally, that one species of (stable) neutrino had a mass in the range 100 to 1 000 000 electron volts, then the total mass of all the predicted neutrinos would have 'closed up' the universe on much less than its present scale.

It would be easy to lengthen this list. The crucial point, however, is that the Big Bang theory has lived dangerously for decades, and survived. I believe we should place 99% confidence in an extrapolation back to the stage when the universe was one second old and at a temperature of ten billion degrees kelvin. However, we have far less confidence about still earlier stages – the first tiny fraction of a second – and I will return to this later. But first let us briefly look forwards rather than backwards – as forecasters rather than fossil hunters.

Futurology

In about five billion years the Sun will die; and the Earth with it. At about the same time (give or take two billion years!) the Andromeda galaxy, already falling towards us, may crash into our own Milky Way. So will the universe go on expanding for ever? Or will the entire firmament eventually collapse to a 'Big Crunch'?

The answer depends on how much the cosmic expansion is being decelerated by the gravitational pull that everything exerts on everything else. It is straightforward to calculate that the expansion can eventually be reversed if there are, on average, more than about five atoms in each cubic metre. That seems very little, but if all the galaxies were dismantled, and their constituent stars and gas spread uniformly through space, they would make an even emptier vacuum – one atom in every ten cubic metres – like one snowflake in the entire volume of the Earth.

Such a concentration is fifty times less than the 'critical density', and at first sight this seems to imply perpetual expansion, by a wide margin. But it is not so straightforward. Astronomers have discovered that galaxies, and even entire clusters of galaxies, would fly apart unless they were held together by the gravitational pull of about ten times more material than we actually see – this is the famous 'dark matter' mystery.

One line of evidence for this dark matter comes from studying the orbits of stars and gas in the outlying parts of galaxies – far outside the region that gives most of the visible light. These orbits are surprisingly fast. It is as though we found that Pluto were orbiting the Sun as fast as the Earth is. Were that the case, we would need to postulate an invisible heavy 'shell' outside the Earth's orbit but inside Pluto's, so that Pluto was 'feeling' the inward gravitational pull of a larger mass. It seems that the luminous parts of galaxies are embedded in a swarm of invisible objects, five to ten times more extensive and contributing five to ten times more total mass. On a still larger scale, entire clusters of galaxies contain dark matter. This is revealed in several ways. One interesting line of evidence comes from gravitational lensing – the bending of light by gravity. This technique was first suggested in the 1930s by the Swiss-American astrophysicist Fritz Zwicky, but has only recently been observationally feasible.

Among the undoubted highlights of the discoveries made with the Hubble Space Telescope are high-resolution pictures of clusters of galaxies. A remarkable image of the cluster Abell2218, a billion light years away, shows many galaxies in the cluster. But the picture also reveals a lot of faint streaks and arcs: each is a remote galaxy, several times further away than the cluster itself, whose image is, as it were, viewed through a distorting lens. Just as a regular pattern on background wallpaper looks distorted when viewed through a curved sheet of glass, the gravity of the cluster of galaxies deflects the light rays passing through it. The visible galaxies in the cluster contain only a tenth as much material as is needed to produce these distorted images – evidence that clusters, as well as individual galaxies, contain ten times more mass than we see.

What could this dark matter be? It is embarrassing that 90% of the universe is unaccounted for! Most cosmologists believe the dark matter is mainly exotic particles left over from the Big Bang, If they are right, we have to take our cosmic modesty one stage further. We are used to the post-Copernican idea that we are not in a special place in the cosmos. But now even 'particle chauvinism' has to go. We are not made of the dominant stuff in the universe. We, the stars, and the galaxies we see are just traces of 'sediment' – almost a seeming afterthought – in a cosmos whose large-scale structure is dominated by particles of a quite different (and still unknown) kind. Checking this is perhaps the number-one problem in the whole subject.

Cosmologists denote the ratio of the actual density to the critical density by the Greek letter omega (Ω). There is certainly enough dark matter around galaxies to make $\Omega = 0.2$ (remember that what we see is only 0.02). Until recently, we could not rule out several times this amount – comprising the full critical density, $\Omega = 1$ – in the space between clusters of galaxies. But it now seems that, in total, atoms and dark matter contribute no more than about $\Omega = 0.3$. The odds, therefore, favour perpetual expansion. The galaxies will fade, as their stars all die, and their material gets locked up in old white dwarfs, neutron stars and black holes. They will recede ever further away, at speeds that may diminish, but never drop to zero. In fact, there is now tantalizing evidence for an extra repulsion force that overwhelms gravity on cosmic scales – what Einstein called the cosmological constant, lambda (λ). The expansion may actually accelerate! If it does, the forecast is an even emptier universe (see Schramm 1998). The American magazine *Science* rated this the most important discovery of 1998, in any field.

The issue of acceleration versus deceleration is important for another issue, the 'age of the universe' – in other words, the time since the Big Bang. If expansion proceeded at an unchanging speed, this would be simply the 'Hubble time' – the inverse of the Hubble constant, which gives the relation between recession speed and distance (for a further explanation of 'Hubble time', see Brush 2001). The Hubble time is still uncertain at the 10% level, but current best-estimates are in the range 13–14 billion years. However, if the universe were decelerating, the average expansion speed would have been faster than the present speed, and the time since the Big Bang consequently shorter than the Hubble time. For example, the age of an 'Einstein–de Sitter' universe, where the matter provides the full critical density (i.e. where $\Omega = 1$), is only two-thirds of the Hubble time. If we were in such a universe, there would be a serious discrepancy with the estimated ages (up to 13 billion years) of globular star clusters. However, in an accelerating universe, the time since the Big Bang can exceed the Hubble time. For the cosmological model which is now most favoured (where the 'cosmical repulsion' is dominant, causing an acceleration, whereas at earlier times, when the matter was denser, there would have been deceleration) the 'age' works out at a value close to the Hubble time. There therefore seems, taking account of the 10% errors in both the Hubble time and in stellar ages, a reassuring concordance between these estimates.

The ultra-early eras

So much for the long-range forecast. Let us now go back to the beginning. People sometimes wonder how our universe can have started off as a hot amorphous fireball and ended up intricately differentiated. Temperatures now range from blazing surfaces of stars (and their even hotter centres) to the night sky only three degrees kelvin above absolute zero. This may seem contrary to a hallowed principle of physics: the second law of thermodynamics. But it is actually a natural outcome of the workings of gravity.

Gravity renders the expanding universe unstable to the growth of structure, in the sense that even very slight initial irregularities would evolve into conspicuous density contrasts. Theorists can now follow a 'virtual universe' in a computer. Slight fluctuations are 'fed in' at the start of the simulation. The calculations can simulate a box containing a few thousand galaxies – large enough to be a fair sample of our universe. As the box expands, regions slightly denser than average lag further and further behind. Eventually, gravitationally bound systems of dark matter would condense out, to become the 'dark halos' of new galaxies. Within these halos, gas cools and condenses into stars. When stars have formed, the negative specific heat of gravitating systems drives things ever further from equilibrium. If the Sun's nuclear fuel were turned off, it would contract, just as Kelvin realized. But it would end up with a hotter centre than before: to provide an enhanced pressure to balance the stronger gravitational force, the centre must, after contraction, get hotter. When a star loses energy, it heats up.

The way slight initial irregularities in the cosmic fireball evolve into galaxies and clusters is in principle as predictable as the orbits of the planets, which have been understood since Newton's time. But to Newton, some features of the solar system were a mystery. Why were the planets 'set up' with their orbits almost in the same plane, all circling the Sun the same way, whereas the comets were not? This is now well understood: the planets have aggregated from smaller bodies within a disc that was 'spun off' from the contracting proto-Sun. Indeed, we have pushed the barrier back from the beginning of the solar system to the first second of the Big Bang. But conceptually we are in no better shape than Newton was. He had to specify the initial trajectories of each planet. We have pushed the causal chain further back by several 'links', but we still reach a stage when we are reduced to saying 'things are as they are, because they were as they were'.

Our calculations of cosmic structure need to specify, at some early time such as one second after the Big Bang, a few numbers:

(i) the cosmic expansion rate;
(ii) the proportions of ordinary atoms, dark matter and radiation in the universe;
(iii) the character of the fluctuations; and, of course
(iv) the basic laws of physics.

Any explanation for these numbers must lie not just within the first second, but within the first tiny fraction of a second. What is the chance, then, of pushing the barrier back still further? The cosmic expansion rate presents a special mystery. The two eschatologies – perpetual expansion, or collapse to a 'crunch' – seem very different. But our universe is still expanding after ten billion years. A universe that collapsed sooner would not have allowed time for stars to evolve, or even to form. On the other hand, if the expansion were too much faster, gravity would have been overwhelmed by kinetic energy and the clouds that developed into galaxies would have been unable to condense out. In Newtonian terms the initial potential and kinetic energies were very closely matched. How did this come about?

I was confident in tracing back to when the universe was a second old. The matter was no denser than air; conventional laboratory physics is applicable and is vindicated by the impressive evidence of the background radiation, helium, and so forth. But for the first trillionth of a second every particle would have had more energy than even CERN's new accelerator will reach (CERN = Organisation Européene pour la Recherche Nucléaire). The further we extrapolate back, the less foothold we have in experiment.

But most cosmologists suspect that the uniformity and expansion rate is a legacy of something remarkable that happened when everything was compressed in scale by 27 powers of ten (and hotter by a similar factor). The expansion would then have been exponentially accelerated, so that an embryo universe could have inflated, homogenized and established the 'fine-tuned' balance between gravitational and kinetic energy when it was only 10^{-38} seconds old (see Turner 2001 for a review). The seeds for galaxies and clusters could then have been tiny quantum fluctuations, imprinted when the entire universe was of microscopic size, and stretched by inflationary expansion.

This generic idea that our universe inflated from something microscopic is compellingly attractive. It looks like 'something for nothing', but really is not. That is because our present vast universe may, in a sense, have zero net energy. Every atom has an energy because of its mass – Einstein's $E = mc^2$. But it has a negative energy due to gravity – we, for instance, are in a state of lower energy on the Earth's surface than if we were up in space. And if we added up the negative potential energy we possess due to the gravitational field of everything else, it could cancel out our rest mass energy. Thus it does not, as it were, cost anything to expand the mass and energy in our universe.

Cosmologists sometimes loosely assert that the universe can essentially arise 'from nothing'. But they should watch their language, especially when talking to philosophers. The physicist's vacuum is latent with particles and forces – it is a far richer construct than the philosopher's 'nothing'. Physicists may, some day, be able to write down fundamental equations governing physical reality. But they will never tell us what 'breathes fire' into the equations, and actualizes them in a real cosmos.

I am uneasy about how cosmology is sometimes presented. The distinction is often blurred between things that are quite well established and those that are still speculative. I have tried to emphasize that as far back as one second after the Big Bang, I regard cosmology as having as firm a base as other historical sciences. But the ultra-early universe is more speculative, and I must offer a special health warning for mentioning it – and a redoubled warning before the brief remarks in the next section.

A multiverse?

In our universe, intricate complexity has unfolded from simple laws – we would not be here if it had not. But simple laws do not necessarily permit complex consequences – one could envisage a set of laws that precluded any emergent structures. To take an analogue, the Mandlebrot set, with its infinite depth of structure, is encoded by a short algorithm. But other algorithms, superficially similar, yield very boring patterns. Why is the physical 'recipe' for our universe, in this analogy, like the Mandelbrot set?

As we have seen, our cosmos could not have evolved its present complexity if it were not expanding at a special rate. And there are other prerequisites for a complex cosmos. We can readily imagine other alterations that would preclude complexity. For instance, if nuclear forces were a few per cent weaker, no atoms other than hydrogen would be stable. The residue of the Big Bang might be entirely dark matter – no ordinary atoms at all. Gravity could be so strong that any large organism would be crushed. Or the number of dimensions might even be different.

This apparent 'fine tuning' could be just a brute fact, but I find another interpretation increasingly compelling: it is that many other universes actually exist. Most would be 'stillborn' because they collapse after a brief existence, or because the physical laws governing them are not rich enough to permit complex consequences. Only some would allow creatures like us to emerge. And we obviously find ourselves in one of that particular subset. The seemingly 'designed' features of our universe need then occasion no surprise. If you go to a clothes shop with a large stock, you're not surprised to find one suit that fits. I have argued in a recent book (Rees 1999) that our universe may not be the only one. Some of the key numbers that characterize ours may take different values in others, which would then be sterile. But this is speculation and it may go the way of Kepler's numerology – he thought there were six planets, in orbits with definite geometrical ratios.

Conclusions

It is helpful to divide cosmic history into three parts.

Part 1 is the first millisecond, a brief but eventful era spanning forty decades of logarithmic time. This is the intellectual habitat of the high-energy theorist and the 'inflationary' or quantum cosmologist. Here, there are uncertainties in the basic physics, which get more serious the further back we go, since we gradually lose our foothold in experiment. But the key features of our universe were imprinted during this era.

Part 2 runs from a millisecond to some millions of years: it is an era when cautious empiricists like myself feel more at home. The physics is well-known, and everything is still smoothly expanding. Theory is corroborated by quantitative evidence: the cosmic helium abundance, the background radiation, etc. Part 2 of cosmic history, though it lies in the remote past, is the easiest to understand. The tractability lasts only as long as the universe remains amorphous and structureless. When the first structures condense out, the first galaxies and stars form and light up – the era studied by traditional astronomers begins. We then witness complex manifestations of well-known basic laws. Gravity, gas dynamics and feedback processes from early stars combine to initiate the complexities around us.

Part 3 is difficult for the same reason that geology and other environmental sciences are difficult. Edwin Hubble concluded his famous book *The Realm of the Nebulae*, with the words: 'Only when empirical resources are exhausted do we reach the dreamy realm of speculation' (Hubble 1936). We still dream and speculate. But there has been astonishing empirical progress since Hubble's time. The last decade has been exceptional, and the crescendo of discovery seems set to continue. Large telescopes on the ground, and the instrument in space that bears Hubble's name, can now view 90% of cosmic history; other techniques can probe right back to the first few seconds of the 'Big Bang'.

There are three great frontiers in science: the very big, the very small and the very complex. Cosmology involves them all. Cosmologists must pin down the basic numbers like the 'density parameter' omega, and find what the dark matter is – I think there is a good chance of achieving this within five years. Second, theorists must elucidate the exotic physics of the very earliest stages, which entails a new synthesis between cosmos and microworld – it would be presumptuous for me to place bets here. But cosmology is also the grandest of the environmental sciences, and its third aim is to understand how a simple fireball evolved, over 10 to 15 billion years, into the complex cosmic habitat we find around us – how, on at least one planet around at least one star, creatures evolved able to wonder about it all. That is the challenge for the new millennium.

I am grateful to H. Huppert and G. Rhee for carefully reading a draft of this article.

References

BRUNO, G. 1584. *On the Infinite Universe and Worlds*.

BRUSH, S. G. 2001. Is the Earth too old? The impact of geochronology on cosmology, 1929–1952. *In*: LEWIS, C. L. E. & KNELL, S. J. (eds) *The Age of the Earth: from 4004 BC to AD 2002*. Geological Society, London, Special Publications, **190**, 157–175.

DALRYMPLE, G. B. 2001. The age of the Earth in the twentieth century: a problem (mostly) solved. *In*: LEWIS, C. L. E. & KNELL, S. J. (eds) *The Age of the Earth: from 4004 BC to AD 2002*. Geological Society, London, Special Publications, **190**, 205–221.

HUBBLE, E. P. 1936. *The Realm of the Nebulae*. Yale University, New Haven.

LEWIS, C. L. E. Arthur Holmes' vision of a geological timescale. *In*: LEWIS, C. L. E. & KNELL, S. J. (eds) *The Age of the Earth: from 4004 BC to AD 2002*. Geological Society, London, Special Publications, **190**, 121–138.

MARCY, G. W. & BUTLER, P. R. 2000. Planets around other suns. *Publications of the Astronomical Society of the Pacific*, **112**, 137–140

REES, M. J. 1999. *Just Six Numbers*. Weidenfeld and Nicolson, London.

SALPETER, E. E. 1999. Stellar nucleosynthesis. *Reviews of Modern Physics*, **71**, S220–222.

SCHRAMM, D. N. (ed.). 1998. Conference proceedings on Age of the Universe, Dark Matter and Structure Formation. *Proceedings of National Academy of Sciences*, **95**, 1–84.

SHIPLEY, B. C. 2001. 'Had Lord Kelvin a right?': John Perry, natural selection and the age of the Earth, 1894–1895. *In*: LEWIS, C. L. E. & KNELL, S. J. (eds) *The Age of the Earth: from 4004 BC to AD 2002*. Geological Society, London, Special Publications, **190**, 91–105.

THOMSON, W. (LORD KELVIN). 1862. On the Age of the Sun's Heat. *Macmillan's Magazine*, **5**, 388–393.

TURNER, M. (ed.) 2001. *Inflationary Cosmology*, Chicago University, Chicago.

Index

Note: Page numbers in *italic* type refer to illustrations; those in **bold** type refer to tables.

absolute age 126, 152–153
absolute timescale 248
Acasta gneiss (Canada) dating 180, 195
Ackroyd, William 110
Age of the Earth Committee (NRC) 129–130, 150, 160, 210
ages of the Earth
  main contributors of dates
    Barnabus 17
    Becker, G. F. 146, 208
    Buffon, Comte de 4, 41, 43
    Chamberlin, T.C. 160
    De Luc, J. A. 56, 59
    Gerling, E. K. 211–212
    Hales, W. 3
    Halley, E. 3
    Haughton, S. 112, 206
    Holmes, A. 9, 126, 128, 150
    Houtermans, F. 212
    Jeffreys, H. 160, 209
    Joly, J. 4, 110, 111, 116
    Kelvin, Lord (Thomson, W.) 7, 91, 101, 108, 205, 206
    King, C. 142, 205, 206
    Maillet, B. de 33
    McGee, W. J. 141, 142, 206
    Nier, A. 152
    Patterson, C. C. 10, 152, 213–214
    Perry, J. 99
    Phillips, J. 88, 96
    Reade, T. M. 144
    Russell, H. N. 148, 160, 209
    Rutherford, E. 160
    Sollas, W. 112
    Swan, J. 18
    Theophilus of Antioch 2
    Upham, W. 144
    Ussher, J. 3, 20
    Walcott, C. D. 112
    Zoroaster 2
  methodology summary **207**
Akilia association (Greenland) 184–185, 195
American Association for the Advancement of Science (AAAS) 141, 143
American Philosophical Society 148
Ames, Joseph 149–150, 151
ammonites as guide fossils 244–248
Amud human fossil site 270
Andromeda galaxy 277, 278, 279
Antarctica and age of Earth site 194
Antediluvian period 86
apparent velocity 165
Ar/Ar dating 265
Archaean 143, 253
  dating *see* zircons
Arduino, Giovanni 32, 34
Aston, Frederick 133, 152, 210
Atapuerca (Spain), human fossils 265, 268–269
atomic physics, leading workers **159**
atoms, creation of 276
Aurignacian 269
Australia
  Archaean eukaryotes 255
  human fossils 268, 271
  zircons for age of earth 178, 193, 233
Auvergne basalts, Desmarest study 41, *43*, 44
Ayrton, W.S. 93

Baade, Walter 157, **158**, 169
bacteria and early life 254
banded ironstone formation 186, 195
Banks, Sir Joseph 66, 78
Barberton Greenstone Belt 193

Barnabas, General Epistle of 16–17
Barr, J. 2, 3
Barrell, Joseph 146, 147, 150, **207**
Barus, Carl 142
basalt use in building timescale 133
Bath Hot Springs 69
Becher, Johann Joachim 28
Becker, George Ferdinand 110, 111, 113, 145, 146, 148, 151, 205–206, **207**, 208, 209, 219
Becquerel, Henri 7, 113, 207
Bergman, Torbern Olof 34
Bethe, Hans 158, **159**
Bevan, Benjamin 66
Bible 1–2, 16, 18–20, 20–22
Big Bang 168, 170, 278–279
biodiversity 261
biogenic carbon, Archaean 186
biohorizons 245
biostratigraphy 240
Blumenbach, Johann Friedrich 55
Boltwood, Bertram 113, 125, 126, 140, 145, 147, 148, 208
Bondi, Hermann **159**, 168, 278
Bourguet, Louis 32
Bretz, J. Harlen 258
Brewer, A. K. **207**
British Association for the Advancement of Science (BAAS) 87
  annual meetings 5, 23, 91, 92, 110, 123, 129, 148, 163, 257
Brogden MP, James 75
Brogniart, Alexandre 4
Brown, Ernest W. 129, 150, 160
Bruno, Giordano 277
Buch, Leopold von 5
Buckland, William 5, 86
Buckman, S.S. 245, 248
Buffon, Comte de *see* Leclerc
bulk silicate earth (BSE) 227–228
Burgess Shale 255, *256*
Burnet, Thomas 30

caesium clock 239
Cainozoic (Cenozoic) 85, 143
calendar
  geological 240, *241*, *242*, 243
  historical 238–239
Cambrian Period 152
Canada Archaean sites 180, 192–193, 195
Cardioceras martini 244–245, *246*
cathode rays 210
Cecil Marquis of Salisbury, Robert 91, 93, 94, 95, 104
Cenozoic *see* Cainozoic
Chamberlin, Thomas Chrowder 144, 149, 150, 152, **159**, 160, 206
Chicxulub (Mexico) 262
China, sites for dating 194, 255, 271
chronology and geochronology 52–53
chronostratigraphy 240, **243**
Clarke, John Mason 149
Cleveland Dyke first dated 132
clocks, primary and secondary 238, 239
coal prospecting, Smith's work 68–74
Cockburn, William 87
Coke, Thomas William 66
comet, role in Earth formation 43, *44*, 47
Committee for the Age of the Earth *see* Age of the Earth Committee (NRC)
Committee for the Measurement of Geological Time by Atomic Disintegration 129, 130, 149
complexity, evolution of 259
conformable lead method 224, 228–230
conodonts as guide fossils 250
convergent evolution 260

cooling earth model for age determination
  Barus 142
  Becker comments on 205–206
  Descartes 26
  Kelvin 88–89, 91–92, 94
  Perry 93–104
  Phillips 87
cosmological constant 280
cosmology and Earth age 157
  expanding universe 161–163, 165–167, 168–170
  leading workers **159**
  steady state 167–168, 168–169
  terrestrial time problem 163–165
creation dates proposed 2, *20*
creationists 157, 158, 170, 171, 257
Cro-Magnons 268–269
Crook, Thomas 66, 67
Cruse, Jeremiah 65
Cunnington, William 66, 74
Curie, Marie and Pierre 123, 159, 207–208
Cuvier, Georges 4, 58
cyclical processes, orders of 249–250

Dana, James Dwight 143, 144
dark matter 279–280
Darwin, Charles 2, 6, 86, 88, 139, 151
  debate with Thomson 139–140, 157
  problem of time 256
  theory of evolution by natural selection 91, 94, 98
Darwin, G. **207**
De La Beche, Henry Thomas *5*
De Luc, Jean André 4, 6, 54, 55–58, 59
Deluge *see* Noah's flood
Dempster, A. J. 210
Descartes, René 26, 27
Desmarest, Nicolas 4, 6, 40–45, *41*, 46–48
deuterium 279
Devonian Period 85, 114, 125
Diluvian period 86
Dirac, Paul 164, 170–171
DNA mapping and evolution 254, 272
Doppler shift 162, 163, 164, 165, 166

Eddington, Arthur **159**, 160, 161, 163, 164
Ediacara 255
Einstein, Albert **159**, 161
electron spin resonance (ESR) dating 268
elements, creation of 276
Elizabeth I, Queen of England 16, 23
energy balance in universe 281–282
époque 40, 41, 42, 53
erosion ageing method 141–142, 144
eukaryotes *255*
Evans, Robley 133
evolution *v.* creationism 257
expanding universe theory 161–163
  Hubble retracts 165–167
  paradox resolved 163–164
  reinstated 168–179
extinction *see* mass extinction
extraterrestrial seeding of life on Earth 254–255, 257

falsifiability criterion 168
Farey, John 4, 66, 69, 70, 72, 73, 76, 77, 79, 80
Fell, John 22
fergusonite 7, 124, 208
Fisher, Osmond 94, 110, 142
Fletcher, Arnold Lockhart 113
fossil humans 265–272
fossils and time 240, 244–248, 250
Fourier, Joseph 6
Freidmann, Aleksandr 161

Gaia hypothesis 261
galaxies, distance to 278
galenas for age 217
Gamow, George **159**, 169, 278

Geike, Archibald 4, 6, 92, 101, 102, 103, 126, *141*, 143, 144
Genesis story of creation 25–26, 51–52, 52–53
  see also creation; Noah's flood
geochron 214, 215, *226*, 227, 231, 233
geochrone 143
geological maps of Smith 78, 79
Geological Society of London 74
geological timescale, construction of 135–136, *135*
Gerling, E. K. **207**, 211–212, 223
Gerling–Holmes–Houtermans model 10, 224
Giraud, Jean Louis (Abbé Soulavie) 4, 6, 54
globular clusters 276
Gold, Thomas **159**, 168, 278
goniatites as guide fossils 250
Gould, Stephen Jay 2, 3, 258
gradualism v. punctuated equilibrium 258
granite, first radioactive dating 114
granitoid gneisses and U-Pb dating 178, 180–195
graptolites as guide fossils 250
gravitational lensing 280
Greenough, George Bellas 5, *5*, 62, 75, 78
Greenland Archaean sites 178, 180–192, 195
Grosseteste, Robert 17, *18*
guide fossils 240, 244–248, 250
Gunflint Formation *241*, 243

Haldane, John 164
Hale, George Ellery 149, **158**
Hales, William 3, 20
Halley, Edmond 3, 109
Haughton, Samuel 112, 206, **207**
Heaviside, Oliver 98, 100
helium analytical methods 132, 133
Helmholtz, H. L. F. von **207**
hemera 245–246
Herschel, John 77
Hertzsprung, Ejnar **158**
Hertzsprung–Russell diagram 161
Hess, Viktor 129
Hevesy, Georg von 129
Hf in age determination 234
Hirne, Urban 28
hibernium 114
Holmes, Arthur 8, 9, 12, 111, 113, 114, 115, 116, 147, 149, 150, 152, **159**, 160, 169, **207**, 208–209, 209–210, 223, 257
  awards and distinctions 121
  dating work 125–127, 129–130
  early years 121–122
  father of geological timescales 136–137, *136*
  first book written 128
  first isotope studies 129
  geological timescale, construction of 135–136, *135*
  initial ratios, invention of 130–132, *131*
  Mozambique work 127–128
  U/Pb studies 133–134
Holmes–Houtermans model 10, 223, *226*, 227, 234
Homo spp. *262*, 268, 271
Hooke, Robert 34
Horner, Leonard 6, 15, 21, 75
horse evolution 258
hot springs and early life 254
Houtermans, Fritz 9, 212, 223
Hoyle, Fred **159**, 168, 278
Hubble, Edwin 157, **158**, 162, *162*, 165–167, 169
human fossils 265–272
Humason, Milton **158**, 162
Hutton, James 4, 6, 53
Huxley, Thomas H. 1, 102, 206, 207, 256

I in age determination 234
initial ratios 9, 122, 180, 184
  invention of 130–132, *131*
isochron 178, 180, 181, 182, 212, *226*, 227
isotopes discovered 129
Isua greenstone belt (Greenland) 185–192, 195
Itsaq orthogneiss (Greenland) dating 180–184

INDEX

James, William 75–76
Jeans, James **159**, 161, 164
Jeffreys, Harold **159**, 160, 164, 167, 169, **207**, 209
Joly, John 4, 8, 145, 149, **207**
  academic career 107–108
  age of Earth methods
    radioactivity 112–115
    sediments 112
    sodium 108–112
  awards and distinctions 116
  early life 107
  isostasy 115
  orogeny 115
  tectonics 115
  thermal cycles theory 115
Joly process of colour photography 108
Judd, John Wesley 102
Jurassic timescale 248, *249*

K–T boundary event 261–262
K/Ar dating 265
Kaapvaal craton 193
Kelvin *see* Thomson, William
Kenrick, John 87
Kepplerites keppleri 244, *245*
Kidd, John 75
kinematic relativity 164
King, Clarence 142, 205, 206
Kircher, Atahanasius 26, *27*
Knopf, Adolph 129, 150, **207**
Kosmoceras 246, *247*, 248
Kovarik, Alois 129, 150

Labrador Archaean rocks 192–193
Lane, Alfred Church 129, 146, 149
Lawson, Andrew Cowper 149
Lawson, Bob 127, 129
Le Moustier 269
lead isotopes 133–134, 215–217, 223–227, 230–233
Leavitt, Henrietta **158**, 161–162
Leclerc, Georges-Louis (Comte de Buffon) 4, 6, 40–45, 46–48, 53, **207**
Lehmann, Johann Gottlob 34
Leibniz, Gottfried Wilhelm 26, 30
Leinster granite 114
Lemaître, Georges **159**, 161, 278
Lhwyd, John 34
life, origins of 253–255, 257, 277
light bending 280
Lightfoot, John 19
Linné, Carl von 33
lithostratigraphy 240
Lloyd, William 22
Lodge, Oliver 94, 96, 100, 101, 104
Lonsdale, William 62
Lowell, Percival 158
Lu-Hf ratios 184, 194, 196
lunar age 233
lunar tides age method 206
Lyell, Charles *5*, 6, 58

McGee, Wilber John 141, 141–142, 142–143, 151, 206, **207**
Maclure, William 78
magnetostratigraphy 250
Maillet, Benoit de 33–34
Manfred complex 193
Mantell, Gideon 77
Maplet, John 15, 17
Marquis of Salisbury *see* Cecil, Robert
Mars and early life 254
Marsili, Luigi Ferdinando 32, 34
mass extinction 261–262, 263
mass spectrograph/spectrometer 210
Massoretic text 3, 18
Meldola, Raphael 101
Mesozoic 85, 143
meteorite bombardment 177, 261–262

meteorite Pb analyses 212–213, 213–215
Mexico and K–T boundary 261–262
mid ocean ridge, and early life 253
Milanković cyclicity 250
Miller, Hugh 6
Milne, Edward 164
mineralogy, ordering of 15
Monti, Giuseppe 32
MORB lead isotopes 227
Moro, Anton Lazzaro 33
Mosaic flood story *see* Noah's flood
Mourne Mts granite 114
Muirhead, Lockhart 75
Murchison, Roderick Impey *5*, 85
Murchison Medal 116
Murthy, V. R. 214–215

National Academy of Sciences 149
National Research Council 149
natural selection *see under* Darwin
natural selection *v.* contingency 259–260
nature's chronometers 57–58
Neanderthals 268, 269
Newcomb, S. **207**
Newton, Sir Isaac 3
Nier, Alfred 133–134, 136, 152, **159**, 211
Noah's flood 19, 26, *31*, 32, 33, 56, 86

ocean chemistry (Na) method 206, **207**
ocean sediment Pb analyses 213, 214
Omer, Guy C. 167
Oort, Jan Hendrik 162–163
optically stimulated luminescence (OSL) dating 268
orbital physics-lunar tides method 206, **207**
order, the Elizabethan quest for 15–16
Order of Strata in the Bath Area 65
orders of cyclicity 249–250
orthogneisses *see* granitoid gneisses
Owen, Richard 5

Palaeolithic, Mid/Upper boundary 268, 269
Palaeozoic 85, 143
Paneth, Fritz 132
parallel evolution 260
Parkinson, James 62
Patterson, Claire C. 9, 10, 11, 152, **159**, 205, 212, 219, 223, 225, 233, 234
  age of Earth calculation 213–215
Payne, Cecilia **158**
Pb *see* lead isotopes
Peking Man 271
Penrose Medal 121, 132
Perry, John 7, *5*, *92*, 93–104, 206
Phillips, John *5*, 5, 6, 62, 80, 85, 86–88, 88–89, 96, 112, 151
Phillips, William 76, 78
Pickering, Edward 158
Pilbara craton 193
planetary geology, leading workers **159**
planetary systems 276–277
Planetesimal Hypothesis 144, 160
Playfair, John 6
pleochroic haloes 113, *114*
Popper, Karl 168
positive rays 210
Postdiluvian period 86
Poulton, Edward 94–95
Powell, John Wesley *141*, 142
primordeal lead 212
Principle of Biosynchroneity 240
Principle of Superposition 240
prokaryote fossils 243–244, *255*
Pu in age determination 234
punctuated equilibrium 258

radioactivity
  beginnings of age dating 145–148
  decay principles 123–124

radioactivity (*continued*)
  discovered 1, 123
  named 208
  use in dating 112–115, 145–151, **207**, 208–210
radiocarbon 267
radiometric timescale for America 150
Ray, John 3, 34
Rayleigh
  Third Baron *see* Strutt, John
  Fourth Baron *see* Strutt, Robert
Rb isotopes 230
Re-Os studies 194
Reade, T. Mellard 109–110, 112, 144, 207
Riftia 255
Ritter, A. **207**
rock-time duality 241–242, **243**
Röntgen, William 7, 123
Russell, Henry Norris 148, **158**, 160, **207**, 209, 210, 211
Russell, R.D. 215–216, 226
Rutherford, Ernest 7, 8, 113, 114, 123, 124, 125, 129, 145, **159**, 160, **207**, 208, 209, 210
  discovery of ^{235}U 133
  spontaneous decay of atoms 123–124
  work with Boltwood 125

Saint-Césaire 269
Salas, José González de 28
salt as age measure 4, 8, 108–112, 150
Scheuchzer, Johann Jakob 31, 32
Schlotheim, Ernest Friedrich, Baron von 5
Schuchert, Charles 129, 146–147, 150, 152
Scott, William 148
Scriven, Michael 168
Sedgwick, Adam 5, 87
sediment accumulation age method 112, 144, 206, **207**
See, Thomas 146
Septuagint 3, 18
sequence stratigraphy 249–250
series, introduction of 244
Shapley, Harlow **158**, 164–165, 166–167
siècles 39, 53
Silurian Period 85
Sino-Korean craton 194
Sirius Passet 255
Sitter, Willem de **159**, 161, 163, 164
Sliper, Vesto Melvin **158**, 161
Sm-Nd studies 192, 196, 230
Smith, William 4, 5, 20–21, 240
  coal trials 68–74
  early career 63–65
  engineering activities 65–67
  life in London 74–77
  mineral prospecting 67–68
  relationship with Phillips 86
  stratigraphical development 77–79
  view of stratigraphy and time 79–81
Soddy, Frederick 10, 123, 129, 208
sodium salt method of dating 3, 7, 108–112, 150
solar origins for Earth 43, *44*, 47
solar system origin, encounter theory 163–164, 164–165
Sollas, William 108, 110, 112, 123, **207**
Soulavie, Abbé *see* Giraud
South Africa 193, 268, 270, 272
Speculum Mundi 17–18
Sr isotopes 230, 248–249
stable isotope ratio stratigraphy 250
stage, introduction of 244
Stanton, R. L. 215–216, 223
stars 158, 276
steady state universe 167–168, 278
Steno, Nicolaus 28–29
Stone Age 270
Strachey, John 63
Strutt, John (Third Baron Rayleigh) 124
Strutt, Robert (Fourth Baron Rayleigh) 10, 113, 124–125, 126, 160, 208

sun
  age 160–161
  composition 158
  end of life 279
  life cycle 275
supernovae 277
Swan, John 17–18
Swedenborg, Emanuel 33

Tait, P. G. 96, 97, 99, 103, **207**
terrestrial rock maximum age 178
Theophilus of Antioch 2
thermal (cooling) dating method 206, **207**
thermal cycles theory 115
thermoluminescence (TL) dating 268
Thomson, J. J. 123, 145, 159, 210
Thomson, William (Lord Kelvin) 6–7, 86, 88–89, 205, 206, **207**
  attacks on uniformitarianism 151
  debate with Chamberlin 144
  debate with Darwin 139–140
  debate with Perry 91, *92*, 93–104
  method of dating 108
  snub to biology 256
  on sun's age 275
time interval, importance of 249
  geochronology divisions **243**
time planes 243
time and timeclocks 238
Tolman, Richard 165, 167
Townsend, Joseph 65, 67, 76
Tozzetti, Giovanni Targioni 32
trilobites as guide fossils 250
troilite in meteorites 212, 213, 214, 216–217
trondhjemite-tonalite-granodiorite (TTG) gneiss 195

U enrichment isotope factors 230
U series dating 267–268
U-Pb zircon dates 177, 178, *179*, 180–197, 233
uniformitarians and uniformitarianism 6, 94
universe
  age problem 157–158, 280
  energy balance 281–282
  expansion of 161–167, 168–170, 278, 280
  initiation of 280–281
  steady state theory 167–168
Upham, Warren 144
Urry, William 133
USA 140–153, 194
Ussher, James *2*, 3, 19, 20, 52

Vallisneri, Antonio 32, 33, 34
Valmont de Bomare, Jacques-Christophe 45–46
Vetlesen Prize 121
Vulgate (Latin) Bible 18

W in age determination 234
Walcott, Charles Doolittle 112, 121, *141*, 142, 143, 151, **207**
Wallace, Alfred 102
  on timescale problem 256
Weald erosion 139, 151
Webster, Thomas 4
Wheeler, Lynde Phelps 145
Whin Sill first dated 132
Whiston, William 30
Williams, Henry Shaler 143, 144
Winchell, A. **207**
Wollaston Medal 62, 1221
Woodward, John 30–32

Xe in age determination 234

zircons and U-Pb dating 177, 178, *179*, 180–197, 233
zone, introduction of 244
Zoroaster 2
Zwicky, Fritz 164, 165